NONSMOOTH AND DISCONTINUOUS PROBLEMS OF CONTROL AND OPTIMIZATION
(NDPCO'98)

A Proceedings volume from the IFAC Workshop,
Chelyabinsk, Russia, 17 - 20 June 1998

Edited by

V.D. BATUKHTIN
Chelyabinsk State University, Chelyabinsk, Russia

F.M. KIRILLOVA
Institute of Mathematics, National Academy of Sciences of Belarus, Minsk, Belarus

and

V.I. UKHOBOTOV
Chelyabinsk State University, Chelyabinsk, Russia

Published for the

INTERNATIONAL FEDERATION OF AUTOMATIC CONTROL

by

PERGAMON
An Imprint of Elsevier Science

UK	Elsevier Science Ltd, The Boulevard, Langford Lane, Kidlington, Oxford, OX5 1GB, UK
USA	Elsevier Science Inc., 660 White Plains Road, Tarrytown, New York 10591-5153, USA
JAPAN	Elsevier Science Japan, Tsunashima Building Annex, 3-20-12 Yushima, Bunkyo-ku, Tokyo 113, Japan

First edition 1999

Library of Congress Cataloging in Publication Data

A catalogue record for this book is available from the Library of Congress

British Library Cataloguing in Publication Data

A catalogue record for this book is available from the British Library

ISBN 0-08-043237 9

Transferred to Digital Printing 2008.

IFAC WORKSHOP ON NONSMOOTH AND DISCONTINUOUS PROBLEMS OF CONTROL AND OPTIMIZATION

Sponsored by
International Federation of Automatic Control (IFAC)
- Technical Committee on Optimal Control

Co-sponsored by
IFAC Technical Committees on
- Control Design
- Nonlinear Systems
Russian Foundation of Fundamental Investigation

Organized by
Russian National Committee on Automatic Control
Belarus National Association on Automatic Control
Chelyabinsk State University

International Programme Committee (IPC)
Kirillova, F.M (Belarus) (Chairperson)
Pallaschke, D. (D) (Vice-Chairperson)
Bardy, M. (I)
Barron, E.N. (USA)
Batukhtin, V.D. (Russia)
Bulirsch, R. (D)
Chernousko, F.L. (Russia)
Clarke, F.H. (CDN)
Demyanov, V.F. (Russia)
Furuta, K. (J)
Gabasov, R. (Belarus)
Gasilov, V.L. (Russia)
Isidori, A. (I)
Krasovskii, N.N. (Russia)
Kryazhimskii, A.V. (Russia)
Maiboroda, L.A. (Russia)
Malanovskii, K. (PL)
Mordukhovich, B. (USA)
Oettly, W. (D)
Olsder, G.J. (NL)
Osipov, Ju.S. (Russia)
Salukvadze, M.E. (Georgia)
Zubov, V.I. (Russia)

National Organizing Committee (NOC)
Batukhtin, V.D. (Chairperson)
Kleimenov, A.F. (Vice-Chairperson)
Ukhobotov, V.I. (Secretary)
Kovalev, Yu.M.
Krasovskii, A.N.
Rolshchikov, V.E.
Shiryaev, V.I.

CONTENTS

*Plenary paper

*Plenary Paper

FINITELY ADDITIVE MEASURES AND PROBLEMS OF ASYMPTOTIC ANALYSIS [1]

Chentsov, A.G. *

* *Institute of Mathematics and Mechanics, 16 Kovalevskaya St., 620219 Ekaterinburg, Russia*

Abstract: Questions of asymptotic attainability under perturbations of integral constraints are considered. The generalized representation in the class of vector finitely additive measures is investigated. Different variants of the weakening of initial conditions are compared. In addition, the unboundedness of the space of controls is assumed. General scheme of analysis of attraction sets is considered.

Keywords: vector finitely additive measure, extension, attraction set, integral constraints.

1. INTRODUCTION

In many control problems distinct integral constraints (IC) arise. Highly often these IC have an indirect character. Such IC are realized by an "recalculation" of other constraints (phase constraints, boundary conditions). This "recalculation" is especially natural in the case of a linear control system under the employment of the Cauchy formula. Of course, very often in control problems the "direct" IC (having a resource character) are used. But, sometimes it is advisable to remove these "direct" IC of the resource character for the attainment of the following goal. Namely, by the above-mentioned step we obtain the interesting possibility in the part of the investigation of a peculiar "real" controllability under other (nonresource) constraints. In this case often unbounded (in a natural sense) admissible sets of controls arise; in addition, the corresponding admissibility of controls is characterized in terms of indirect conditions. So, it is possible to consider the following linear system

$$\dot{x}(t) = A(t)x(t) + B(t)f(t), \quad x(t_0) = x_0, \qquad (1.1)$$

in n-dimensional phase space $\mathbb{R}^n$; $t_0 \in \mathbb{R}$, $x_0 \in \mathbb{R}^n$, $t_0 \le t \le \vartheta_0$, where $\vartheta_0 \in]t_0, \infty[$. In (1.1) $f = (f(t) \in \mathbb{R}^r, t_0 \le t < \vartheta_0)$ is nonnegative (component-wise) vector control. Now postulate that $f \in F$, where F is the set of all such nonnegative piecewise constant and continuous from the right r-vector-functions $f = (f_1, ..., f_r)$ on $I \triangleq [t_0, \vartheta_0[$ (below a more general case is considered). Under $f \in F$ we interprete $f_i(t) \in [0, \infty[$ as the intensity of an power supply in the i-channal (of control system) at the moment $t \in I$. Suppose that in (1.1) $A(\cdot)$ is a $n \times n$-matriciant on $I_0 \triangleq [t_0, \vartheta_0]$ with continious components $A_{i,j}(\cdot)$, $i \in \overline{1,n}$, $j \in \overline{1,n}$. Let $B(\cdot)$ be a $n \times r$-matriciant on I; suppose that all components $B_{i,j}(\cdot)$, $i \in \overline{1,n}$, $j \in \overline{1,r}$, of $B(\cdot)$ are uniform limits of piecewise constant and continuous from the right real-valued (r.-v.) functions on I. For the system (1.1) with a "positive" control f the "direct" IC have the following sense: the sum of all integrals of components of f is bounded above by a number $c, c \ge 0$; in addition, this number c is fixed. The concrete indirect condition can be obtained by the following requirement: $h'(t)x(t) \ge d(t)$ under $t \in I^0$; here the prime denotes the operation of transposition, I^0 is a nonempty subset of I_0, $h(\cdot)$ is a given r-vector-function on I^0 and $d(\cdot)$ is a r.-v. function on I^0.

[1] Supported by the Russian Foundation of Fundamental Researches, project no. 97-01-00458.

We consider distinct variants of weakening of the last constraints restricting us now for simplicity, to the consideration of the case of the finite set $I^0 : I^0 = \{t_i : i \in \overline{1,N}\}$. In this case we obtain the natural system of constraints in the form of a finite system of integral inequalities. Under the weakening of this condition it is possible to relax all or not all inequalities. In general case it is possible to use many other variants of a weakening of the initial system of constraints. Note that by virtue of the natural representation in terms of the Cauchy formula it is possible to transform the above-mentioned constraints system to an integral form. As a result, we obtain a variant of the general system (Chentsov, 1996a; Chentsov, 1997) of IC. Therefore, for the asymptotic analysis of limits of real trajectories of (1.1) it is possible to use the constructions (Chentsov, 1996a; Chentsov, 1997) of a weakening of IC (see, for example, (Chentsov, 1997, ch.1,2)).

2. EXTENSION IN THE CLASS OF VECTOR FINITELY ADDITIVE MEASURES

Consider an abstract version of (potentially unbounded) problem connected with the observance of IC. Fix a nonempty set E and a semialgebra (Chentsov, 1996a; Chentsov, 1997; Neveu, 1964) $\mathcal{L}$ of subsets of E. Let $\mathbb{A}(\mathcal{L})$ be the set of all r.-v. finitelly additive measures (FAM) with a bounded variation on $\mathcal{L}$; the linear space $\mathbb{A}(\mathcal{L})$ (here and below linear operations, product and order in spaces of functionals are defined as pointwise) is generated by the cone $(add)_+[\mathcal{L}]$ of all nonnegative r.-v. FAM on $\mathcal{L}$. The strong norm of $\mathbb{A}(\mathcal{L})$ is defined by the complete variation (see (Chentsov, 1996a; Chentsov, 1997). Denote by $B_0(E,\mathcal{L})$ the set of all r.-v. $\mathcal{L}$-step-functions on E. Let $B(E,\mathcal{L})$ be the closure of $B_0(E,\mathcal{L})$ in the topology of the sup-norm $\|\cdot\|$ (Danford and Schwartz, 1958, p.261) of Banach space $\mathbb{B}(E)$ of all bounded r.-v. functons on E. Then $B(E,\mathcal{L})$ (as a subspace of $(\mathbb{B}(E), \|\cdot\|)$) is Banach space for which $\mathbb{A}(\mathcal{L})$ equipped with the strong norm-variation defines the topologically conjugate space (Chentsov, 1996a, p.71),(Chentsov, 1997, p.41), (Danford and Schwartz, 1958, ch.IV). We equip $\mathbb{A}(\mathcal{L})$ with the topologies $\tau_*(\mathcal{L})$, $\tau_0(\mathcal{L})$ (Chentsov, 1996a, p.70),(Chentsov, 1997, p.41) and consider $(add)_+[\mathcal{L}]$ as the corresponding subspace; $\tau_*^+(\mathcal{L})$ and $\tau_0^+(\mathcal{L})$ are the natural relative topologies of $(add)_+[\mathcal{L}]$ (see (Chentsov, 1996a, p.81)). Under $\mu \in \mathbb{A}(\mathcal{L})$ and $f \in B(E,\mathcal{L})$ we introduce $f * \mu \in \mathbb{A}(\mathcal{L})$ (Chentsov, 1996a, p.69) (the indefinite μ-integral of f). For each set T and $k \in \mathcal{N} \triangleq \{1;2;...\}$ (as usually) we exploit T^k in the traditional capacity of the k-multiple product of the samples of T (see (Chentsov, 1997, p.36)); if τ is a topology of T, then $\otimes^k[\tau]$ corresponds to (Chentsov, 1997, p.36) (the natural topology of product). Equip the real line $\mathbb{R}$ with the "ordinary" $|\cdot|$-topology $\tau_{\mathbb{R}}$ and with discrete topology τ_∂. For each nonempty set T and each topology τ of $\mathbb{R}$ we denote by $\otimes^T(\tau)$ the topology of Tichonoff product of samples $(\mathbb{R},\tau)$ with index set T (see (Chentsov, 1997, p.35)). Denote by $B_0^+(E,\mathcal{L})$ and $B^+(E,\mathcal{L})$ the "positive" (nonnegative) cones of $B_0(E,\mathcal{L})$ and $B(E,\mathcal{L})$ respectively. Fix $\eta \in (add)_+[\mathcal{L}]$ and introduce the cone $(add)^+[\mathcal{L};\eta]$ (see (Chentsov, 1996a, p.79); (Chentsov, 1997, p.43)), $(add)^+[\mathcal{L};\eta] \subset (add)_+[\mathcal{L}]$, of all weakly η-continuous nonnegative FAM on $\mathcal{L}$. Use vector-functions and vector measures in the correspondence with (Chentsov, 1998). Fix $r \in \mathcal{N}$ and understand $(add)_r^+[\mathcal{L}]$, $(add)_r^+[\mathcal{L} \mid \eta]$, $B_r^+[E;\mathcal{L}]$, $B_{0,r}^+[E;\mathcal{L}]$ and $\mathbf{N}_r^+(\mathcal{L})$ in the correspondence with (Chentsov, 1998) (for example: $(add)_r^+[\mathcal{L}] = (add)_+[\mathcal{L}]^r$). Introduce (see (Chentsov, 1998)) the operator $\mathbb{I}$ from $\mathbb{M}_r^+(\mathcal{L}) \triangleq (add)_+[\mathcal{L}] \times B_r^+[E;\mathcal{L}]$ into $(add)_r^+[\mathcal{L}]$ by the condition $\mathbb{I}(\mu,(f_i)_{i\in\overline{1,r}}) \triangleq (f_i * \mu)_{i\in\overline{1,r}}$. Use $\mathbf{i}_\eta \triangleq (\mathbb{I}(\eta,\cdot) \mid B_{0,r}^+[E;\mathcal{L}])$ (Chentsov, 1998), obtaining the mapping from $B_{0,r}^+[E;\mathcal{L}]$ into $(add)_r^+[\mathcal{L} \mid \eta]$.

Introduce the abstract system of IC. Fix a nonempty set Γ (below, $\mathbb{R}^\Gamma$ is the set of all functionals on Γ), an operator

$$(\gamma,i) \mapsto S_{\gamma,i} : \Gamma \times \overline{1,r} \to B(E,\mathcal{L})$$

and a closed (in $(\mathbb{R}^\Gamma, \otimes^\Gamma(\tau_{\mathbb{R}}))$) set Y, $Y \subset \mathbb{R}^\Gamma$. Consider the mapping $\mathbb{S}_* : (add)_r^+[\mathcal{L}] \to \mathbb{R}^\Gamma$ for which

$$\mathbb{S}_*((\mu_i)_{i\in\overline{1,r}}) \triangleq (\sum_{i=1}^r \int_E S_{\gamma,i} d\mu_i)_{\gamma\in\Gamma}.$$

Moreover, introduce $\mathbb{S} \triangleq (\mathbb{S}^* \mid (add)_r^+[\mathcal{L} \mid \eta])$ (Chentsov, 1998) for the consideration of the "extended" condition $\mathbb{S}(\mu) \in Y$. Then $\mathbb{S}^{-1}(Y)$ is the set of admissible elements for this condition. Consider $\mathbf{s} \triangleq \mathbb{S} \circ \mathbf{i}_\eta$ as a "realizable" integral operator

$$(f_i)_{i\in\overline{1,r}} \mapsto (\sum_{i=1}^r \int_E S_{\gamma,i} f_i d\eta)_{\gamma\in\Gamma} :$$

$$B_{0,r}^+[E;\mathcal{L}] \to \mathbb{R}^\Gamma. \qquad (2.1)$$

In particular, it is possible to use the concrete version of (2.1) for the representation of trajectories of the system (1.1). This possibility is connected with the Cauchy formula. As a corollary, for many profound problems of section 1 it is possible to represent the corresponding IC in the form of the requirement

$$\mathbf{s}(f) \in Y. \qquad (2.2)$$

It is possible to relate other concrete variants of the last condition (see (Chentsov, 1997, ch.VI)).

Return to the condition (2.2) in its general form. We have (Chentsov, 1996a, ch.I); (Chentsov, 1997, ch.1,2) the very "nonregular" IC (see (2.1), (2.2)). Therefore, we give the basic attention to "regularizations" on the basis of $\mathbb{S}^{-1}(Y)$. And what is more, the corresponding "regularization" is a highly universal with respect to different variants of the weakening of IC (2.1), (2.2). The principle moment is connected with employment of FAM in the capacity of generalized elements (g.e.). Namely, in the space of such g.e. $\mathbb{S}^{-1}(Y)$ defines an "universal" attraction set in the class of appoximate solutions. For the clarification of this "universality" (see (Chentsov, 1998)) we consider only two polar variants of pertubations of (2.2). But, first we recall the version of an attraction set used in (Chentsov, 1996a; Chentsov, 1997), (Chentsov, 1998); in addition, we follow (Chentsov, 1996a, p.39,40); (Chentsov, 1997, p.35). Therefore, consider the following

Remark. Denote by $\mathcal{B}[X]$ the set of all nonempty families $\mathcal{X}$ of subsets of a set X with the property $\forall A \in \mathcal{X} \ \forall B \in \mathcal{X} \ \exists C \in \mathcal{X} : \ C \subset A \cap B$; if $\mathcal{H} \in \mathcal{B}[X]$ such that $\emptyset \notin \mathcal{H}$, then $\mathcal{H}$ is the basis of a filter of X. Denoting by $cl(\cdot, \tau)$ the closure operator in an arbitrary topological space $(\mathbb{T}, \tau)$, we introduce the natural set-theoretical limit. Namely, if the set X is equipped with the family $\mathcal{X} \in \mathcal{B}[X]$, $(\mathbb{T}, \tau)$ is a topological space and s is an operator from X into $\mathbb{T}$, then by definition $(\tau - LIM)[\mathcal{X} \mid s]$ is the intersection of all sets $cl(s^1(U), \tau)$, $U \in \mathcal{X}$, where $s^1(\cdot)$ is used for the designation of the s-image operation. Note the very useful particular case: $\mathcal{X}$ has a countable basis and $(\mathbb{T}, \tau)$ is the space with the first axiom of countability. Then (Chentsov, 1997, p.38) $(\tau - LIM)[\mathcal{X} \mid s]$ is the attraction set in the class of sequential approximate solutions, i.e. in fact, $(\tau - LIM)[\mathcal{X} \mid s]$ is an "ordinary attractor". In the general case of $\mathcal{X}$ and τ for the realization of points of $(\tau - LIM)[\mathcal{X} \mid s]$ it should be used nets.

Return to the basic problem. By $Fin(T)$ denote the family of all nonempty finite subsets of a set T. Let $\Gamma_0 \stackrel{\triangle}{=} \{\gamma \in \Gamma \mid \forall i \in \overline{1,r} : S_{\gamma,i} \in B_0(E, \mathcal{L})\}$. If $y \in \mathbb{R}^\Gamma$, $K \in Fin(\Gamma)$ and $\varepsilon \in]0, \infty[$, then introduce the set $N_0(y, K, \varepsilon)$ of all functionals h on Γ such that $(\forall \gamma \in K \cap \Gamma_0 : \ y(\gamma) = h(\gamma)) \& (\forall \gamma \in K \backslash \Gamma_0 : \mid y(\gamma) - h(\gamma) \mid < \varepsilon)$. Moreover, introduce in the consideration the Hausdorff topology $\mathcal{I}_0$ of $\mathbb{R}^\Gamma$ in the form of the family of all sets G, $G \subset \mathbb{R}^\Gamma$, for each of which $\forall g \in G$ $\exists K \in Fin(\Gamma) \ \exists \varepsilon \in]0, \infty[: \ N_0(g, K, \varepsilon) \subset G$. Of course, $\otimes^\Gamma(\tau_\mathbb{R}) \subset \mathcal{I}_0$. Let $\Gamma_1 \stackrel{\triangle}{=} \Gamma \backslash \Gamma_0$. Under $\Gamma_0 \neq \emptyset$ and $\Gamma_1 \neq \emptyset$ we have, in the form of $(\mathbb{R}^\Gamma, \mathcal{I}_0)$ and $\mathbb{R}^{\Gamma_0} \times \mathbb{R}^{\Gamma_1}$ in the topology of the product of the spaces $(\mathbb{R}^{\Gamma_0}, \otimes^{\Gamma_0}(\tau_\partial))$ and $(\mathbb{R}^{\Gamma_1}, \otimes^{\Gamma_1}(\tau_\mathbb{R}))$, the gomeomorphs. For $\Gamma_0 = \emptyset$ and $\Gamma_0 = \Gamma$ we have $\mathcal{I}_0 = \otimes^\Gamma(\tau_\mathbb{R})$ and $\mathcal{I}_0 = \otimes^\Gamma(\tau_\partial)$ respectively. If

$\Gamma_0 = \Gamma \in Fin(\Gamma)$, then $\mathcal{I}_0$ is the discrete topology of $\mathbb{R}^\Gamma$; if $\Gamma_1 = \Gamma \in Fin(\Gamma)$, then $\mathcal{I}_0 = \otimes^\Gamma(\tau_\mathbb{R})$ is the natural topology of coordinate-wise convergence finite-dimensional space. Denote by $\mathfrak{I}_0$ the family of all neighboinrhoods (Bourbaki, 1968, § 1.2) of Y in space $(\mathbb{R}^\Gamma, \mathcal{I}_0)$. Consider the family $\mathbf{s}^{-1}[\mathfrak{I}_0] \stackrel{\triangle}{=} \{\mathbf{s}^{-1}(H) : H \in \mathfrak{I}_0\} \in \mathcal{B}[B_{0,r}^+[E; \mathcal{L}]]$ in the capacity of the first "rigid" variant of the perturbation of IC (2.1), (2.2). Under this variant many requirements on the choice of a (simplest) control are preserved. Consider the second variant. For $K \in Fin(\Gamma)$ and $\varepsilon \in]0, \infty[$ we denote by $< Y \mid K, \varepsilon >$ the set of all functions (in fact: vectors) $z : K \to \mathbb{R}$ such that $\exists y \in Y \ \forall \gamma \in K : \mid y(\gamma) - z(\gamma) \mid \leq \varepsilon$; moreover, if $\mathcal{K} \in Fin(\mathcal{L})$, then $\Omega(\mathcal{K}, K, \varepsilon)$ is the set of all $(\mu, f) \in \mathbb{M}_+^+(\mathcal{L})$ for each of which

$$((\mu \mid \mathcal{K}) = (\eta \mid \mathcal{K})) \ \& \ (\exists z \in < Y \mid K, \varepsilon > \ \forall \gamma \in K :$$
$$\mid (\mathbb{S}_* \circ \mathbb{I})(\mu, f)(\gamma) - z(\gamma) \mid \leq \varepsilon).$$

Then $\mathfrak{A} \stackrel{\triangle}{=} \{\Omega(\mathcal{K}, K, \varepsilon) : \ (\mathcal{K}, K, \varepsilon) \in Fin(\mathcal{L}) \times Fin(\Gamma) \times]0, \infty[\} \in \mathcal{B}[\mathbb{M}_+^+(\mathcal{L})]$. The following statement is true.

Theorem 2.1. $\forall \tau \in \mathbf{N}_r^+(\mathcal{L}) : \ \mathbb{S}^{-1}(Y) = (\tau - LIM)[\mathbf{s}^{-1}[\mathfrak{I}_0] \mid \mathbf{i}_\eta] = (\tau - LIM)[\mathfrak{A} \mid \mathbb{I}]$.

The basic singularity of Theorem 2.1 is the following property. Namely, we have in the form $\mathbb{S}^{-1}(Y)$ the "universal" regularization of the admissible set $\mathbf{s}^{-1}(Y)$ in conditions of the potentially unbounded (in strong sense) problem. It is possible to supplement Theorem 2.1 by statements about many intermediate variants of the perturbation of IC (2.1), (2.2); see (Chentsov, 1998).

3. ASYMPTOTIC ATTAINABILITY UNDER INTEGRAL CONSTRAINTS

Fix a topological space $(\mathbb{X}, \mathcal{T})$, $\mathbb{X} \neq \emptyset$, and a continuous (in the sense of the topologies $\otimes^r[\tau_*^+(\mathcal{L})]$ and $\mathcal{T}$) operator w from $(add)_r^+[\mathcal{L}]$ into $\mathbb{X}$; let $W \stackrel{\triangle}{=} w \circ \mathbb{I}$ and $\mathcal{W} \stackrel{\triangle}{=} (W(\eta, \cdot) \mid B_{0,r}^+[E; \mathcal{L}])$. Interpret $\mathcal{W}$ as an "operator of a system" (for example, in this capacity it is possible to consider the operator entry — exit of the system (1.1)). In addition,

$$w^1(\mathbb{S}^{-1}(Y)) \subset (\mathcal{T} - LIM)[\mathbf{s}^{-1}[\mathfrak{I}_0] \mid$$
$$\mathcal{W}] \subset (\mathcal{T} - LIM)[\mathfrak{A} \mid W]. \qquad (3.1)$$

Theorem 3.1. If w is perfect (Engelking, 1986, p.277) mapping in the sense of $\otimes^r[\tau_*^+(\mathcal{L})]$ and $\mathcal{T}$, then all three sets in (3.1) coincide.

The corresponding reasoning is analogous to (Chentsov, 1997, p.105). As a corollary, we note that in the case when w is a gomeomorphism in the sense of $\otimes^r[\tau_*^+(\mathcal{L})]$ and $\mathcal{T}$, all three sets in (3.1) coincide. We establish conditions sufficient for the regularization of the problem about constructing the "attainable set" $\mathcal{W}^1(\mathbf{s}^{-1}(Y))$ in terms of $w^1(\mathbb{S}^{-1}(Y))$. Consider a basis of some applications of the above-mentioned corollary of Theorem 3.1. For $h \in B(E,\mathcal{L})$ denote by $\mathcal{J}_h$ the (continuous in the sense of $\tau_*(\mathcal{L})$) operator acting in $\mathbb{A}(\mathcal{L})$ by the rule $\mathcal{J}_h(\mu) \triangleq h * \mu$. Introduce $\mathbf{B}^0(E,\mathcal{L}) \triangleq \{h \in B(E,\mathcal{L}) \mid \exists \varepsilon \in]0,\infty[\ \forall x \in E : \varepsilon \leq \mid h(x) \mid\}$ and $\mathbf{B}^0_+(E,\mathcal{L}) \triangleq \mathbf{B}^0(E,\mathcal{L}) \cap B^+(E,\mathcal{L})$. For $h \in \mathbf{B}^0(E,\mathcal{L})$ the mapping $\mathcal{J}_h$ is the gomeomorphism of $(\mathbb{A}(\mathcal{L}),\tau_*(\mathcal{L}))$ on itselfs; in this case the mapping inverse with respect to $\mathcal{J}_h$ is $\mathcal{J}_l$, where $l \in B(E,\mathcal{L})$ is defined by the rule $l(x) = (h(x))^{-1}$. For $h \in \mathbf{B}^0_+(E,\mathcal{L})$ in the form $(\mathcal{J}_h \mid (add)_+[\mathcal{L}])$ we have the gomeomorphism of $((add)_+[\mathcal{L}],\tau_*^+(\mathcal{L}))$ on itselfs; the structure of the inverse operator is defined by the previous general construction with the employment of the obvious restriction on $(add)_+[\mathcal{L}]$. The vector variant of the last procedure realizes (in particular) some useful representations having the sense of regularizations of the following problems for the system (1.1). Omiting concrete conditions of (Chentsov, 1998, § 5), we note only that for the realization of the above-mentioned "extended" condition in the case of the control problem for the system (1.1) it is advisable to use the construction of the generalized system similar (Chentsov, 1996*a*, p.136). Namely, for the realization of the requirement $\mathbb{S}(\mu) \in Y$ in this case it should be used g.e. (vector FAM) as generalized controls. Let $\Phi(\cdot,\cdot)$ be the matriciant of the system $\dot{x} = A(t)x$ considered on I_0. Here define $\mathcal{L}$ as the semialgebra of all the intervals $[a,b[$, $t_0 \leq a \leq b \leq \vartheta_0$, postulating that $\eta([\alpha,\beta[) \triangleq \beta - \alpha$, $t_0 \leq \alpha \leq \beta \leq \vartheta_0$. Then for $\mu \triangleq (\mu_i)_{i\in\overline{1,r}} \in (add)_r^+[\mathcal{L} \mid \eta]$ we consider the "trajectory" with conponents

$$t \mapsto (\Phi(t,t_0)x_0)_i + \sum_{j=1}^{r} \int_{[t_0,t[} (\Phi(t,\xi)B(\xi))_{i,j}\mu_j(d\xi) : \quad I_0 \to \mathbb{R}, \quad i \in \overline{1,n}. \quad (3.2)$$

In terms of (3.2) it is possible to construct the concrete version of condition $\mathbb{S}(\mu) \in Y$ for the cases when Y-constraints are formed as a system of requirements on trajectory of (1.1). In this connection note settings and constructions of (Krasovskii, 1968).

4. STRUCTURE OF ATTRACTION SETS IN TOPOLOGICAL SPACES

In connection with (3.1) and Theorem 3.1 it is advisable to consider some general questions of the representation of attraction sets in topological spaces. We use designations and notions of Remark of section 2. If (U,τ), $U \neq \emptyset$, and (V,ϑ), $V \neq \emptyset$, are topological spaces, then we denote by $C(U,\tau,V,\vartheta)$ the set of all (τ,ϑ)-continuous mappings from U into V; moreover, we denote by $[C(\tau,\vartheta) - clos]_0(U,V)$ the set of all closed (Engelking, 1986, ch.7) (in the sense of the topologies τ and ϑ) mappings from U into V; $[C(\tau,\vartheta) - clos]_0(U,V) \subset C(U,\tau,V,\vartheta)$. Moreover, for each set X we introduce the set $\mathcal{B}_\mathcal{N}(X)$ of all families $\mathcal{H} \in \mathcal{B}[X]$ with a countable basis in the sense of the relation (3.3.17) of monograph (Chentsov, 1997) (for $\mathcal{X} \in \mathcal{B}_\mathcal{N}(X)$ it is possible to choose a sequence $(U_i)_{i\in\mathcal{N}}$ in $\mathcal{X}$ for which $\forall U \in \mathcal{X}\ \exists k \in \mathcal{N} : U_k \subset U$).

Proposition 4.1. Let: 1)X be a nonempty set equipped with a family $\mathcal{X} \in \mathcal{B}_\mathcal{N}(X)$; 2)$(Y,\tau)$, $Y \neq \emptyset$, be a topological T_1-space (Engelking, 1986); 3)(K,θ), $K \neq \emptyset$, be a countably compact space (Engelking, 1986); 4)m be a mapping from X into K; 5)$g \in [C(\theta,\tau) - clos]_0(K,Y)$. Then

$$(\tau - LIM)[\mathcal{X} \mid g \circ m] = g^1((\theta - LIM)[\mathcal{X} \mid m]). \quad (4.1)$$

In connection with Proposition 4.1 it is advisable to recall the useful Proposition 3.3.1 of (Chentsov, 1997). Really, in the case of the employment in the capacity of topological space (Y,τ) a T_1-space with the first axiom of countability we have in Proposition 4.1 the useful representation of the natural attraction set in the class of sequential approximate solutions. Moreover, we note that each compact space is countably compact (see (Engelking, 1986)). Therefore, the constructions of (Chentsov, 1996*a*; Chentsov, 1997) (used compactifications) are coordinated with Proposition 4.1. The last proposition can be added. We use the traditional definition of a countably compact set $\mathbb{C}$ in an arbitrary topological space (H,τ): the (relative) topology of $\mathbb{C}$, $\mathbb{C} \subset H$, considered as a subspace of (H,τ) converts this set $\mathbb{C}$ in a countably compact space. If (U,τ), $U \neq \emptyset$, and (V,θ), $V \neq \emptyset$, are topological spaces and (moreover) $h \in [C(\tau,\theta) - clos]_0(U,V)$, then one call h a quasiperfect (in the sense of the spaces (U,τ) and (V,θ)) mapping iff for each $v \in V$ the set $h^{-1}(\{v\})$ is countably compact in (U,τ).

Proposition 4.2. Let: 1)X be a nonempty set equipped with a family $\mathcal{X} \in \mathcal{B}_\mathcal{N}(X)$; 2)$(Y,\tau)$, $Y \neq \emptyset$, *and* (H,θ), $H \neq \emptyset$, are arbitrary topological spaces; 3)m be a mapping from X into H; 4)g be a quasiperfect (in the sense of (H,θ) and (Y,τ)) mapping from H into Y. Then the relation (4.1) is true.

Under the proof we use the following practically obvious corollary of the relations (3.3.16) and (3.3.17) of (Chentsov, 1997). Namely, for $\mathcal{X} \in \mathcal{B}_{\mathcal{N}}(X)$ it is possible to indicate a sequence $(U_i)_{i\in\mathcal{N}}$ in $\mathcal{X}$ for which $\forall U \in \mathcal{X}\ \exists k \in \mathcal{N}\ \forall m \in \mathcal{N} : (k \leq m) \Longrightarrow (U_m \subset U)$. Moreover, we use the known property of countable centered systems of closed sets in a countably compact space. It is useful to use Proposition 4.2 in the combination with Proposition 3.3.1 of (Chentsov, 1997).

Proposition 4.3. Let: 1)X be a nonempty set equipped with a family $\mathcal{X} \in \mathcal{B}[X]$; 2)$(Y,\tau)$, $Y \neq \emptyset$, and (H,θ), $H \neq \emptyset$, be topological spaces; 3) m be a mapping from X into H; 4) g be an almost perfect (Engelking, 1986) (in the sense of (H,θ) and (Y,τ)) mapping from H into Y. Then the relation (4.1) is true.

From Proposition 4.3 it is possible to extract Theorem 2.5.2 of (Chentsov, 1996*a*). For this we note that in the case of a compact space (H,θ) and a Hausdorff space (Y,τ) for each $f \in C(H,\theta,Y,\tau)$ we have the following known (Engelking, 1986) property: f is an almost perfect mapping (in the sense of (H,θ) and (Y,τ)). Recall that the above-mentioned Theorem 2.5.2 of (Chentsov, 1996*a*) is used in constructions of extension of extremal problems (see, for example, (Chentsov, 1996*a*, ch.2,7-9) and (Chentsov, 1997, ch.5)). Therefore, in (4.1) we obtain a very useful representation of the attraction set corresponding to the problem of asymptotic attainability on values of the mapping $g \circ m$. We note a very general statement having the sense of an estimate.

Proposition 4.4. Let: 1) X be a nonempty set equipped with a family $\mathcal{X} \in \mathcal{B}[X]$; 2)$(Y,\tau)$, $Y \neq \emptyset$, and (H,θ), $H \neq \emptyset$, be topological spaces; 3) m be a mapping from X into H; 4) $g \in C(H,\theta,Y,\tau)$. Then

$$g^1((\theta - LIM)[\mathcal{X} \mid m]) \subset (\tau - LIM)[\mathcal{X} \mid g \circ m].$$

The proof exploits simplest properties of continuous mappings. Note one useful cotrollary. Namely, with the employment of Proposition 4.4 it is possible to establish the existence of asymptotically attainable elements in the space (Y,τ) in terms of the nonemptyness of the corresponding attraction set in the "auxiliary" space (H,θ). In the capacity of the example of such a type we note the relation (3.1). Namely, the set on the left-hand side of (3.1) is (by virtue of Theorem 2.1) the required lower bound of the basic attraction sets.

In Propositions 4.1—4.4 we have highly general representations for attraction sets (now we omit many generalizations of such a type). Theorems 2.1 and 3.1 are concrete variant of these Propositions (for example, Theorem 3.1 is the concrete variant of Proposition 4.3). Now we note basic statements of a general character permitting to consider in Theorems 2.1 and 3.1 very different variants of a weakening of the initial IC.

Fix three nonempty sets: $\mathbf{F}$, $\mathbf{X}$ and $\mathbf{H}$. Moreover, we fix a set $\mathbf{Y}$, $\mathbf{Y} \subset \mathbf{X}$, and mappings

$$s : \mathbf{F} \longrightarrow \mathbf{X}, \quad h : \mathbf{F} \longrightarrow \mathbf{H}. \tag{4.2}$$

Consider the condition $s(f) \in \mathbf{Y}$ (of course, $f \in \mathbf{F}$) that is an abstract analog of (2.2). Then, by (4.2) $h^1(s^{-1}(\mathbf{Y}))$ is the set of all h-attainable elements of $\mathbf{H}$ under this condition. Of course, $s^{-1}(\mathbf{Y})$ is the set of all admissible elements. We equip $\mathbf{X}$ and $\mathbf{H}$ with the topologies τ and θ respectively. Consider the attainable sets $h^1(s^{-1}(\mathbf{G}))$, where $\mathbf{G}$ is a neighborhood of $\mathbf{Y}$ in $(\mathbf{X},\tau)$. Here and below we call the set M, $M \subset \mathbf{X}$, a τ-neighborhood of the point $x \in \mathbf{X}$ (of the set N, $N \subset \mathbf{X}$) iff it is possible to choose an open set $G \in \tau$ for which $x \in G$ (for which $N \subset G$) and $G \subset M$. Of course, in the given construction $\mathbf{G}$ "converges" to $\mathbf{Y}$. In addition, it is possible to use all or not all neighborhoods of $\mathbf{Y}$. Often it is possible to choose a natural "part" of the family $\mathcal{Y}$ of all τ-neighborhoods of $\mathbf{Y}$; this "part" corresponds to practically interesting variants of the weakening of the basic $\mathbf{Y}$-constraint. But, it is advisable to compare the possibilities in the asymptotic realization of elements of $\mathbf{H}$ under the employment of different subfamilies of $\mathcal{Y}$.

For IC of section 1-3 this comparison is especially important for unbounded (in the strong sense) problems. In the conformity with designations of section 2 for each family $\mathfrak{X}$ of subsets of $\mathbf{X}$ we denote by $s^{-1}[\mathfrak{X}]$ the family of all sets $s^{-1}(M)$, $M \in \mathfrak{X}$; we have $\forall \tilde{\mathfrak{X}} \in \mathcal{B}[\mathbf{X}] : s^{-1}[\tilde{\mathfrak{X}}] \in \mathcal{B}[\mathbf{F}]$. In particular, $s^{-1}[\mathcal{Y}] \in \mathcal{B}[\mathbf{F}]$. Consider one typical variant of a subfamily of $\mathcal{Y}$ using the scheme of an uniformization.

So, we fix a set Q, $Q \neq \emptyset$, and introduce a mapping Λ from $\mathbf{X} \times Q$ into the family of all subsets of $\mathbf{X}$ with the following properties: 1)for each $x \in \mathbf{X}$ and $q \in Q$ the set $\Lambda(x,q)$ is a τ-neighborhood of x in the space $(\mathbf{X},\tau)$; 2)$\forall q_1 \in Q\ \forall q_2 \in Q\ \exists q_3 \in Q\ \forall x \in \mathbf{X}$:

$$\Lambda(x,q_3) \subset \Lambda(x,q_1) \cap \Lambda(x,q_2).$$

If $q \in Q$, then the union of all sets $\Lambda(y,q)$, $y \in \mathbf{Y}$, is an element of $\mathcal{Y}$. And what is more,

$$\mathfrak{Y} \triangleq \{\bigcup_{y\in\mathbf{Y}} \Lambda(y,q) : q \in Q\} \in \mathcal{B}[\mathbf{X}].$$

Theorem 4.1. If the conditions

$$(\exists g \in C(\mathbf{H},\theta,\mathbf{X},\tau) : s = g \circ h) \& (\forall x \in \mathbf{X}\setminus\mathbf{Y}\ \exists q \in Q :$$
$$\Lambda(x,q) \cap (\bigcup_{y\in\mathbf{Y}} \Lambda(y,q)) = \emptyset) \tag{4.3}$$

are correct, then the following equality for attraction sets takes place:

$$(\theta - LIM)[s^{-1}[\mathcal{Y}] \mid h] = (\theta - LIM)[s^{-1}[\mathfrak{Y}] \mid h].$$

Discuss the condition (4.3). The first part of (4.3) is a very typical (for constructions of (Chentsov, 1996*a*; Chentsov, 1997)) condition: see, for example, the above-cited Theorem 2.5.2 of (Chentsov, 1996*a*). In the connection with the second part of (4.3) it is possible to note the following example. Namely, let $\mathbf{X}$ be the set equipped with a pseudometric

$$\rho: \mathbf{X} \times \mathbf{X} \longrightarrow [0, \infty[.$$

Moreover, let τ be the topology of $\mathbf{X}$ generated by ρ (τ is the ρ-topology of $\mathbf{X}$). Let $Q =]0, \infty[$. Suppose that for each $x \in \mathbf{X}$ and $q \in Q$ the representation

$$\Lambda(x, q) = \{\tilde{x} \in \mathbf{X} \mid \rho(x, \tilde{x}) < q\}$$

takes place. Finally, consider the natural case: $\mathbf{Y}$ is a closed (in $(\mathbf{X}, \tau)$) subset of $\mathbf{X}$. Then the second part of (4.3) is correct. With the considered particular case the following profound circumstance is connected. Namely, consider the widely known case $\mathbf{X} = \mathbb{R}^m$ where $m \in \mathcal{N}$. Suppose that ρ is the Euclidean metric of $\mathbf{X}$ and $\mathbf{Y}$ is a closed subset of $\mathbf{X}$. Then traditional ε-neighborhoods of $\mathbf{Y}$ (under the enumeration of all $\varepsilon \in]0, \infty[$) not compose, generally speaking, a fundamental system of neighborhoods of $\mathbf{Y}$. But, Theorem 4.1 permits to investigate this case without the special supposition about a boudedness of the s-image of the set $\mathbf{F}$.

Remark. Note that in Theorem 2.1 the particular variant of Theorem 4.1 is realized. This circumstance is highly essential since for IC considered in sections 2, 3 we assume the case of the unboundedness in the strong sense (by the variation). Therefore, the given case (see sections 2, 3) does not admit, generally speaking, the employment of compactifications in the sense of (Chentsov, 1996*a*, ch.2). But, we note that $\otimes^r[\tau_*^+(\mathcal{L})]$ is the locally compact topology of $(add)_r^+[\mathcal{L}]$.

5. ASYMPTOTIC ATTAINABILITY IN THE CLASS OF CONTROLS WITH ALTERNATING SIGNS

In this section we refuse the reguirement of the nonnegativity of controls that in sections 1-3 was used. As a result, some singularities of the topological character (see (Chentsov, 1996*a*, ch.IV)) arise. Therefore, we several change the construction of attraction sets adding to the used construction the natural stabilizer in the form of the strong norm-variation. Note that controls with alternating signs are used in control problems very often. The following construction is the development of scheme of (Chentsov, 1997, ch.6). We are restricted to the consideration of the case of scalar controls. Under $\mu \in \mathbb{A}(\mathcal{L})$ we denote by v_μ the variation of μ as a set function on $\mathcal{L}$ (see (Chentsov, 1996*a*, p.61), (Chentsov, 1997, p.39)), $v_\mu \in (add)_+[\mathcal{L}]$, and consider the functional

$$\mu \mapsto v_\mu(E): \quad \mathbb{A}(\mathcal{L}) \to [0, \infty[\tag{5.1}$$

as the strong norm of $\mathbb{A}(\mathcal{L})$. In terms of the norm (5.1) we introduce the natural bounded $*$-weak topology $\tau_{\mathbb{B}}^*(\mathcal{L})$; in addition, we follow (Chentsov, 1997, p.45). Let $\mathfrak{M}_*(\mathcal{L}) \triangleq \{\tau_*(\mathcal{L});\ \tau_{\mathbb{B}}^*(\mathcal{L})\}$ and $\mathfrak{M}(\mathcal{L}) \triangleq \mathfrak{M}_*(\mathcal{L}) \cup \{\tau_\otimes(\mathcal{L});\ \tau_0(\mathcal{L})\} = \{\tau_*(\mathcal{L});\ \tau_{\mathbb{B}}^*(\mathcal{L});\ \tau_\otimes(\mathcal{L});\ \tau_0(\mathcal{L})\}$ (see the designations for topologies in (Chentsov, 1996*a*, p.80), (Chentsov, 1997, p.44)). Introduce

$$\mathbb{A}_\eta[\mathcal{L}] \triangleq \{\mu \in \mathbb{A}(\mathcal{L}) \mid \forall L \in \mathcal{L}:$$
$$(\eta(L) = 0) \Longrightarrow (\mu(L) = 0)\}$$

obtaining the linear space generated by the cone $(add)^+[\mathcal{L}; \eta]$. Suppose $\mathbb{M}(\mathcal{L}) \triangleq (add)_+[\mathcal{L}] \times B(E, \mathcal{L})$ and denote by $\mathcal{I}$ the mapping

$$(\mu, f) \mapsto f * \mu: \quad \mathbb{M}(\mathcal{L}) \to \mathbb{A}(\mathcal{L}).$$

Let $\mathbf{j}_\eta \triangleq (\mathcal{I}(\eta, \cdot) \mid B_0(E, \mathcal{L}))$; so, $\mathbf{j}_\eta(f) = f * \eta \in \mathbb{A}_\eta[\mathcal{L}]$ for $f \in B_0(E, \mathcal{L})$. In the following along with the sets Γ and Y corresponding to the conditions of section 2 we fix a nonempty set Ω and mappings:

$$\gamma \mapsto S_\gamma: \quad \Gamma \to B(E, \mathcal{L}), \tag{5.2}$$

$$\omega \mapsto L_\omega: \quad \Omega \to \mathcal{L}, \tag{5.3}$$

$$\omega \mapsto c_\omega: \quad \Omega \to [0, \infty[. \tag{5.4}$$

It should be distinguish (5.2) away from the basic operator of section 2. We use (5.2) in a "new" Y-constraint; recall that Y is the closed set in $(\mathbb{R}^\Gamma, \otimes^\Gamma(\tau_\mathbb{R}))$. The mappings (5.3) and (5.4) are used in partial "resource" constraints. For $f \in B_0(E, \mathcal{L})$ consider the following system of conditions corresponding to (Chentsov, 1997, p.242):

$$((\int_E S_\gamma f\, d\eta)_{\gamma \in \Gamma} \in Y)\ \&(\forall \omega \in \Omega:$$

$$\int_{L_\omega} \mid f \mid d\eta \leq c_\omega). \tag{5.5}$$

Note that even in the case of the finite set Ω the corresponding admissible set for the condition (5.5) is not, generally speaking, bounded in the sense of values of η-integrals for $\mid f \mid$ on the set E for elements f (of this admissible set), since the sets L_ω, $\omega \in \Omega$, defined by (5.3), can not generate a cover of E. As a corollary, under the natural immersion of the set of "ordinary controls" in the space of generalized elements we do not obtain, generally speaking, the boundedness of the corresponding image (of the admissible set) in the sense of the strong norm (5.1). Introduce some generalized problem. Let $\mathbf{S}_*$ be the mapping from $\mathbb{A}(\mathcal{L})$ into $\mathbb{R}^\Gamma$ for which

$$\mathbf{S}_*(\mu) \triangleq \left(\int_E S_\gamma d\mu\right)_{\gamma\in\Gamma} \tag{5.6}$$

under $\mu \in \mathbb{A}(\mathcal{L})$. Of course (see section 4), $\mathbf{S}_* \in C(\mathbb{A}(\mathcal{L}), \tau_*(\mathcal{L}), \mathbb{R}^\Gamma, \otimes^\Gamma(\tau_\mathbb{R}))$. It is useful to introduce $\mathbf{S} \triangleq (\mathbf{S}_* \mid \mathbb{A}_\eta[\mathcal{L}])$; the obtained operator from $\mathbb{A}_\eta[\mathcal{L}]$ into $\mathbb{R}^\Gamma$ takes values (5.6). In conformity with (Chentsov, 1997, p.244) suppose

$$\tilde{\mathbb{A}}_0 \triangleq \{\mu \in \mathbf{S}^{-1}(Y) \mid \forall \omega \in \Omega : v_\mu(L_\omega) \leq c_\omega\}. \tag{5.7}$$

Follow the scheme of (Chentsov, 1997, ch.6), adding some new statements connected with a highly "universal" (in the sense of different variants of the perturbations introduction) character of the regularization (5.7) of the admissible set for the initial problem. As in sections 2 and 3, here only two polar versions of the weakening of the conditions (5.5) are considered. We use the topology $\mathcal{I}_0$ of section 2 and the mapping $\mathcal{S}$ from $B_0(E,\mathcal{L})$ into $\mathbb{R}^\Gamma$ such that $\forall f \in B_0(E,\mathcal{L})$:

$$\mathcal{S}(f) \triangleq (\int_E S_\gamma f d\eta)_{\gamma\in\Gamma}. \tag{5.8}$$

Of course, it is possible to consider $\mathbf{S}$ in the capacity of extension of $\mathcal{S}$ (5.8); in this connection see (Chentsov, 1996*a*, ch.4) and (Chentsov, 1997, ch.3,6). By analogy with constructions of (Chentsov, 1997, ch.6) for each sets H, $H \subset \mathbb{R}^\Gamma$, and Q, $Q \subset \Omega$, the set $(Adm)[H;Q]$ is defined as the set of all $f \in \mathcal{S}^{-1}(H)$ such that

$$\forall \omega \in Q : \int_{L_\omega} | f | d\eta \leq c_\omega.$$

It is advisable to change one previous definition. Namely, in the following we postulate that

$$\Gamma_0 \triangleq \{\gamma \in \Gamma \mid S_\gamma \in B_0(E,\mathcal{L})\}.$$

In this terms the following designations $\mathcal{I}_0$ and $\mathfrak{Z}_0$ correspond (below) to section 2. Moreover, we use

$$N_0(y,K,\varepsilon),\ y \in \mathbb{R}^\Gamma,\ K \in Fin(\Gamma),\ \varepsilon \in]0,\infty[,$$

in the correspondence with the new definition of Γ_0. Denote by $\mathfrak{A}_0$ the family of all sets

$$(Adm)[H;Q],\ H \in \mathfrak{Z}_0,\ Q \in Fin(\Omega);$$

then $\mathfrak{A}_0 \in \mathcal{B}[B_0(E,\mathcal{L})]$. For $K \in Fin(\Gamma)$ and $\varepsilon \in]0,\infty[$ the union $N_Y^0(K,\varepsilon)$ of all sets $N_0(y,K,\varepsilon)$, $y \in Y$, is an element of $\mathfrak{Z}_0$. The given type of neighborhoods of Y corresponds to the above-mentioned construction of the uniformization connected with Theorem 4.1. But, along with this uniformization it is advisable to consider some additional possibilities in the part of the weakening of the Y-constraint. Namely, introduce an analog of the family $\mathfrak{A}$ of sections 2 and 3. So (like (Chentsov, 1997, ch.6)) for $\mathcal{K} \in Fin(\mathcal{L})$, $P \in Fin(\Gamma)$, $Q \in Fin(\Omega)$ and $\varepsilon \in]0,\infty[$ denote by $(\mathbb{A}dm)[\mathcal{K};P;Q;\varepsilon]$ the set of all $(\mu,f) \in \mathbb{M}(\mathcal{L})$ such that

$$((\mu \mid \mathcal{K}) = (\eta \mid \mathcal{K}))\&(\exists y \in Y\ \forall \gamma \in P : | \int_E S_\gamma f d\mu - y(\gamma) | \leq \varepsilon)\&(\forall \omega \in Q : \int_{L_\omega} | f | d\mu \leq c_\omega + \varepsilon).$$

In addition, the family $\mathfrak{U}$ of all sets

$$(\mathbb{A}dm)[\mathcal{K};P;Q;\varepsilon],\ (\mathcal{K},P,Q,\varepsilon) \in Fin(\mathcal{L}) \times Fin(\Gamma) \times Fin(\Omega) \times]0,\infty[,$$

is the element of $\mathcal{B}[\mathbb{M}(\mathcal{L})]$. The following theorems are (in fact) natural corollaries of statements of (Chentsov, 1998).

Theorem 5.1. Under the equipment of $\mathbb{A}(\mathcal{L})$ with the topologies of $\mathfrak{M}_*(\mathcal{L})$ the set $\tilde{\mathbb{A}}_0$ is the common attraction set with respect to the families $\mathfrak{A}_0$ and $\mathfrak{U}$. Namely, $\forall \tau \in \mathfrak{M}_*(\mathcal{L}) : \tilde{\mathbb{A}}_0 = (\tau - LIM)[\mathfrak{A}_0 \mid \mathrm{j}_\eta] = (\tau - LIM)[\mathfrak{U} \mid \mathcal{I}]$.

Theorem 5.2. Let $\forall \gamma \in \Gamma : S_\gamma \in B_0(E,\mathcal{L})$. Then $\forall \tau \in \mathfrak{M}(\mathcal{L})$:

$$\tilde{\mathbb{A}}_0 = (\tau - LIM)[\mathfrak{A}_0 \mid \mathrm{j}_\eta] = (\tau - LIM)[\mathfrak{U} \mid \mathcal{I}].$$

Theorem 5.2 realizes the property of a very broad asymptotic equivalence of the generalized admissible set with respect to different variants of the constraints weakening and the topological equipment of the FAM space.

Again introduce in the consideration the topological space $(\mathbb{X},\mathcal{T})$ of section 3. We return to the problem about the asymptotic attainability in $(\mathbb{X},\mathcal{T})$, using another construction of attraction sets. Namely, we exploit the strong norm (5.1) in the capacity of a some stabilizer. In the capacity of approximate solutions here it should be used (in general case) nets (Engelking, 1986). Under the consideration of nets and the Moore-Smith convergence we follow symbolics of (Chentsov, 1997, p.33,34).

If $\mathbf{K}$ is a nonempty set and κ is an operator from $\mathbf{K}$ into $(add)_+[\mathcal{L}]$, then introduce (Chentsov, 1997, ch.6) the set $\mathbb{B}[\mathbf{K};E;\mathcal{L};\kappa]$ of all mappings h from $\mathbf{K}$ into $B(E,\mathcal{L})$ for each of which the functional

$$k \longmapsto \int_E \mid h(k) \mid d\kappa(k) : \mathbf{K} \longrightarrow [0, \infty[\quad (5.9)$$

is bounded. Moreover, for a set $\mathbf{K}$, $\mathbf{K} \neq \emptyset$, denote by $\mathbb{B}(\mathbf{K}, E, \mathcal{L}, \eta)$ the set $\mathbb{B}[\mathbf{K}; E; \mathcal{L}; \kappa]$ in the case when the mapping κ from $\mathbf{K}$ into $(add)_+[\mathcal{L}]$ is defined by the condition

$$\forall k \in \mathbf{K} : \ \kappa(k) \triangleq \eta;$$

then $\mathbb{B}_0(\mathbf{K}, E, \mathcal{L}, \eta) \triangleq \{h \in \mathbb{B}(\mathbf{K}, E, \mathcal{L}, \eta) \mid \forall k \in \mathbf{K} : \ h(k) \in B_0(E, \mathcal{L})\}$ is the set of all mappings p from $\mathbf{K}$ into $B_0(E, \mathcal{L})$ for each of which the functional

$$k \longmapsto \int_E \mid p(k) \mid d\eta : \mathbf{K} \longrightarrow [0, \infty[$$

is bounded. For each nonempty set $\mathbf{T}$ and arbitrary mappings g from $\mathbf{T}$ into $(add)_+[\mathcal{L}]$ and $h \in \mathbb{B}[\mathbf{T}; E; \mathcal{L}; g]$ the operator

$$t \longmapsto (g(t), h(t)) : \mathbf{T} \longrightarrow \mathbb{M}(\mathcal{L})$$

is designated by $\langle g, h \rangle$ and, as a corollary, $\mathcal{I} \circ \langle g, h \rangle$ is the mapping

$$t \longmapsto h(t) * g(t) : \mathbf{T} \longrightarrow \mathbb{A}(\mathcal{L}). \quad (5.10)$$

Of course, under conditions defining (5.10) we have (under $t \in \mathbf{T}$) the coincidence

$$v_{h(t)*g(t)}(E) = \int_E \mid h(t) \mid dg(t).$$

Using this property with respect to (5.9) under a natural change in designations, we obtain a verification in the part of the employment of the stabilizer in the form of the strong norm for the considered approximate construction.

Introduce some designations. If $(D, \preceq, h)$ is a net in a set H, then the filter (Engelking, 1986) $(H - ass)[D; \preceq; h]$ associated with the net $(D, \preceq, h)$ is defined as the family of all sets U, $U \subset H$, with the property

$$\exists \delta \in D \ \forall d \in D : \ (\delta \preceq d) \Longrightarrow (h(d) \in U).$$

Recall that designations connected with nets are used in the correspondence with (Chentsov, 1997, p.32-34). Fix

$$\mathbf{w} \in C(\mathbb{A}(\mathcal{L}), \tau_*(\mathcal{L}), \mathbb{X}, \mathcal{T})$$

and suppose $\mathbb{W} \triangleq \mathbf{w} \circ \mathcal{I}$. Finally,

$$\mathfrak{W} \triangleq (\mathbb{W}(\eta, \cdot) \mid B_0(E, \mathcal{L}))$$

is the following mapping

$$f \longmapsto \mathbf{w}(f * \eta) : B_0(E, \mathcal{L}) \longrightarrow \mathbb{X}.$$

Below questions of asymptotic attainability (in $(\mathbb{X}, \mathcal{T})$) on values of the mappings $\mathbb{W}$ and $\mathfrak{W}$ are considered. Moreover, the attainable set in the sense of possibilities defined by $\mathbf{w}$ is used. In the correspondence with the approach of (Chentsov, 1997, ch.6) the following definition is introduced: if $\mathcal{H} \in \mathcal{B}[\mathbb{M}(\mathcal{L})]$ then

$$(\mathbf{B}\mathbb{W} - Att)[\mathcal{H}]$$

is identified with the set of all $\mathbf{x} \in \mathbb{X}$ for each of which it is possible to choose a net $(\mathbf{K}, \preceq, g)$ in the set $(add)_+[\mathcal{L}]$ and a mapping $h \in \mathbb{B}[\mathbf{K}; E; \mathcal{L}; g]$ with the following properties

$$(\mathcal{H} \subset (\mathbb{M}(\mathcal{L}) - ass)[\mathbf{K}; \preceq; \langle g, h \rangle])\&((\mathbf{K}, \preceq, \mathbb{W} \circ \langle g, h \rangle) \xrightarrow{\mathcal{T}} \mathbf{x}).$$

Moreover, by analogy with (Chentsov, 1997, ch.6) suppose that

$$(\mathbf{B}\mathfrak{W} - att)[\mathfrak{A}_0]$$

is the set of all $\mathbf{x} \in \mathbb{X}$ for each of which it is possible to choose a nonempty directed set $(\mathbf{K}, \preceq)$ and an operator $h \in \mathbb{B}_0(\mathbf{K}, E, \mathcal{L}, \eta)$ such that

$$(\mathfrak{A}_0 \subset (B_0(E, \mathcal{L}) - ass)[\mathbf{K}; \preceq; h])\&((\mathbf{K}, \preceq, \mathfrak{W} \circ h) \xrightarrow{\mathcal{T}} \mathbf{x}).$$

Finally, we consider one more variant of an attraction set, oriented towards the problem about the validity of IC under the inexact realization of values of the mapping (5.2). If $K \in (Fin)[\Gamma]$ and $\varepsilon \in]0, \infty[$, then $\mathcal{O}_K(\varepsilon)$ is defined as the set of all mappings

$$(\tilde{S}_\gamma)_{\gamma \in K} : K \longrightarrow B(E, \mathcal{L})$$

for each of which $\forall \gamma \in K$:

$$||\tilde{S}_\gamma - S_\gamma|| \leq \varepsilon.$$

In this relation we use a model of the corresponding fragment of the operator (5.2). If $\mathcal{K} \in Fin(\mathcal{L})$, $P \in Fin(\Gamma)$, $Q \in Fin(\Omega)$ and $\varepsilon \in]0, \infty[$, then denote by $(ADM)[\mathcal{K}; P; Q; \varepsilon]$ the set of all $(\mu, f) \in \mathbb{M}(\mathcal{L})$ for each of which

$$((\mu \mid \mathcal{K}) = (\eta \mid \mathcal{K}))\&(\exists (\tilde{S}_\gamma)_{\gamma \in P} \in \mathcal{O}_P(\varepsilon)$$

$$\exists y \in Y \ \forall \gamma \in P : \mid \int_E \tilde{S}_\gamma f d\mu - y(\gamma) \mid \leq \varepsilon)\&(\forall \omega \in Q :$$

$$\int_{L_\omega} \mid f \mid d\mu \leq c_\omega + \varepsilon).$$

Introduce in the consideration the family

$$\mathcal{A} \triangleq \{(ADM)[\mathcal{K}; P; Q; \varepsilon] : \ (\mathcal{K}, P, Q, \varepsilon)$$

$$\in Fin(\mathcal{L}) \times Fin(\Gamma) \times Fin(\Omega) \times]0, \infty[\} \in \mathcal{B}[\mathbb{M}(\mathcal{L})],$$

obtaining a new type of the asymptotics of attainable sets. This asymptotics corresponds (in idea) to the employment of inexact models of (5.2). It is possible to verify that always

$$\mathbf{w}^1(\tilde{\mathbb{A}}_0) \subset (\mathbf{B}\mathfrak{W} - att)[\mathfrak{A}_0]$$

$$\subset (\mathbf{B}\mathbb{W} - Att)[\mathfrak{U}] \subset (\mathbf{B}\mathbb{W} - Att)[\mathcal{A}].$$

Theorem 5.3. Let $(\mathbb{X}, \mathcal{T})$ be a Hausdorff space. Then

$$\mathbf{w}^1(\tilde{\mathbb{A}}_0) = (\mathbf{B}\mathfrak{W} - att)[\mathfrak{A}_0] = (\mathbf{B}\mathbb{W} - Att)[\mathfrak{U}] = (\mathbf{B}\mathbb{W} - Att)[\mathcal{A}].$$

The proof is analogous in essence to verifications of (Chentsov, 1997, ch.6). This statement determines the fact of a very broad asymptotic equivalence of different variants of perturbations of (5.5). The above-mentioned equivalence established under minimal suppositions is the property of "attractors of a bounded convergence" (this term corresponds to (Chentsov, 1997)). In connection with Theorem 5.3 it is useful to recall the properties similar to relation (3.9.22) of (Chentsov, 1997). The given construction (with the employment of the stabilizer) realising Theorem 5.3 is "nearer" to compactifications of (Chentsov, 1996*a*, ch.2). Note the important fact: in (Chentsov, 1997, p.244,245) the constructive variant of the approximate realization of generalized elements defined as FAM with the property of the weak absolute η-continuity is given. It is useful to note that the above-mentioned constructive procedure of the determination of approximate solutions is coordinated with the stabilizer in the form of the strong norm (see (Chentsov, 1997, p.245)).

6. COMPACTIFIABLE PROBLEMS WITH INTEGRAL CONSTRAINTS

We continue the consideration of the problem of section 5, imposing one additional condition typical for control theory. Namely, in the given section the case of the integrally bounded problem is investigated. This corresponds to the employment of a natural constraint of a resource character. The last requirement is widely known in concrete problems of control. Consider the mapping (5.3). Suppose until the end of the given section that

$$\exists \omega \in \Omega : L_\omega = E. \tag{6.1}$$

By (6.1) an analog of the natural resource condition on the choice of controls was introduced. Note that here $\tilde{\mathbb{A}}_0$ is a compact set in the locally convex σ-compactum

$$(\mathbb{A}(\mathcal{L}), \tau_*(\mathcal{L})). \tag{6.2}$$

The space (6.2) is used by analogy with (Chentsov, 1997, ch.6) as a basic space.

Introduce some designations connected with finite sets. If $\mathbb{T}$ is a set and $K \in Fin(\mathbb{T})$, then denote by $(Fin)[\mathbb{T} \mid K]$ the family of all sets $\tilde{K} \in Fin(\mathbb{T})$ such that $K \subset \tilde{K}$. Moreover, by (6.1) the family $(FIN)[\Omega]$ of all sets $Q \in Fin(\Omega)$ for each of which $\{\omega \in Q \mid L_\omega = E\} \neq \emptyset$ is not empty. Let $K_* \in (FIN)[\Omega]$. It is obvious that

$$\begin{aligned}\mathbf{A} \triangleq \{(ADM)[\mathcal{K}; P; Q; \varepsilon] :\ & (\mathcal{K}, P, Q, \varepsilon) \\ & \in Fin(\mathcal{L}) \times Fin(\Gamma) \\ & \times (Fin)[\Omega \mid K_*] \times]0,1]\} \in \mathcal{B}[\mathbb{M}(\mathcal{L})]\end{aligned} \tag{6.3}$$

is fundamental in $\mathcal{A}$. Namely, $\mathbf{A} \subset \mathcal{A}$ and

$$\forall A_1 \in \mathcal{A}\ \exists A_2 \in \mathbf{A} : A_2 \subset A_1. \tag{6.4}$$

Note that $\forall Q_1 \in Fin(\Omega)\ \exists Q_2 \in (Fin)[\Omega \mid K_*] : Q_1 \subset Q_2$.

Suppose until the end of the present section that $(\mathbb{X}, \mathcal{T})$ is a Hausdorff space. Then $\mathbf{w}^1(\tilde{\mathbb{A}}_0)$ is a compact and (in particular) closed set in $(\mathbb{X}, \mathcal{T})$. From Proposition 4.3, (6.3) and (6.4) the equalities

$$\begin{aligned}&(\mathcal{T} - LIM)[\mathcal{A} \mid \mathbb{W}] \\ &= (\mathcal{T} - LIM)[\mathbf{A} \mid \mathbb{W}] = \mathbf{w}^1(\tilde{\mathbb{A}}_0)\end{aligned} \tag{6.5}$$

follows. In connection with the concrete variant of parameters in Proposition 4.3 we note that $\mathbf{A}$ (6.3) is an element of $\mathcal{B}[X]$ under

$$X \triangleq \{(\mu, f) \in \mathbb{M}(\mathcal{L}) \mid \int_E \mid f \mid d\mu \leq c\} \tag{6.6}$$

for some $c \in [0, \infty[$. Then $(\mathcal{I} \mid X)$, where X correspondends to (6.6), transform X in the ball (more exactly: in the c-ball) of $\mathbb{A}(\mathcal{L})$ in the strong norm. The above-mentioned ball in the topology of a subspace of the space (6.2) is a compactum by Alaoglu, theorem. As a corollary, for the proof of (6.5) it is possible (in fact) to use Theorem 2.5.2 of (Chentsov, 1996*a*) under an unessential version of the latter. Moreover, see (Chentsov, 1996*b*, p.195). From Proposition 4.4 and Theorem 5.1 we have the inclusion

$$\begin{aligned}\mathbf{w}^1(\tilde{\mathbb{A}}_0) &= w^1((\tau_*(\mathcal{L}) - LIM)[\mathfrak{A}_0 \mid \mathbf{j}_\eta]) \\ &\subset (\mathcal{T} - LIM)[\mathfrak{A}_0 \mid \mathfrak{W}].\end{aligned} \tag{6.7}$$

In addition, it should be used obvious property $\mathfrak{W} = \mathbf{w} \circ \mathbf{j}_\eta$. On the other hand, the inclusion

$$\begin{aligned}&(\mathcal{T} - LIM)[\mathfrak{A}_0 \mid \mathfrak{W}] \\ &\subset (\mathcal{T} - LIM)[\mathcal{A} \mid \mathbb{W}]\end{aligned} \tag{6.8}$$

takes place. For the verification of (6.7) fix $\mathcal{K}_* \in Fin(\mathcal{L})$, $P_* \in Fin(\Gamma)$, $Q_* \in (Fin)[\Omega]$ and $\varepsilon_* \in]0, \infty[$, obtaining $N_Y^0(P_*, \varepsilon_*) \in \mathfrak{Z}_0$ in correspondence with definitions of section 5. Then

$$\begin{aligned}D_* \triangleq (Adm)[N_Y^0(P_*, \varepsilon_*); Q_*] \subset \{f \in B_0(E, \mathcal{L}) \mid \\ (\exists y \in Y\ \forall \gamma \in P_* : \mid \int_E S_\gamma f d\eta - y(\gamma) \mid \leq \varepsilon_*) \\ \&(\forall \omega \in Q_* : \int_E \mid f \mid d\eta \leq c_\omega)\}.\end{aligned}$$

Therefore, we obtain the inclusion

$$\{\eta\} \times D_* \subset (ADM)[\mathcal{K}_*; P_*; Q_*; \varepsilon_*].$$

But, $D_* \in \mathfrak{A}_0$. So, if $H_1 \in \mathcal{A}$, then it is possible to choose $H_2 \in \mathfrak{A}_0$ for which $\{\eta\} \times H_2 \subset H_1$ and, as a corollary,

$$(\mathcal{T} - LIM)[\mathfrak{A}_0 \mid \mathfrak{W}] \subset cl(\mathbb{W}^1(H_1), \mathcal{T}).$$

Since H_1 was choosen arbitrary, the inclusion (6.8) is true. From (6.5), (6.7) and (6.8) the equalities

$$\mathbf{w}^1(\tilde{\mathbb{A}}_0) = (\mathcal{T} - LIM)[\mathfrak{A}_0 \mid \mathfrak{W}] = (\mathcal{T} - LIM)[\mathcal{A} \mid \mathbb{W}] = (\mathcal{T} - LIM)[\mathbf{A} \mid \mathbb{W}] \quad (6.9)$$

follow. In (6.9) one more "broad" asymptotic equivalence of different variants of perturbations of IC was established. On the basis of (6.9) it is possible to obtain the statements about conditions of an asymptotic nonsensitivity in terms of neighborhoods. From the definition of $\mathbb{W}$ and from properties of section 4 it is follows that for each $\mathcal{K} \in Fin(\mathcal{L})$, $P \in Fin(\Gamma)$, $Q \in (FIN)[\Omega]$ and $\varepsilon \in]0, \infty[$ the set

$$cl(\mathbb{W}^1((ADM)[\mathcal{K}; P; Q; \varepsilon]), \mathcal{T})$$

is compact in $(\mathbb{X}, \mathcal{T})$. In this verification the known relation (3.7.20) of (Chentsov, 1997) is used. From the last property and from the representation (6.9) the useful statement similar to (Chentsov, 1996*b*, p.199) follows.

Theorem 6.1. Let $G_0 \in \mathcal{T}$ be such that $\mathbf{w}^1(\tilde{\mathbb{A}}_0) \subset G_0$. Then

$$\exists \mathcal{K}_0 \in Fin(\mathcal{L})\ \exists P_0 \in Fin(\Gamma)\ \exists Q_0 \in (FIN)[\Omega]$$
$$\exists \varepsilon_0 \in]0, \infty[\ \forall \mathcal{K} \in (Fin)[\mathcal{L} \mid \mathcal{K}_0]\ \forall P \in (Fin)[\Gamma \mid P_0]$$
$$\forall Q \in (Fin)[\Omega \mid Q_0]\ \forall \varepsilon \in]0, \varepsilon_0] :$$
$$\mathbf{w}^1(\tilde{\mathbb{A}}_0) \subset cl(\mathbb{W}^1((ADM)[\mathcal{K}; P; Q; \varepsilon]), \mathcal{T}) \subset G_0.$$

By definitions of section 5 we obtain that $\forall \mathcal{K} \in Fin(\mathcal{L})\ \forall P \in Fin(\Gamma)\ \forall Q \in Fin(\Omega)\ \forall \varepsilon \in]0, \infty[$:

$$\mathfrak{W}^1((Adm)[N_Y^0(P, \varepsilon); Q]) \subset \mathbb{W}^1((ADM)[\mathcal{K}; P; Q; \varepsilon]). \quad (6.10)$$

From (6.10) the corresponding inclusion for closures of sets on the left and right parts of (6.10) follows.

Suppose until the end of the given section that $(\mathbb{X}, \mathcal{T})$ is a metrizable space. Let ρ be the metric of $\mathbb{X}$ generating the topology $\mathcal{T}$. If M is a subset of $\mathbb{X}$ and $\alpha \in]0, \infty[$, then

$$U_\rho^0(M, \alpha) \triangleq \{\mathbf{x} \in \mathbb{X} \mid \exists m \in M : \rho(m, \mathbf{x}) < \alpha\} \in \mathcal{T}.$$

Theorem 6.2. The following property of a rougness takes place. Namely,

$$\forall \alpha \in]0, \infty[\ \exists \mathcal{K}_0 \in Fin(\mathcal{L})\ \exists P_0 \in Fin(\Gamma)$$
$$\exists Q_0 \in (FIN)[\Omega]\ \exists \varepsilon_0 \in]0, \infty[\ \forall \mathcal{K} \in (Fin)[\mathcal{L} \mid \mathcal{K}_0]$$
$$\forall P \in (Fin)[\Gamma \mid P_0]\ \forall Q \in (Fin)[\Omega \mid Q_0]\ \forall \varepsilon \in]0, \varepsilon_0] :$$
$$\mathfrak{W}^1((Adm)[N_Y^0(P, \varepsilon); Q]) \subset \mathbb{W}^1((ADM)[\mathcal{K}; P; Q; \varepsilon]) \subset U_\rho^0(\mathfrak{W}^1((Adm)[N_Y^0(P, \varepsilon), Q]), \alpha).$$

The proof is obvious. If $\alpha \in]0, \infty[$, then by Theorem 6.1 choose $\mathcal{K}_0 \in Fin(\mathcal{L})$, $P_0 \in Fin(\Gamma)$, $Q_0 \in (FIN)[\Omega]$ and $\varepsilon_0 \in]0, \infty[$ such that $\forall \mathcal{K} \in (Fin)[\mathcal{L} \mid \mathcal{K}_0]$ $\forall P \in (Fin)[\Gamma \mid P_0]$ $\forall Q \in (Fin)[\Omega \mid Q_0]$ $\forall \varepsilon \in]0, \varepsilon_0]$:

$$\mathbf{w}^1(\tilde{\mathbb{A}}_0) \subset cl(\mathbb{W}^1((ADM)[\mathcal{K}; P; Q; \varepsilon]), \mathcal{T}) \subset U_\rho^0(\mathbf{w}^1(\tilde{\mathbb{A}}_0), \frac{\alpha}{2}). \quad (6.11)$$

Fix $\mathcal{K} \in (Fin)[\mathcal{L} \mid \mathcal{K}_0]$, $P \in (Fin)[\Gamma \mid P_0]$, $Q \in (Fin)[\Omega \mid Q_0]$ and $\varepsilon \in]0, \varepsilon_0]$. As a corollary, (6.11) is correct. On the other hand, $N_Y^0(P, \varepsilon) \in \mathfrak{J}_0$. Therefore, $(Adm)[N_Y^0(P, \varepsilon); Q] \in \mathfrak{A}_0$ and by (6.9):

$$\mathbf{w}^1(\tilde{\mathbb{A}}_0) \subset cl(\mathfrak{W}^1((Adm)[N_Y^0(P, \varepsilon); Q]), \mathcal{T}).$$

Therefore, from (6.11) we have the inclusion

$$\mathbb{W}^1((ADM)[\mathcal{K}; P; Q; \varepsilon]) \subset U_\rho^0(\mathfrak{W}^1((Adm)[N_Y^0(P, \varepsilon); Q]), \alpha).$$

Using (6.10), we obtain the statement of the given Theorem.

Theorem 6.2 determines the property of the asymptotic nonsensitivity under the perturbation of a part of constraints in terms of neighborhoods. Here a highly natural with the practical point of view characterization of "directions of a roughness" is given. For the definition of such directions the families $\mathfrak{A}_0$ and $\mathcal{A}$ should be compared. So, we supplement the limit representation (6.9).

7. THE REGULARIZATION OF ATTRACTION SETS

Consider the case when fragments of (5.2) are known not exactly. So, it is required to observe the constraints under the conditions of an inexact information. Of course, the validity of constraints is approximate. The question about an "coordination" of errors arises naturally. For the realization of the given "coordination" a procedure of the regularization is used. We take into account the following circumstances. Namely, it is impossible to use the objective operator (5.2) itself. But, estimates of its fragments are known, since we know possibilities of the corresponding "instruments". And what is more, here it is possible to raise the precision of this "instruments". The similar questions are considered in theory of ill-posed problems (Tichonoff and Arsenin, 1979). In the given case singularities arise. In particular, here the regularization of asymptotic objects is considered. Namely, consider the regularization of attraction sets. In these sets we take into account different effects connected with the weakening of constraints (in particular, of IC). New perturbations have another character. For example, these perturbations have not, generally speaking, the property of monotonicity.

Consider precise definitions. Below postulate that (6.1) is correct. If $\mathcal{K} \in Fin(\mathcal{L})$, $P \in Fin(\Gamma)$, $Q \in (FIN)[\Omega]$, $(\tilde{S}_\gamma)_{\gamma \in P}$ is a mapping from P into $B(E, \mathcal{L})$, $t \in [0, \infty[$ and $\varepsilon \in]0, \infty[$, then denote by

$$(\mathbb{A}dm)^*[\mathcal{K}; P; Q; (\tilde{S}_\gamma)_{\gamma \in P}; t; \varepsilon]$$

the set of all $(\mu, f) \in \mathbb{M}(\mathcal{L})$ for each of which

$$((\mu \mid \mathcal{K}) = (\eta \mid \mathcal{K})) \,\&(\exists y \in Y \; \forall \gamma \in P :$$

$$\mid \int_E \tilde{S}_\gamma f d\mu - y(\gamma) \mid \leq t + \varepsilon) \,\&(\forall \omega \in Q :$$

$$\int_{L_\omega} \mid f \mid d\mu \leq c_\omega + \varepsilon). \tag{7.1}$$

Moreover, if $P \in Fin(\Gamma)$, $Q \in (FIN)[\Omega]$, $(\tilde{S}_\gamma)_{\gamma \in P}$ is a mapping from P into $B_0(E, \mathcal{L})$, $t \in [0, \infty[$ and $\varepsilon \in]0, \infty[$, then denote by

$$(adm)_0^*[P; Q; (\tilde{S}_\gamma)_{\gamma \in P}; t; \varepsilon]$$

the set of all $f \in B_0(E, \mathcal{L})$ for each of which

$$(\exists y \in Y : \; (\forall \gamma \in P \cap \Gamma_0 : \mid \int_E \tilde{S}_\gamma f d\eta - y(\gamma) \mid \leq t)$$

$$\&(\forall \gamma \in P \setminus \Gamma_0 : \mid \int_E \tilde{S}_\gamma f d\eta - y(\gamma) \mid \leq t + \varepsilon))$$

$$\&(\forall \omega \in Q : \int_{L_\omega} \mid f \mid d\eta \leq c_\omega). \tag{7.2}$$

In (7.1), (7.2) definitions of (Chentsov, 1996*b*, p.201) are used. If $Q \in (FIN)[\Omega]$, then $\mathbb{L}_0[Q] \triangleq \{\omega \in Q \mid L_\omega = E\} \in (FIN)[\Omega]$ and

$$c_0[Q] \triangleq inf(\{c_\omega : \; \omega \in \mathbb{L}_0[Q]\}) \in [0, \infty[.$$

We use $\delta c_0[Q]$, where $\delta \in [0, \infty[$ is a parameter of the precision, in the capacity of t in (7.1), (7.2). From the relation (5.4) of (Chentsov, 1996*b*) the useful property follows. Namely, $\forall P \in Fin(\Gamma)$ $\forall Q \in (FIN)[\Omega]$ $\forall \delta \in [0, \infty[$ $\forall (\tilde{S}_\gamma)_{\gamma \in P} \in \mathcal{O}_P(\delta)$ $\forall \varepsilon \in]0, \infty[$:

$$\mathbf{w}^1(\tilde{\mathbb{A}}_0) \subset cl(\mathfrak{W}^1((adm)_0^*[P; Q; (\tilde{S}_\gamma)_{\gamma \in P}; \delta c_0[Q]; \varepsilon]), \mathcal{T}). \tag{7.3}$$

Let (until the end of this section) $(\mathbb{X}, \mathcal{T})$ be a Hausdorff space.

Theorem 7.1. Let $G \in \mathcal{T}$ be such that $\mathbf{w}^1(\tilde{\mathbb{A}}_0) \subset G$. Then

$$\exists \mathcal{K}_0 \in Fin(\mathcal{L}) \; \exists P_0 \in Fin(\Gamma) \; \exists Q_0 \in (FIN)[\Omega]$$
$$\exists \varepsilon_0 \in]0, \infty[\; \forall \mathcal{K} \in (Fin)[\mathcal{L} \mid \mathcal{K}_0]$$
$$\forall P \in (Fin)[\Gamma \mid P_0] \; \forall Q \in (Fin)[\Omega \mid Q_0] \; \forall \delta \in [0, \varepsilon_0]$$
$$\forall (\tilde{S}_\gamma)_{\gamma \in P} \in \mathcal{O}_P(\delta) \; \forall \varepsilon \in]0, \varepsilon_0] :$$

$$\mathbf{w}^1(\tilde{\mathbb{A}}_0) \subset cl(\mathbb{W}^1((\mathbb{A}dm)^*[\mathcal{K}; P; Q; (\tilde{S}_\gamma)_{\gamma \in P}; \delta c_0[Q]; \varepsilon]), \mathcal{T}) \subset G.$$

Theorem 7.2. Let $G \in \mathcal{T}$ be such that $\mathbf{w}^1(\tilde{\mathbb{A}}_0) \subset G$. Then $\exists P_0 \in Fin(\Gamma)$ $\exists Q_0 \in (FIN)[\Omega]$ $\exists \varepsilon_0 \in]0, \infty[$ $\forall P \in (Fin)[\Gamma \mid P_0]$ $\forall Q \in (Fin)[\Omega \mid Q_0]$ $\forall \delta \in [0, \varepsilon_0]$ $\forall (\tilde{S}_\gamma)_{\gamma \in P} \in \mathcal{O}_P(\delta)$ $\forall \varepsilon \in]0, \varepsilon_0]$:

$$w^1(\tilde{\mathbb{A}}_0) \subset cl(\mathfrak{W}^1((adm)_0^*[P; Q; (\tilde{S}_\gamma)_{\gamma \in P}; \delta c_0[Q]; \varepsilon]), \mathcal{T}) \subset G.$$

Until the end of this section suppose that $(\mathbb{X}, \mathcal{T})$ is a metrizable space, using as in section 6, the metric ρ that generates $\mathcal{T}$. We exploit the designation for ε-neighborhoods in $(\mathbb{X}, \rho)$ used in section 6.

Theorem 7.3. Let $\alpha \in]0, \infty[$. Then

$$\exists \mathcal{K}_0 \in Fin(\mathcal{L}) \quad \exists P_0 \in Fin(\Gamma) \quad \exists Q_0 \in (FIN)[\Omega]$$
$$\exists \varepsilon_0 \in]0, \infty[\quad \forall \mathcal{K} \in (Fin)[\mathcal{L} \mid \mathcal{K}_0]$$
$$\forall P \in (Fin)[\Gamma \mid P_0] \quad \forall Q \in (Fin)[\Omega \mid Q_0]$$
$$\forall \delta \in [0, \varepsilon_0] \quad \forall (\tilde{S}_\gamma)_{\gamma \in P} \in \mathcal{O}_P(\delta) \quad \forall \varepsilon \in]0, \varepsilon_0] :$$
$$\mathfrak{W}^1((adm)_0^*[P; Q; (\tilde{S}_\gamma)_{\gamma \in P}; \delta c_0[Q]; \varepsilon]) \subset$$
$$\mathbb{W}^1((\mathbb{A}dm)^*[\mathcal{K}; P; Q; (\tilde{S}_\gamma)_{\gamma \in P}; \delta c_0[Q]; \varepsilon]) \subset$$
$$U_\rho^0(\mathfrak{W}^1((adm)_0^*[P; Q; (\tilde{S}_\gamma)_{\gamma \in P}; \delta c_0[Q]; \varepsilon]), \alpha).$$

For the proof we use Theorem 7.1 under $G = U_\rho^0(\mathbf{w}^1(\tilde{\mathbb{A}}_0), \alpha/2)$. Choose $\mathcal{K}_0 \in Fin(\mathcal{L})$, $P_0 \in Fin(\Gamma)$, $Q_0 \in (FIN)[\Omega]$ and $\varepsilon_0 \in]0, \infty[$ such that under $\mathcal{K} \in (Fin)[\mathcal{L} \mid \mathcal{K}_0]$, $P \in (Fin)[\Gamma \mid P_0]$, $Q \in (Fin)[\Omega \mid Q_0]$, $\delta \in [0, \varepsilon_0]$, $(\tilde{S}_\gamma)_{\gamma \in P} \in \mathcal{O}_P(\delta)$ and $\varepsilon \in]0, \varepsilon_0]$ the inclusions

$$\mathbf{w}^1(\tilde{\mathbb{A}}_0) \subset cl(\mathbb{W}^1((\mathbb{A}dm)^*[\mathcal{K}; P; Q; (\tilde{S}_\gamma)_{\gamma \in P}; \delta c_0[Q]; \varepsilon]), \mathcal{T}) \subset U_\rho^0(w^1(\tilde{\mathbb{A}}_0), \alpha/2) \tag{7.4}$$

is true. From (7.3) and (7.4) we obtain the useful estimate

$$\mathbb{W}^1((\mathbb{A}dm)^*[\mathcal{K}; P; Q; (\tilde{S}_\gamma)_{\gamma \in P}; \delta c_0[Q]; \varepsilon]) \subset U_\rho^0(\mathfrak{W}^1((adm)_0^*[P; Q; (\tilde{S}_\gamma)_{\gamma \in P}; \delta c_0[Q]; \varepsilon]), \alpha). \tag{7.5}$$

On the other hand, from (7.1) and (7.2) we have $\{\eta\} \times (adm)_0^*[P; Q; (\tilde{S}_\gamma)_{\gamma \in P};$

$$\delta c_0[Q]; \varepsilon] \subset (\mathbb{A}dm)^*[\mathcal{K}; P; Q; (\tilde{S}_\gamma)_{\gamma \in P}; \delta c_0[Q]; \varepsilon].$$

Then by definitions of section 5,

$$\mathfrak{W}^1((adm)_0^*[P;Q;(\tilde{S}_\gamma)_{\gamma\in P};\delta c_0[Q];\varepsilon]) \subset$$
$$\mathbb{W}^1((\mathbb{A}dm)^*[\mathcal{K};P;Q;(\tilde{S}_\gamma)_{\gamma\in P};\delta c_0[Q];\varepsilon]). \quad (7.6)$$

From (7.5) and (7.6) we obtain the following chain of inclusions

$$\mathfrak{W}^1((adm)_0^*[P;Q;(\tilde{S}_\gamma)_{\gamma\in P};\delta c_0[Q];\varepsilon]) \subset$$
$$\mathbb{W}^1((\mathbb{A}dm)^*[\mathcal{K};P;Q;(\tilde{S}_\gamma)_{\gamma\in P};\delta c_0[Q];\varepsilon]) \subset$$
$$U_\rho^0(\mathfrak{W}^1((adm)_0^*[P;Q;(\tilde{S}_\gamma)_{\gamma\in P};\delta c_0[Q];\varepsilon]),\alpha).$$

Theorem 7.3 is the basic statement of this section. This theorem is a regularized analog of Theorem 6.2. From Theorem 7.3 it is seen that after the corresponding regularization it is possible to obtain again the important property of an asymptotic nonsensitivity under the perturbation of a part of constraints.

Conclusion. The problem connected with extensions and relaxations can be considered as an instrument of the regularization (of unstable problems) in the broad sense. The essence of the given method is constructing more "perfect" space of solutions. Usually, this operation of the passage to more "perfect" space is realized by a compactification of the space of solutions of initial problem. Of course, this compactification having some difference from that in general topology exists in not any case (see example (Chentsov, 1997, p.156)). Therefore, it is advisable to find other constructions. Useful properties supplied of these constructions depend in the essential part on the choice of the space of generalized solutions under the employment of this space by analogy with the scheme of a compactification. In the given investigation FAM are used in the capacity of generalized elements. As a result, a new quality even in "unbounded" problems is attained (see sections 3,5). Namely, by such application of FAM with the specific property of the weak absolute continuity one can obtain a common representation of attraction sets for very different variants of the weakening of conditions. This representation is the natural analog of attainability domain in the class of generalized controls (Warga, 1972) that were used in nonlinear control problems with geometric constraints on the choice of an instantaneous control (Pontryagin *et al.*, 1983) (moreover see monographs (Krasovskii and Subbotin, 1988); (Gamkrelidze, 1977); finally, see the extensive bibliography of (Warga, 1972)). In the given investigations in the capacity of generalized elements countably additive Radon measures or measure-valued functions (with the values in the space of normed Borel measures) were used. Moreover, in many settings the solutions of differential inclusions were used for immediate constructing generalized trajectories. In problems with IC new singularities arise. In particular, here some effects have the sense of product of a discontinuous function and a generalized function. It seems that FAM determines some (although incomplete) analog of the above-meationed product. Of course, such FAM must have the above-cited property of the weak absolute continuity with respect to the given measure space. In this connection see theorems of (Chentsov, 1996*a*),(Chentsov, 1997) about the density property of sets composed from indefinite integrals of usual controls. These topological constructions of (Chentsov, 1996*a*),(Chentsov, 1997) generate the basis of the considered method of investigation.

8. REFERENCES

Bourbaki, N. (1968). *General topology.* Nauka. Moscow (in Russian).

Chentsov, A.G. (1996*a*). *Finitely additive measures and relaxation of extremal problems.* Plenum Publishing Corporation. New York, London and Moscow.

Chentsov, A.G. (1996*b*). To the question about conditions of asymptotics nonsensitivity of an attainable set under the perturbation of a part of constraints. *Vestnik Chelyabinskogo univ. Ser. Matematika i mekhanika* **3**, 189–205 (in Russian).

Chentsov, A.G. (1997). *Asymptotic Attainability.* Kluwer Academic Publishers. Dordrecht.

Chentsov, A.G. (1998). The relaxation of integral constraints in the class of vector finitely additive measures. *Doklady Akademii Nauk* **358**, 609–613.

Danford, N. and J.T. Schwartz (1958). *Linear operators. Vol.1.* Interscience. New York.

Engelking, R. (1986). *General topology.* Mir. Moscow (in Russian).

Gamkrelidze, R.V. (1977). *Foundations of optimal control theory.* Izdat.Tbilissi.Univ.. Tbilissi (in Russian).

Krasovskii, N.N. (1968). *The theory of control of motion.* Nauka. Moscow (in Russian).

Krasovskii, N.N. and A.I. Subbotin (1988). *Game-theoretical control problems.* Springer-Verlag. Berlin.

Neveu, J. (1964). *Bases mathematiques du calcul des probabilites.* Masson. Paris.

Pontryagin, L.S., V.G. Boltjanskii, R.V. Gamkrelidze and E.F. Mishchenko (1983). *The mathematical theory of optimal processes.* Nauka. Moscow (in Russian).

Tichonoff, A.N. and V.M. Arsenin (1979). *Solving methods for ill-posed problems.* Nauka. Moscow (in Russian).

Warga, J. (1972). *Optimal control of differential and functional equations.* Academic Press. New York.

EXISTENCE OF EQUILIBRIA FOR G-MONOTONE MAPPINGS

W. Oettli and D. Schläger

*Lehrstuhl für Mathematik VII, Universität Mannheim,
68131 Mannheim, Germany*

Abstract: Existence results are given for a class of vectorial equilibrium problems involving g-monotone multivalued mappings. *Copyright © 1998 IFAC*

Keywords: Equilibrium, monotonicity, multivalued mapping.

1. INTRODUCTION

Let K be a convex set in some real topological vector space X, and let $f\colon K \times K \to \mathbb{R}$ be a given function such that $f(x,x) \geq 0$ for all $x \in K$. The scalar equilibrium problem (Blum and Oettli, 1994) deals with the existence of

$$\bar{x} \in K \quad \text{such that} \quad f(\bar{x},y) \geq 0 \quad \forall y \in K. \quad (1)$$

This problem subsumes in particular optimization problems, Nash equilibria in noncooperative games, and variational inequalities. In the latter one is given a mapping $T\colon K \to X^*$, and one asks for the existence of

$$\bar{x} \in K \quad \text{such that} \quad \langle T(\bar{x}), y - \bar{x}\rangle \geq 0 \quad \forall y \in K. \quad (2)$$

In connection with problem (1) the function $f(\cdot,\cdot)$ is said to be pseudomonotone (Bianchi et al., 1997) iff, for all $x, y \in K$,

$$f(x,y) \geq 0 \implies f(y,x) \leq 0, \quad (3)$$

and maximal pseudomonotone (Ansari et al., 1997) iff, in addition, for all $x \in K$,

$$f(y,x) \leq 0 \ \forall y \in K \implies f(x,y) \geq 0 \ \forall y \in K. \quad (4)$$

Pseudomonotonicity is interesting since it permits to weaken the continuity requirements needed to prove the existence of a solution to (1). Existence of a solution under (3) and (4) has been studied by several authors. Usually, (3) is made an explicit requirement, whereas (4) is enforced by other conditions, and thus does not occur explicitly.

In order to unify the general case and the pseudomonotone case, Oettli and Schläger (1998) introduced another function $g\colon K \times K \to \mathbb{R}$, and defined f to be g-monotone iff, for all $x, y \in K$,

$$f(x,y) \geq 0 \implies g(x,y) \leq 0, \quad (5)$$

and maximal g-monotone iff, in addition, for all $x \in K$,

$$g(x,y) \leq 0 \ \forall y \in K \implies f(x,y) \geq 0 \ \forall y \in K. \quad (6)$$

For $g(x,y) = -f(x,y)$, (5) and (6) are trivially satisfied, whereas for $g(x,y) = f(y,x)$ one obtains (3) and (4). Still other choices of g are possible.

Here this approach is carried over to multivalued vectorial equilibrium problems, which can be formulated as follows: There is given a multivalued mapping $f\colon K \times K \rightrightarrows Z$, Z a real topological vector space. For each $x \in K$ there is given a closed convex cone $P(x) \subseteq Z$ with $P(x) \neq Z$ and $\operatorname{int} P(x) \neq \emptyset$. Then one asks for the existence of

$$\bar{x} \in K \quad \text{such that} \quad f(\bar{x},y) \not\subseteq -\operatorname{int} P(\bar{x}) \ \forall y \in K. \quad (7)$$

For technical reasons the following notion of g-monotonicity is used: for all $x, y \in K$,

$$f(x,y) \not\subseteq -\operatorname{int} P(x) \implies g(x,y) \subseteq -P(y), \quad (8)$$

where $g\colon K \times K \rightrightarrows Z$ is another multivalued mapping. The associated maximal g-monotonicity requires, in addition, that for all $x \in K$,

$$g(x,y) \subseteq -P(y) \ \forall y \in K \implies f(x,y) \not\subseteq -\operatorname{int} P(x) \ \forall y \in K. \quad (9)$$

Clearly (7), (8), (9) constitute legitimate extensions of (1), (5), (6) respectively. But whereas (5) and (6) are trivially satisfied under the choice $g(x,y) = -f(x,y)$, this is no longer true for (8) and (9). Therefore it is interesting to note that the convex cones $-\operatorname{int} P(x)$ and $-P(y)$ occurring in (7), (8), (9) can be replaced by more general sets $C(x)$ and $D(y)$. This is done in Theorem 1. Then, in Corollary 3, we replace the requirement (9) by other, more practical, conditions.

In the second part we consider a combination of equilibria and variational inequalities, called mixed equilibria. Here we are given in addition a multivalued mapping $T\colon K \rightrightarrows L(X,Z)$, and we are concerned with the existence of

$$\bar{x} \in K \quad \text{and} \quad \bar{\xi} \in T(\bar{x}) \quad \text{such that}$$
$$f(\bar{x},y) + \langle \bar{\xi}, y - \bar{x}\rangle \not\subseteq -\operatorname{int} P(\bar{x}) \ \forall y \in K. \quad (10)$$

The monotonicity requirement in this case does not refer to the mapping $f(x,y)$ itself, but rather to the family of mappings $f(x,y) + \langle \xi, y - x\rangle$ for all $\xi \in T(K)$.

2. EQUILIBRIA

In what follows, X and Z are real topological vector spaces, and K is a nonempty, convex subset of X. For every $x \in K$, let $P(x) \subseteq Z$ be a closed convex cone (not necessarily pointed) with nonempty interior and $P(x) \neq Z$.

Theorem 1. Let K be nonempty, convex, and compact. Let $C, D\colon K \rightrightarrows Z$ and $f, g\colon K \times K \rightrightarrows Z$ satisfy the following conditions:

(i) for all $x, y \in K$, $f(x,y) \not\subseteq C(x)$ implies $g(x,y) \subseteq D(y)$;
(ii) for all $y \in K$, $\{x \in K \mid g(x,y) \subseteq D(y)\}$ is closed in K;
(iii) for all $x \in K$, $\{y \in K \mid f(x,y) \subseteq C(x)\}$ is convex;
(iv) for all $x \in K$, if $g(x,y) \subseteq D(y) \ \forall y \in K$, then $f(x,y) \not\subseteq C(x) \ \forall y \in K$;
(v) for all $x \in K$, $f(x,x) \not\subseteq C(x)$.

Then there exists $\bar{x} \in K$ such that $f(\bar{x},y) \not\subseteq C(\bar{x})$ for all $y \in K$.

PROOF. For every $y \in K$, let

$$S(y) := \{x \in K \mid g(x,y) \subseteq D(y)\}.$$

Then $S(\cdot)$ is a KKM-map, i.e., for every finite subset $\{y_1, \dots, y_n\}$ of K there holds

$$\operatorname{conv}\{y_1, \dots, y_n\} \subseteq \bigcup_{i=1}^{n} S(y_i).$$

In fact, let $x \in \operatorname{conv}\{y_1, \dots, y_n\}$ and assume, for contradiction, that $x \notin S(y_i)$ for all i. Then $g(x,y_i) \not\subseteq D(y_i)$ for all i, hence from (i) $f(x,y_i) \subseteq C(x)$ for all i, hence from (iii) $f(x,x) \subseteq C(x)$, contradicting (v). Thus $S(\cdot)$ is a KKM-map, and $S(y)$ is closed from (ii). Since K is convex and compact, it follows from the KKM-Lemma, as extended by Fan (1961), that there exists $\bar{x} \in K$ such that $\bar{x} \in S(y)$ for all $y \in K$, i.e., $g(\bar{x},y) \subseteq D(y)$ for all $y \in K$. From (iv) follows $f(\bar{x},y) \not\subseteq C(\bar{x})$ for all $y \in K$. □

Remark 2. Theorem 1 remains true if compactness of K is replaced by the following coercivity condition:

(vi) there exist a nonempty compact set $A \subseteq K$ and a compact, convex set $B \subseteq K$ such that, for every $x \in K \setminus A$, there exists $y \in B$ with $g(x,y) \not\subseteq D(y)$.

To see this, let $y_1, \dots, y_n \in K$, and set

$$K' := \operatorname{conv}(B \cup \{y_1, \dots, y_n\}).$$

K' is compact, since B is convex and compact. From the proof of Theorem 1 there exists $x' \in K'$ such that $g(x',y) \subseteq D(y)$ for all $y \in K'$. From $B \subseteq K'$ and (vi) it follows $x' \in A$, and from $y_i \in K'$ it follows $g(x',y_i) \subseteq D(y_i)$. Hence every finite subfamily of the family of sets

$$S(y) := \{x \in A \mid g(x,y) \subseteq D(y)\},$$

$y \in K$, has nonempty intersection. Since A is compact and the sets $S(y)$ are closed, there exists $\bar{x} \in A$ which belongs to all $S(y)$, $y \in K$; using condition (iv) completes the proof. □

Condition (ii) of Theorem 1 is fulfilled, if, for all $y \in K$, the mapping $g(\cdot,y)$ is lower semicontinuous and $D(y)$ is closed. In fact, $\{x \in K \mid g(x,y) \cap (Z \setminus D(y)) \neq \emptyset\}$ is open, being the lower inverse of an open set under a lower semicontinuous mapping; see Berge (1966).

A similar fact about upper semicontinuity is used in the sequel: If, given $y \in K$, the mapping $f(\cdot,y)$ is upper semicontinuous with compact values, and the mapping $C(\cdot)$ has open graph in $K \times Z$, then

$$R := \{u \in K \mid f(u,y) \not\subseteq C(u)\}$$

is closed in K. To see this, consider a net $\{u_i\}$ in R such that $u_i \to u \in K$. For every i there exists $z_i \in f(u_i,y)$ with $z_i \notin C(u_i)$. From the upper semicontinuity of $f(\cdot,y)$, $f(u,y) \neq \emptyset$. It suffices to find a subnet $\{z_j\}$ which converges to some $z \in f(u,y)$; then $(u_j, z_j) \to (u,z)$, and it follows $z \notin C(u)$, since the graph of $C(\cdot)$ is open, hence $u \in R$. Now assume that such a subnet does not exist. Then, for every $w \in f(u,y)$, there exist an open set $V(w) \ni w$ and an index $j(w)$ with $z_i \notin V(w)$ for all $i \geq j(w)$. The sets $V(w)$ cover the compact $f(u,y)$, hence there exist $w_1, \ldots, w_n \in f(u,y)$ with

$$f(u,y) \subseteq V(w_1) \cup \ldots \cup V(w_n) =: V.$$

Then $z_i \notin V$ and therefore $f(u_i,y) \not\subseteq V$ whenever $i \geq j(w_1), \ldots, j(w_n)$. But from $u_i \to u$ and the upper semicontinuity of $f(\cdot,y)$ we see that $f(u_i,y) \subseteq V$ for sufficiently large i, a contradiction.

Given a multivalued mapping $h\colon K \rightrightarrows Z$ and a convex cone $P \subseteq Z$, we say that h is *right P-convex* iff, for all $x,y \in K$ and $\alpha \in [0,1]$,

$$h(\alpha x + (1-\alpha)y) \subseteq \alpha h(x) + (1-\alpha)h(y) - P.$$

Similarly, h is said to be *left P-convex* iff

$$h(\alpha x + (1-\alpha)y) + P \supseteq \alpha h(x) + (1-\alpha)h(y).$$

If h is single-valued, then both notions coincide with ordinary P-convexity.

We shall apply Theorem 1 to $g(x,y) := f(y,x)$, $C(x) := -\operatorname{int} P(x)$, $D(x) := -P(x)$. Then condition (iii) of Theorem 1 is fulfilled if, for every $x \in K$, $f(x,\cdot)$ is right $P(x)$-convex (use $P(x) + \operatorname{int} P(x) = \operatorname{int} P(x)$).

Corollary 3. Let K be nonempty and convex. Let $f\colon K \times K \rightrightarrows Z$ be given such that

(i) for all $x,y \in K$, $f(x,y) \not\subseteq -\operatorname{int} P(x)$ implies $f(y,x) \subseteq -P(y)$;
(ii) for all $y \in K$, $f(y,\cdot)$ is lower semicontinuous;
(iii) for all $x \in K$, $f(x,\cdot)$ is right $P(x)$-convex;
(iv) the mapping $\operatorname{int} P(\cdot)$ has open graph in $K \times Z$;
(v) for all $x,y \in K$, $f(\cdot,y)$ is upper semicontinuous and compact-valued on $[x,y]$;
(vi) for all $x \in K$, $f(x,x) \not\subseteq -\operatorname{int} P(x)$;
(vii) there exist a nonempty, compact set $A \subseteq K$ and a compact, convex set $B \subseteq K$ such that for every $x \in K \setminus A$ there exists $y \in B$ with $f(x,y) \subseteq -\operatorname{int} P(x)$.

Then there exists $\overline{x} \in A$ such that $f(\overline{x},y) \not\subseteq -\operatorname{int} P(\overline{x})$ for all $y \in K$.

PROOF. Let $C(x) := -\operatorname{int} P(x)$, $D(x) := -P(x)$, and $g(x,y) := f(y,x)$. We show first that condition (iv) of Theorem 1 holds on every convex subset $K' \subseteq K$. Let $x \in K'$ satisfy

$$g(x,y) \subseteq -P(y) \quad \forall y \in K'.$$

Assume that $f(x,y) \subseteq -\operatorname{int} P(x)$ for some $y \in K'$. Then $y \neq x$ from (vi). From (iv) and (v),

$$\{u \in [x,y] \mid f(u,y) \subseteq -\operatorname{int} P(u)\}$$

is open in $[x,y] \subseteq K'$. Hence there exists $u \in \,]x,y[$ such that $f(u,y) \subseteq -\operatorname{int} P(u)$. Since $f(u,x) = g(x,u) \subseteq -P(u)$, it follows from (iii), that $f(u,u) \subseteq -\operatorname{int} P(u)$, contradicting (vi). Hence

$$f(x,y) \not\subseteq -\operatorname{int} P(x) \quad \forall y \in K'.$$

Now let $y_1, \ldots, y_m \in K$, and

$$K' := \operatorname{conv}(B \cup \{y_1, \ldots, y_m\}).$$

Then K' is compact, since B is compact and convex. Hence all conditions of Theorem 1 are satisfied on K', so there exists $x' \in K'$ such that

$$f(x',y) \not\subseteq -\operatorname{int} P(x') \quad \forall y \in K'.$$

From $B \subseteq K'$ and (vii) it follows $x' \in A$, and from $y_i \in K'$ and (i) it follows

$$f(y_j,x') \subseteq -P(y_j) \quad \forall j,$$

i.e., every finite subfamily of the family of sets

$$S(y) := \{x \in A \mid f(y,x) \subseteq -P(y)\},$$

$y \in K$, has nonempty intersection. These are closed subsets of the compact A, hence there exists $\overline{x}$ belonging to all $S(y)$, i.e., $\overline{x} \in A$ and $f(y,\overline{x}) \subseteq -P(y)$ for all $y \in K$. Applying the first part of the proof with $K' := K$ yields $f(\overline{x},y) \not\subseteq -\operatorname{int} P(\overline{x})$ for all $y \in K$. □

The coercivity condition (vii) of Corollary 3 is automatically satisfied if K is compact, by choosing $A := K$. Note that, in view of the monotonicity property, (vii) is weaker than the coercivity condition given in Remark 2.

3. MIXED EQUILIBRIA

In this section, let $L(X,Z)$ denote the space of all continuous linear operators $X \to Z$. For $\phi \in L(X,Z)$ we write $\langle \phi, x \rangle := \phi(x)$, and for $\Phi \subseteq L(X,Z)$ we write $\langle \Phi, x \rangle := \{\langle \phi, x \rangle \mid \phi \in \Phi\}$. $L(X,Z)$ is supposed to be topologized in such a way that it is locally convex and $\langle \cdot, \cdot \rangle$ is continuous on $M \times X$ whenever $M \subseteq L(X,Z)$ is compact.

The mapping $P\colon K \rightrightarrows Z$ remains as before.

Theorem 4. Let K be nonempty, convex, and compact. Let $M \subseteq L(X,Z)$ be nonempty, compact, convex. Let $T\colon K \rightrightarrows M$ and $f,g\colon K \times K \rightrightarrows Z$ be given such that

(i) T is upper semicontinuous with nonempty, closed, convex values;
(ii) for all $y \in K$, $g(\cdot,y)$ is lower semicontinuous and compact-valued;
(iii) for all $x \in K$, $f(x,\cdot)$ and $g(\cdot,x)$ are right $P(x)$-convex;
(iv) for all $x,y \in K$ and $\xi \in M$, if $f(x,y) + \langle \xi, y-x \rangle \not\subseteq -\operatorname{int} P(x)$, then $g(x,y) - \langle \xi, y - x \rangle \subseteq -P(y)$;

(v) for all $x \in K$ and $\xi \in M$, if $g(x,y) - \langle \xi, y - x\rangle \subseteq -P(y)$ for all $y \in K$, then $f(x,y) + \langle \xi, y-x\rangle \not\subseteq -\operatorname{int} P(x)$ for all $y \in K$;

(vi) for all $x \in K$, $f(x,x) \not\subseteq -\operatorname{int} P(x)$.

Then there exist $\overline{x} \in K$ and $\overline{\xi} \in T(\overline{x})$ such that $f(\overline{x}, y) + \langle \overline{\xi}, y - \overline{x}\rangle \not\subseteq -\operatorname{int} P(\overline{x})$ for all $y \in K$.

PROOF. By condition (v) it suffices to show the existence of $(\overline{x}, \overline{\xi})$ satisfying

$$\begin{aligned} &\overline{x} \in K, \quad \overline{\xi} \in T(\overline{x}), \\ &g(\overline{x}, y) - \langle \overline{\xi}, y - \overline{x}\rangle \subseteq -P(y) \quad \forall y \in K. \end{aligned} \tag{11}$$

Since $-P(y) = \bigcap\{z - P(y) \mid z \in \operatorname{int} P(y)\}$, (11) holds if, for all $y \in K$ and $z \in \operatorname{int} P(y)$, $(\overline{x}, \overline{\xi})$ is contained in

$$\begin{aligned} R(y,z) := \{(x,\xi) \in K \times M \mid \xi \in T(x),& \\ g(x,y) - \langle \xi, y - x\rangle \subseteq z - P(y)\}.& \end{aligned}$$

These are closed subsets of $K \times M$, since T has closed graph and the mapping

$$(x,\xi) \mapsto g(x,y) - \langle \xi, y - x\rangle$$

is lower semicontinuous. But $K \times M$ is compact, so it suffices to show $\bigcap_{j=1}^{m} R(y_j, z_j) \neq \emptyset$ for all $y_1, \ldots, y_m \in K$ and $z_j \in \operatorname{int} P(y_j)$. To this end, fix $\xi \in M$, and consider the sets

$$C_j := \{x \in K \mid g(x, y_j) - \langle \xi, y_j - x\rangle \subseteq -P(y_j)\},$$

which are closed in K. For every nonempty $J \subseteq \{1, \ldots, m\}$ there holds

$$\operatorname{conv}\{y_j \mid j \in J\} \subseteq \bigcup_{j\in J} C_j.$$

In fact, assume for contradiction that there exists $x = \sum_{j\in J} \mu_j y_j$ with $\mu_j \geq 0$, $\sum_{j \in J} \mu_j = 1$, and $x \notin C_j$ for all j. Then $g(x, y_j) - \langle \xi, y_j - x\rangle \not\subseteq -P(y_j)$, hence from (iv), $f(x,y_j) + \langle \xi, y_j - x\rangle \subseteq -\operatorname{int} P(x)$, hence from (iii),

$$\begin{aligned} f(x,x) &\subseteq \sum_{j\in J} \mu_j f(x, y_j) - P(x) \\ &= \sum_{j\in J} \mu_j (f(x,y_j) + \langle \xi, y_j - x\rangle) - P(x) \\ &\subseteq -\operatorname{int} P(x) - P(x) \\ &\subseteq -\operatorname{int} P(x), \end{aligned}$$

contradicting (vi). So from the KKM-Lemma it follows that $\bigcap_{j=1}^{m} C_j \neq \emptyset$, i.e., there exists $x \in K$ such that $g(x,y_j) - \langle \xi, y_j - x\rangle \subseteq -P(y_j) \subseteq z_j - \operatorname{int} P(y_j)$ for all j. Hence M is covered by the sets

$$U(x) := \bigcap_{j=1}^{m} \{\xi \in M \mid g(x, y_j) - \langle \xi, y_j - x\rangle \subseteq z_j - \operatorname{int} P(y_j)\},$$

$x \in K$. But M is compact, and $U(x)$ is an open subset of M, since the mapping

$$\xi \mapsto g(x,y_j) - \langle \xi, y_j - x\rangle,$$

due to compactness of $g(x, y_j)$, is upper semicontinuous. Hence there exist a finite subcover $U(x_1), \ldots, U(x_n)$ and a continuous partition of unity $\beta_1, \ldots, \beta_n$ subordinate to this subcover; i.e., nonnegative continuous functions $\beta_i\colon M \to \mathbb{R}$ with $\sum_{i=1}^{n} \beta_i(\xi) = 1$ for all $\xi \in M$ and $\xi \in U(x_i)$ whenever $\beta_i(\xi) > 0$. Then $p(\xi) := \sum_{i=1}^{n} \beta_i(\xi) x_i$ defines a continuous $p\colon M \to K$. From (i), $T \circ p\colon M \rightrightarrows M$ is upper semicontinuous with nonempty, closed, convex values, hence has a fixed point $\xi' \in T(p(\xi'))$. Let $I := \{i \mid \beta_i(\xi') > 0\}$ and $x' := p(\xi') = \sum_{i\in I} \beta_i(\xi') x_i$. Then, from (iii) and $\xi' \in U(x_i)$ for all $i \in I$ it follows for all j that

$$\begin{aligned} &g(x', y_j) - \langle \xi', y_j - x'\rangle \\ &\subseteq \sum_{i\in I} \beta_i(\xi')(g(x_i, y_j) - \langle \xi', y_j - x_i\rangle) - P(y_j) \\ &\subseteq z_j - \operatorname{int} P(y_j) - P(y_j) \subseteq z_j - P(y_j), \end{aligned}$$

hence $(x', \xi') \in R(y_j, z_j)$, and this proves (11). □

Remark 5. Theorem 4 remains true if compactness of K is replaced by the following coercivity condition:

(vi) there exist a nonempty, compact set $A \subseteq K$ and a compact, convex set $B \subseteq K$ such that for every $x \in K \setminus A$ and $\xi \in T(x)$ there exists $y \in B$ with $g(x,y) - \langle \xi, y - x\rangle \not\subseteq -P(y)$.

The proof is similar to that of Remark 2

Corollary 6. Let K be nonempty and convex. Let $M \subseteq L(X,Z)$ be nonempty, compact, convex. Let $T\colon K \rightrightarrows M$ and $f\colon K \times K \rightrightarrows Z$ be given such that

(i) T is upper semicontinuous with nonempty, closed, convex values;

(ii) for all $y \in K$, $f(y, \cdot)$ is lower semicontinuous and compact-valued;

(iii) for all $x \in K$, $f(x, \cdot)$ is right $P(x)$-convex;

(iv) for all $x, y \in K$ and $\xi \in M$, if $f(x,y) + \langle \xi, y - x\rangle \not\subseteq -\operatorname{int} P(x)$, then $f(y,x) - \langle \xi, y - x\rangle \subseteq -P(y)$;

(v) for all $x, y \in K$, $f(\cdot, y)$ is upper semicontinuous on $[x,y]$, and $\operatorname{int} P(\cdot)$ has open graph;

(vi) for all $x \in K$, $f(x,x) \not\subseteq -\operatorname{int} P(x)$;

(vii) there exist a nonempty, compact set $A \subseteq K$ and a compact, convex set $B \subseteq K$ such that for every $x \in K \setminus A$ and $\xi \in T(x)$ there exists $y \in B$ with $f(x,y) + \langle \xi, y - x\rangle \subseteq -\operatorname{int} P(x)$.

Then there exist $\overline{x} \in A$ and $\overline{\xi} \in T(\overline{x})$ such that $f(\overline{x}, y) + \langle \overline{\xi}, y - \overline{x}\rangle \not\subseteq -\operatorname{int} P(\overline{x})$ for all $y \in K$.

PROOF. Let $g(x,y) := f(y,x)$. We show first that condition (v) of Theorem 4 holds on every convex subset $K' \subseteq K$. Let $x \in K'$, $\xi \in M$ satisfy

$$g(x,y) - \langle \xi, y - x\rangle \subseteq -P(y) \quad \forall y \in K'.$$

Assume that $f(x,y) + \langle \xi, y - x\rangle \subseteq -\operatorname{int} P(x)$ for some $y \in K'$. Then $y \neq x$ from (vi). From (v),

$$\{u \in [x,y] \mid f(u,y) + \langle \xi, y-u \rangle \subseteq -\operatorname{int} P(u)\}$$

is open in $[x,y]$. Hence there exists $u \in\,]x,y[$ such that $f(u,y) + \langle \xi, y-u \rangle \subseteq -\operatorname{int} P(u)$. Since $f(u,x) + \langle \xi, x-u \rangle = g(x,u) - \langle \xi, u-x \rangle \subseteq -P(u)$, it follows from (iii), that $f(u,u) \subseteq -\operatorname{int} P(u)$, contradicting (vi). Hence

$$f(x,y) + \langle \xi, y-x \rangle \not\subseteq -\operatorname{int} P(x) \quad \forall y \in K'.$$

Now let $y_1, \dots, y_m \in K$, and

$$K' := \operatorname{conv}(B \cup \{y_1, \dots, y_m\}).$$

Then K' is compact, since B is compact and convex. From Theorem 4 there exist $x' \in K'$ and $\xi' \in T(x')$ such that

$$f(x',y) + \langle \xi', y-x' \rangle \not\subseteq -\operatorname{int} P(x') \quad \forall y \in K'.$$

From $B \subseteq K'$ and (vii) it follows $x' \in A$, and from $y_i \in K'$ and (iv) it follows

$$f(y_j, x') - \langle \xi', y_j - x' \rangle \subseteq -P(y_j) \quad \forall j,$$

i.e., every finite subfamily of the family of sets

$$S(y) := \{(x,\xi) \in A \times M \mid \xi \in T(x), \\ f(y,x) - \langle \xi, y-x \rangle \subseteq -P(y)\},$$

$y \in K$, has nonempty intersection. These are closed subsets of the compact $A \times M$, hence there exists $(\overline{x}, \overline{\xi})$ belonging to all $S(y)$, i.e., $\overline{x} \in A$, $\overline{\xi} \in T(\overline{x})$ and $f(y,\overline{x}) - \langle \overline{\xi}, y - \overline{x} \rangle \subseteq -P(y)$ for all $y \in K$. Applying the first part of the proof to $K' := K$ yields $f(\overline{x}, y) + \langle \overline{\xi}, y - \overline{x} \rangle \not\subseteq -\operatorname{int} P(\overline{x})$ for all $y \in K$. □

References

Ansari, Q.H., W. Oettli and D. Schläger (1997). A generalization of vectorial equilibria. *Math. Methods Oper. Res.*, **46**, 147–152.

Berge, C. (1966). *Espaces topologiques – Fonctions multivoques.* Dunod, Paris.

Bianchi, M., N. Hadjisavvas and S. Schaible (1997). Vector equilibrium problems with generalized monotone bifunctions. *J. Optim. Theory Appl.*, **92**, 531–546.

Blum, E. and W. Oettli (1994). From optimization and variational inequalities to equilibrium problems. *Math. Student*, **63**, 123–145.

Fan, K. (1961), A generalization of Tychonoff's fixed point theorem. *Math. Ann.*, **142**, 305–310.

Oettli, W. and D. Schläger (1998). Generalized vectorial equilibria and generalized monotonicity. In: *Functional Analysis with Current Applications* (M. Brokate and A.H. Siddiqi (Ed)), 145–154. Longman, London.

TIME-OPTIMAL CONTROL IN A THIRD-ORDER SYSTEM

F. L. Chernousko and A. M. Shmatkov*

Institute for Problems in Mechanics of RAS pr. Vernadskogo 101-1, Moscow, 117526

Abstract. A third-order controlled system which simulates the motion of an inertial object acted upon by a control force with a bounded rate of change is considered. The time-optimal open-loop control of the system is constructed. The feedback optimal control is given in closed form.

Key Words. Optimal control; open-loop control; feedback control; maximum principle

1. FORMULATION OF THE PROBLEM

A system with a single degree of freedom is considered, described by the equations

$$\dot{x}_1 = x_2, \quad m\dot{x}_2 = F \tag{1}$$

where x_1 is a generalized coordinate, x_2 is a generalized velocity, m is a constant inertial characteristic (the mass or moment of inertia), F is the control (the force or the moment of the force) and dots denote derivatives with respect to the time t.

When formulating optimal control problems, it is usually assumed that the absolute magnitude of the force F is bounded by a constant F_0, that is, $|F| \le F_0$. In the case of a time-optimal control problem it is well known (Pontryagin, *et al.*, 1983) that this constraint leads to the bang-bang form of optimal control. In this case, the force $F(t)$ takes limiting values $\pm F_0$ and instantaneously switches from one of these values to the other. Such a control is not always practicable, for example, when an electric drive is used to realize the control.

Let's assume that there is a more realistic constraint on the rate of change of the control force of the form

$$|\dot{F}| \le v_0 \tag{2}$$

where $v_0 > 0$ is a specified constant, and that the bound on the absolute magnitude of the force is not attained and $|F(t)| < F_0$ always.

Making the change of variables

$$x_1 = (v_0/m)x, \quad x_2 = (v_0/m)y, \quad F = v_0 z$$

one can reduce (1) and constraint (2) to the form

$$\dot{x} = y, \quad \dot{y} = z, \quad \dot{z} = u, \quad |u| \le 1 \tag{3}$$

Here, the variables x, y and z are phase coordinates and u plays the role of a bounded control.

The initial conditions for system (3) are specified in the form

$$x(0) = x_0, \quad y(0) = y_0, \quad z(0) = z_0 \tag{4}$$

where the initial instant of time is assumed to be equal to zero without any loss in generality.

It is possible now to formulate the problem of constructing a control $u(t)$ which satisfies the constraint $|u(t)| \le 1$ when $t \ge 0$ and which transfers system (3) from an arbitrary initial state (4) to a specified terminal manifold

$$x(T) = 0, \quad y(T) = 0 \tag{5}$$

for arbitrary $z(T)$ after the shortest time T.

In addition to determining the open-loop control, the problem of the feedback time-optimal control for system (3) will also be solved. This control $u(x, y, z)$, which is expressed as a function of the current (or initial) phase coordinates x, y, z, ensures that system (3) is brought to the specified terminal manifold (5) after the shortest time.

2. THE MAXIMUM PRINCIPLE

Let's apply the maximum principle (Pontryagin, *et al.*, 1983) to the time-optimal control problem (3) - (5). The Hamiltonian function can be written in the form

$$H = p_x y + p_y z + p_z u \tag{6}$$

and it leads to the conjugate equations

$$\dot{p}_x = 0, \quad \dot{p}_y = -p_x, \quad \dot{p}_z = -p_y \tag{7}$$

Here p_x, p_y, p_z are the conjugate variables. System (7) is integrated subject to the transversality condition $p_z(T) = 0$, which corresponds to the condition that $z(T)$ is not fixed, and one can obtain

$$p_x = c_x,\ p_y = c_y + c_x\tau,\ p_z = c_y\tau + \frac{c_x\tau^2}{2} \quad (8)$$

Here $\tau = T - t$ is the time measured from the end of the process (the "inverse" time), and c_x and c_y are arbitrary constants. The condition for the Hamiltonian (6) to be a maximum with respect to u subject to the constraint $|u| \leq 1$ from (3) gives $u(t) = \operatorname{sign} p_z(t)$. It follows from formula (8) for p_z that the function $p_z(t)$ changes sign not more than once when $t \leq T$, $\tau \geq 0$. Consequently, the optimal control $u(t) = \pm 1$ has not more than one switching when $t \leq T$.

3. OPEN-LOOP CONTROL

Let's denote the lengths of the two possible segments of constancy of the control $u(t)$ by θ_1 and θ_2 and the value of $u(t)$ in the first of these segments by $\sigma = \pm 1$. The optimal control can then be represented in the form

$$u(t) = \begin{cases} \sigma; & t \in (0, \theta_1) \\ -\sigma; & t \in (\theta_1, T),\ \theta_1 + \theta_2 = T \end{cases} \quad (9)$$

Now control (9) can be substituted into system (3) and integrated subject to the initial conditions (4). One can obtain

$$\begin{aligned} x(t) &= x_0 + y_0 t + z_0 t^2/2 + \sigma t^3/6 \\ y(t) &= y_0 + z_0 t + \sigma t^2/2 \\ z(t) &= z_0 + \sigma t \quad \text{when} \quad t \in (0, \theta_1) \end{aligned}$$

$$\begin{aligned} x(t) &= x_0 + y_0\theta_1 + z_0\theta_1^2/2 + \sigma\theta_1^3/6 + \\ &\quad (y_0 + z_0\theta_1 + \sigma\theta_1^2/2)(t - \theta_1) + \\ &\quad (z_0 + \sigma\theta_1)(t - \theta_1)^2/2 - \sigma(t - \theta_1)^3/6 \\ y(t) &= y_0 + z_0\theta_1 + \sigma\theta_1^2/2 + \\ &\quad (z_0 + \sigma\theta_1)(t - \theta_1) - \sigma(t - \theta_1)^2/2 \\ z(t) &= z_0 + \sigma\theta_1 - \sigma(t - \theta_1) \text{ when } t \in (\theta_1, T) \end{aligned} \quad (10)$$

Substituting solution (10) into condition (5), two relations can be obtained

$$\begin{aligned} & x_0 + y_0(\theta_1 + \theta_2) + z_0(\theta_1 + \theta_2)^2/2 + \\ &\quad \sigma(\theta_1^3 + 3\theta_1^2\theta_2 + 3\theta_1\theta_2^2 - \theta_2^3)/6 = 0 \\ & y_0 + z_0(\theta_1 + \theta_2) + \sigma(\theta_1^2 + 2\theta_1\theta_2 - \theta_2^2)/2 = 0 \end{aligned}$$

Then, on solving these for x_0 and y_0 one obtains

$$\begin{aligned} x_0 &= z_0 T^2/2 + \sigma(\theta_1^3 + 3\theta_1^2\theta_2 - \theta_2^3)/3 \\ y_0 &= -z_0 T - \sigma(\theta_1^2 + 2\theta_1\theta_2 - \theta_2^2)/2 \end{aligned} \quad (11)$$

The following notation is introduced

$$\begin{aligned} & \xi = z_0^{-3} x_0,\ \eta = z_0^{-1}|z_0|^{-1} y_0,\ \zeta = \operatorname{sign} z_0 \\ & s = |z_0|^{-1} T,\ \lambda = \theta_2 T^{-1}\ (z_0 \neq 0) \quad (12) \\ & X(\lambda) = (1 - 3\lambda^2 + \lambda^3)/3,\ Y(\lambda) = \lambda^2 - 1/2 \end{aligned}$$

Relations (11) then takes the form

$$\begin{aligned} \zeta(\xi s^{-3} - s^{-1}/2) &= \sigma X(\lambda) \\ \zeta(\eta s^{-2} + s^{-1}) &= \sigma Y(\lambda) \end{aligned} \quad (13)$$

When $z_0 = 0$, relations (11) gives

$$x_0 T^{-3} = \sigma X(\lambda), \quad y_0 T^{-2} = \sigma Y(\lambda) \quad (14)$$

When the parameter λ changes from 0 to 1, a point with coordinates $X(\lambda)$, $Y(\lambda)$ traverses the arc of the curve which joints points A_1 and A_2 with coordinates $(1/3, -1/2)$, $(-1/3, 1/2)$. When $\lambda \in [0, 1]$ and $\sigma = \pm 1$, points with coordinates $\sigma X(\lambda)$, $\sigma Y(\lambda)$ form a closed curve Γ which is symmetric about the origin of coordinates and has corner points A_1 and A_2, (see Fig. 1). The curve Γ bounds a convex domain containing the origin of the system of coordinates.

The solution of the time-optimal open-loop control problem (3) – (5) can then be represented as follows.

Let's initially assume that $z_0 \neq 0$ and determine ξ, η, ζ in accordance with (12) from (5) using the specified initial data x_0, y_0, z_0. The left-hand sides of relations (13) specify the coordinates of a certain point P, which depends on the parameter $s \in [0, \infty)$. As s changes from ∞ to 0, P moves along a smooth semi-infinite curve from the origin of the system of coordinates (when $s \to \infty$) to infinity (when $s \to 0$). This point falls at least once on the closed curve Γ, which encircles the origin of the system of coordinates. The least value of $s = s_*$ for which $P \in \Gamma$ is found numerically. According to (12), the optimal time is equal to $T = |z_0|s_*$. The position of the point P on the curve Γ, when $s = s_*$, determines the values of the parameters $\sigma = \pm 1$ and $\lambda \in [0, 1]$. By virtue of (12), the lengths of the segments of constancy of the control are equal to $\theta_1 = (1 - \lambda)T$ and $\theta_2 = \lambda T$.

When $z_0 = 0$, equalities (14) will be considered instead of (13). The left-hand sides of these equalities specify the coordinates of the point P which

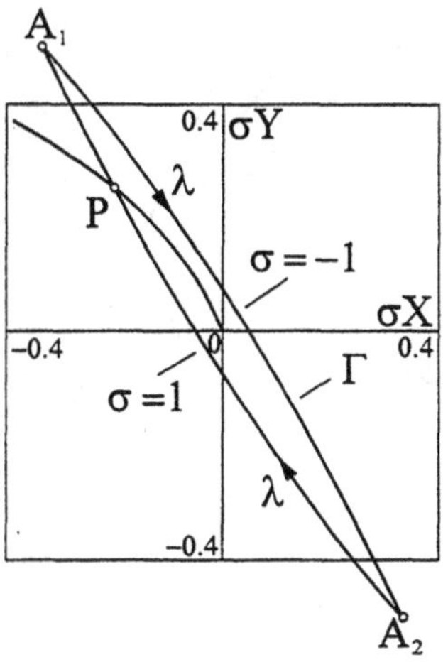

Fig. 1. The example of open-loop control problem solution

depends on the parameter T. When T changes from ∞ to 0, the point P moves along a semicubic parabola from the origin of the system of coordinates (when $T \to \infty$) to infinity (when $T \to 0$). The least value of the parameter T for which $P \in \Gamma$ is the optimal time. The values of parameters σ, λ, θ_1, θ_2 are determined from the position of the point P on Γ as in the case when $z_0 \neq 0$.

When the quantities σ, θ_1, θ_2 have been determined, the optimal control $u(t)$ and the corresponding optimal trajectory are specified by equalities (9) and (10). The proposed algorithm completely determines the solution of the time-optimal open-loop control problem. According to the construction, this solution is unique.

As an example, the results of the determination of the optimal control for the initial data are presented

$$x_0 = -72 + 27\sqrt{3} \approx -25.2, \ y_0 = 3, \ z_0 = 1$$

In this case, one obtains

$$T = s = 6, \ \sigma = 1$$

$$\theta_1 = 6 - 3\sqrt{3} \approx 0.80, \qquad \theta_2 = 3\sqrt{3} \approx 5.20$$

The corresponding trajectory of the point P when T changes from ∞ to 0 is shown in Fig. 1.

4. FEEDBACK OPTIMAL CONTROL

In order to construct a feedback optimal control it suffices to find the switching surfaces in the phase space xyz on which the sign of the control $u = \pm 1$ changes. On these surfaces, the length of one of the segments of constancy vanishes, that is, $\theta_1 = 0$ or $\theta_2 = 0$. From (12), here $\lambda = 0$ or $\lambda = 1$. According to (13), values of X and Y equal to $\pm 1/3$ and $\mp 1/2$ correspond to these values of λ respectively. From (13), one obtains the conditions

$$\begin{aligned} \zeta(\xi s^{-3} - s^{-1}/2) &= \pm\sigma/3 \\ \zeta(\eta s^{-2} + s^{-1}) &= \mp\sigma/2 \end{aligned} \tag{15}$$

which are satisfied in the $\xi\eta$ plane on the switching curves when $z_0 \neq 0$. However, relations (15) are insufficient for determining the switching curves: for this, a direct analysis of relations (13) is required which will be carried out below.

Note that, in the feedback control, the initial conditions x_0, y_0, z_0 can be treated as the current values of the phase coordinates x, y, z. Relations (12)can be considered as formulae for the change of variables

$$\xi = z^{-3}x, \quad \eta = z^{-1}|z|^{-1}y, \quad \zeta = \operatorname{sign} z \tag{16}$$

in phase state. This change of variables, which introduces the self-similar variables ξ and η, enables one, when $z \neq 0$, to reduce the dimension of the phase space by one and to construct the feedback optimal control in the $\xi\eta$ plane.

Let's first consider the case when $z = 0$ separately. By analogy with (15), one obtains the conditions

$$xT^{-3} = \pm\sigma/3, \quad yT^{-2} = \mp\sigma/2 \tag{17}$$

from (14). These conditions are satisfied at the intersection of the switching surfaces with the $z = 0$ plane. When $z = 0$, conditions (17) define two halves of the semicubic parabolae which form the switching curve (SC) in the $z = 0$ plane, described by the equation

$$\gamma(x, y) \equiv 3x + 2y|y|^{3/2} = 0 \tag{18}$$

The analysis of the signs of σ on the branches of the SC (18) enables one to determine the signs of the controls on the different sides of the switching curve. As a result, the feedback optimal control when $z = 0$ is obtained in the form

$$u(x, y, 0) = \begin{cases} -\operatorname{sign}\gamma(x, y), & \gamma \neq 0 \\ \operatorname{sign} x = -\operatorname{sign} y, & \gamma = 0 \end{cases} \tag{19}$$

When $z \neq 0$, the change of variables (16) transforms the first two equations of (3) to the form

$$\dot{\xi} = |z|^{-1}(\eta - 3u\zeta\xi), \ \dot{\eta} = |z|^{-1}(1 - 2u\zeta\eta) \tag{20}$$

On dividing the first equation of (20) by the second, the linear equation in ξ is obtained

$$\frac{d\xi}{d\eta} = \frac{\eta - 3\alpha\xi}{1 - 2\alpha\eta}, \quad \alpha = u\zeta = \pm 1 \tag{21}$$

The parameter α retains a constant value along

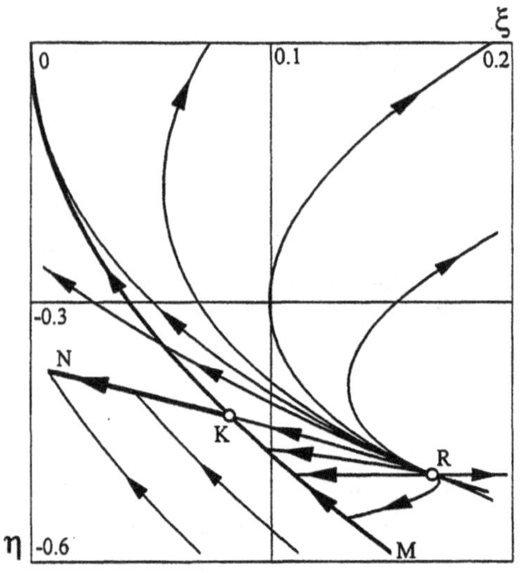

Fig. 2. Feedback optimal control: a large scale

the optimal trajectories which do not intersect the $z = 0$ plane. On integrating (21) in the case of constant α, its general solution can be found

$$\xi = \Phi(\eta, \alpha, A) \equiv \alpha\eta - \frac{1}{3} + A|1 - 2\alpha\eta|^{3/2} \quad (22)$$

where A is an arbitrary constant. Note that the second equation (20) enables one to determine the direction of motion along the optimal trajectories. If $\alpha = 1$, the motion occurs in the direction of an increase in η when $\eta < 1/2$ and in the direction of a decrease in η when $\eta > 1/2$. If, however, $\alpha = -1$, the motion occurs in the direction of a decrease in η when $\eta < -1/2$ and an increase in η when $\eta > -1/2$.

Let's completely present the feedback optimal control. To be specific, one can take $z > 0$ and $\zeta = 1$. The switching curve in the $\xi\eta$ plane is defined by the equalities

$$\xi = f(\eta) = \begin{cases} \Phi(\eta, 1, 1/3), & \eta \le \eta^* \\ \Phi(\eta, -1, -1/3), & \eta > \eta^* \end{cases} \quad (23)$$

where the notation of (22) is used, and $\eta^* = -\sqrt{3}/4$. The switching curve is continuous and has a kink at the point K with the coordinates $\xi^* = 1/12$, $\eta^* = -\sqrt{3}/4$. This curve is represented by the solid line in Fig. 2 and Fig. 3.

On account of the fact that the scale in Fig. 3 is smaller than that in Fig. 2, the points K и R shown in Fig. 2 are practically indistinguishable in Fig. 3 and are therefore not labelled. On the other hand, the scale the scale used in Fig. 3 enables us to depict all the characteristic phase trajectories, the important part of which is missing in Fig. 2. The rest of the notation employed in Fig. 2 and Fig. 3 is identical. To be specific, let's henceforth mainly refer to Fig. 2.

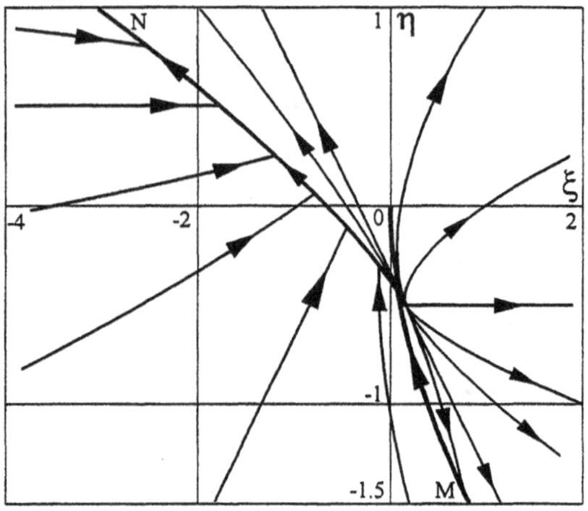

Fig. 3. Feedback optimal control: a small scale

The branches of the switching curve corresponding to $\eta < \eta^*$ and $\eta > \eta^*$ are denoted by the letters M and N respectively. In the $\xi\eta$ plane, one has

$$u = \begin{cases} 1, & \xi < f(\eta) \\ 1, & \xi = \Phi(\eta, 1, 1/3), \quad \eta \le 0 \\ -1, & \text{at the remaining points} \end{cases} \quad (24)$$

Hence $u = 1$ to the left of and below the switching curve (23), on its segment MK and also on the arc of the curve $\xi = \Phi(\eta, 1, 1/3)$ which joins the origin of the system of coordinates and the point K, see Fig. 2, where this arc is a part of the switching curve. In the remaining part of the $\xi\eta$ plane, one has $u = -1$.

When $z < 0$, $\zeta = -1$ the switching curve remains the same and one simply has to interchange the positions of the set of points $\xi\eta$ where $u = 1$ and $u = -1$ in relations (24). So, the synthesis of the optimal control $u(x, y, z)$ is completely determined by relations (16), (18), (19), (22), (23), (24) for all x, y, z.

Let's now describe the set of optimal trajectories which, in the variables ξ and η, consist of arcs of the curves (22). Suppose that the initial point x, y, z is specified and, to be specific, let's assume that $z > 0$. According to formulae (12), one can find that ξ, η and $\zeta = 1$.

If a point $\xi\eta$ lies on the curve $\xi = \Phi(\eta, 1, 1/3)$, where $\eta \le 0$, then motion occurs along this curve $MK0$ with a control $u = 1$ until it reaches the origin of the system of coordinates.

All the remaining optimal trajectories also arrive at the origin of the system of coordinates along this curve. An exception is the segment $R0$ of the curve $\xi = \Phi(\eta, -1, 1/3)$ when $\eta \in [-1/2, 0]$: this segment is a phase trajectory for $u = -1$ which begins at the point R with coordinates $(1/6, -1/2)$ and reaches the origin of the system of coordinates. The point R represents the whole phase trajectory for $u = -1$ except the final point which is the origin of the system of coordinates in the case when $z = 0$. Phase trajectories are denoted by thin lines in Fig. 2 and Fig. 3 and arrows indicate the direction of the motion.

If the initial point lies in the curvilinear corner

$$\eta \leq 0, \quad \Phi(\eta, 1, 1/3) < \xi < \Phi(\eta, -1, 1/3) \quad (25)$$

then the optimal trajectory consists of the segment with $u = -1$ until it reaches the curve $\xi = \Phi(\eta, 1, 1/3)$ and of the subsequent motion along this curve with $u = 1$.

If the initial $\xi < f(\eta)$, the motion initially occurs with $u = 1$ until it intersects the curve $\xi = \Phi(\eta, -1, -1/3)$ which is the part KN of the switching curve (23) (see Fig. 2) and then with $u = -1$ along this curve which departs to infinity. By (12), one has $z = 0$ at an infinitely distant point of the $\xi\eta$ plane. At infinity, z changes sign and, then, $z < 0$, $\zeta = -1$. The phase trajectory continues, arriving, when $u = -1$, from infinity along the curve $\xi = \Phi(\eta, 1, 1/3)$ and arrives along this curve at the origin of the system of coordinates. Note that motion through an infinitely distant point occurs without a change in the control and takes a finite time.

It remains to consider initial points in the domain $\xi > f(\eta)$ but outside of the curvilinear corner (25). Here, one initially has $u = -1$ and the trajectory $\xi = \Phi(\eta, -1, A)$ departs to infinity, where $A > -1/3$. When $\zeta = -1$ and $u = -1$, motion subsequently occurs along the curves $\xi = \Phi(\eta, 1, -A)$ with a change in the sign of A. These curves lie in the domain $\xi < f(\eta)$ and persist in the branch KN of the switching curve $\xi = \Phi(\eta, -1, -1/3)$. A trajectory with $u = 1$ departs to infinity along this curve, where the sign of z changes again. Later, when $\zeta = 1$, motion occurs, when $u = 1$, along the curve $\xi = \Phi(\eta, 1, 1/3)$ until it reaches the origin of the system of coordinates.

Note that certain phase trajectories contain segments of the lines $\xi = \pm\eta - 1/3$ and $\eta = \pm(2\alpha)^{-1}$, which correspond to the values $A = 0$ and $A = \infty$ in (22) respectively. On departing to infinity along these lines the variable x (in the case of the straight line with $A = 0$) or the variable y (in the case of the straight line with $A = \infty$) simultaneously vanishes together with z, as is easily shown using (12). In other respects these lines are treated in the same manner as the remaining trajectories (22).

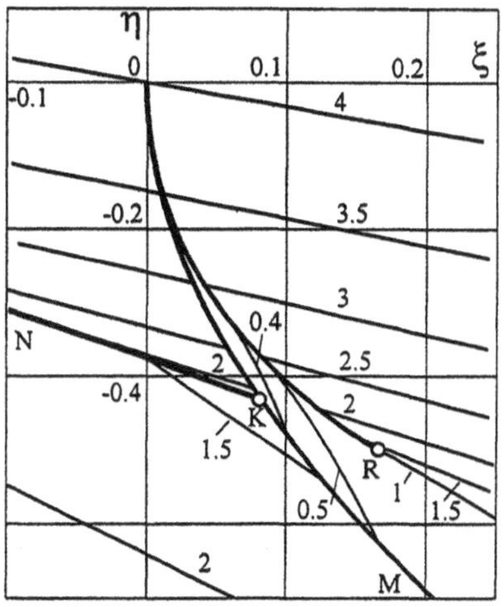

Fig. 4. Level lines of normalized optimal time $s(\xi, \eta)$

Hence for any initial point x, y, z, the motion is completely described by the trajectories of Fig. 2, Fig. 3 and contains not more than two segments where the control is constant. In this case, the sign of z cannot change more than twice.

The results of an investigation of the normalized optimal time s as a function of ξ and η are presented in Fig. 4. In Fig. 4, the thin lines are level lines of the function $s(\xi, \eta)$ and the bold lines are the lines of discontinuity of this function. The rest of the notation is the same as in Fig. 2.

This research was supported financially by the Russian Foundation for Basic Research (96–01–01137).

5. REFERENCES

Pontryagin, L. S., Boltyanskii, V. G., Gamkrelidze, R. V. and Mishchenko, Ye. F. (1983). *Mathematical Theory of Optimal Processes.* Nauka, Moscow.

APPROXIMATE GRADIENT METHODS AND THE NECESSARY CONDITIONS FOR THE EXTREMUM OF DISCONTINUOUS FUNCTIONS

V. D. Batukhtin, S. I. Bigil'deev, T. B. Bigil'deeva

Chelyabinsk State University
129, Br. Kashirinyikh str, Chelyabinsk, 454021, Russia

Abstract: The theory of solving extremal problems is developed and its practical applications are discussed. On the basis of approximate gradient theory (Batukhtin and Maiboroda, 1984,1995; Batukhtin, 1993) a multivalued mapping is constructed that makes it possible to study the extremum of functions integrated with the square in Lebesgue measure. The connection with the subdifferential F.H.Clarke is established. Numerical algorithms that practically realize this approach are given. The work of algorithms is illustrated with examples. *Copyright ©1998 IFAC*

Keywords: Mathematical programming, Discontinuous function, Integrals, Numerical methods, Computer experiments.

1. INTRODUCTION

In the monographs (Batukhtin and Maiboroda, 1984, 1995) the concept of approximate gradient was introduced and on its basis an approach to solving discontinuous extremal problems was developed. Approximate gradient being an integral operator, is a vector that can be considered as gradient of differentiable functions analogue. Yet, it is defined not only for differentiable functions but for discontinuous ones as well. The interesting interpretation of this approach is founded in the papers (Bigil'deev, 1996; Bigil'deev and Rolshchikov, 1997; Batukhtin, et al., 1997).

The methods of approximate gradient are studying the functions for the extremum via constructing the sequence of points at each of which the approximate gradient is zero vector and the diameter of the integration domain converges to zero. Such approach proved to be fairly productive in numerical solutions of extremal problems.

At the same time, some rather complex problems are encountered on this way. First and foremost while constructing the numerical methods. It is, to a great extent, connected with the complex behaviour of the studied objects, that is discontinuous functions. Particularly, one of the important problems is to create effective methods of constructing a sequence of points where approximate gradient equals to zero at a proper choice of a weight function. The point is that, when the diameter of the integration domain converges to zero on a function discontinuity the approximate gradient is infinitely growing in the norm to impede the numerical realization.

The intention of this paper is to consider the approximate gradient methods from a more general point of view, that, as the authors believe, would result in a further step in the understanding of the problem and would encourage the development of the approximate gradient theory. The theoretical part of the paper is based on the representation of the approximate gradient in a more general form and the study of a set of approximate gradients on all possible weight functions as well as the study of the limiting properties of the set.

When using approximate gradient, the problem of conditional optimization of the function $F : U \to \mathbf{R}$ on set $U \subset \mathbf{R}^n$ may be consider as the problem of unconditional optimization of an extended function on all space $\mathbf{R}^n$:

$$f(x) = \begin{cases} F(x) \ , \text{if } x \in U \\ \Psi(x) \ , \text{otherwise} \end{cases}$$

The various ways of function extention were considered by authors in detail at the papers (Batukhtin and Maiboroda, 1984, 1995; Batukhtin, et al., 1997). Note, that the problem of function extention does not involve any principle difficulties, that is a consequence of wideness of the studied functions class.

Let $f : \mathbf{R}^n \to \mathbf{R}$ be a square integrable in Lebesgue measure μ function. Consider the problem of unconditional minimization:

$$f(x) \longrightarrow \min_{x \in \mathbf{R}^n} \tag{1}$$

Following (Liusternik and Sobolev, 1965), let us introduce the concept of substantial minimum of the function. Let $B_\delta(x)$ be a sphere with the radius δ and the center at the point x. Denote the class $\mathbf{X}$ of the sets X of the measure 0 that belong to $B_\delta(x)$ and consider the following function on $\mathbf{X}$:

$$\inf_{y \in B_\delta(x) \setminus X} f(y) = \alpha(X)$$

Let $\alpha_0 = \sup_{X \subset \mathbf{X}} \alpha(X)$. In addition, if the function $\alpha(X)$ is finite for some $X \subset \mathbf{X}$, then it takes the maximal value on a certain set $X_\alpha \subset \mathbf{X}$ and $\alpha_0 > -\infty$ (Liusternik and Sobolev, 1965).

The α_0 is called the substantial minimum of the function f on the sphere $B_\delta(x)$ and is denoted :

$$vrai \inf_{y \in B_\delta(x)} f(y) = \sup_{X \subset \mathbf{X}} \left\{ \inf_{y \in B_\delta(x) \setminus X} f(y) \right\}.$$

Here the following limit

$$vr\, f(x) = \lim_{\delta \to +0} vrai \inf_{y \in B_\delta(x)} f(y) \tag{2}$$

is called the substantial value of the function f at the point x.

Definition 1. The point $x^* \in \mathbf{R}^n$ is called a point of substantial local minimum of the function f if there exists $\varepsilon > 0$ such that,

$$f(x) \geq vr\, f(x^*) \tag{3}$$

on $B_\varepsilon(x^*)$ almost everywhere (Kolmogorov and Fomin, 1989). If the prefix *vrai* is absent in (2) and (3) holds for all x from $B_\varepsilon(x^*)$, then the point x^* is called just a point of local minimum.

It is evident that for the continuous function the sets of minimum points and substantial minimum points coincide.

Definition 2. The nonnegative square integrable in Lebesgue measure μ function $p_\delta : \mathbf{R}^n \to [0; +\infty)$ is called a weight one if it is other than zero only in a neighborhood Ω_δ of the origin that is contained in the sphere of the radius $\delta > 0$ and takes positive values there on the set of non zero measure.

Definition 3. The following integral operator

$$a(x; \delta, p_\delta; f)$$
$$= D(p_\delta)^{-1} \int_{\Omega_\delta} (s - \bar{s}_\delta) f(x + s)\, p_\delta(s) \mu(ds), \tag{4}$$

where

$$\bar{s}_\delta = \int_{\Omega_\delta} s p_\delta(s) \mu(ds) / \int_{\Omega_\delta} p_\delta(s) \mu(ds),$$

$$D(p_\delta) = \int_{\Omega_\delta} (s - \bar{s}_\delta)(s - \bar{s}_\delta)^T p_\delta(s) \mu(ds)$$

is the positively defined matrix due to the definition of the weght function p_δ, is called the approximate gradient $a(x; \delta, p_\delta; f)$ of the function f at the point x.

NOTE. In the mentioned papers (Batukhtin and Maiboroda, 1984,1995; Batukhtin 1993) the approximate gradient is defined on a weight function that represents a probability distribution density function of the vector s the components of which are centered noncorrelated random values.

In order to construct the algorithms for solving the problem (1) numerically, let us consider the properties of the limiting points of the set of approximate gradients (4) as $\delta \to +0$ on the set of all weight functions. Here, this set of limiting points is called the approximate subdifferential of the function f at the point x and is denoted by $\partial_a f(x)$.

Similarly (Batukhtin and Maiboroda, 1984) it is not difficult to establish that in the case of the differentiable function the approximate subdifferential consists of a single point that coinsides with the function gradient. However, if no additional conditions are imposed on weight functions, then the approximate subdifferential may not be contained in the subdifferential even for the convex function f (Bigil'deev and Rolshchikov, 1997). Hence, further we shall confine ourselves with the weight functions being other than zero in the sphere $B_r(\bar{s}_\delta) \subseteq \Omega_\delta$, where $r > 0$ that depend only on the distance from points $\bar{s}_\delta$. As the result, the weight function for each δ will be given by a certain point from the neighborhood Ω_δ and a function of one variable on the segment $[0; r]$. We shall denote the mentioned point $\bar{s}_\delta$ from Ω_δ by $\bar{s}_r$

and the weight function by $p_r(s) = p_r(\| s - \bar{s}_r \|)$. Such weight functions are called symmetrical, and in the case $\bar{s}_r = 0$ - centered. Denote by Q the set of the symmetrical weight functions.

When symmetrical weight functions are used approximate gradient (4) takes a simpler form. In this case the set Ω_δ may be replaced by $B_r(\bar{s}_r)$ and the matrix $D(p_r)$ becomes diagonal with the same elements on the diagonal:

$$d_r(p_r) = \frac{1}{n} \int\limits_{B_r(\bar{s}_r)} \| s - \bar{s}_r \|^2 p_r(\| s - \bar{s}_r \|)\mu(ds)$$

$$= \frac{1}{n} \int\limits_{B_r} \| \nu \|^2 p_r(\| \nu \|)\mu(d\nu) > 0,$$

where $\nu = s - \bar{s}_r$, n is the space dimension, B_r is the sphere of the radius r with the center at the origin.

As the vector $\bar{s}_r$ is defined arbitrarily, let us include it in the arguments of the approximate gradient, then Formula (4) will be rewritten as :

$$a(x; r, p_r, \bar{s}_r; f) =$$

$$\frac{1}{d_r(p_r)} \int\limits_{B_r} \nu f(x + \bar{s}_r + \nu)\, p_r(\| \nu \|)\mu(d\nu) \quad (5)$$

Approximate gradient (5) on symmetrical weight functions is continuous in $\bar{s}_r$ due to its continuity in x on the centered weight functions (Batukhtin and Maiboroda, 1984)

By definition for the symmetrical weight functions

$$\partial_a f(x) = \left\{ \lim_{\substack{r \to +0 \\ \bar{s}_r \to 0 \\ p_r \in Q}} a(x; r, p_r, \bar{s}_r; f) \right\}, \quad (6)$$

where for each $r > 0$ the weight function $p_r(\| s - \bar{s}_r \|)$ is given arbitrarily.

Let us consider the properties of the set.

It can be unbounded even for the function of one variable (Bigil'deev, 1996;). By definition, this set is closed. Besides, the multivalued mapping $x \mapsto \partial_a f(x)$ is upper semicontinuous in inclusion. The latter follows from the fact that in the definition of approximate gradient (5) $\bar{s}_r$ gives only a parallel transfer of the variable x. At the same time there can be any order of r and $\bar{s}_r$ converging to zero in (6). Hence, directing first r then $\bar{s}_r$ to zero we obtain the definition of upper semicontinuity in x.

As $\partial_a f(x)$ is closed its convex hull $co\, \partial_a f(x)$ is a closed convex set.

Let us define the function $f^a(x; u)$ as the upper limit of the following type:

$$f^a(x; u) = \limsup_{\substack{r \to +0 \\ \bar{s}_r \to 0 \\ p_r \in Q}} \langle a(x; r, p_r, \bar{s}_r; f),\, u \rangle \quad (7)$$

for $\forall u \in \mathbf{R}^n$.

As in (Clarke, 1983), it can be established that the function $u \mapsto f^a(x; u)$ is positively uniform and subadditive. Upper semicontinuity of $f^a(x; u)$ as the function of x follows from upper semicontinuity in inclusion of the multivalued mapping $x \mapsto \partial_a f(x)$.

On the other hand, by Formulas (7) and (6) for the function $f^a(x; u)$ and the set $\partial_a f(x)$, $\xi \in co\, \partial_a f(x)$ if and only if $\langle \xi, u \rangle \le f^a(x; u)$ for all $u \in \mathbf{R}^n$. Hence, $f^a(x; u)$ is the supporting function of $co\, \partial_a f(x)$ (Pschenichny, 1980).

Following (Clarke, 1983) we call the following function as the generalized derivative of the function f in the direction $u \in \mathbf{R}^n$ at the point x :

$$f^o(x; u) = \limsup_{\substack{y \to x \\ t \to +0}} \frac{f(y + tu) - f(y)}{t}, \quad (8)$$

which, as well as the function $f^a(x; u)$, is the positively uniform and subadditive function in u.

Theorem 4. For any point x and for all $u \in \mathbf{R}^n$ the following inequality is true

$$f^a(x; u) \le f^o(x; u)$$

■

PROOF. Due to positive uniformity of functions (7), (8) it is sufficient to consider $u \in \mathbf{R}^n$ with $\| u \| = 1$. The following equality is true for this vector

$$\langle a(x; r, p_r, \bar{s}_r; f),\, u \rangle$$

$$= \frac{1}{d_r(p_r)} \int\limits_{B_r} \langle \nu, u \rangle f(x + \bar{s}_r + \nu)\, p_r(\| \nu \|)\mu(d\nu) =$$

$$\frac{1}{d_r(p_r)} \int\limits_{B_r} \langle \nu, u \rangle [f(y_r + \nu) - f(y_r + \nu^\perp)] p_r(\| \nu \|)\mu(d\nu),$$

where $y_r = x + \bar{s}_r$, $\nu^\perp = \nu - \langle \nu, u \rangle u$ is the vector ν constituent orthogonal to the vector u. We have

$$\int\limits_{B_r} \langle \nu, u \rangle f(y_r + \nu^\perp)\, p_r(\| \nu \|)\mu(d\nu) = 0$$

as the function $f(y_r + \nu^\perp) = const$ for all points ν with the same consituent $\nu^\perp$, that is on the straight lines paralled to the vector u.

Let X be the set of points ν from the sphere B_r for which $\langle \nu, u\rangle = 0$. As $\mu(X) = 0$ we obtain

$$\langle a(x;r,p_r,\bar{s}_r;f),u\rangle$$

$$= \int_{B_r\setminus X} \frac{f(y_r+\nu)-f(y_r+\nu^{\perp})}{\langle \nu,u\rangle} q_r(\nu)\mu(d\nu),$$

where

$$q_r(\nu) = \frac{1}{d_r(p_r)}\langle \nu,u\rangle^2 p_r(\|\nu\|) \geq 0$$

and

$$\int_{B_r\setminus X} q_r(\nu)\mu(d\nu) = \frac{1}{d_r(p_r)}\int_{B_r}\langle \nu,u\rangle^2 p_r(\|\nu\|)\mu(d\nu)$$

$$= \frac{1}{nd_r(p_r)}\int_{B_r}\|\nu\|^2 p_r(\|\nu\|)\mu(d\nu) = 1$$

due to the symmetry of the weight function p_r.

As the result we obtain

$$\langle a(x;r,p_r,\bar{s}_r;f),u\rangle$$

$$\leq \sup_{\nu\in B_r\setminus X} \frac{f(y_r+\nu)-f(y_r+\nu^{\perp})}{\langle \nu,u\rangle}$$

$$= \sup\{\sup 1; \sup 2\}$$

$$\leq \sup_{\substack{0<t_r\leq r\\ \|z_r-y_r\|\leq r}} \frac{f(z_r+t_r u)-f(z_r)}{t_r},$$

where

$$\sup 1 = \sup_{\substack{\langle \nu,u\rangle>0\\ \|\nu\|\leq r}} \frac{f(z_r'+\langle \nu,u\rangle u)-f(z_r')}{\langle \nu,u\rangle},$$

$$\sup 2 = \sup_{\substack{\langle \nu,u\rangle<0\\ \|\nu\|\leq r}} \frac{f(z_r''-\langle \nu,u\rangle u)-f(z_r'')}{-\langle \nu,u\rangle},$$

$z_r' = y_r+\nu^{\perp}$, $z_r'' = y_r+\nu$.

Consequently,

$$f^a(x;u) \leq \limsup_{\substack{z\to x\\ t\to+0}} \frac{f(z+tu)-f(z)}{t} = f^o(x;u).$$

Theorem 4 is proved ■

Theorem 5. If the function f is locally Lipschitzian in a neighborhood of the point x, then the F.H.Clarke subdifferential $\partial_{Cl}f(x) = co\,\partial_a f(x)$ on symmetrical weight functions (Bigil'deev and Rolshchikov, 1997). ■

PROOF. This statement follows from the fact, that the extreme points of the set $\partial_{Cl}f(x)$ are limit points of the sequence of gradients $f'(y)$ as $y\to x$ on points y of the differentiability of the function f (Clarke, 1983). But the extreme points are contained in $\partial_a f(x)$, too, because when selecting $\bar{s}_r = y-x$ for all $r>0$ $a(x;r,p_r,y-x;f)\to f'(y)$ as $r\to 0$.

Theorem 5 is proved ■

2. STRUCTURE OF THE APPROXIMATE SUBDIFFERENTIAL

The idea on the structure of the set $\partial_a f(x)$ for the square integrable in Lebesgue measure μ function f results from the following theorems.

Theorem 6. At fixed $r>0$ and $\bar{s}_r$ the set of approximate gradients $\{a(x;r,p_r,\bar{s}_r;f)\}$ on the symmetrical weight functions is a convex set. ■

PROOF. For any $\lambda\in[0;1]$ and any weight functions $p_r'(\|s-\bar{s}_r\|)$, $p_r''(\|s-\bar{s}_r\|)$

$$\lambda a(x;r,p_r',\bar{s}_r;f)+(1-\lambda)a(x;r,p_r'',\bar{s}_r;f)$$

$$= \lambda\frac{1}{d_r(p_r')}\int_{B_r}\nu f(y_r+\nu)p_r'(\|\nu\|)\mu(d\nu)$$

$$+(1-\lambda)\frac{1}{d_r(p_r'')}\int_{B_r}\nu f(y_r+\nu)p_r''(\|\nu\|)\mu(d\nu)$$

$$= \frac{1}{d_r(p_r)}\int_{B_r}\nu f(y_r+\nu)\,p_r(\|\nu\|)\mu(d\nu)$$

$$= a(x;r,p_r,\bar{s}_r;f),$$

where $y_r = x+\bar{s}_r$,

$$p_r(\|\nu\|) = \gamma p_r'(\|\nu\|)+(1-\gamma)p_r''(\|\nu\|),$$

$$\gamma = \frac{\lambda d_r(p_r'')}{\lambda d_r(p_r'')+(1-\lambda)d_r(p_r')} \in [0;1].$$

Theorem 6 is proved ■

Theorem 7. The set of the points of the approximate subdifferential $\partial_a f(x)$ that can be obtained on the set of centered symmetrical weight functions is a convex set. ■

PROOF. Let us fix $\bar{s}_r = \bar{s}$ for all $r>0$ and consider

the set $\left\{\lim_{\substack{r\to+0\\ p_r\in Q}} a(x;r,p_r,\bar{s};f)\right\}$.

If $r_1\leq r_2$, then

$$\{a(x;r_1,p_{r_1},\bar{s};f)\}\subseteq\{a(x;r_2,p_{r_2},\bar{s};f)\}$$

as for any weight function p_{r_1}, in the sphere B_{r_1}, there exists a weight function

$$p_{r_2}(\| \nu \|) = \begin{cases} p_{r_1}(\| \nu \|) & , \text{if } \| \nu \| \leq r_1 \\ 0 & , \text{if } \| \nu \| > r_1 \end{cases}$$

Hence, at fixed $\bar{s}$ the set

$$\left\{ \begin{array}{l} \lim\limits_{\substack{r \to +0 \\ p_r \in Q}} a(x; r, p_r, \bar{s}; f) \end{array} \right\}$$

is an intersection of convex sets.

For the case $\bar{s} = 0$ we obtain the statement ■

Let us fix an arbitrary sequence of points $X = \{x_i\} \subseteq \mathbf{R}^n$ converging to the point x at $i \to \infty$. Consider at $r \to +0$ the set ∂_X of limiting points of the sequences of the approximate gradients computed at the points x_i on the centered symmetrical weight functions. Unlike the points of the set in the Theorem 7 the points of the set ∂_X result from the changing r and the functions p_r at the transition from one point of the sequence X to another. At the same time, the set considered in the Theorem 7 is the set ∂_X for the sequence X all points of which coincide with the point x. Denote it by ∂_0.

Theorem 8. On any sequence $X = \{x_i\} \subseteq \mathbf{R}^n$ converging to the point x the set

$$\partial_X = \left\{ \begin{array}{l} \lim\limits_{\substack{r \to +0 \\ x_i \to x \\ p_r \in Q}} a(x_i; r, p_r, 0; f) \end{array} \right\}$$

of the limiting points at $r \to +0$ of approximate gradients on the centered symmetrical weight functions is a convex closed set that contains the set ∂_0. In addition, the set $\partial_a f(x)$ is the union of the sets ∂_X in all sequences $X = \{x_i\}$ converging to the point x ■

PROOF. Let $\{\xi_j\}_{j=1}^{\infty}$ be a sequence of points of the set ∂_X converqing to a point ξ as $j \to +\infty$. Then, from the definition of the set ∂_X for each point ξ_j there exists a sequence of numbers $r_{j_i} \to +0$ and weight functions $p_{r_{j_i}}$ such that $a(x_i; r_{j_i}, p_{r_{j_i}}, 0; f) \to \xi_j$ as $i \to +\infty$.

For each index $j > 0$ the index $i^* = i(j)$ is selected so that

$$\| a(x_{i^*}; r_{j_{i^*}}, p_{r_{j_{i^*}}}, 0; f) - \xi_j \| < \frac{1}{j}.$$

As the result, we obtain the sequence of approximate gradients

$$\left\{ a(x_{i^*}; r_{j_{i^*}}, p_{r_{j_{i^*}}}, 0; f) \right\}$$

converqing to the point ξ as $j \to +\infty$. Consequently, $\xi \in \partial_X$ and the set is closed.

The convexity of the set ∂_X follows from it being the limiting points of the sequences of convex sets. Indeed, for any two points ξ and η from ∂_X there exist sequences of approximate gradients $\xi_i = a(x_i; \delta_i, p'_{\delta_i}, 0; f)$ and $\eta_i = a(x_i; \rho_i, p''_{\rho_i}, 0; f)$ such that $\xi_i \to \xi$, $\eta_i \to \eta$ as $i \to +\infty$. For any $\lambda \in [0; 1]$ $\lambda\xi_i + (1 - \lambda)\eta_i = a(x_i; r_i, p_{r_i}, 0; f)$ for $r_i = \max\{\delta_i, \rho_i\}$ and a certain centered weight function p_{r_i} due to Theorem 6. Hence,

$$\lambda\xi + (1 - \lambda)\eta = \lim_{\substack{i \to +\infty \\ p_{r_i} \in Q}} a(x_i; r_i, p_{r_i}, 0; f)$$

and any point that lies on the segment connecting points ξ and η belongs to the set ∂_X.

To prove the inclusion $\partial_0 \subseteq \partial_X$ let us take an arbitrary single vector $u \in \mathbf{R}^n$. For it

$$\limsup_{\substack{r \to +0 \\ p_r \in Q \\ x_i \to x}} \langle a(x_i; r, p_r, 0; f), u \rangle$$

$$\geq \limsup_{\substack{r \to +0 \\ p_r \in Q}} \left\{ \limsup_{x_i \to x} \langle a(x_i; r, p_r, 0; f), u \rangle \right\}$$

$$= \limsup_{\substack{r \to +0 \\ p_r \in Q}} \langle a(x; r, p_r, 0; f), u \rangle$$

due to the continuity of approximate gradient in x at fixed $r > 0$ and weight function p_r.

As the result, it is shown that supporting function of the convex closed set ∂_X can not be less than the supporting function of the convex closed set ∂_0. Hence, $\partial_0 \subseteq \partial_X$.

The inclusion of the set ∂_X into the approximate subdifferential $\partial_a f(x)$ follows from (see Formula (5))

$$a(x_i; r, p_r, 0; f) = a(x; r, p_r, x_i - x; f)$$

for any point x_i converging to x sequence. On the other hand, any element of the set $\partial_a f(x)$ is

$$\lim_{\substack{r \to +0 \\ \bar{s}_r \to 0 \\ p_r \in Q}} a(x; r, p_r, \bar{s}_r; f)$$

that is the element of a set ∂_X.

Theorem 8 is proved ■

Theorem 9. The approximate subdifferential $\partial_a f(x)$ of the square integrable in Lebesgue measure function f is a convex set on symmetrical weight functions. ■

PROOF. Consider two arbitrary elements ξ and η of the set $\partial_a f(x)$. Let $\xi \in \partial_X$, $\eta \in \partial_Y$ and the sets ∂_X, $\partial_Y \subset \partial_a f(x)$ are obtained, correspondingly, on the sequences $\{x_i\}$, $\{y_i\}$ converging to the point x. The sequence $\{z_j\}$ with $z_j = x_i$ for $j = 2i$ and with $z_j = y_i$ for $j = 2i + 1$ also converges to the point x. In addition, the convex set $\partial_Z \subset \partial_a f(x)$, that corresponds to this sequence, will include sets ∂_X and ∂_Y, as for any single vector $u \in \mathbf{R}^n$ and the sequence $\{z_j\} \supseteq \{\{x_i\} \cup \{y_i\}\}$

$$\limsup_{\substack{r \to +0 \\ p_r \in Q \\ z_j \to x}} \langle a(z_j; r, p_r, 0; f), u \rangle \geq sup \{ls1; ls2\},$$

where

$$ls1 = \limsup_{\substack{r \to +0 \\ p_r \in Q \\ x_i \to x}} \langle a(x_i; r, p_r, 0; f), u \rangle,$$

$$ls2 = \limsup_{\substack{r \to +0 \\ p_r \in Q \\ y_i \to x}} \langle a(y_i; r, p_r, 0; f), u \rangle.$$

Theorem 9 is proved ∎

Corollary 10. If the function f is locally Lipschitzian in the neighborhood of the point x, then, the F.H.Clarke subdifferential $\partial_{Cl} f(x) = \partial_a f(x)$ on symmetrical weight functions. ∎

PROOF. This statement repeats the Theorem 5 with due account of convexity of the approximate subdifferential ∎

3. THE NECESSARY CONDITION OF OPTIMALITY

Definition 11. The point $x_0 \in \mathbf{R}^n$ is called a substantially stationary point of the square integrable in Lebesgue measure μ function f if

$$0 \in \partial_a f(x_0). \qquad (9)$$

It is known that the necessary minimum condition for a differentiable function is its gradient being equal to zero. Hence, a point of a substantial local minimum of the function that is equivalent (Kolmogorov and Fomin, 1989) to a differentiable function is a substantially stationary point.

It follows from the previous section that the approximate gradient methods (Batukhtin and Maiboroda, 1984, 1995) can be interpreted as the methods of constructing a substantially stationary point. Indeed, for the point x to be a substantially stationary one it is sufficient that it should be a generalized stationary point (Batukhtin and Maiboroda, 1995) that is when $r \to +0$ there exist points x_r, at each of which the approximate gradient should be equal to zero for a certain weight function, converging to the point x.

Condition (9) introduced here is a necessary condition of minimum for different classes of functions. For convex functions f it is a necessary as well as a sufficient condition of minimum that zero belongs to approximate subdifferential. It is a necessary condition of minimum for discontinuous functions obtained by a vertical shift in hyperplanes of parts of epigraphs of convex functions. These two statements follows from properties of the approximate gradient which state by V.E.Rolshchikov (Batukhtin and Maiboroda, 1995).

It follows from the Theorem 9 that (9) is a necessary condition of minimum for locally Lipschitzian functions (Clarke, 1983).

It has not yet been established in the general case for the class of the square integrable in Lebesgue measure μ functions that (9) is a necessary condition of substantial minimum. As the generalized stationary point is at the same time the substantially stationary point then for these functions that satisfy the special condition (Batukhtin and Maiboroda, 1995) this statement is true (Theorem 13). Particularly, strongly quasiconvex functions that can be discontinuous satisfy this special condition (Bazaraa and Shetty, 1979).

To formulate the above condition, let us introduce the following notation:
$l_{y,u} = \{x \in \mathbf{R}^n : x = y + \alpha u, \alpha \geq 0\}$ is a beam originating from the point y in the direction of the vector $u \neq 0$;
$S(y, R) = \{x \in \mathbf{R}^n : \| x - y \| = R\}$ is a sphere of the radius R with the center at the point y.

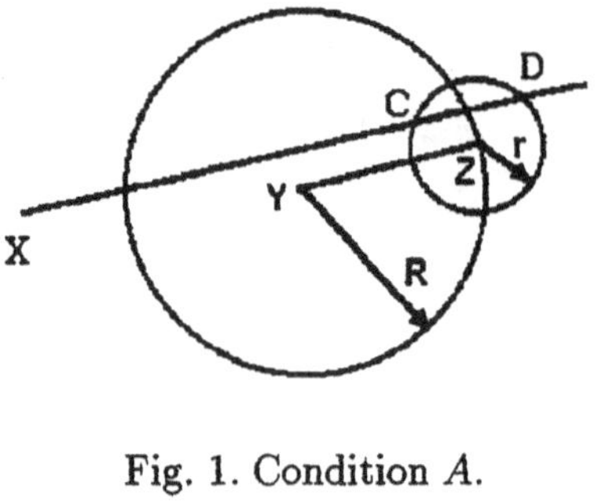

Fig. 1. Condition A.

Definition 12. The function f satisfies condition A at the point y if for any sufficiently small $R > 0$ there exists $r > 0$ such that, for any $x \in \mathbf{R}^n$ and $z \in S(y, R)$ the function $f(x+\alpha(z-y))$ for $\alpha \geq 0$ is a nondecreasing function in α on the segment $l_{x,z-y} \cap B_r(z)$ (segment CD on Fig.1).

Theorem 13. If the square integrable in Lebesgue measure μ lower semicontinuous function f satisfies condition A at the point of local minimum $x^\star$ then inclusion (9) is satisfied at this point (Batukhtin and Maiboroda, 1995). ■

Definition 14. The function f is called strongly quasiconvex if for any $x,\ y \in \mathbf{R}^n,\ x \neq y$ and any $\alpha \in (0;1)$

$$f(\alpha x + (1-\alpha) y) \ < \ max\, \{f(x),\ f(y)\}\,.$$

Theorem 15. Let f be a strongly quasiconvex square integrable in Lebesgue measure μ lower semicontinuous function and $x^\star$ is its point of minimum at which it is continuous. Then, it is satisfy condition A at the point $x^\star$ and (9) is fulfilled there. ■

PROOF. As f is a strongly quasiconvex function, then $x^\star$ is the single point of minimum of this function. Here $M_c(f) = \{x \in \mathbf{R}^n \,:\, f(x) \leq c\}$ is the nonempty and convex set for any $c \geq f(x^\star)$.

Let us fix an arbitrary $R > 0$ and consider the points of the sphere $B_R(x^\star)$. As the lower semicontinuous function f on the compact reaches its least value, then denote $c^o = \min_{x \in S(x^\star, R/2)} f(x) > f(x^\star)$.

Assume $\varepsilon = c^o - f(x^\star) > 0$. Due to the continuity of the function f at the point $x^\star$ there exists $\delta > 0$ such that $B_\delta(x^\star) \subseteq M_{co}(f)$. Consequently, $x^\star \in intM_{c^o}(f)$. In addition, $\delta < R/2$ and for any point $x \in B_{\delta/2}(x^\star) \quad x \in intM_{c^o}(f)$. But then it follows from the properties of the convex sets (Vasilyev, 1988) that any beam $l_{x,u}$ for any direction $u \in \mathbf{R}^n$ intersects the boundary of the sets $M_c(f)$ at $c \geq c^o$ at no more than one point. It means that for $\forall x \in B_{\delta/2}(x^\star)$ and $\forall u \in \mathbf{R}^n$ the function f increases along the beam $l_{x,u}$ on its parts outside $M_{c^o}(f)$ and, hence, outside $B_{R/2}(x^\star)$. In addition, due to strong quasiconvexity on the beams there are no constant parts of function f.

Assume $r = \delta/2$. Then, for any point $y \in S(x^\star, R)$ the function f increases in the sphere $B_r(y)$ along every beam $l_{x,y-x^\star}$ for $x \in B_{\delta/2}(x^\star)$. In addition, $l_{x,y-x^\star} \cap B_r(y) = \emptyset$ if $x \notin B_{\delta/2}(x^\star)$.

Hence, the function f satisfies condition A at the point $x^\star$ and statement (9) is true ■

Theorem 16. Let the function f satisfy the conditions of Theorem 15 and φ is an increasing function of one variable for which the superposition $h(x) = \varphi(f(x))$ is a square integrable in Lebesgue measure μ function. Then, the point of minimum of the function $h(x)$ satisfies (9). ■

PROOF. Due to strong monotonicity of the function φ the point $x^\star$ is the single point of minimum as well as for the function h, for which condition A is fulfilled at the point of minimum.

Theorem 16 is proved ■

To conclude this section note, that all the above mentioned theorems are true for the functions equivalent to the discussed ones. Hence, numerical search for the point of substantial minimum of the square integrable in Lebesgue measure μ function can be effected by finding an substantially stationary point with the help of approximate gradient like it is done in the convex and smooth cases.

4. NUMERICAL METHODS, HOW THEY WORK

The described theoretical results make it possible to construct numerical methods of solving discontinuous extremal problems. Approximate gradient (taken with the minus) can be used as a motion direction when constructing the minimizing sequence.

The main scheme of approximate gradient methods is constructing the sequence of points $\{x^{(k)}\}$, $x^{(k)} \in \mathbf{R}^n$ in accordance with the iteration procedure:

$$x^{(k+1)} = x^{(k)} - \alpha_k a(x^{(k)}; r^{(k)}, p^{(k)}, \bar{s}^{(k)}; f),$$

where $p^{(k)} = p_{r^{(k)}}(s)$, $k = 0, 1, 2, \ldots$ and the step α_k is chosen either by means of one-dimensional minimization:

$$\alpha_k = arg \min_{\alpha > 0} f(x^{(k)} - \alpha a(x^{(k)}; r^{(k)}, p^{(k)}, \bar{s}^{(k)}; f))$$

or from the condition defined by the inequality

$$f(x^{(k)}) - f(x^{(k+1)})$$

$$\geq \sigma \alpha_k \parallel a(x^{(k)}; r^{(k)}, p^{(k)}, \bar{s}^{(k)}; f) \parallel^2,$$

$$0 < \sigma < 0{,}5.$$

If $\parallel a(x^{(k)}; r^{(k)}, p^{(k)}, \bar{s}^{(k)}; f) \parallel < \varepsilon$ where ε is a sufficiently small positive number or the function f is not decreasing in the direction $-a(x^{(k)}; r^{(k)}, p^{(k)}, \bar{s}^{(k)}; f)$, then, the value of $r^{(k)}$ is decreasing. The process is stopped if $r^{(k)} < r_{min}$. The convergence of these methods is proved for nonsmooth and discontinuous functions. In addition, the main condition imposed on the function

f is that it belongs to the class of functions satisfying the generalized Lagrange mean-value theorem (Batukhtin and Maiboroda, 1995).

These methods appear good when solving nonsmooth and discontinuous optimization problems. The authors presented computational results at international workshops as well as published them in some papers (Batukhtin and Maiboroda, 1984, 1995; Batukhtin, et al., 1997; Bigil'deeva and Rolshchikov, 1994).

In the main scheme of the approximate gradient methods computation $a(x^{(k)}; r^{(k)}, p^{(k)}, \bar{s}^{(k)}; f)$ is performed in a certain uniform lattice in the integration domain. Here, function values at all nodes are taken as given with (most often equal) weights. However, in this case it is not always possible to find a fairly accurate solution of discontinuous problems where: a) a point of minimum is on a discontinuity surface, b) the function is decreasing along the discontinuity. This is conditioned by the fact that the approximate gradient norm converges to infinity at the discontinuity point if the radius of the domain converges to zero. Hence, at such points the approximate gradient "turns back" in the direction perpendicular to the discontinuity line hampering the motion along it.

That is why some adaptive algorithms were constructed.The main idea of adaptive algorithms is to correct the initial approach of the approximate gradient by adding new nodes. Each new (m+1)-th node is choosen with due account of the direction of the obtained approximate gradient approach on m nodes. Then, the approximate gradient is recomputed for the (m+1) points that defines the choice of the next node.

The procedure of accumulating the nodes and approximate gradient correction is continued until one of the following three conditions is observed

(1) an approximate gradient approach gives a node at which the function value is not less than at the previous one;
(2) the approximate gradient norm less than a beforehand given number;
(3) the node number reaches the maximum permissible one.

Otherwise, the adaptive algorithms are similar to the main scheme of the approximate gradient methods. If for the given r the accumulation of nodes does not lead to constructing the direction of the function f decrease, then r is subdivided until it reaches the minimum permissible value or the desired direction is found. Finding the point in the direction of function decrease is defined by one-dimensional minimization.

Let us draw computational formulae for one of the nodes accumulation methods. For this, let us replace the approximate gradient $a(x^{(k)}; r^{(k)}, p^{(k)}, \bar{s}^{(k)}; f)$ at the point x at $r > 0$ with the following integral sum:

$$a(x^{(k)}; r^{(k)}, p^{(k)}, \bar{s}^{(k)}; f) \approx a^{(m)} = d_m^{-1} \sum_{i=0}^{m} (s^{(i)} - \bar{s}^{(m)}) f_i \lambda_i,$$

where

$s^{(i)}$ are nodes in the sphere B_r $(0 \le i \le m)$; $f_i = f(x + s^{(i)})$;

$\lambda_i = \left(\int_{B_r} p_r(s)\mu(ds)\right)^{-1} \int_{\Omega_i} p_r(s)\mu(ds) \ge 0$

are weight coefficients,

$\sum_{i=0}^{m} \lambda_i = 1;$

Ω_i are some subdomains B_r such that, $\bigcup_{i=0}^{m} \Omega_i = B_r$ and $\Omega_i \cap \Omega_j = \emptyset$ for $i \ne j$ $(0 \le i \le m,\ 0 \le j \le m)$;

$\bar{s}^{(m)} = \sum_{i=0}^{m} \lambda_i s^{(i)}; \quad d_m = \sum_{i=0}^{m} \| s^{(i)} - \bar{s}^{(m)} \| \lambda_i.$

Let $\bar{f}^{(m)} = \sum_{i=0}^{m} \lambda_i f_i$ and $g^{(m)} = \sum_{i=0}^{m} \lambda_i f_i s^{(i)}$.

Then,

$$a^{(m)} = d_m^{-1}(g^{(m)} - \bar{f}^{(m)} \bar{s}^{(m)}).$$

The initial approach of the approximate gradient is computed after $(n+1)$ points where n is the space dimension situated on coordinate axes, that is

$s^{(0)} = (0; ...; 0)^T$, $s^{(1)} = (r; 0; ...; 0)^T$, ...,
$s^{(n)} = (0; ...; 0; r)^T$.

The weight coefficients λ_i are choosen equal to $\frac{1}{n+1}$. The next node is choosen on the surface of the sphere B_r so that

$$s^{(m+1)} = -r \frac{a^{(m)}}{\| a^{(m)} \|}.$$

When computing a new approach of the approximate gradient the weight coefficient for an additional node is assumed equal to one and the same number β $(0 < \beta < 1)$. Then,

$$\bar{s}^{(m+1)} = (1 - \beta)\bar{s}^{(m)} + \beta s^{(m+1)};$$

$$\bar{f}^{(m+1)} = (1 - \beta)\bar{f}^{(m)} + \beta f_{(m+1)};$$

$$g^{(m+1)} = (1 - \beta) g^{(m)} + \beta f_{(m+1)} s^{(m+1)};$$

$$d_{m+1} = (1-\beta)[d_m + \beta \parallel s^{(m+1)} - \bar{s}^{(m)} \parallel^2].$$

As a result $(1-\beta)$ plays the role of a certain "memory" coefficient of mean values $\bar{s}^{(m)}$, $\bar{f}^{(m)}$, d_m and $g^{(m)}$.

It should be noted that though adaptive algorithms contain heuristic elements and there is no proof as yet of their convergence the idea used in them provides the lacking flexibility to the main scheme of approximate gradient methods. It is first of all connected with the fact that the procedure of node accumulating makes it possible to construct the approaches for $\bar{s}_r$ and the weight function $p_r(s)$ for which the approximate gradient norm is minimal.

At the same time in the adaptive algorithms the initial approach of approximate gradient can be fairly rough and contain only information on function values in lineary independent directions. A special feature of this kind of algorithm also is that it is sufficient to preserve only the averaged characteristics of the nodes set to correct the direction of the approximate gradient .

Experience shows that adaptive algorithms appear more economical as compared to the main scheme in the number of function computations and provide more accurate problem solutions.

Let us consider some examples to illustrate the work of adaptive algorithms.

We shall use the following notation:
x^0 is a point of minimum;
x is an obtained approximation of the point of minimum;
$\delta_x = \parallel x - x^0 \parallel$ is an obtained solution accuracy in the argument;
$\delta_f = \mid f(x) - f(x^0) \mid$ is an obtained solution accuracy in the function values;
k_f is the number of a minimized function computations.

So, for example, for the discontinuous function

$f(x_1, x_2) = 2(x_1 - 0,1)^2 + (x_2 - 4,5)^2$,

if $x_1 + x_2 \leq 5$, $0 \leq x_1$, $0 \leq x_2$ and

$f(x_1, x_2) = 3(x_1 - 0,1)^2 + 2(x_2 - 4,5)^2$ otherwise

the main scheme provides : $\delta_f = 10^{-6}$, $\delta_x = 10^{-3}$, $k_f = 1258$ and the adaptive algorithm provides: $\delta_f = 10^{-9}$, $\delta_x = 10^{-4}$, $k_f = 776$.

For the smooth function

$f(x_1, x_2) = 0$, if $x_1^2 + x_2^2 = 0$ and

$f(x_1, x_2) = 5 \cdot 10^{-5}(x_1^2 + x_2^2)(99 \sin \frac{100}{\sqrt{x_1^2 + x_2^2}} + 101)$,

if $x_1^2 + x_2^2 \neq 0$

at a sufficiently big initial r the adaptive algorithm with a good accuracy finds the absolute minimum $(0;0)^T$: $\delta_f = 10^{-9}$, $\delta_x = 6 \cdot 10^{-3}$, $k_f = 1953$. At the same time for the smooth methods points of local minimum that are concentrated in a neighborhood of the point $(0;0)^T$ appear to be "traps". For example, from the same initial point the conjugate gradient method terminated its work with $\delta_x = 1,3544$.

As in the previous example, the point of absolute minimum of the discontinuous function

$f(x_1, x_2) = 0$, if $x_1^2 + x_2^2 = 0$ and

$f(x_1, x_2) = -\sqrt{x_1^2 + x_2^2} + 2j + 1$,

if $j \leq \sqrt{x_1^2 + x_2^2} < j + 1$

$f(x_1, x_2) = -\sqrt{x_1^2 + x_2^2} + \frac{2j+1}{j(j+1)}$,

if $(j+1)^{-1} \leq \sqrt{x_1^2 + x_2^2} < j^{-1}$ for $j = 1, 2, ...$

is the origin, in the neighborhood of which the discontinuities are concentrated. Moreover, each point of function discontinuity is a point of substantial local minimum as well as maximum.

For this problem the adaptive algorithm method with initial $r = 1$ finds the point of absolute minimum with the following accuracy: $\delta_f = 10^{-6}$, $\delta_x = 10^{-6}$, $k_f = 4558$. Fig. 2 shows the algorithm motion trajectory.

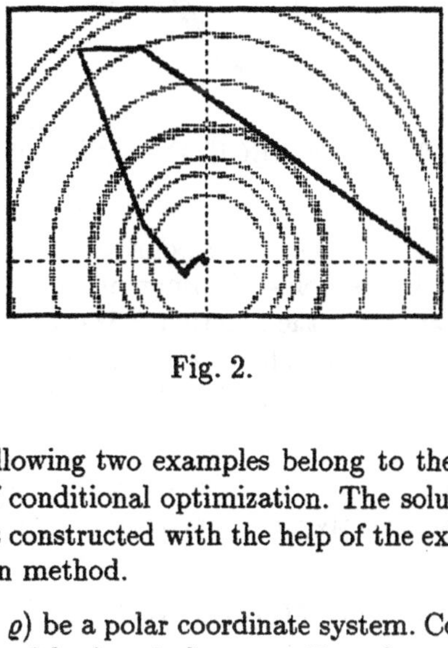

Fig. 2.

The following two examples belong to the problems of conditional optimization. The solution in them is constructed with the help of the extended function method.

Let (φ, ϱ) be a polar coordinate system. Consider two logarithmic spirals $\varrho_1 = \ell^{\varphi}$ and $\varrho_2 = \ell^{2\varphi}$. If for a fixed angle φ the point $x_1 = \varrho \cdot cos\varphi$, $x_2 = \varrho \cdot sin\varphi$ has the value $\varrho \in [\varrho_1; \varrho_2]$, we say that the point $(x_1, x_2)^T$ belongs to the set U.

Consider the problem of conditional minimization

$$F(x_1, x_2) = x_1^2 + x_2^2 \longrightarrow \min_{(x_1, x_2)^T \in U}$$

To solve it with the adaptive algorithm the function F was extended in the following way

$$f(x_1, x_2) = \begin{cases} F(x_1, x_2) & , \text{if } (x_1, x_2)^T \in U \\ F(x_1, x_2) + 1000 & , \text{otherwise} \end{cases}$$

And the minimum of the latter was found with the result: $\delta_f = 10^{-6}$, $\delta_x = 10^{-4}$, $k_f = 3193$.

Fig. 3 and Fig. 4 show level lines of the function f as well as the whole trajectory of algorithm motion and its fragment in the neighborhood of the point of minimum in the bottom of the twisting ravine.

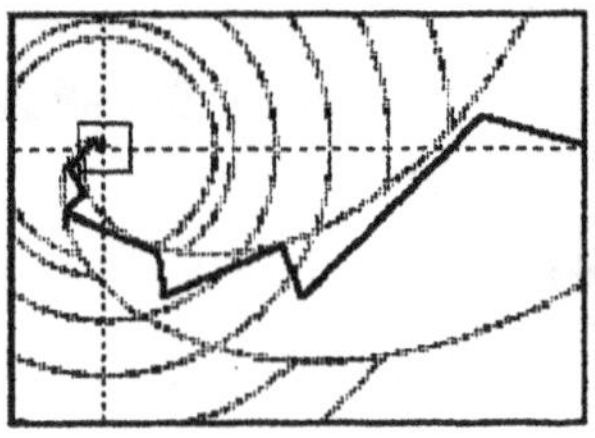

Fig. 3.

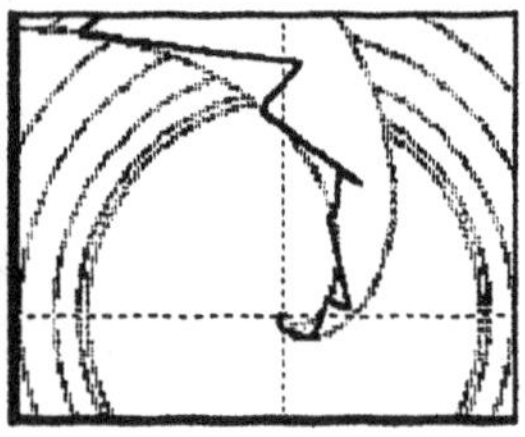

Fig. 4.

In the second example of the problem of conditional minimization it is required to minimize the function $F(x_1, x_2) = x_2^3$ in the domain U, pressed between the circle and the ellipse at $x_1 \geq 0$ (see Fig. 5).

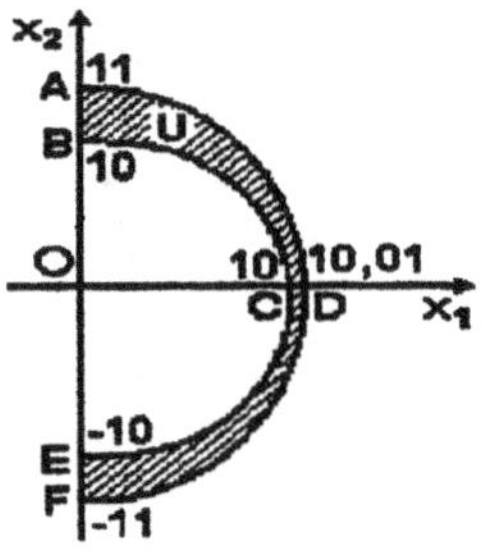

Fig. 5. The curve ABCEFDA is the boundary U.

The point (0;11) is the point of maximum and (0;-11) is the point of minimum of the function F. In addition, the function F is constant on the straight lines parallel to the axis Ox_1 and on the axis Ox_1 (CD segment on Fig. 5) it has the point of its inflection.

Outside the domain U let us extend the function F by sufficiently great constants and the parabolic function:

$f(x_1, x_2) = x_2^3$, if $(x_1, x_2)^T \in U$;

$f(x_1, x_2) = 11x_2^2 + 1$,

if $(\frac{x_1}{10{,}01})^2 + (\frac{x_2}{11})^2 > 1$ and $x_1 \geq 0$;

$f(x_1, x_2) = 1100$, if $x_1^2 + x_2^2 < 100$ and $x_1 \geq 0$;

$f(x_1, x_2) = 1010$, if $x_1 < 0$.

At such extension of the function F the points of the curves ABC and DFE (Fig. 5) are saddle points for the function f.

Fig. 6 shows the general view of the adaptive algorithm motion trajectory along the boundary of the domain U from the point of maximum of the function F and the level line f. Fig. 7 shows a fragment of the algorithm passing the narrow part of the domain U. For this problem $\delta_f = 10^{-3}$, $\delta_x = 10^{-2}$, $k_f = 8853$.

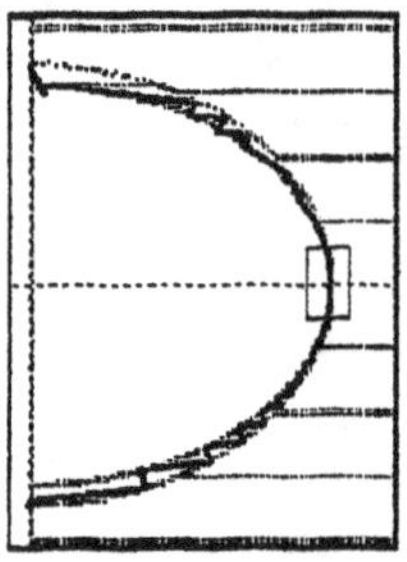

Fig. 6.

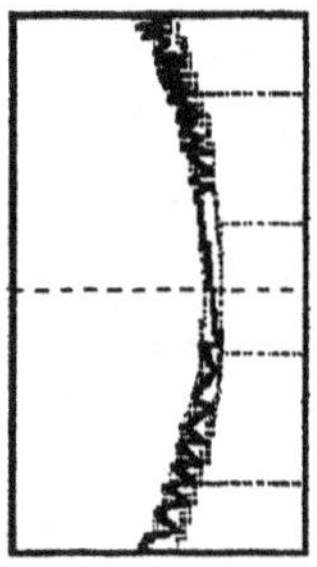

Fig. 7.

REFERENCES

Batukhtin, V.D. and L.A. Maiboroda (1984). *Optimization of Discontinuous Functions.* Nauka, Moskow.

Batukhtin, V.D. and L.A. Maiboroda (1995). *Discontinuous Extremal Problems.* Gippocrat, St.-Peterburg.

Batukhtin, V.D. (1993). On Solving Discontinuous Extremal Problems. *Journal of Optimization Theory and Applications,* **77**, 575-589.

Bigil'deev, S.I. (1996). Approximated Derivative as a Multivalued Mapping. *Vestn. Chelyabinsk State University,* **1**, 21-33.

Bigil'deev, S.I. and V.E. Rolshchikov (1997). Properties of Approximate Gradient in Depending on Weight Functions. *Journal of Computer and System Sciences,* **4**, 89-94.

Batukhtin, V.D., S.I. Bigil'deev and T.B. Bigil'deeva (1997). Numerical Methods for Solution of Discontinuous Extremal Problems. *Journal of Computer and System Sciences,* **3**, 113-120.

Liusternik, L.A. and V.I. Sobolev (1965). *Elements of Functional Analysis.* Nauka, Moskow.

Kolmogorov, A.N. and S.V. Fomin (1989). *Elements of the Theory of Functions and Functional Analysis.* Nauka, Moskow.

Clarke, F.H. (1983). *Optimization and Nonsmooth Analysis.* Wiley, New York.

Pschenichny, B.N. (1980). *Convex Analysis and Extremal Problems.* Nauka, Moskow.

Bazaraa, M.S. and C.M. Shetty (1979). *Nonlinear Programing. Theory and Algorithms.* Wiley, New York.

Vasilyev, F.P. (1988). *Numerical Methods of Solving Extremal Problems.* Nauka, Moskow.

Bigil'deeva, T.B. and V.E. Rolshchikov (1994). Numerical Methods of Optimization Discontinuous Functions. *News Russian Academy of Science, Technical Cybernetics,* **3**, 47-54.

STABILIZATION OF DYNAMICAL SYSTEMS WITH THE HELP OF OPTIMIZATION METHODS

R. Gabasov[1], F.M.Kirillova[2], E.A.Ruzhitskaya[3]

[1]*Faculty of Applied Mathematics and Informatics,*
Department of Optimal Control Methods,
Byelorussian State University, F.Scorina prosp. 4, Minsk, 220080, Belarus
[2]*Institute of Mathematics, National Academy of Sciences of Belarus,*
Surganov str. 11, Minsk, 220072, Belarus
[3] *Faculty of Mathematics, Gomel State University*
Sovjetskaya str. 104, 246699 Gomel, Belarus
e-mail: [1,2,3] *kirill@nsys.minsk.by*

Abstract: Schemes of realization of bounded stabilizing feedbacks based on optimal control methods and linear programming are under consideration. Both approaches use the principle of correction of current solutions in real–time mode. The methods elaborated are used for robust stabilization and stabilization under additional conditions on transients, such as degree of stability, degree of oscillation, degree of overcontrol, monotonicity of transients. The result are illustrated by nontrivial examples. *Copyright ©1998 IFAC*

Keywords: stabilization, optimal control, feedback, real time, algorithm

1. INTRODUCTION

Stabilization problem is one of the important problems of the control theory (Aizerman, 1958; Malkin, 1966). With the growth of dimensions of stabilized systems and the rise of demands to quality of transient processes optimal control methods (Pontryagin, *et al.*, 1983) together with modern computers have been becoming powerfull tools while stabilizing dynamical systems. The first results in this directions were obtained in the theory of linear–quadratic problems of optimal control with infinite horizon (Kalman, 1961; Letov, 1960) and later on with a finite horizon (Kwon and Pearson, 1977; Mayne and Michalska, 1990). Both (Kalman, 1961; Letov, 1960) and (Kwon and Pearson, 1977; Mayne and Michalska, 1990) deal with explicit form of positional solutions of linear–quadratic problems of optimal control. It had been a consequence of "no" restrictions on controls because of what problems in questions did not leave the frame of calculus of variations. In the papers (Gabasov, *et al.*, 1991; Gabasov, *et al.*, 1992) methods of optimal control which take into account geometrical restrictions on control functions and allows to consruct bounded stabilizing feedbacks are suggested. As the construction of bounded stabilizing feedback in explicit form represent an especially difficult problem, in (Gabasov, *et al.*, 1991) methods of realization of positional solutions of optimal control problems by modern computer tools are justified. This approach was developed for stabilization of dynamic systems in (Gabasov, *et al.*, 1992). In the paper we at first describe principles of constructing bounded stabilizing feedbacks for dynamic systems. Then we investigate problems of robust stabilization and dynamic systems stabilization with additional properties of transients.

The main features of the approach under considereation consist in the following: 1) structure of feedbacks is not given beforehand; 2) stabilizing controls are restricted; 3) auxiliary constrained optimization problems are introduced to be solved for the initial position of the dynamic system; 4) principle of continuous correcting current solutions is used during a finite period of time.

2. PROBLEM STATEMENT

Consider dynamical system on the interval $t \geq 0$

$$\dot{x} = Ax + bu \ (x \in R^n, \ u \in R) \tag{1}$$

where A, b a given matrix and a vector.

Let G be a vicinity of $x = 0$ of (1) ($u = 0$), L, $0 \leq L < \infty$ be a given number.

A function

$$u = u(x), \ x \in G \tag{2}$$

is said to be a bounded stabilizing feedback for (1) if: 1) $u(0) = 0$; 2) $|u(x)| \leq L$, $x \in G$; 3) the equation

$$\dot{x} = Ax + bu(x) \tag{3}$$

obtained after closing (1) by (2) has a solution $x(t)$, $t \geq 0$, for all $x(0) \in G$; 4) the trivial solution $x(t) = 0$, $t \geq 0$, of (3) is asymptotically stable in G.

The problem consists in constructing such function (2) for which the domain of attraction G of asymptotically stable solution $x(t)$, $t \geq 0$, of equation (3) are close to maximum one. The construction of $u(x)$, $x \in G$ in explicit form is a very difficult problem. By analogy with (Gabasov, *et al*, 1992) one can inrtoduce a new statement of the problem which is based on the use of modern computer technology.

We assume that a bounded stabilizing feedback (2) has been constructed. Consider the behavior $x^*(t)$, $t \geq 0$, of closed system (3) in some particular process of stabilization having started from $x^*(0) = x_0^*$, $x^* \in G$:

$$\dot{x}^*(t) = Ax^*(t) + bu(x^*(t)), \ x^*(0) = x_0^*. \tag{4}$$

According to (4) in this particular process the control

$$u^*(t) = u(x^*(t)), \ t \geq 0 \tag{5}$$

is fed to input of the dynamic system, i.e. in a particular process of stabilization feedback (2) is not used as a whole. It is needed only its meaning along isolated continuous curve $x^*(t)$, $t \geq 0$. This values is supposed to be used not beforhand but at every current moment $\tau \geq 0$.

Function (5) is said to be a realization of the feedback for a particular process of stabilization. A device which is capable to calculate its values in the course of stabilization process is said to be Stabilizer.

Thuth, the problem of stabilization is reduced to the construction for Stabilizer an real-time algorithm.

The basic idea of the stated approach consists in introduction auxiliary (accompanying) problem of optimal control, in construction, following (Gabasov, *et al.*, 1991), an algorithm of work of an optimal regulator and in the proof that the optimal regulator appears to be Stabilizer of dynamic system (1).

3. STABILIZATION OF DYNAMICAL SYSTEM USING A MINIMUM INTENSITY CONTROL PROBLEM

Assume that system (1) is controllable:

$$\text{rank}(b, Ab, \ldots, A^{n-1}b) = n.$$

Choose a positive number Θ (parameter of a method) and in the class of piecewise continuous functions $u(t)$, $t \in T = [0, \Theta]$, consider the accompanying optimal control problem

$$\rho(z) = \min_{u} \max_{t \in T} |u(t)|,$$

$$\dot{x} = Ax + bu, \ x(0) = z, \ (z \in R^n) \tag{6}$$

$$x(\Theta) = 0, \ |u(t)| \leq \rho, \ t \in T.$$

Designate: $u^0(t|z)$, $t \in T$, is optimal open loop control (6) for an initial condition $z \in R^n$

$$u^0(t|z) = \rho(z)\text{sign}\Delta^0(t), \ z \in R^n,$$

where $\Delta^0(t) = \Delta^0(t|z) = \psi'(t)b = -y'(z)F(\vartheta - t)b$ is the optimal cocontrol, $y = y(z)$ is the optimal n-vector of potentials, $F(t)$ is the fundamental matrix of solutions to system $\dot{x} = Ax$; $\rho = y'F(\Theta)z$ is the optimal intensity of control. $G(\Theta)$ is the set of all condition z, for which problem (6) has a solution.

A function

$$u^0(z) = u^0(0|z), \ z \in G(\Theta), \tag{7}$$

is said to be an optimal start feedback for (6).

It can be proved that the function $u^0(x) = u^0(0|x)$, $x \in R^n$, is the bounded stabilizing feedback for system (1) (Balashevich, *et al*, 1994).

Two ways of realization of the bounded stabilizing feedback are possible:

1) "continuous", at which in the optimal open loop control a finite set of defining elements (structure of optimal control $u^0(t|x^*(\tau))$, $t \in T$) is extracted, and for it the system of defining equations is introduced that for problem (6) has the form:

$$\Delta^0(t_i(\tau) + 0) = 0, \ x(\Theta) = 0$$

or

$$\sum_{i=0}^{p} \int_{t_i(\tau)}^{t_{i+1}(\tau)} F(\Theta - t)bdtk_i\rho(\tau) + F(\Theta)x^*(\tau) = 0$$

$$-y'_\tau F(\Theta - t_i(\tau))b = 0, \ i = \overline{1,p} \tag{8}$$

$$y'_r(\tau) = F(\Theta)x^*(\tau) = \rho(x^*(\tau)) = \rho(\tau))$$
$$t \in [0, \Theta), \ t_0(\tau) = 0, \ t_{p+1}(\tau) = \Theta,$$
$$k_i = \text{sign}\Delta^0(t_i(\tau) + 0),$$

where $t_i(\tau), i = 1, \ldots, p$ (point of switching of the optimal open loop control), $y(\tau)$ (optimal n-vector of potentials), $\rho(\tau)$ (optimal intensity of the control) are unknowns under rather general conditions. The Jacobi matrix for (8) is nonsingular. An algorithm of solution of (8) (Gabasov, *et al.*, 1992), allowing in the course of process to calculate realizations $u^*(\tau)$, $\tau \geq 0$, of the bounded stabilizing feedback (7) is described.

2) a "discrete" way, at which the accompanying problem is considered in the class piecewise constant controls with the constant period of quantization $h > 0$, $h = \Theta/N : \ N < +\infty$; $u(t) = u_k$, $t \in [kh], k+1)h[$, $k = 0, 1, \ldots, N-1$. The optimal open loop control $u^0(t|x^*(kh))$, $t = 0, h, \ldots, (N-1)h$, in the current moment $\tau = kh$ is under construction by a dual method of linear programming by correction of the control $u^0(t|x^*((k-1)h)0, h, \ldots, (N-1)h)$, constructed on the previous step $\tau - h = (k-1)h$. At each of the mentioned ways used methods of correction allow quickly to calculate values $u^*(\tau)$, $\tau \geq 0$.

Using the discrete way of realization of the bounded stabilizing feedback, robust stabilization problem and stabilization of dynamic systems with of additional properties of transients are solved below.

4. ROBUST STABILIZATION OF DYNAMIC SYSTEM WITH THE HELP OF BOUNDED CONTROLS

Assume that the accessible information on parameters A, b of system (1) inexact: $n \times n$-matrix A and n-vector b are those that

$$A = A_0 + \Delta A, \ b = b_0 + \Delta b,$$

where A_0, b_0 are an known $n \times n$-matrix and an n-vector accordingly, ΔA, Δb are unknowns satisfying the inequalities:

$$||\Delta A|| \leq \alpha, \ ||\Delta b|| \leq \beta \ (\alpha, \ \beta > 0).$$

At a given $\varepsilon > 0$ and a fixed numbers $\nu > 0$, $L > 0$, a function

$$u(t, x), \ x \in G, \ t \in [0, \nu[, \qquad (9)$$

is said to be a robust bounded stabilizing open-closed loop control of system (1) in G if 1) $u(t, 0) = 0$, $t \in [0, \nu[$; 2) $|u(t, x)| \leq L$, $x \in G$, $t \in [0, \nu[$; 3) the trajectory of the closed system

$$\dot{x} = Ax + bu(t, x), \ x(0) = x_0, \ x_0 \in G, \qquad (10)$$

is a continuous solution of the equation $\dot{x} = Ax + bu(t)$, $x(0) = x_0$, at $u(t) = u(t - k\nu, x(k\nu))$, $t \in [k\nu, (k+1)\nu[$, $k = 0, 1, \ldots$; 4) system (10) at $A \equiv A_0$, $b \equiv b_0$ is assymototically stable in G; 5)there exists a finite number $t(\varepsilon) > 0$, such that every solution $x(t)$, $t \geq 0$, of system (10) satisfies the condition $||x(t)|| \leq \varepsilon$, $t \geq t(\varepsilon)$.

Choose natural numbers $N, m(N > m > n)$, a real number $h > 0$. Assume $\nu = mh, \Theta = Nh$.

In the class piecewise constant functions $u(t)$, $t \in T = [0, \Theta]$, satisfying the restriction $|u(t)| \leq L$, $t \in T$, we shall consider the accompanying problem of optimal control:

$$\rho(z) = \min \rho,$$
$$\dot{x} = A_0 x + b_0 u, \ x(0) = z, \ (z \in R^n) \qquad (11)$$
$$x(\Theta) = 0, \ |u(t)| \leq \rho, \ t \in T.$$

Optimal start open–closed loop control is defined by the equality

$$u^0(t, z) = u^0(t|z), \ t \in [0, \nu[, \ z \in R^n.$$

Introduce the set
$G_\Theta = \{z \in R^n : \ |u^0(t, z)| \leq L, \ t \in [0, \nu]\}$.

For any $\varepsilon > 0$ there exists such $\Theta > 0$ that an ε-vicinity of G_Θ contains all states of (1) which can be transfered to $x = 0$ for a finite time. Using the Lyapunov function method (Barbashin, 1967; Bromberg, 1967) it is possible to show that at given ε, G, α, β and an appropriate choice of parameters $\Theta > 0$, $\nu > 0$, $h > 0$ of problem (11) the feedback $u(t, x) = u^0(t, x)$, $x \in G_\Theta$, $t \in [0, \nu[$, will satisfy all requirements of definition of the robust bounded stabilizing open-closed loop control with $G = G_\Theta$. As the Lyapunov function one can take the optimal values of the criterion of quality $\rho(z), z \in G_\Theta$, of problem (11).

Consider stabilization problem of a mathematical pendulum in the upper unstable state of equilibrium (Malkin, 1966). The mathematical model of such system has the form

$$\dot{x}_1 = x_2, \ \dot{x}_2 = x_1 + x_3, \ \dot{x}_3 = u \qquad (12)$$

where x_1 is the angle of deviation of the pendulum from the vertical, x_2 is the angular speed of the pendulum, x_3 is the moment enclosed to the pendulum.

Let at the initial moment $t = 0$ system (12) is in the condition $x_1(0) = 0.3$, $x_2(0) = 1.0$, $x_3(0) = -1.2$. It is required to stabilize it in the upper vertical state $x_1 = 0$, $x_2 = 0$, $x_3 = 0$.

The accompanying problem is

$$\rho \to \min$$

$$\dot{x}_1 = x_2, \ \dot{x}_2 = x_1 + x_3, \ \dot{x}_3 = u$$

$$x_1(0) = x_1^*(\tau),\ x_2(0) = x_2^*(\tau),\ x_3(0) = x_3^*(\tau)$$

$$x_1(\Theta) = 0,\ x_2(\Theta) = 0,\ x_3(\Theta) = 0$$

$$|u(t)| \le \rho,\ t \in T = [0, \Theta]$$

where $x^*(\tau) = (x_1^*(\tau), x_2^*(\tau), x_3^*(\tau))$ is the condition of system (12) in the current moment τ.

For the solution of the accompanying problem the following parameters were chosen: $\Theta = 1$, $h = 0.025$, $\nu = 5h$.

During work of Stabilizer the coefficient at x_3 is changed (Fig. 1) (curves 1 corresponds to the value $1x_3$, the curves 2, 3 correspond to the values $0.5x_3$ and $1.5x_3$).

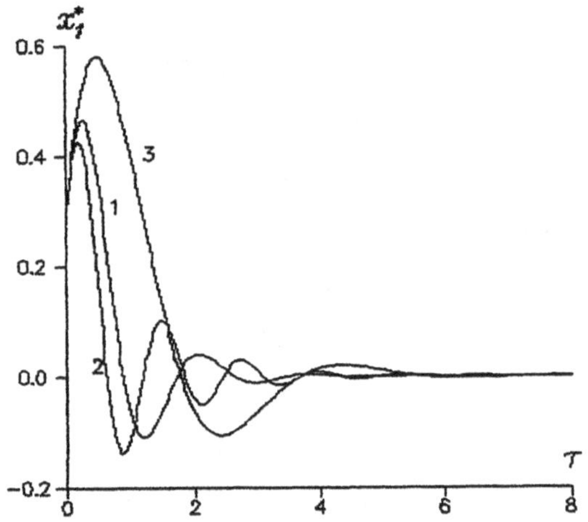

Fig. 1. The trajectory of system (12).

5. STABILIZING FEEDBACKS PROVIDING GIVEN PROPERTIES OF TRANSIENT PROCESSES

Consider system (1). Let $y = Hx$, $H \in R^{m\times n}$, an m-vector of output. Introduce the sets

$$Y(t) = \{y \in R^m : g_*(t) \le y \le g^*(t)\},\ t \ge 0,$$

where $g_*(t)$, $g^*(t)$, $-\infty < g_*(t) \le g^*(t) < \infty$, $t \ge 0$, are given continuous m-vector functions.

Definition 1. At fixed numbers $\nu > 0, h > 0, L > 0$ a function

$$u_s(t, x),\ s \in [0, \nu[,\ t \ge 0,\ x \in G, \qquad (13)$$

is said to be an bounded stabilizing open-closed loop control of (1) in G, if 1) $u_s(t, 0) = 0$, $s \in [0, \nu[$, $t \ge 0$; 2) $|u_s(t, x)| \le L$, $s \in [0, \nu[$, $t \ge 0$, $x \in G$; 3) a trajectory of the closed system

$$\dot{x} = Ax + bu_s(t, x),\ x(0) = x_0,\ x_0 \in G, \qquad (14)$$

is a continuous solution of the equation

$$\dot{x} = Ax + bu(t),\ x(0) = x_0,$$

at $u(t) = u_s(k\nu, x(k\nu))$, $s \in [0, \nu[$, $t \in [k\nu, (k+1)\nu[$, $k = 0, 1, \ldots$; 4) system (14) is assymototically stable in G.

Definition 2. A bounded stabilizing open-closed control (13) is said to be a stabilizing open-closed loop control with the property A if the output signal $y(t)$, $t \ge 0$, of system (14) is contained in set $Y(t)$, $t \ge 0$, where $g_*(t) = -a\exp(-\alpha t)$, $g^*(t) = a\exp(-\alpha t)$, $\alpha > 0$. The number $\alpha > 0$ is called degree of stability of a transient.

For construction of the bounded stabilizing feedback with the property A two accompanying problems were considered.

The first problem is the optimal control problem

$$B(\tau, z) = \min \int_0^\Theta |u(t)| dt,$$

$$\dot{x} = Ax + bu,\ x(0) = z,\ x(\Theta) = 0, \qquad (15)$$

$$g_*(t + \tau) \le Hx(t) \le g^*(t + \tau),$$

$$|u(t)| \le 1,\ t \in T = [0, \Theta].$$

Let $G(\Theta, \tau)$ is the set of all $z \in R^n$ for which problem (15) has a solution.

An optimal start open–slosed loop control is defined by the equality

$$u_s^0(\tau, z) = u^0(s|\tau, z),\ s \in T_h,\ z \in G(\Theta, \tau),\ \tau \in R_\nu.$$

By the Lyapunov function method it is possible to show that at given $g_*(t)$, $g^*(t)$, $t \ge 0$, and appropriate choice of parameters $\Theta > 0$, $\nu > 0$, $h > 0$ of problem (15) the feedback $u_s(t, x) = u_s^0(t, x)$, $s \in T_h$, $x \in G(\Theta, t)$, $t \in R_\nu$, will satisfy all requirements of definition of bounded stabilizing open-closed loop controls with the property A. As the Lyapunov function we shall take optimal values of the criterion of quality of problem (15).

Illustrate the obtained results on an example of stabilization of oscillatory system (Sussmann, *et al.*, 1994)

$$\dot{x}_1 = x_2,\ \dot{x}_2 = -x_1 + x_3, \qquad (16)$$

$$\dot{x}_3 = x_4,\ \dot{x}_4 = -x_3 + u,$$

where (x_1, x_2, x_3, x_4) is a state of system (16), u is a control.

In the work Sussmann'a, *et al.*, 1994 the bounded stabilization feedback:

$$u = -\varepsilon \mathrm{sat}\left(\frac{x_4}{\varepsilon} + \frac{1}{29}\mathrm{sat}\left(\frac{29}{\varepsilon}(-x_1 + x_3 + x_4)\right)\right),$$

$$|u(t)| \le \varepsilon,\ \mathrm{sat}(s) = \mathrm{sign}(s)\min\{|s|, 1\} \qquad (17)$$

is constructed.

For comparison of results we at first shall construct the bounded stabilizing feedback by the above method for the case when the restriction are not imposed on the output $y(t)$, $t \ge 0$. The accompanying problem of optimal control in this case is taken as

$$\int_0^{\Theta} |u(t)|dt \to \min,$$

$$\dot{x}_1 = x_2,\ \dot{x}_2 = -x_1 + x_3,\ \dot{x}_3 = x_4,\ \dot{x}_4 = -x_3 + u,$$

$$x_1(0) = x_1^*(\tau),\ x_2(0) = x_2^*(\tau), \qquad (18)$$

$$x_3(0) = x_3^*(\tau),\ x_4(0) = x_4^*(\tau),$$

$$x_1(\Theta) = 0,\ x_2(\Theta) = 0,\ x_3(\Theta) = 0,\ x_4(\Theta) = 0,$$

$$|u(t)| \le 1,\ t \in T = [0, \Theta].$$

It was solved at the following values of parameters: $\Theta = 8$, $N = 25$, $\nu = 0.32$. As an initial condition the vector $x_0^* = (0.1, 0.1, 0.1, 0.1)$ was taken.

In Fig. 2 the transients of system (16) are given with feedback (17) at $\varepsilon = 1$ (curve 1) and with the feedback constructed on accompanying problem (18) (curve 2).

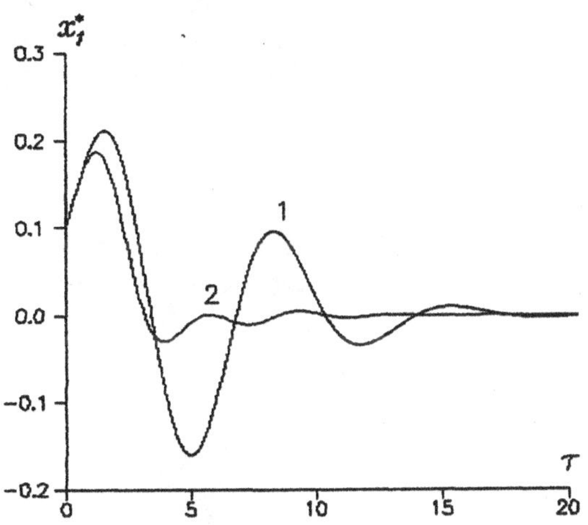

Fig. 2. Transients without restrictions on output.

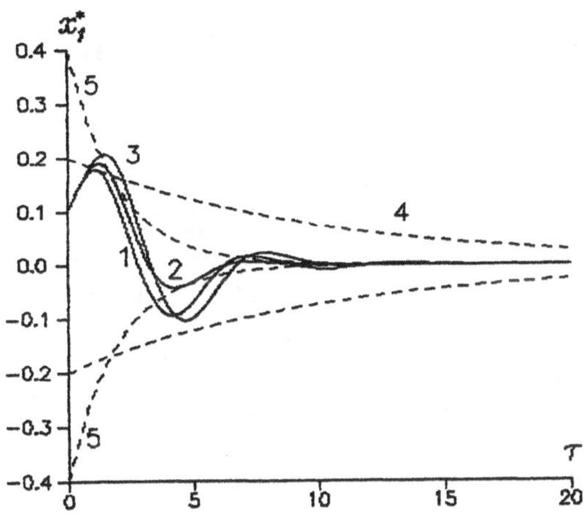

Fig. 3. Influence of parameters α, a to behaviour of the system.

Now construct the bounded stabilizing feedback for system (16) in view of restrictions on output. As the output we shall take $y(t) = x_1(t), t \ge 0$. Assume $g_*(t) = -a\exp(-\alpha t), g^*(t) = a\exp(-\alpha t)$, $t \ge 0$.

The accompanying problem of optimal control:

$$\int_0^{\Theta} |u(t)|dt \to \min,$$

$$\dot{x}_1 = x_2,\ \dot{x}_2 = -x_1 + x_3,\ \dot{x}_3 = x_4,\ \dot{x}_4 = -x_3 + u,$$

$$x_1(0) = x_1^*(\tau),\ x_2(0) = x_2^*(\tau), \qquad (19)$$

$$x_3(0) = x_3^*(\tau),\ x_4(0) = x_4^*(\tau),$$

$$x_1(\Theta) = 0,\ x_2(\Theta) = 0,\ x_3(\Theta) = 0,\ x_4(\Theta) = 0,$$

$$-a\exp(-\alpha t) \le x_1(t) \le a\exp(-\alpha t),$$

$$|u(t)| \le 1,\ t \in T = [0, \Theta].$$

At the solution of problem (19) the following paramaters were chosen: $\Theta = 8, h = 0.4, \nu = 5h, x_0^* = (0.1, 0.1, 0.1, 0.1)$. In Fig. 3 the behaviour of the output $y(\tau) = x_1^*(\tau)$, $\tau \ge 0$, for various values α, a have been shown. The curve 1 corresponds to $\alpha = 0.1, a = 0.2$, the curve 2 stands for $\alpha = 0.5, a = 0.4$. For comparison the transient process is shown when on the output $y(t) = x_1(t)$, $t \ge 0$, the restrictions (curve 3) are not imposed. The dotted lines stand for the case of restrictions on values of output. The curves 4, 5 correspond to $g_*(t) = -0.2\exp(-0.1t)$, $g^*(t) = 0.2\exp(-0.1t)$; $g_*(t) = -0.4\exp(-0.5t)$, $g^*(t) = 0.4\exp(-0.5t)$ respectively.

The second problem in question has been the accompanying problem of minimization of intensity of control:

$$\rho \to \min,$$

$$\dot{x}_1 = x_2,\ \dot{x}_2 = -x_1 + x_3,\ \dot{x}_3 = x_4,\ \dot{x}_4 = -x_3 + u,$$

$$x_1(0) = x_1^*(\tau),\ x_2(0) = x_2^*(\tau),$$

$$x_3(0) = x_3^*(\tau),\ x_4(0) = x_4^*(\tau), \qquad (20)$$

$$x_1(\Theta) = 0,\ x_2(\Theta) = 0,\ x_3(\Theta) = 0,\ x_4(\Theta) = 0,$$

$$-a\exp(-\alpha t) \le x_1(t) \le a\exp(-\alpha t),$$

$$|u(t)| \le \rho,\ t \in T = [0, \Theta].$$

At the solution of problem (20) the following parameters were chosen: $\Theta = 2$, $h = 0.08$, $\nu = 5h$, $\alpha = 0.1$, $a = 0.11$.

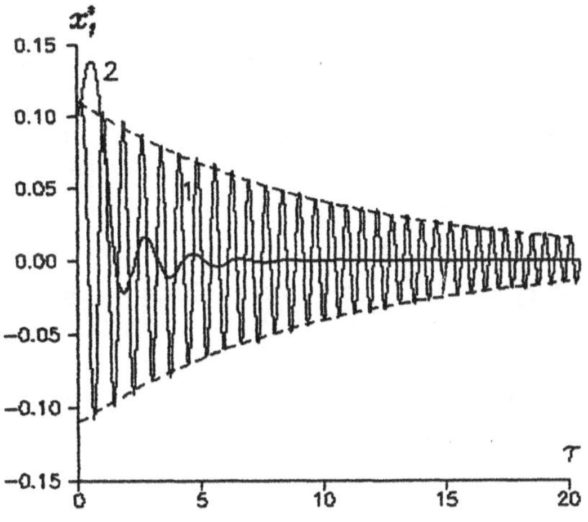

Fig. 4. The transient in view of restrictions on output.

In Fig. 4. the results of calculations are presented. The curve 1 corresponds to the solution of (20). The dotted lines show restrictions on a output. For comparison, the transitive process is shown when on output of the system the restrictions (curve 2) are not imposed.

Definition 3. A bounded stabilizing open–closed control (13) is said to be a stabilizing open–closed loop control with the property B if derivatives of the output $y(t)$, $t \geq 0$, corresponding to the solution $x(t)$, $t \geq 0$, of systems (14), change values in given moments of time $t = k/q$, $k = 1, 2, \ldots$. The number q is degree of oscillation.

For system (16) one can construct the stabilizing open–closed control with the property B ($y = x_1$, $\dot{y} = x_2$).

Let l be an integer of periods of constant signs of length $1/q$ on the interval $[0, \Theta]$, $N_1 = \{1, 3, 5, 7, \ldots\}$, $N_2 = \{2, 4, 6, 8, \ldots\}$.

The accompanying problem of optimal control has the form

$$\rho \to \min,$$

$$\dot{x}_1 = x_2,\ \dot{x}_2 = -x_1 + x_3,\ \dot{x}_3 = x_4,\ \dot{x}_4 = -x_3 + u,$$

$$x_1(0) = x_1^*(\tau),\ x_2(0) = x_2^*(\tau),$$

$$x_3(0) = x_3^*(\tau),\ x_4(0) = x_4^*(\tau), \qquad (21)$$

$$x_1(\Theta) = 0,\ x_2(\Theta) = 0,\ x_3(\Theta) = 0,\ x_4(\Theta) = 0,$$

$$|u(t)| \leq \rho,\ t \in T.$$

Let $\Theta < 1/q$.

If $(k-1)/q \leq \tau \leq k/q - \Theta$, $k \in N_1$, then $x_2(t) \geq 0$, $t \in T = [0, \Theta]$;

if $(k-1)/q \leq \tau \leq k/q - \Theta$, $k \in N_2$, then $x_2(t) \leq 0$, $t \in T$;

if $k/q - \Theta < \tau < k/q + \Theta$, $k \in N_1$, then $x_2(t) \geq 0$, $t \in [0, k/q - \tau]$, $x_2(t) \leq 0$, $t \in [k/q - \tau, \Theta]$;

if $k/q - \Theta < \tau < k/q + \Theta$, $k \in N_2$, then $x_2(t) \leq 0$, $t \in [0, k/q - \tau]$, $x_2(t) \geq 0$, $t \in [k/q - \tau, \Theta]$.

Let $\Theta \geq 1/q$.

If $\tau = (k-1)/q$, $k \in N_1$, then $x_2(t) \geq 0$,

$$t \in T_1 = \bigcup_{s=0,2,4,\ldots,l-1} [s/q, (s+1)/q] \bigcup_{l-even} [l/q, \Theta],$$

$x_2(t) \leq 0$,

$$t \in T_2 = \bigcup_{s=1,3,5,\ldots,l-1} [s/q, (s+1)/q] \bigcup_{l-odd} [l/q, \Theta];$$

if $\tau = (k-1)/q$, $k \in N_2$, then $x_2(t) \leq 0$, $t \in T_1$, $x_2(t) \leq 0$, $t \in T_2$;

if $(k-1)/q < \tau < k/q$, $k \in N_1$, then $x_2(t) \geq 0$,

$$t \in T_3 = [0, k/q - \tau] \bigcup_{s=1,3,5,\ldots,l-1} [k/q - \tau + s/q, k/q - \tau + (s+1)/q] \bigcup_{l-even} [(l-1)/q + k/q - \tau, \Theta],$$

$x_2(t) \leq 0$,

$$t \in T_4 = \bigcup_{s=0,2,4,\ldots,l-1} [k/q - \tau + s/q, k/q - \tau + (s+1)/q] \bigcup_{l-odd} [(l-1)/q + k/q - \tau, \Theta];$$

if $(k-1)/q < \tau < k/q$, $k \in N_2$, then $x_2(t) \leq 0$, $t \in T_3$, $x_2(t) \geq 0$, $t \in T_4$.

At the solution of (21) the following values of parameters were chosen: $\Theta = 2$, $h = 0.1$, $\nu = 5h$. As a initial condition the vector $x_0^* = (-0.1, 0.1, 0.1, 0.1)$ was taken.

In Fig. 5 the trajectories for three processes of stabilization are presented corresponding to various degrees of oscillation: 1) on the interval $[0, \Theta]$ there are two periods of constant signs of speed x_2 (curves 1); 2) on an interval $[0, \Theta]$ there are four periods of constants signs of speed x_2 (curves 2); 3) the period of constant signs of speed x_2 is equal to Θ (curves 3).

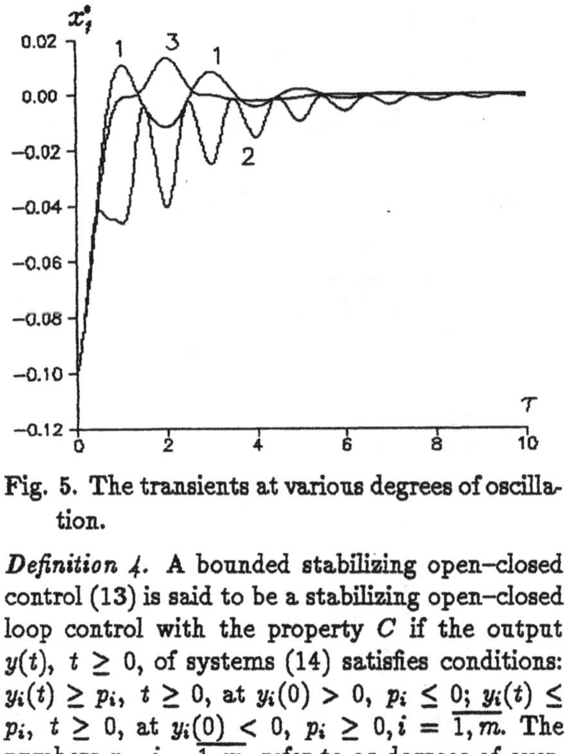

Fig. 5. The transients at various degrees of oscillation.

Definition 4. A bounded stabilizing open–closed control (13) is said to be a stabilizing open–closed loop control with the property C if the output $y(t)$, $t \geq 0$, of systems (14) satisfies conditions: $y_i(t) \geq p_i$, $t \geq 0$, at $y_i(0) > 0$, $p_i \leq 0$; $y_i(t) \leq p_i$, $t \geq 0$, at $y_i(0) < 0$, $p_i \geq 0, i = \overline{1, m}$. The numbers p_i, $i = \overline{1, m}$, refer to as degrees of overcontrol. If $p_i = 0$, $i = \overline{1, m}$, then in the closed system is no any overcontrol.

Construct the stabilizing open–close control without overcontrol ($p_1 = 0$) for system (16) with the output $y(t) = x_1(t)$, $t \geq 0$.

The accompanying problem has the form

$$\rho \to \min,$$

$$\dot{x}_1 = x_2,\ \dot{x}_2 = -x_1 + x_3,\ \dot{x}_3 = x_4,\ \dot{x}_4 = -x_3 + u,$$

$$x_1(0) = x_1^*(\tau),\ x_2(0) = x_2^*(\tau),$$
$$x_3(0) = x_3^*(\tau),\ x_4(0) = x_4^*(\tau), \qquad (22)$$
$$x_1(\Theta) = 0,\ x_2(\Theta) = 0,\ x_3(\Theta) = 0,\ x_4(\Theta) = 0,$$
$$x_1(t) \geq 0,\ |u(t)| \leq \rho,\ t \in T = [0, \Theta].$$

For the solution of (22) the following values of parameters were chosen: $\Theta = 2$, $h = 0.08$, $\nu = 5h$. As an initial condition the vector $x_0^* = (0.1, 0.1, 0.1, 0.1)$ was taken.

In Fig. 6. the output is given for the process of stabilization (curve 1).

Definition 5. A bounded stabilizing open–closed control (13) is said to be a stabilizing open–closed loop control with the property D if the derivative of the output $y(t)$, $t \geq 0$, of systems (14) keep constant signs on an interval $t \geq 0$.

Let construct for system (16) a stabilizing open–closed loop control with property D (the output $y = x_1$). The accompanying problem

$$\rho \to \min,$$
$$\dot{x}_1 = x_2,\ \dot{x}_2 = -x_1 + x_3,\ \dot{x}_3 = x_4,\ \dot{x}_4 = -x_3 + u,$$
$$x_1(0) = x_1^*(\tau),\ x_2(0) = x_2^*(\tau),$$
$$x_3(0) = x_3^*(\tau),\ x_4(0) = x_4^*(\tau), \qquad (23)$$
$$x_1(\Theta) = 0,\ x_2(\Theta) = 0,\ x_3(\Theta) = 0,\ x_4(\Theta) = 0,$$
$$x_2(t) \leq 0,\ |u(t)| \leq \rho,\ t \in T = [0, \Theta].$$

As an initial condition the vector $x_0^* = (0.1, -0.1, 0.1, 0.1)$ was taken. The following values of parameters of problem (23) were taken: $\Theta = 2$, $h = 0.08$, $\nu = 5h$. In Fig. 6 the output given (curve 2).

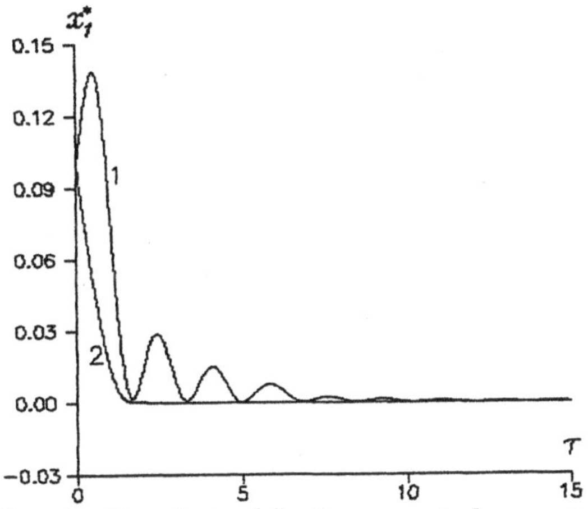

Fig. 6. Transient of "no" overcontrol; monotone transient.

ACKNOWLEDGEMENTS

This research was party supported by the Byelorussian Republican Foundation of Basic Researches under Grant F97M–139.

REFERENCES

Aizerman, M.A. (1958). *Lectures on the theory of automatic control*, Nauka, Moscow [in Russian].

Balashevich, N.V., R.Gabasov and F.M.Kirillova (1994). An optimal damper for dynamical systems. *Automation and Remote Control*, **55**, pp. 615–622.

Barbashin, E.A. (1967). *Introduction in the theory of stability.* Nauka, Moscow [in Russian].

Bromberg, P.V. (1967). *Matrix methods in the theory of a relay pulse regulator.* Nauka, Moscow [in Russian].

Gabasov, R., F.M. Kirillova and O.I. Kostyukova (1991). Constructing of optimal controls of feedback type in a linear problem. *Soviet Math. Dokl.*, **320, No. 2**, pp. 1294–1299.

Gabasov, R., F.M. Kirillova and O.I. Kostyukova (1992). Optimization of linear control system in real time. *Cybernetics*, **No. 4**, pp. 3–19 [in Russian].

Gabasov, R., F.M. Kirillova and S. V. Prischepova (1995). *Optimal Feedback Control.* Lecture Notes in Control and Information Sciences. (Thoma M. ed.) Springer. **207**.

Kalman, R.E (1961). The general theory of control systems. In: *Proc. Ist of the IFAC*, **2**, pp. 521–547. Izd. Akad. Nauk SSSR.

Kwon, W.H. and A.E. Pearson (1977). A modified quadratic cost problem and feedback stabilization of a linear system. *IEEE Trans. Automatic Control.* **CA - 22, No. 5**, pp. 838–842.

Letov, A.M. (1960). Analytical construction of controllers. *Avtomatika i Telemekhanika*, **21**, pp. 436–441 [in Russian].

Malkin, I.G. (1966). *Theory of the stability of motion*, pp. 475 - 514. Nauka, Moscow [in Russian].

Mayne, D.Q., and H. Michalska (1990). Receding horizon control of nonlinear systems. *IEEE Trans. Aut. Control.* **AC - 35, No. 7**, pp. 814–824.

Pontryagin, L.S. et al. (1964). *The mathematical theory of optimal processes*, Macmillan, USA.

Sussmann, H.J., E.D. Sontag and Y. Yang (1994). A general result on the stabilization of linear systems using bounded control. *IEEE Transaction on Automatic Control*, **39, No. 12**, pp. 2411–2425.

ASYMPTOTICS FOR SINGULARLY PERTURBED DIFFERENTIAL GAMES

N.N.Subbotina

Institute of Mathematics and Mechanics, Ural Branch Russ. Acad. Sci., S.Kovalevskoi str.,16, 620219 Ekaterinburg, Russia e-mail: subb@uran.ru

Abstract: Singularly perturbed differential games with "fast" and "slow" motions and the Bolza type payoff functionals are considered. Sufficient conditions are obtained for the value functions of the games to converge to the value function of the asymptotic unperturbed game as a parameter of singularity ε tends to 0.

Keywords: Attractors, characteristic equations, invariance, minimax techniques, non-linear systems, perturbation analysis

1. INTRODUCTION

Mathematical models of dynamical systems with "fast" and "slow" motions occur quite frequently in various problems of economics, engineering, mechanics, biology and other applications.

The paper deals with singularly perturbed differential games $\mathbf{G}^\varepsilon$ whose dynamics are described in the following way

$$\dot{x} = f^\varepsilon(t, x, y, u, v),$$

$$\varepsilon \dot{y} = k^\varepsilon(t, x, y, u, v) + h^\varepsilon(t, x, u, v, \alpha, \beta) \qquad (1)$$

where $\varepsilon > 0$ is a small parameter, time $t \in [0, \theta]$ $(x, y) \in R^n \times R^k$ is the phase vector, x is the "slow" variable, y is the "fast" one. These "attributes" underline a difference between dynamics of variables x and y. The velocity $\dot{y}$ is of order $1/\varepsilon$. Hence y can rapidly vary with respect to time.

It is assumed that controls of the first player (u, α) and controls of the second player (v, β) are restricted with the constraints

$$u \in P \subset R^{m1}, \; v \in Q \subset R^{m2},$$

$$\alpha \in A \subset R^{r1}, \; \beta \in B \subset R^{r2}, \qquad (2)$$

where P, Q, A, B are compacta.

The considered games have the fixed end time θ and a pay-off functional of the Bolza type

$$\gamma^\varepsilon_{t,x,y}(x(\cdot), y(\cdot)) = \sigma^\varepsilon(x(\theta)) +$$

$$+ \int_t^\theta g^\varepsilon(\tau, x(\tau), y(\tau), u(\tau), v(\tau)) d\tau \qquad (3)$$

where $(x(\cdot), y(\cdot)) : [t, \theta] \mapsto R^n \times R^k$ are the trajectories of the equations (1) started at the point $(x(t), y(t)) = (x, y)$, $t \in [0, \theta]$ under controls $(u(\cdot), v(\cdot), \alpha(\cdot), \beta(\cdot)) : [t, \theta] \mapsto P \times Q \times A \times B$. The first player has the aim to minimized the pay-off functional and the second player tries to maximized γ^ε.

The paper deals with conditions under which the value functions $\omega^\varepsilon(t, x, y)$ of the singularly perturbed games $\mathbf{G}^\varepsilon$ converge to the value function $\omega(t, x)$ of an unperturbed game $\mathbf{G}$, as $\varepsilon \downarrow 0$. The unperturbed game $\mathbf{G}$ is the asymptotics for $\mathbf{G}^\varepsilon$.

Researches of singularly perturbed control problems are wide represented in literature, see, for example, (Kokotovic, 1984, Bensoussan, 1989) and references within.

The classical approach of obtaining asymptotics (Tikhonov, 1952; Vasil'eva and Butuzov, 1973; O'Malley, 1974) contains

• equating ε to 0 in relations (1)-(3);
• solving the asymptotic algebraic equation in formulae (1) relative to y and
• substituting the expression for y into the asymptotic equation in relations (1) and the asymptotic pay-off functional (3).

The other well known approach is a decomposition of the problem to
• an optimization problem for "fast" variables under fixed "slow" ones and
• an asymptotic control problem for "slow" variables taking into account results of previous calculations having been undepended on "fast" variables.
See to get acquainted with (Pervozvansky and Gaitsgory, 1988) and references within.

Two mentioned approaches consider convergence problems in the whole of "fast" variables subspace. When a convergence problem in optimal control theory for singularly perturbed data and state constraints on "fast" variables is under consideration it makes sense to consider the convergence only within the constraints. In this case controllability conditions for "fast" variables play an essential role. References on this topic one can find in (Bagagiolo et al., 1997; Bardi and Capuzzo-Dolcetta, 1997).

In the paper an different approach based on existence of attractors in the subspace of "fast" variables is presented. Here main results are obtained due to constructions of the minimax solutions to the first-order PDEs, namely, to the Isaacs–Bellman equations (Subbotin 1980, 1991, 1995). The approach is close to two mentioned above ones considering convergence problems in the whole of "fast" variable subspace.

2. PRELIMINARIES

2.1 Value function as minimax solution of the Isaacs equation

It is known that the value function $\omega^\varepsilon(t,x,y)$ of the game $\mathbf{G}^\varepsilon$ is the minimax (Subbotin, 1980, 1991, 1995) or/and viscosity (Crandall and Lions, 1983; Crandall et al., 1984) generalized solution to the following Cauchy problem $\mathbf{P}^\varepsilon$

$$\partial\omega^\varepsilon/\partial t + H^\varepsilon(t,x,y,D_x\omega^\varepsilon,D_y\omega^\varepsilon) = 0, \quad (4)$$

$$t \in [0,\theta), \quad x \in R^n, \quad y \in R^k;$$

$$\omega^\varepsilon(\theta,x,y) = \sigma^\varepsilon(x), \quad x \in R^n, \ y \in R^k, \quad (5)$$

where vectors $p = D_x\omega^\varepsilon$, $q = D_y\omega^\varepsilon$ are gradients of $\omega^\varepsilon(t,x,y)$ relative to x and y, namely, $D_x\omega^\varepsilon = (\partial\omega^\varepsilon/\partial x_1,...,\partial\omega^\varepsilon/\partial x_n)$, $D_y\omega^\varepsilon = (\partial\omega^\varepsilon/\partial y_1,...,\partial\omega^\varepsilon/\partial y_k)$ and $H^\varepsilon(t,x,y,p,q)$ is the Hamiltonian

$$\min_{u\in P,\alpha\in A}\ \max_{v\in Q,\beta\in B}[\langle f^\varepsilon(t,x,y,u,v),p\rangle +$$
$$+ g^\varepsilon(t,x,y,u,v) + 1/\varepsilon\langle k^\varepsilon(t,x,y,u,v),q\rangle +$$
$$+ 1/\varepsilon\langle h^\varepsilon(t,x,u,v,\alpha,\beta),q\rangle] =$$
$$\max_{v\in Q,\beta\in B}\ \min_{u\in P,\alpha\in A}[\langle f^\varepsilon(t,x,y,u,v),p\rangle +$$
$$+ g^\varepsilon(t,x,y,u,v) + 1/\varepsilon\langle k^\varepsilon(t,x,y,u,v),q\rangle +$$
$$+ 1/\varepsilon\langle h^\varepsilon(t,x,u,v,\alpha,\beta),q\rangle] =$$
$$= H^\varepsilon(t,x,y,p,q) \quad (6)$$

The equation (4) is called the Isaacs equation (Isaacs R., 1965) for the singularly perturbed differential game $\mathbf{G}^\varepsilon$ (1)-(3). It is singularly perturbed because it contains terms with coefficients $1/\varepsilon$ where ε is a small parameter. These terms are positive homogeneous with respect to impulse "fast" variables $q = D_y\omega^\varepsilon$.

Definitions of minimax solutions to the first-order PDE will be essentially used in the paper to obtain sufficient conditions of convergence $\omega^\varepsilon(t,x,y)$ as $\varepsilon \downarrow 0$. Hence, let us consider this concept.

It should be mentioned that the notion of a minimax solution to the first-order PDE may be defined in a variety of ways; among other things, one can use the tools of nonsmooth analysis: directional derivatives, tangent or contingent cones, sub- and super-differentials (the definitions and proof of equivalence of all the definitions can be found, for example, in (Subbotin, 1995)).

The present paper deals with the definition that can be viewed as a generalization of the classical method of characteristics. In this definition the characteristic system is replaced by a family of "characteristic" differential inclusions. The graph of the minimax solution is weakly invariant with respect to these differential inclusions. The role of the characteristic inclusions to the Cauchy problem $\mathbf{P}^\varepsilon$ (4)-(5) can play inclusions which define u-stability and v-stability properties of the value function. Let us remain the definition.

Let S be a nonempty set and M^ε a multivalued mapping

$$[0,\theta] \times R^n \times R^k \times S \ \ni (t,x,y,s) \mapsto$$
$$\mapsto M^\varepsilon(t,x,y,s) \subset R^n \times R^k \times R \quad (7)$$

The pair (S, M^ε) is called *a characteristic complex* (or, briefly, a complex) to the singularly perturbed problem $\mathbf{P}^\varepsilon$ (4)-(5) if the following conditions hold:

1°*a*) for any $(t,x,y) \in [0,\theta] \times R^n \times R^k$ and $s \in S$ the set $M^\varepsilon(t,x,y,s) = \{f,d,g\} \subset R^n \times R^k \times R$ is nonempty, convex and closed. Components $(f,d,g) \in M^\varepsilon(t,x,y,s)$ satisfy the inequalities

$$\|f\| \le \mu^\varepsilon(1 + \|x\| + \|y\|);$$

$$\|d\| \le \mu^\varepsilon(1 + \|x\| + \|y\|);$$

$$\|g\| \leq \mu^\varepsilon(1 + \|x\| + \|y\|);$$

where $\mu^\varepsilon > 0$ are constant. The multivalued mappings $(t, x, y) \mapsto M^\varepsilon(t, x, y, s)$ are upper semicontinuous for all $s \in S$;

2°a) for any $(t, x, y) \in [0, \theta] \times R^n \times R^k$ and $(p, q) \in R^n \times R^k$

$$\max_{s \in S} \min\{\langle f, p\rangle + 1/\varepsilon\langle d, q\rangle - g : \ (f, d, g) \in$$

$$\in M^\varepsilon(t, x, y, s)\} = H^\varepsilon(t, x, y, p, q);$$

2°b) for any $(t, x, y) \in [0, \theta] \times R^n \times R^k$ and $(p, q) \in R^n \times R^k$

$$\min_{s \in S} \max\{\langle f, p\rangle + 1/\varepsilon\langle d, q\rangle - g : \ (f, d, g) \in$$

$$\in M^\varepsilon(t, x, y, s)\} = H^\varepsilon(t, x, y, p, q)$$

The symbols $\|\cdot\|$, and $\langle\cdot,\cdot\rangle$ denote the Euclidean norms and inner products respectively.

The set of all characteristic complexes (S, M^ε) will be denoted by the symbol $C(H^\varepsilon)$.

A pair (S^*, M^ε) $((S_*, M^\varepsilon))$ is called *an upper (lower) characteristic complex* if conditions 1°a) and 2°a) (1°a) and 2°b)) are satisfied. The set of upper (lower) characteristic complexes (S^*, M^ε) $((S_*, M^\varepsilon))$ will be denoted by $C^\uparrow(H^\varepsilon)$ $(C^\downarrow(H^\varepsilon))$.

It is easy to check that the following complexes can be considered as upper and lower characteristic ones for the Isaacs equation (4)

$$S^* = Q \times B \ni s^* = (v^*, \beta^*),$$

$$M^\varepsilon(t, x, y, s^*) = \overline{co}\ \{f^\varepsilon(t, x, y, P, v^*),$$

$$k^\varepsilon(t, x, y, P, v^*) + h^\varepsilon(t, x, P, v^*, A, \beta^*),$$

$$- g^\varepsilon(t, x, y, P, v^*)\}; \qquad (10)$$

$$S_* = P \times A \ni s_* = (u_*, \alpha_*),$$

$$M^\varepsilon(t, x, y, s_*) = \overline{co}\ \{f^\varepsilon(t, x, y, u_*, Q),$$

$$k^\varepsilon(t, x, y, u_*, Q) + h^\varepsilon(t, x, u_*, Q, \alpha_*, B),$$

$$- g^\varepsilon(t, x, y, u_*, Q)\}; \qquad (11)$$

The symbol $\overline{co}\ D$ denotes the closed convex hull of the set D.

For any $(S^*, M^\varepsilon) \in C^\uparrow(H^\varepsilon)$ $((S_*, M^\varepsilon) \in C^\downarrow(H^\varepsilon))$ and $s = s^* \in S^*$ $(s = s_* \in S_*)$ the symbol Sol (t_0, x_0, y_0, z_0, s) will denote the set of absolutely continuous functions $(x(\cdot), y(\cdot), z(\cdot)) : [0, \theta] \mapsto R^n \times R^k \times R$, that satisfy the condition $(x(t_0), y(t_0), z(t_0)) = (x_0, y_0, z_0)$ and the *upper (lower) characteristic differential inclusion*

$$(\dot{x}(t), \varepsilon\dot{y}(t), \dot{z}(t)) = M^\varepsilon(t, x(t), y(t), s) \qquad (12)$$

Definition I. A lower (upper) semicontinuous function $[0, \theta] \times R^n \times R^k \ni (t, x) \mapsto u^\varepsilon(t, x, y) \in R$ is called *an upper (lower) minimax solution* of the singularly perturbed Hamilton-Jacobi equation (4) if the following invariance conditions hold:

- for any $(t_0, x_0, y_0, z_0) \in \text{epi}\ u^\varepsilon$ $((t_0, x_0, y_0, z_0) \in \text{hypo}\ u^\varepsilon)$, $s = s^* \in S^*$ $(s = s_* \in S_*)$ a trajectory $(x(\cdot), y(\cdot), z(\cdot)) \in$ Sol (t_0, x_0, y_0, z_0, s) exists such that $(\tau, x(\tau), y(\tau), z(\tau)) \in \text{epi}\ u^\varepsilon$ $((\tau, x(\tau), y(\tau), z(\tau)) \in \text{hypo}\ u^\varepsilon)$ for all $\tau \in [t_0, \theta]$.

Here $(S^*, M^\varepsilon) \in C^\uparrow(H^\varepsilon)$ $((S_*, N^\varepsilon) \in C^\downarrow(H^\varepsilon))$. The symbols epi u^ε, hypo u^ε and gr u^ε denote the epigraph, the hypograph and the graph of the function $u^\varepsilon(t, x, y)$ respectively, i.e. the sets

$$\text{epi}\ u^\varepsilon = \{(t, x, y, z) : \ z \leq u^\varepsilon(t, x, y)\},$$

$$\text{hypo}\ u^\varepsilon = \{(t, x, y, z) : \ z \geq u^\varepsilon(t, x, y)\},$$

$$\text{gr}\ u^\varepsilon = \{(t, x, y, z) : \ z = u^\varepsilon(t, x, y)\}.$$

Definition II. A continuous function $[0, \theta] \times R^n \times R^k \ni (t, x) \mapsto u^\varepsilon(t, x, y) \in R$ is called *a minimax solution* of the singularly perturbed Hamilton-Jacobi equation (4) iff it is an upper minimax solution and a lower minimax solution simultaneously.

It is known that for the minimax solution the following invariance condition holds.

- For any $(t_0, x_0, y_0, z_0) \in \text{gr}\ u^\varepsilon$, $s \in S$, $(S, M^\varepsilon) \in C(H^\varepsilon)$ a trajectory $(x(\cdot), y(\cdot), z(\cdot)) \in$ Sol (t_0, x_0, y_0, z_0, s) exists such that $(\tau, x(\tau), y(\tau), z(\tau)) \in \text{gr}\ u^\varepsilon$ for all $\tau \in [t_0, \theta]$.

It should be noted that the definitions of minimax and upper (lower) minimax solutions do not depend on which of complexes $(S, M^\varepsilon) \in C(H^\varepsilon)$ and $(S^*, M^\varepsilon) \in C^\uparrow(H^\varepsilon)$ $((S_*, M^\varepsilon) \in C^\downarrow(H^\varepsilon))$ are utilized.

2.2 Nonsmooth analysis tools

Let us choose among wide variety of nonsmooth analysis tools those ones are applicable in the constructions below. These are the following notions of proximal calculus (see Clarke et al, 1998).

Let S be a nonempty closed set in a finite dimensional space. Suppose that x is a point not lying in S. Suppose further that there exist a point s in S whose distance to x is minimal. Then s is called a *closest point* or a *projection* of x onto S.

The vector $x - s$ determines what is called a *proximal normal direction* to S at s. Any nonnegative multiply $n = \lambda(x - s)$, $\lambda \geq 0$, of such a vector is called a *proximal normal* (or a P-normal) to S at s. The set of all n obtainable in this manner is termed the *proximal normal cone* to S at s, and is denoted by $N_S^P(s)$. Note that for any point s belonging to the interior of the set S the proximal normal cone is a singleton. It consists of the unique null element.

The other important thing is the notion of *strong invariance.* Let S be a nonempty closed set in

$R\times R^n$ and its section at the moment t be denoted by S_t. Let F be a multivalued mapping $R\times R^n$ to the subsets of R^n. Consider a differential inclusion

$$\dot{x}(t) \in F(t, x(t)), \quad t \in [t_0, \theta] \tag{13}$$

Definition III. The pair (S, F) is said to be *strongly invariant* if *every* trajectory $x(\cdot)$ of (13) on $[t_0, \theta], t_0 \in [0, \theta]$ for which $x(t_0) \in S_{t_0}$ is such that $x(t) \in S_t$ for all $t \in [t_0, \theta]$.

2.3 Basic assumptions and main result

It is assumed for data of singularly perturbed games $\mathbf{G}^\varepsilon$ (1)-(3) to satisfy the following conditions

1) functions $f^\varepsilon(t,x,y,u,v)$, $g^\varepsilon(t,x,y,u,v)$, $\sigma^\varepsilon(x)$, $k^\varepsilon(t,x,y,u,v)$, $h^\varepsilon(t,x,u,v,\alpha,\beta)$, are continuous relative to all variables and parameter ε when $t \in [0,\theta]$, $x \in R^n$, $y \in R^k$, $u \in P$, $v \in Q$, $\alpha \in A$, $\beta \in B$, $\varepsilon \in [0,1]$;

2) there exist constants $\mu^\varepsilon > 0$, $\nu^\varepsilon \in (0, \mu^\varepsilon/2)$ such that

$$\|f^\varepsilon(t,x,y,u,v)\| \leq \mu^\varepsilon(1 + \|x\| + \|y\|);$$

$$\|k^\varepsilon(t,x,y,u,v)\| \leq \nu^\varepsilon(1 + \|x\| + \|y\|);$$

$$\|h^\varepsilon(t,x,u,v,\alpha,\beta)\| \leq \mu^\varepsilon/2(1 + \|x\|);$$

$$|g^\varepsilon(t,x,y,u,v)| \leq \mu^\varepsilon(1 + \|x\| + \|y\|);$$

Note that constants μ^ε are the same as in condition 1°a);

3) for any compactum $C \in [0,\theta] \times R^n \times R^k$ and any $u \in P$, $v \in Q$, $\alpha \in A$, $\beta \in B$ there exist constants $L^\varepsilon > 0$, $N^\varepsilon \in (0, L^\varepsilon/2)$ such that

$$\|f^\varepsilon(t',x',y',u,v) - f^\varepsilon(t'',x'',y'',u,v)\| \leq$$
$$\leq L^\varepsilon(|t'-t''| + \|x'-x''\| + \|y'-y''\|);$$
$$\|k^\varepsilon(t',x',y',u,v) - k^\varepsilon(t'',x'',y'',u,v)\| \leq$$
$$\leq N^\varepsilon(|t'-t''| + \|x'-x''\| + \|y'-y''\|);$$
$$\|h^\varepsilon(t',x',u,v,\alpha,\beta) - h^\varepsilon(t'',x'',u,v,\alpha,\beta)\| \leq$$
$$\leq L^\varepsilon/2(|t'-t''| + \|x'-x''\|);$$
$$|g^\varepsilon(t',x',y',u,v) - g^\varepsilon(t'',x'',y'',u,v)| \leq$$
$$\leq L^\varepsilon(|t'-t''| + \|x'-x''\| + \|y'-y''\|);$$

4) for any $(t,x,y) \in [0,\theta] \times R^n \times R^k$, $(p,q) \in R^n \times R^k$ the Isaacs condition (6) is valid.

It is known (Krasovskii and Subbotin, 1974, 1985; et al.), that conditions 1)–4) guarantee existence of the value functions $\omega^\varepsilon(t,x,y)$ in the games $\mathbf{G}^\varepsilon$ for any $\varepsilon \in (0,\theta]$. To provide a convergence property for the value functions $\omega^\varepsilon(t,x,y)$ as ε tends to zero the following constructions and assumptions are added.

Let sets Y^ε, Y^ε_{up}, Y^ε_{lo} define via the relations

$$Y^\varepsilon = Y^\varepsilon(t,x,u,v,\alpha,\beta) = \{y :$$

$$k^\varepsilon(t,x,y,u,v) + h^\varepsilon(t,x,u,v,\alpha,\beta) = 0\}; \tag{14}$$

$$Y^\varepsilon_{up} = Y^\varepsilon_{up}(t,x,v^*,\beta^*) =$$
$$= \bigcup_{u\in P,\alpha\in A} Y^\varepsilon(t,x,u,v^*,\alpha,\beta^*); \tag{15}$$

$$Y^\varepsilon_{lo} = Y^\varepsilon_{lo}(t,x,u_*,\alpha_*) =$$
$$= \bigcup_{v\in Q,\beta\in B} Y^\varepsilon(t,x,u_*,v,\alpha_*,\beta); \tag{16}$$

Assume that

5) for any $(t,x) \in [0,\theta]\times R^n$, $s^* \in S^*$, $s_* \in S_*$ the sets Y^ε, $Y^\varepsilon_{up}(t,x,s^*)$, $Y^\varepsilon_{lo}(t,x,s_*)$ are nonempty and restricted

$$\forall y \in Y^\varepsilon_{up}(t,x,s^*) : \|y\| \leq \chi^\varepsilon(1 + \|x\|), \tag{17}$$

$$\forall y \in Y^\varepsilon_{lo}(t,x,s_*) : \|y\| \leq \chi^\varepsilon(1 + \|x\|), \tag{18}$$

where χ^ε are constant, $\chi^\varepsilon \in (0, \mu^\varepsilon]$;

6) for any $(t',x') \in [0,\theta] \times R^n$, $(t'',x'') \in [0,\theta] \times R^n$, $s^* \in S^*$, $s_* \in S_*$ the following Lipschitz conditions are valid

$$\text{dist}\ (Y^\varepsilon_{up}(t'x's^*),\ Y^\varepsilon_{up}(t'',x'',s^*)) \leq$$
$$\leq K^\varepsilon(|t'-t''| + \|x'-x''\|), \tag{19}$$
$$\text{dist}\ (Y^\varepsilon_{lo}(t'x's_*),\ Y^\varepsilon_{lo}(t'',x'',s_*)) \leq$$
$$\leq K^\varepsilon(|t'-t''| + \|x'-x''\|), \tag{20}$$

where $K^\varepsilon > 0$ are constant and the symbol dist (Y^1, Y^2) denotes the Hausdorff distance between the sets Y^1 and Y^2;

7) the maps $(t,x) \mapsto Y^\varepsilon_{up}(t,x,v^*,\beta^*)$, $(t,x) \mapsto Y^\varepsilon_{lo}(t,x,u_*,\alpha_*)$ are upper semicontinuous for any $(v^*,\beta^*) = s^* \in S^* = Q\times B$, $(u_*,\alpha_*) = s_* \in S_* = P\times A$;

8) for any compacta $D, D^0, D_0 : D \supset [0,\theta] \times R^n$ and

$$R^k \supset D^0 \supset \bigcup_{\varepsilon\in[0,1]} \bigcup_{(t_0,x_0)\in D, s^*\in S^*} Y^\varepsilon(t_0,x_0,s^*),$$

$$R^k \supset D_0 \supset \bigcup_{\varepsilon\in[0,1]} \bigcup_{(t_0,x_0)\in D, s_*\in S_*} Y^\varepsilon(t_0,x_0,s_*)$$

there exist numbers $\kappa^\varepsilon > 0$ such that the following inequalities are valid

$$\max_{u\in P}\langle y^* - y^0,\ k^\varepsilon(t,x,y^*,u,v^*) - k^\varepsilon(t,x,y^0,u,v^*)\rangle$$

$$\leq -\kappa^\varepsilon \text{dist}\ ^2(y^*, Y^\varepsilon_{up}(t,x,s^*)) \tag{21}$$

for any $y^* \notin Y^\varepsilon_{up}(t,x,s^*)$, $y^0 \in Y^\varepsilon_{up}(t,x,s^*)$, $s^* \in S^*$, $(t,x) \in D\times \exp(\mu^\varepsilon(\theta - t_0)(1 + \chi^\varepsilon + 2d^0_y))$, $d^0_y = \max_{y'\in D^0, y''\in D^0} \|y' - y''\|$;

analogously,

$$\max_{v\in Q}\langle y_* - y_0,\ k^\varepsilon(t,x,y_*,u_*,v) - k^\varepsilon(t,x,y_0,u_*,v)\rangle$$

$$\leq -\kappa^\varepsilon \text{dist}\,^2(y_*, Y^\varepsilon_{lo}(t,x,s_*)) \quad (22)$$

for any $y_* \notin Y^\varepsilon_{lo}(t,x,s_*)$, $y_0 \in Y^\varepsilon_{lo}(t,x,s_*)$, $s_* \in S_*$, $(t,x) \in D \times \exp(\mu^\varepsilon(\theta - t_0)(1+\chi^\varepsilon + 2d_{0,y}))$, $d_{0,y} = \max_{y' \in D_0, y'' \in D_0} \|y' - y''\|$,

(the symbol dist (y,Y) denotes the distance between point y and set Y, i.e.

$$\text{dist}\,(y,Y) = \inf_{y' \in Y} \|y - y'\| \quad);$$

9) for any $y^0 \in Y^\varepsilon_{up}(t,x,s^*), s^* \in S^*$ and a P-normal $n^*(y^0)$ to the set $Y^\varepsilon_{up}(t,x,s^*)$ at the point y^0 the equality holds

$$\max_{u \in P, \alpha \in A} \langle n^*(y^0), k^\varepsilon(t,x,y^0,u,v^*) +$$

$$+ h^\varepsilon(t,x,u,v^*,\alpha,\beta^*)\rangle = 0; \quad (23)$$

analogously, for any $y_0 \in Y^\varepsilon_{lo}(t,x,s_*), s_* \in S_*$ and a P-normal $n_*(y_0)$ to the set $Y^\varepsilon_{lo}(t,x,s_*)$ at the point y_0 the equality holds

$$\max_{v \in Q, \beta \in B} \langle n_*(y_0), k^\varepsilon(t,x,y_0,u_*,v) +$$

$$+ h^\varepsilon(t,x,u_*,v,\alpha_*,\beta)\rangle = 0 \quad (24)$$

10) constants μ^ε, ν^ε, χ^ε, K^ε, L^ε, N^ε, κ^ε depend continuously on parameter $\varepsilon \in [0,1]$;

Let us introduce the "upper" H^ε_{up} and "lower" H^ε_{lo} Hamiltonians via the formulae

$$H^\varepsilon_{up}(t,x,s) = \max_{s^* = (v^*, \beta^*) \in S^*} \min[\langle f, s\rangle - g :$$

$$(f,g) \in \overline{co}\,\{f^\varepsilon(t,x,Y^\varepsilon_{up}(t,x,s^*),P,v^*),$$

$$- g^\varepsilon(t,x,Y^\varepsilon_{up}(t,x,s^*),P,v^*)\}, \quad (25)$$

$$H^\varepsilon_{lo}(t,x,s) = \min_{s_* = (u_*, \alpha_*) \in S_*} \max[\langle f, s\rangle - g :$$

$$(f,g) \in \overline{co}\,\{f^\varepsilon(t,x,Y^\varepsilon_{lo}(t,x,s_*),u_*,Q),$$

$$- g^\varepsilon(t,x,Y^\varepsilon_{lo}(t,x,s_*),u_*,Q)\}, \quad (26)$$

11) Assume that

$$|H^\varepsilon_{up}(t,x,s) - H^\varepsilon_{lo}(t,x,s)| \leq \delta(\varepsilon), \quad (27)$$

where $\delta(\varepsilon) \downarrow 0$ as $\varepsilon \downarrow 0$.

Let us denote by $H^0(t,x,s)$ the limits

$$H^0(t,x,s) = \lim_{\varepsilon \downarrow 0} H^\varepsilon_{up}(t,x,s) =$$

$$\lim_{\varepsilon \downarrow 0} H^\varepsilon_{lo}(t,x,s) \quad (28)$$

which will be considered as the Hamiltonian in the unperturbed Cauchy problem **P**

$$\partial\omega/\partial t + H^0(t,x,D_x\omega) = 0, \quad (29)$$

$$t \in [0,\theta), \quad x \in R^n,$$

$$\omega(\theta,x) = \sigma^0(x), \quad x \in R^n, \quad (30)$$

where $\sigma^0(x) = \lim_{\varepsilon \downarrow 0} \sigma^\varepsilon(x)$.

The main result of the paper is the following

Theorem I. Let conditions 1) -11) in the singularly perturbed differential games $\mathbf{G}^\varepsilon$ (1) -(3) be satisfied. Then on any compactum $C \subset [0,\theta] \times R^n \times R^k$ the value function $\omega^\varepsilon(t,x,y)$ converges uniformly to the value function $\omega(t,x)$ of the following unperturbed game **G**

$$\dot{x} \in f^0(t,x,Y^0(t,x,u,v,\alpha,\beta),u,v), \quad (31)$$

$$(t,x) \in [0,\theta] \times R^n;$$

$$(u,\alpha) \in P \times A, \quad (v,\beta) \in Q \times B; \quad (32)$$

$$\gamma_{t_0,x_0}(x(\cdot)) \in \sigma^0(x(\theta)) + \int_{t_0}^{\theta} g^0(\tau, x(\tau), Y^0(\tau,$$

$$x(\tau),u(\tau),v(\tau),\alpha(\tau),\beta(\tau)),u(\tau),v(\tau))d\tau \quad (33)$$

where $(x(\cdot)) : [t,\theta] \mapsto R^n$ is a trajectory of the differential inclusion (31) started at the point $x(t_0) = x_0$, $t_0 \in [0,\theta]$ under controls $(u(\cdot),v(\cdot),\alpha(\cdot),\beta(\cdot)) : [t_0,\theta] \mapsto P \times Q \times A \times B$.

The functions $f^0(\cdot), g^0(\cdot), \sigma^0(\cdot)$ are obtained from the data of the singularly perturbed games $\mathbf{G}^\varepsilon$ if $\varepsilon = 0$. The sets $Y^0(t,x,u,v,\alpha,\beta)$ are the limits of the sets $Y^\varepsilon(t,x,u,v,\alpha,\beta)$ in the Hausdorff metric as $\varepsilon \downarrow 0$.

The unperturbed game **G** (31)-(33) is called the asymptotics for the singularly perturbed differential games $\mathbf{G}^\varepsilon$ (1)-(3).

The value function $\omega(t,x)$ is the unique minimax solution of the problem **P** (29),(30).

3. SUFFICIENT CONDITIONS FOR CONVERGENCY OF THE VALUE FUNCTIONS

The Theorem I provides sufficient conditions for convergency of the value functions to the singularly perturbed differential games. Its proof is based on the results of the papers (Subbotina, 1996, 1998) where convergence of the minimax solutions to the singularly perturbed Hamilton–Jacobi equations (4) was studed.

The Hamiltonians for the Isaacs equations arising in the differential games theory have the special form (6). Also the specific form of characteristic complexes (10) and (11) will be used below to construct sufficient conditions for convergence of the value functions of the differential games with fast and slow motions. These complexes were wide utilized in researches on u-stability and v-stability properties of the value functions in the differential games theory (see Krasovskii and Subbotin, 1974, 1985).

3.1 Properties of the sets Y^{ε}_{up}, Y^{ε}_{lo}

It is easy to see from the conditions 1), 5), 7), 10) that for any $(t,x) \in [0,\theta] \times R^n$, $s^* = (v^*,\beta^*) \in S^* = Q \times B$, $s_* = (u_*,\alpha_*) \in S_* = P \times A$ the sets Y^{ε}, Y^{ε}_{up}, Y^{ε}_{lo} are compacta and

$$Y^{\varepsilon}(t,x,u,v,\alpha,\beta) \to Y^0(t,x,u,v,\alpha,\beta)$$

$$Y^{\varepsilon}_{up}(t,x,v^*,\beta^*) \to Y^0_{up}(t,x,v^*,\beta^*)$$

$$Y^{\varepsilon}_{lo}(t,x,u_*,\alpha_*) \to Y^0_{lo}(t,x,u_*,\alpha_*) \qquad (34)$$

as $\varepsilon \downarrow 0$ in the Hausdorff metric. The sets $Y^0_{up}(t,x,s^*)$, $Y^0_{lo}(t,x,s_*)$, $Y^0(t,x,u,v,\alpha,\beta)$ are compacta and the maps

$$(t,x) \mapsto Y^0(t,x,u_*,v^*,\alpha_*,\beta^*),$$

$$(t,x) \mapsto Y^0_{up}(t,x,v^*,\beta^*),$$

$$(t,x) \mapsto Y^0_{lo}(t,x,u_*,\alpha_*)$$

are uniformly Lipschitz continuous for any $(v^*,\beta^*) = s^* \in S^* = Q \times B$, $(u_*,\alpha_*) = s_* \in S_* = P \times A$;

Conditions 9) imply (see Subbotin, 1995; Clarke et al., 1997) that for any continuous bounded function $x(\cdot) : [0,\theta] \mapsto R^n$, $s* \in S^*$, $s_* \in S_*$, $\varepsilon > 0$ the sets $Y^{\varepsilon}_{up}(t,x(t),s^*)$ are strongly invariant with respect to the differential inclusions

$$\varepsilon\dot{y}^0(t) \in \bar{co}\,\{k^{\varepsilon}(t,x(t),y^0(t),P,v^*) + \\ + h^{\varepsilon}(t,x(t),P,v^*,A,\beta^*)\}, \qquad (35)$$

and sets $Y^{\varepsilon}_{lo}(t,x(t),s_*)$ are strongly invariant with respect to the differential inclusions

$$\varepsilon\dot{y}_0(t) \in \bar{co}\,\{k^{\varepsilon}(t,x(t),y_0(t),u_*,Q) + \\ + h^{\varepsilon}(t,x(t),u_*,Q,\alpha_*,B)\}, \qquad (36)$$

• It means that all trajectories of the differential inclusions started at the points on the corresponding sets stay in these sets all the time up to θ.

In order to be specific below important properties will be obtained for the *upper* sets Y^{ε}_{up} and the *upper* characteristic inclusions (10). All the conclusions are valid for the *lower* sets Y^{ε}_{lo} and the *lower* characteristic inclusions too.

Let $x^{\varepsilon}(\cdot), y^{\varepsilon}(\cdot), z^{\varepsilon}(\cdot) : [t_0,\theta] \mapsto R^n \times R^k \times R$, $(x^{\varepsilon}(t_0), y^{\varepsilon}(t_0), z^{\varepsilon}(t_0)) = (x_0,y_0,z_0)$ be a solution of the upper characteristic inclusion (10) corresponding to a parameter $s^* = (v^*,\beta^*) \in S^*$, i.e. $(x^{\varepsilon}(\cdot), y^{\varepsilon}(\cdot), z^{\varepsilon}(\cdot)) \in \mathrm{Sol}\,(t_0,x_0,y_0,z_0,s^*)$. Let us estimate dist $^2(y^{\varepsilon}(t), Y^{\varepsilon}_{up}(t,x^{\varepsilon}(t),s^*))$.

Let us choose

$$y_0 \notin Y^{\varepsilon}_{up}(t_0,x_0,s^*)$$

and define

$$\tilde{y}_0 \in Y^{\varepsilon}_{up}(t_0,x,s^*)$$

$$\|y_0 - \tilde{y}_0\| = \mathrm{dist}\,(y_0, Y^{\varepsilon}_{up}(t_0,x_0,s^*)) = d_0 > 0$$

Consider a solution of the differential inclusion

$$\varepsilon\dot{y}^0(t) \in \bar{co}\,\{k^{\varepsilon}(t,x^{\varepsilon}(t),y^0(t),P,v^*) + \\ + h^{\varepsilon}(t,x^{\varepsilon}(t),P,v^*,A,\beta^*)\}, \qquad (37)$$

$$y^0(t_0) = \tilde{y}_0$$

The strong invariance property of the sets $Y^{\varepsilon}_{up}(t,x^{\varepsilon}(t),s^*)$ with respect to the inclusion (37) implies that $y^0(t) \in Y^{\varepsilon}_{up}(t,x^{\varepsilon}(t),s^*)$ for all $t \in [t_0,\theta]$.

Remember that the dynamics of $y^{\varepsilon}(t)$ describes in the follows way

$$\varepsilon\dot{y}^{\varepsilon}(t) \in \bar{co}\,\{k^{\varepsilon}(t,x^{\varepsilon}(t),y^{\varepsilon}(t),P,v^*) + \\ + h^{\varepsilon}(t,x^{\varepsilon}(t),P,v^*,A,\beta^*)\}, \qquad (38)$$

$$y^{\varepsilon}(t_0) = y_0$$

And let the terms h^{ε} undepended relative to y in the right-hand sides of the inclusions (37) and (38) coincide.

Hence the following estimates are valid

$$\mathrm{dist}\,(y^{\varepsilon}(t), Y^{\varepsilon}_{up}(t,x^{\varepsilon}(t),s^*)) \le \|y^{\varepsilon}(t) - y^0(t)\|;$$

$$\frac{d\|y^{\varepsilon}(t) - y^0(t)\|^2}{dt} = 2\langle y^{\varepsilon}(t) - y^0(t), \frac{dy^{\varepsilon}(t)}{dt} - \frac{dy^0(t)}{dt}\rangle$$

$$\le 2/\varepsilon \max_{u \in P}\langle y^{\varepsilon}(t) - y^0(t), k^{\varepsilon}(t,x^{\varepsilon}(t),y^{\varepsilon}(t),u,v^*) - \\ - k^{\varepsilon}(t,x^{\varepsilon}(t),y^0(t),u,v^*)\rangle$$

Using these relations and the condition 8) one can obtain that

$$\mathrm{dist}\,(y^{\varepsilon}(t), Y^{\varepsilon}_{up}(t,x^{\varepsilon}(t),s^*)) \le d_0 - \\ - \int_{t_0}^{t} \frac{2\kappa^{\varepsilon}}{\varepsilon}\mathrm{dist}\,(y^{\varepsilon}(\tau), Y^{\varepsilon}_{up}(\tau,x^{\varepsilon}(\tau),s^*))d\tau;$$

$$\mathrm{dist}\,(y^{\varepsilon}(t), Y^{\varepsilon}_{up}(t,x^{\varepsilon}(t),s^*)) \le \\ \le d_0 e^{-\frac{\kappa^{\varepsilon}}{\varepsilon}(t-t_0)} \le d_0 \qquad (39)$$

So the fast components $y^{\varepsilon}(t)$ of the upper characteristics (38) go fast to the corresponding sets $Y^{\varepsilon}_{up}(t,x^{\varepsilon}(t),s^*)$. Analogous property holds for the lower characteristics and the corresponding lower sets Y^{ε}_{lo}.

Hence, the sets $Y^{\varepsilon}_{up}, Y^{\varepsilon}_{lo}$ play roles of attractors for the considered characteristic inclusions. The conditions 8) and 9) imply that any d_0–neighbourhood ($d_0 > 0$) of the sets is strongly invariant relative to the corresponding characteristic inclusions.

At last let us show that the following relation takes place for any $(t,x) \in [0,\theta] \times R^n$, $s^* \in S^*$, $s_* \in S_*$

$$Y^0_{up}(t,x,s^*) \bigcap Y^0_{lo}(t,x,s_*) \neq \emptyset \qquad (40)$$

It is known (Subbotin, 1991, 1995) that the analogous conditions are valid for the right-hand side sets in the characteristic inclusions, namely

$$M^\varepsilon_{up}(t,x,y,s^*) \bigcap M^\varepsilon_{lo}(t,x,y,s_*) = \qquad (41)$$

$$= M^\varepsilon(t,x,y) \neq \emptyset; \qquad (41)$$

for all $(t,x,y,s^*,s_*,\varepsilon) \in [0,\theta] \times R^n \times R^k \times S^* \times S_* \times (0,1]$.

Consider the differential inclusion

$$(\dot{x}^\varepsilon(t), \varepsilon\dot{y}^\varepsilon(t), \dot{z}^\varepsilon(t)) = M^\varepsilon(t, x(t), y(t)) \qquad (42)$$

$$(x^\varepsilon(t_0), y^\varepsilon(t_0), z^\varepsilon(t_0)) = (x_0, y_0, z_0) \qquad (43)$$

It is easy to see (Filippov, 1985) that there exists a solution of the inclusion.

From the above properties of the sets $Y^\varepsilon_{up}, Y^\varepsilon_{lo}$ it follows that for any $\eta > 0$ there exist $\varepsilon(\eta) > 0$, $\delta(\varepsilon) > 0, (\delta(\varepsilon)) \downarrow 0$ as $\varepsilon \downarrow 0)$ such that for any $\varepsilon \leq \varepsilon(\eta), \tau \geq t_0 + \delta(\varepsilon)$ the inclusion holds

$$y^\varepsilon(\tau) \in \{(Y^\varepsilon_{up}(\tau, x^\varepsilon(\tau), s^*) + \eta B\} \bigcap$$

$$\bigcap \{Y^\varepsilon_{up}(\tau, x^\varepsilon(\tau), s_*) + \eta B\} \qquad (44)$$

where the symbol B denotes the unit ball in R^k.

Choosing $\eta_i \downarrow 0$, $\varepsilon_i = \varepsilon(\eta_i) \downarrow 0$, $\delta_i = \delta(\varepsilon_i) \downarrow 0$ one can obtain in the limit the last required property

$$Y^0_{up}(t_0,x_0,s^*) \bigcap Y^0_{up}(t_0,x_0,s_*) \neq \emptyset \qquad (45)$$

3.2. Proof of the main result.

To prove the Theorem I. it will be shown that the conditions 8),9),11) are analogous to the conditions 3 and 4 in the paper (Subbotina, 1996) and guarantee convergence $\omega^\varepsilon(t,x,y)$ to $\omega(t,x)$.

For the sake of definity all the constructions below will be done for the upper characteristic complexes and their attractors. The similar results for the lower characteristics and attractors one can easy obtain in the same way.

The following fact of the theory of differential inclusions will be useful for the forthcoming estimates.

Let $(t,x,z) \mapsto F_i(t,x,z) \subset R^n \times R$: $[t_0,\theta] \times R^n \times R \mapsto 2^{R^n \times R}, i = 1,2$ – to be two multivalued mappings which have convex, compact values. For the mappings are assumed to be upper semicontinuous.

Let $x_0 \in R^n, z_0 \in R$. Consider the differential inclusions

$$(\dot{x}_i(t), \dot{z}_i(t)) \in F_i(t, x_i(t), z_i(t)) \quad t \in [t_0, \theta], \qquad (46)$$

$$(x_i(t_0), z_i(t_0)) = (x_0, z_0), \quad i = 1,2. \qquad (47)$$

The set of all solutions $(x_i(\cdot), z_i(\cdot))$ of the differential inclusion (46), (47) with number "i" will be denote as Sol $_i(t_0,x_0,z_0)$. The following proposition is valid (see Filippov, 1985).

Proposition 1. For any solution $(x_1(\cdot), z_1(\cdot)) \in$ Sol $_1(t_0,x_0,z_0)$ there exists a solution $(x_2(\cdot), z_2(\cdot)) \in$ Sol $_2(t_0,x_0,z_0)$ such that the following estimates take place for all $t \in [t_0, \theta]$

$$\|x_1(t) - x_2(t)\| \leq \int_{t_0}^{t} \text{dist } (F_1(\tau,$$

$$x_1(\tau), z_1(\tau)), F_2(\tau, x_2(\tau), z_2(\tau)))d\tau; \qquad (48)$$

$$\|z_1(t) - z_2(t)\| \leq \int_{t_0}^{t} \text{dist } (F_1(\tau,$$

$$x_1(\tau), z_1(\tau)), F_2(\tau, x_2(\tau), z_2(\tau)))d\tau \qquad (49)$$

Let $(t_0,x_0,y_0) \in D \times D_0$, $z_0 \in R^1$, $\varepsilon \in (0,1]$ and $Y^\varepsilon_{up}(t,x,s^*)$, $s^* \in S^*$, (S,M^ε) be defined by the relations (15), (10). Fix a solution $(x^\varepsilon(\cdot), y^\varepsilon(\cdot), z^\varepsilon(\cdot)) \in$ Sol (t_0,x_0,y_0,z_0,s^*), $s^* \in S^*$, i.e.

$$(\dot{x}^\varepsilon(t), \varepsilon\dot{y}^\varepsilon(t), \dot{z}^\varepsilon(t)) \in M^\varepsilon_{up}(t, x^\varepsilon(t), y^\varepsilon(t), s^*)$$

$$(x^\varepsilon(t_0), y^\varepsilon(t_0), z^\varepsilon(t_0)) = (x_0, y_0, z_0)$$

For the function $y^\varepsilon(\cdot) : [t_0, \theta] \mapsto R^k$ let us construct the multivalued mapping

$$(t,x) \mapsto Y^\varepsilon_0(t,x,s^*) \subset Y^\varepsilon_{up}(t,x,s^*)$$

$$Y^\varepsilon_0(t,x,s^*) = \{y_0 \in Y^\varepsilon_{up}(t,x,s^*):$$

$$\text{dist } (y^\varepsilon(t), Y^\varepsilon_{up}(t,x,s^*)) = \|y^\varepsilon(t) - y_0\|\} \neq \emptyset \quad (50)$$

One can easy check that for any s^* the mapping $(t,x) \mapsto Y^\varepsilon_0(t,x,s^*)$ has compact values and it is upper semicontinuous . Hence, these properties are transferring to the mapping

$$(t,x) \mapsto \bar{co} \; \{f^\varepsilon(t,x,Y^\varepsilon_0(t,x,s^*),s^*),$$

$$- \; g^\varepsilon(t,x,Y^\varepsilon_0(t,x,s^*),s^*)\} \qquad (51)$$

Consider the following differential inclusion corresponding to the mapping (51)

$$(\dot{x}^\varepsilon_0(t), \dot{z}^\varepsilon_0(t)) \in \bar{co} \; \{f^\varepsilon(t, x^\varepsilon_0(t), Y^\varepsilon_0(t, x^\varepsilon_0(t), s^*), s^*)$$

$$- g^\varepsilon(t, x^\varepsilon_0(t), Y^\varepsilon_0(t, x^\varepsilon_0(t), s^*), s^*)\} \qquad (52)$$

$$x^\varepsilon_0(t_0) = x_0, \qquad z^\varepsilon_0(t_0) = z_0 \qquad (53)$$

According to the theory of differential inclusions (see Filippov, 1985) a solution of (52), (53) exists for $t \in [t_0, \theta]$. Denote the set of all solutions $(x_0^\varepsilon(\cdot), z_0^\varepsilon(\cdot))$ of (52), (53) by the symbol Sol $_0^\varepsilon(t_0, x_0, z_0, s^*)$. The symbol Sol $^\varepsilon(t_0, x_0, z_0, s^*)$ will be denoted the set of all solutions $(x_\varepsilon(\cdot), z_\varepsilon(\cdot))$ of the differential inclusion

$$(\dot{x}_\varepsilon(t), \dot{z}_\varepsilon(t)) \in \bar{co}\,\{f^\varepsilon(t, x_\varepsilon(t), Y_{up}^\varepsilon(t, x_\varepsilon(t), s^*), s^*),$$

$$- g^\varepsilon(t, x_\varepsilon(t), Y_{up}^\varepsilon(t, x_\varepsilon(t), s^*), s^*)\} \qquad (54)$$

$$x_\varepsilon(t_0) = x_0, \qquad z_\varepsilon(t_0) = z_0 \qquad (55)$$

The inclusion is obvious

$$\text{Sol}\ _0^\varepsilon(t_0, x_0, z_0, s^*) \subset \text{Sol}\ ^\varepsilon(t_0, x_0, z_0, s^*.) \qquad (56)$$

For the chosen trajectory $(x^\varepsilon(\cdot), y^\varepsilon(\cdot), z^\varepsilon(\cdot)) \in$ Sol $(t_0, x_0, y_0, z_0, s^*)$ let us estimate the distance between $(x^\varepsilon(t), z^\varepsilon(t))$ and $(x_\varepsilon(t), z_\varepsilon(t))$ which is a point of the trajectory $(x_\varepsilon(\cdot), z_\varepsilon(\cdot)) \in$ Sol $^\varepsilon(t_0, x_0, z_0, s^*)$, closest to $(x^\varepsilon(\cdot), z^\varepsilon(\cdot))$. Here $t \in [t_0, \theta]$.

It is follows from the definitions that the mentioned distance is not more than the distance between $(x^\varepsilon(t), z^\varepsilon(t))$ and $(x_0^\varepsilon(t), z_0^\varepsilon(t))$ which is a point on the trajectory $(x_0^\varepsilon(\cdot), z_0^\varepsilon(\cdot)) \in$ Sol $_0^\varepsilon(t_0, x_0, z_0, s)$ closest to $(x^\varepsilon(\cdot), z^\varepsilon(\cdot))$.

Using the Proposition 1. one can obtain the following relations

$$\|x^\varepsilon(t) - x_\varepsilon(t)\| \le \|x^\varepsilon(t) - x_0^\varepsilon(t)\| \le$$

$$\int_{t_0}^{t} \|\dot{x}^\varepsilon(\tau) - \dot{x}_0^\varepsilon(\tau)\| d\tau \le$$

$$\le \int_{t_0}^{t} \text{dist}\ (\bar{co}\ f^\varepsilon(\tau, x^\varepsilon(\tau), y^\varepsilon(\tau), P, v^*),\ \bar{co}\ f^\varepsilon(\tau,$$

$$x_0^\varepsilon(\tau), Y_0^\varepsilon(\tau, x_0^\varepsilon(\tau), P, v^*, A, \beta^*), P, v^*)) d\tau$$

$$\le \int_{t_0}^{t} \max_{u \in P} \|f^\varepsilon(\tau, x^\varepsilon(\tau), y^\varepsilon(\tau), u, v^*) -$$

$$- f^\varepsilon(\tau, x_0^\varepsilon(\tau), y_0^\varepsilon(\tau), u, v^*)\| d\tau\ ; \qquad (57)$$

$$\|z^\varepsilon(t) - z_\varepsilon(t)\| \le$$

$$\|z^\varepsilon(t) - z_0^\varepsilon(t)\| \le \int_{t_0}^{t} \|\dot{z}^\varepsilon(\tau) - \dot{z}_0^\varepsilon(\tau)\| d\tau \le$$

$$\le \int_{t_0}^{t} \max_{u \in P} \|g^\varepsilon(\tau, x^\varepsilon(\tau), y^\varepsilon(\tau), u, v^*) -$$

$$- g^\varepsilon(\tau, x_0^\varepsilon(\tau), y_0^\varepsilon(\tau), u, v^*))\| d\tau \qquad (58)$$

where $y_0^\varepsilon(\cdot) : [t_0, \tau] \mapsto D_0 : t \mapsto y_0^\varepsilon(t) \in Y_0^\varepsilon(t, x_0^\varepsilon(t), s^*)$ is a measurable function satisfying in accordance with (50)

$$\|y^\varepsilon(t) - y_0^\varepsilon\| = \text{dist}\ (y^\varepsilon(t), Y_{up}^\varepsilon(t, x_0^\varepsilon(t), s^*))$$

Taking into account the assumed Lipschitz conditions and properties of the operation "dist " one can continue the inequalities (60).

$$\|x^\varepsilon(t) - x_0^\varepsilon(t)\| \le \int_{t_0}^{t} L^\varepsilon\{\|x^\varepsilon(\tau) - x_0^\varepsilon(\tau)\| + \|y^\varepsilon(\tau) -$$

$$- y_0^\varepsilon(\tau)\|\} d\tau \le \int_{t_0}^{t} L^\varepsilon\{\|x^\varepsilon(\tau) - x_0^\varepsilon(\tau)\| + \text{dist}\ (y^\varepsilon(\tau),$$

$$Y_{up}^\varepsilon(\tau, x^\varepsilon(\tau), s^*)) + \text{dist}\ (Y_{up}^\varepsilon(\tau, x^\varepsilon(\tau), s^*),$$

$$Y_{up}^\varepsilon(\tau, x_0^\varepsilon(\tau), s^*)\} d\tau \le \int_{t_0}^{t} L^\varepsilon\{(1 + \nu^\varepsilon)\|x^\varepsilon(\tau) -$$

$$- x_0^\varepsilon(\tau)\| + 2\text{dist}\ (y^\varepsilon(\tau), Y_{up}^\varepsilon(\tau, x^\varepsilon(\tau), s^*))\} d\tau \qquad (59)$$

Using the exponential estimates (39) for the distance between the fast components of the characteristics and the corresponding attractors and also the Gronwall inequality (see Warga, 1976) the relations (59) are completing

$$\|x^\varepsilon(t) - x_0^\varepsilon(t)\| \le \int_{t_0}^{\theta} L^\varepsilon(1 + \nu^\varepsilon)\|x^\varepsilon(\tau) -$$

$$x_0^\varepsilon(\tau)\| d\tau\ +\ 2\int_{t_0}^{\theta} exp\{-\frac{\kappa^\varepsilon}{\varepsilon}(\tau - t_0)\} d_0\ d\tau;$$

$$\|x^\varepsilon(t) - x_0^\varepsilon(t)\| \le 2\frac{\varepsilon d_0}{\kappa^\varepsilon}\left(1 - exp\{-\frac{\kappa^\varepsilon}{\varepsilon}(t - t_0)\}\right)$$

$$+ 2L^\varepsilon(1 + \nu^\varepsilon)\frac{\varepsilon}{\kappa^\varepsilon} exp\{L^\varepsilon(1 + \nu^\varepsilon)(\theta - t_0)\} \times$$

$$\times [(t - t_0) + \left(1 - exp\{-\frac{\kappa^\varepsilon}{\varepsilon}(t - t_0)\}\right)] = \rho(\varepsilon) \qquad (60)$$

One can see that $\rho(\varepsilon) \downarrow 0$ as $\varepsilon \downarrow 0$. Analogous estimate can be obtain for $\|z^\varepsilon(t) - z_0^\varepsilon\|$.

Let us denote by $G_\varepsilon(t_0, \tau, x_0, y_0, z_0, s^*)$ – the projection of the attainability set for the system (12) at the moment τ to the subspace of variables x, z and by $G_\varepsilon^{\rho(\varepsilon)}(t_0, \tau, x_0, z_0, s^*)$ – a closed $\rho(\varepsilon)$ -neighbourhood of the attainability set for the system (54),(55)

Estimates obtained above for upper (and analogous estimates for lower) characteristic complexes imply that the following proposition is truth.
Proposition 2 For any compacta D, D_0 from the condition 8) there exist the mappings $(0, 1] \mapsto R_+ \times R_+$, $\varepsilon \mapsto (\alpha(\varepsilon), \rho(\varepsilon))$, such that $\alpha(\varepsilon) \downarrow 0$, $\rho(\varepsilon) \downarrow 0$ as $\varepsilon \downarrow 0$, and for any $(t_0, x_0) \in D$, $y_0 \in D_0$, $z_0 \in R$, $s' = s^* \in S^*$, $(s' = s_* \in S_*)$, $\varepsilon \in (0, 1]$, $(x^\varepsilon(\cdot), y^\varepsilon(\cdot), z^\varepsilon(\cdot)) \in$ Sol (t_0, x_0, y_0, z_0, s')

there exist such $(x_\varepsilon(\cdot), z_\varepsilon(\cdot)) \in \mathrm{Sol}^{\,\varepsilon}(t_0, x_0, z_0, s')$, that for $\tau \in [t_0 + \delta(\varepsilon), \theta]$, the following relations are valid

$$\|x^\varepsilon(\tau) - x_\varepsilon(\tau)\| \le \rho(\varepsilon), \quad \|z^\varepsilon(\tau) - z_\varepsilon(\tau)\| \le \rho(\varepsilon) \tag{61}$$

The proposition imply that the situation is close to that which was studed in papers (Subbotina, 1996, 1998). Modifying proofs of these papers the main result of this paper can be successfully proved.

3.3 Example

Consider the following singularly perturbed differential game

$$\dot{x}(t) = f(t, x(t), y(t)), \tag{62}$$

$$\varepsilon\dot{y}(t) = k(y) + \xi(t, x, \alpha, \beta) \tag{63}$$

where $k(y)$ has the form

$$k(y) = \begin{cases} -y, & if \quad y \ge 0, \\ -2y, & if \quad y \le 0 \end{cases} \tag{64}$$

$$\alpha \in A, \quad \beta \in B, \tag{65}$$

sets A and B are compacta. Let the Isaacs condition holds. The pay-off functional has the form

$$\gamma_{t_0,x_0,y_0}(x(\cdot), y(\cdot)) = \sigma(x(\theta)) +$$

$$+ \int_{t_0}^{\theta} g(\tau, x(\tau), y(\tau))d\tau \tag{66}$$

Here the upper characteristic complexes and the corresponding attractors satisfying the assumptions 1)-11) are the following

$$s^* = \beta, \quad S^* = B$$

$$M^\varepsilon_{up}(t, x, y, \beta) = \overline{co}\ \{(f(t, x, y),\ \frac{1}{\varepsilon}(\xi - y),$$

$$g(t, x, y)):\quad \xi \in \overline{co}\ \xi(t, x, A, \beta)\}; \tag{67}$$

$$M^0_{up}(t, x, \beta) = \overline{co}\ \{(f(t, x, \xi),\ g(t, x, \xi)):$$

$$\xi \in Y(t, x, \beta)\} \tag{68}$$

where

$$Y(t, x, \beta) = \varphi(\xi) \bigcup_{\alpha \in A} \xi(t, x, \alpha, \beta) \tag{69}$$

and the function $\varphi(\xi)$ is defined as

$$\varphi(\xi) = \begin{cases} 1, & if \quad \xi \ge 0, \\ 1/2, & if \quad \xi \le 0 \end{cases} \tag{70}$$

The upper attractors Y^ε_{up} and Y^0_{up} are obtained by the formulae

$$Y^\varepsilon_{up}(t, x, \beta) = Y(t, x, \beta)^\varepsilon,$$

$$Y^0_{up}(t, x, \beta) = Y(t, x, \beta) \tag{71}$$

where the symbol $Y(t, x, \beta)^\varepsilon$ denotes the closed ε -neighbourhood of the set $Y(t, x, \beta)$.

To obtain the lower characteristic complexes and corresponding arrtactors one can change places β and α, B and A.

The asymptotic game has the form

$$\dot{x}(t) \;=\; f(t, x(t), Y(t, x(t), \alpha, \beta)), \tag{72}$$

$$\alpha \in A, \quad \beta \in B; \tag{73}$$

$$\gamma_{t_0,x_0}(x(\cdot)) \;=\; \sigma(x(\theta)) +$$

$$+ \int_{t_0}^{\theta} g(\tau, x(\tau), Y(\tau, x(\tau), \alpha, \beta))d\tau \tag{75}$$

where

$$Y(t, x, \alpha, \beta) = \varphi(\xi)\xi(t, x, \alpha, \beta) \tag{76}$$

4. CONCLUSION

In this paper, differential games with fast and slow motions are considered. There are obtained sufficient conditions for the value functions in singularly perturbed differential games to converge to the value function in the asymptotics game. It was shown that the convergence property is based on the existence of attractors in the subspace of fast variables.

5. ACKNOWLEDGEMENTS

The research was supported by the Russian Fund of Fundamental Researches under Grants N96-01-00219, N96-15-96245 and N97-01-00371.

REFERENCES

Bagagiolo, F., M. Bardi and I. Capuzzo–Dolcetta (1997). A Viscosity Solutions Approach to Some Asymptotic Problems in Optimal Control. In: *Partial differential equation methods in control and shape analysis* (Da Prato, G. and J. P. Zolésio. (Ed)), 29–39. Dekker, Inc., New York.

Bardi, M. and I. Capuzzo-Dolcetta (1997). *Optimal Control and Viscosity Solutions of Hamilton–Jacobi–Bellman Equations*, Birkhäuser, Boston.

Barron, E. N., L. C. Evans and R. Jensen (1984). Viscosity solutions of Isaacs' equations and differential games with Lipschitz controls. *J. Different. Equat.*, **53**(2), 213–233.

Bensoussan, A. (1988). *Perturbation Methods in Optimal Control Problems*, Wiley–Gautier, New York.

Clarke, F. H., Yu. S. Ledyaev, R. J. Stern and P. R. Wolenski (1998). *Nonsmooth Analysis and Control Theory*, Springer–Verlag, New York.

Crandall M. G. and P. L. Lions (1983). Viscosity solutions of Hamilton-Jacobi equations. *Trans. Amer. Math. Soc.*, **277**, 1–42.

Crandall M. G., L. C. Evans and P. L. Lions (1984). Some properties of viscosity solutions of Hamilton-Jacobi equations. *Trans. Amer. Math. Soc.*, **282**, 487–502.

Filippov, A. F. (1985). *Differential Equations with Discontinuous Right-hand Side*, Nauka, Moscow [in Russian].

Gaitsgory, V. G. (1996). Limit Hamilton–Jacobi–Isaacs Equations for Singularly Perturbed Zero–Sum Differential Games. *J. Math. Anal. Appl.*, **202**, 862-899.

Isaacs, R. (1965). *Differential Games*. Wiley, New York.

Krasovskii, N.N. and A.I. Subbotin (1974). *Positional Differential Games*. Nauka, Moscow [in Russian].

Krasovskii, N.N. and A.I. Subbotin (1988). *Game-Theoretical Control Problems*. Springer-Verlag, New York.

Kokotovic, P. V. (1984). Applications of singular perturbations techniques to control problems. *SIAM Rev.*, **26**, 501-550.

O'Malley, R. E. (1974). *Introduction to Singular Perturbations*, Academic Press, New York.

Pervozvansky, A. A. and V. G. Gaitsgory (1988). *Theory of Suboptimal Solutions*, Kluwer Academic, Dordrecht.

Subbotin, A. I. (1980). Generalization of the main equation of the theory of differential games. *Soviet. Math. Dokl.*, **22**(2), 358-362.

Subbotin, A. I. (1991) *Minimax inequalities and Hamilton-Jacobi equations*, Nauka, Moscow [in Russian].

Subbotin, A. I. (1995) *Generalized Solutions of First-Order PDEs. The Dynamical Optimization perspective*, Birkhäuser, Boston.

Subbotina, N. N. (1996). Asymptotic properties of minimax solutions of Isaacs–Bellman equations in differential games with fast and slow motions. *J. Appl. Maths Mechs*, **60**(6), 883-890.

Subbotina, N. N. (1998). Asymptotics for singularly perturbed Hamilton–Jacobi equations. *Prikl. Mat. Mech.*, [in Russian], (to appear).

Tikhonov, A. N. (1952). Systems of differential equations containing small parameters near derivatives. *Mat. Sbornik*, **31**(3), 575-586 [in Russian].

Vasil'eva, A. B. and A. F. Butuzov (1973). *Asymptotic Expansions of Solutions to Singularly Perturbed Equations*, Nauka, Moscow [in Russian].

Warga, J. (1976). *Optimal Control of Differential and Functional Equations*, Academic Press, New York.

OPTIMAL CONTROL OF A SYSTEM UNDER DISTURBANCE

V.M. Alexandrov

S.L.Sobolev Institute of Mathematics, Siberian Branch of the Russian Academy of Sciences
630090 Novosibirsk, av. of acad. Koptyug, 4 e-mail: alexegor@math.nsc.ru

Abstract. It has been obtained a system of linear algebraic equations to connect the increments of initial conditions of a normed conjugate system and that of the final control moment with the deviations of phase coordinates of a controlled system. The deviations are caused by various disturbances the object is subjected to. The control of the disturbed system is carried out according to its linear model. It has been proposed a method of sequential correction of the switching moments and the final moment in following up the representative point of the optimal time control. The method is based on the Caushy initial-value problem and the system of linear algebraic equations. The features of motion of the controlled system about the varieties of the switching moments have been considered.

Keywords. Optimal control, disturbance, switching, multivarible systems, differential equations, computational method.

1. INTRODUCTION.

The problem of synthesis of optimal control is of great theoretical and practical interest since only when synthesizing a controlled system is insensitive to: external and parametrical disturbances, both either random or determinate; discrepancy between the object and the used mathematical model; inaccuracy the switching moments to be realized with; restrictions to phase coordinates and so on. The problem of synthesis is highly complicated and up to now there are no solutions to have been found for high-order systems with restricted control.

In the present paper it has been proposed a method of sequential synthesis of time optimal control using the linear model of an object and closeness of the phase tra-

2. THE STATEMENT OF PROBLEM.

Let a controlled system be described by the differential equation

$$\dot{x} = A(t)x + B(t)u + C(t)F, x(t_i) \in D. \quad (1)$$

Here x is an n-dimensional vector of phase state; $A(t)$, $B(t)$ and $C(t)$ are continuously differentiable matrix-functions of dimensions $n \times n$, $n \times m$ and $n \times l$, respectively; u is an m-dimensional control vector with its components being piecewise continuous functions satisfying

$$|u_j| \leq M_j, M_j > 0, j = \overline{1,m}; \quad (2)$$

D is a bounded set of initial conditions belonging to a controllability set $V(t)$; F is an l-dimensional vector of external and parametric disturbances. This is a bounded piecewise unknown (uncontrollable) disturbance so that $|F_\omega| \leq F_\omega^\circ$, $\omega = \overline{1,l}$. In a special case $F = F(x,t)$. The linear model of system (1) is supposed to be completely controllable

$$\dot{x}_A = A(t)x_A + B(t)u, x_A(t_i) \in D. \quad (3)$$

Problem. *Given: The optimal control* $u^0(t), t \in [t_i, t_k^{(i)}]$, *moving linear system* (3) *from an initial state* $x_A(t_i)$ *to the origin* $x_A(t_k^{(i)}) = 0$ *in a minimal time* $T^{(i)} = t_k^{(i)} - t_i$. *This control acts simultaneously on system* (1) *being* $\Delta\tilde{x}(t_i)$ *distant from* (3) *at the moment* t_i : $\Delta\tilde{x}(t_i) = x(t_i) - x_A(t_i)$. *To find: At the moment* $t_{i+1} = t_i + \Delta t$ *the optimal time control* $u^\circ(t)$, $t \in [t_i, t_k^{(i)} + \Delta t_k^{(i)}]$ *moving system* (1) *under* $F \equiv 0$ *from point* $x(t_i)$ *to the origin* $x(t_k^{(i)} + \Delta t_k^{(i)}) = 0$.

The method considered, in essence, is this. Given is an optimal control moving linear system (3) from some support (reference) point $x_A(t_i)$ to the origin. The very same control is fed into the system under disturbance (1) which is $\Delta\tilde{x}(t_i)$ distant from the linear system. After computing the optimal control, in time $\Delta t = t_{i+1} - t_i$ corresponding to linear system (3) with new turbed system (1) will get over to the point $x(t_{i+1})$. If we take $F \equiv 0$, $t \in [t_i, t_{i+1}]$ then system (1), being, in this case, under the updated optimal control, gets over to the point $x_A(t_{i+1})$ distant by $\Delta\tilde{x}(t_{i+1})$ from the point $x(t_{i+1})$. The optimal control moving linear system (3) from the point $x_A(t_{i+1})$ to the origin is known, since it has been computed by the sampling instant t_{i+1}. Known as well at the moment t_{i+1} is the deviation $\Delta\tilde{x}(t_{i+1}) = x(t_{i+1}) - x_A(t_{i+1})$ caused by disturbance. Next computing iteration starts in following up the representative point. In this way, on the basis of its linear model (3) sequential correction of the optimal control of the disturbed system (1) takes place.

3. SOLVING TECHNIQUE.

3.1. *Finding Deviations of Phase Coordinates Caused by Disturbance.*

Since F is supposed to be an unknown function (the instance of the most practical and theoretical interest), the deviations are determined by using of the value of phase state $x(t_{i+1})$ measured and that of coordinate point $x_A(t_{i+1})$ computed:

$$\Delta\tilde{x}(t_{i+1}) = x(t_{i+1}) - \Big[\Phi(t_{i+1}, t_i)x(t_i) + \\ + \int_{t_i}^{t_{i+1}} \Phi(t_{i+1}, \tau)B(\tau)u^\circ(\tau)d\tau\Big].$$

In moving system (1) distortionlessly ($u(t) \equiv 0$, $F \equiv 0$), increment $\Delta\tilde{x}(t_i)$ at the left end of the phase trajectory causes at the right one its deviation from the origin at the moment $t_k^{(i)}$: $\Delta\tilde{x}(t_k^{(i)}) = \Phi(t_k^{(i)}, t_i)\Delta\tilde{x}(t_i)$. The deviation needs to be balanced by changing all the switching moments, including that of the final moment $t_k^{(i)}$.

3.2. *Variation of the Control Switching Moments.*

and the final one for piecewise continuous control $u(t)$ components of which are switched at the moments $t = \nu_j^p(t_i)$, $j = \overline{1,m}$; $p = \overline{1,r_j}$ and take values $u_j(t) = \hat{u}_j^p$, $t \in [\nu_j^{p-1}(t_i), \nu_j^p(t_i)]$, cause the following changings of the phase coordinates at the final moment $t_k^{(i)}$:

$$\Delta\hat{x}(t_k^{(i)}) \cong \sum_{j=1}^{m}\sum_{p=1}^{r_j-1} \Phi(t_k^{(i)}, \nu_j^p(t_i))B_j(\nu_j^p(t_i))\times$$

$$\times[\hat{u}_j^p - \hat{u}_j^{p+1}]\Delta\nu_j^p + \sum_{j=1}^{m} \Phi(t_k^{(i)}, t_k^{(i)})B_j(t_k^{(i)})\times$$

$$\times \hat{u}_j^{r_j}\Delta t_k^{(i)}. \qquad (4)$$

Optimal control moving a linear system from a state $x(t_i)$ to the origin $x(t_k^{(i)}) = 0$ is formed on the basis of principle of the maximum according to the algorithm $u_j^\circ(t) = M_j \operatorname{sign}[B_j(t)]^*\psi(t), j = \overline{1,m}$, where $\psi(t)$ is a solution of the conjugate system $\dot{\psi} = -A^*(t)\psi$. Let us introduce the designation $S_j(p) = \operatorname{sign}[B_j(t)]^*\psi(t)$, $t \in [\nu_j^{p-1}(t_i), \nu_j^p(t_i)]$. The control parameters take values $\hat{u}_j^p = M_jS_j(p)$, $\hat{u}_j^{p+1} = M_jS_j(p+1)$. Since $S_j(p+ +1) = -S_j(p)$, deviations of phase coordinates (4) are written

$$\Delta\hat{x}(t_k^{(i)}) \cong 2\sum_{j=1}^{m}\sum_{p=1}^{r_j-1} \Phi(t_k^{(i)}, \nu_j^p(t_i))B_j(\nu_j^p(t_i))\times$$

$$\times M_jS_j(p)\Delta\nu_j^p(t_i) + \sum_{j=1}^{m} \Phi(t_k^{(i)}, t_k^{(i)})B_j(t_k^{(i)})\times$$

$$\times M_jS_j(r_j)\Delta t_k^{(i)}. \qquad (5)$$

3.3. *Computation Error of Optimal Control.*

Under disturbance F the phase trajectory of motion of system (1) may considerably differ from that of its linear model (3). Computation of increments of switching moments, including that of the final control moment, based on closeness of the phase trajectories, becomes less and less exact and, altogether, the breakdown point of the method is likely to occur. To avoid this substantial disadvantage and computation we should abandon the conventional concept of the fixed reference trajectory and turn to a "sliding" reference trajectory. For this purpose at every step the phase trajectories of linear system (3) start from points $x(t_i)$, $i = 1, 2, 3, \ldots$, belonging to the phase trajectory of system (1) and being placed at Δt intervals. The error of moving the linear system to the origin $\Delta\overline{x}(t_k^{(i)})$ can be presented by means of a corresponding initial condition $\Delta\overline{x}(t_i)$ for homogeneous system $\dot{x} = A(t)x$ in the following way:

$$\Delta\bar{x}(t_k^{(i)}) = \Phi(t_k^{(i)}, t_i)\Delta\bar{x}(t_i). \qquad (6)$$

3.4. *Key Equation of Balance of the Deviations.*

Thus, deviation $\Delta\tilde{x}(t_k^{(i)})$, caused by action of the disturbance F on system (1) and deviation $\Delta\bar{x}(t_k^{(i)})$ (6), due to the control failing to be computed accurately must be balanced by deviation of phase coordinates $\Delta\hat{x}(t_k^{(i)})$ (5) produced by changing the switching moments and the final moment of optimal control. The key equation of balance of the deviations is required to be fulfilled

$$\Delta\tilde{x}(t_k^{(i)}) + \Delta\bar{x}(t_k^{(i)}) + \Delta\hat{x}(t_k^{(i)}) = 0. \qquad (7)$$

Substitute Exp. $\Delta\tilde{x}(t_k^{(i)})$, (5), (6) in (7). Taking into account that the matrix of fundamental solutions $\Phi(t_k^{(i)}, t_i)$ is nonsingular, we obtain the system of n linear algebraic equations relating increments of the switching moments to deviations of the phase coordinates at the moments t_i:

$$\Delta\tilde{x}(t_i) + \Delta\bar{x}(t_i) + 2\sum_{j=1}^{m}\sum_{p=1}^{r_j-1} \Phi^{-1}(\nu_j^p(t_i), t_i)\times$$

$$\times B_j(\nu_j^p(t_i))M_jS_j(p)\Delta\nu_j^p(t_i) + \sum_{j=1}^{m} \Phi^{-1}(t_k^{(i)},$$

$$t_i)B_j(t_k^{(i)})M_jS_j(r_j)\Delta t_k^{(i)} = 0. \qquad (8)$$

To find optimal control (rather than simply an admissible one) it is necessary to

ative position of the switching moments using the conjugate system. With this in mind, let us determine relation between $\Delta \nu_j^p(t_i)$ and increments of initial conditions of the normed conjugate system $\Delta \hat{\psi}(t_i)$.

3.5. Relation between Increments of Switching Moments and Increments of Initial Conditions of Normed Conjugate System.

A solution of the conjugate system may be written as $\psi(t) = \left[\Phi^{-1}(t, t_i) \right]^* \psi(t_i)$. The switching moments of the components of the optimal control and their number on the interval $t \in [t_i, t_k^{(i)}]$ depend uniquely on the switching functions $[B_j(t)]^* \psi(t)$, $j = \overline{1, m}$, and can be determined provided that $\psi(t)$ is known, i.e. the initial conditions of the conjugate system $\psi_\xi(t_i)$, $\xi = \overline{1, n}$ are known.

The problem has been specified by giving an optimal control at the initial moment t_i, i.e. given are $\psi_\xi(t_i)$, $\xi = \overline{1, n}$. Since Hamiltonian function $H(\psi(t), x(t), u(t), t) = (\psi(t), A(t)x) + (\psi(t), B(t)u)$ is homogeneous relative to $|\psi(t)|$ we suppose $|\psi_\alpha(t_i)| = 1$, $\alpha \in [1, n]$. Here $\psi_\alpha(t_i)$ is an initial condition of α-th phase coordinate being nonzero at the moment t_i where α is any number out of set $[1, n]$, characterized by $\psi_\alpha(t_i) \neq 0$. Let us introduce a designation $\hat{\psi}(t_i) = \psi(t_i)/\psi_\alpha(t_i)$ and substitute it into the switching functions

$$[B_j(\nu_j^p(t_i))]^* [\Phi^{-1}(\nu_j^p(t_i), t_i)]^* \hat{\psi}(t_i) = 0, \quad (9)$$
$$j = \overline{1, m}, \ p = \overline{1, r_j - 1},$$

Now we have the system of $(n - 1)$ linear algebraic equations with $(n-1)$ unknowns (since $\hat{\psi}_\alpha(t_i) = 1$). Let $\hat{\psi}(t_i)$ be changed by $\Delta \hat{\psi}(t_i)$. This results in a change of $\Delta \nu_j^p(t_i)$ in $\nu_j^p(t_i)$:

$$[B_j(\nu_j^p(t_i) + \Delta \nu_j^p(t_i))]^* [\Phi^{-1}(\nu_j^p(t_i) + \Delta \nu_j^p(t_i),$$
$$t_i)]^* (\hat{\psi}(t_i) + \Delta \hat{\psi}(t_i)) = 0, \quad j = \overline{1, m};$$
$$p = \overline{1, r_j - 1}.$$

lor series and confine ourselves only to the linear terms. Eventually, we get the following expression (Alexandrov, V.M., 1988)

$$\Delta \nu_j^p(t_i) \cong \left\{ \left\{ [B_j(\nu_j^p(t_i))]^* A^*(\nu_j^p(t_i)) - \right. \right.$$
$$- [\dot{B}_j(\nu_j^p(t_i))]^* \right\} [\Phi^{-1}(\nu_j^p(t_i), t_i)]^* \hat{\psi}(t_i) \right\}^{-1} \times$$
$$\times [B_j(\nu_j^p(t_i))]^* [\Phi^{-1}(\nu_j^p(t_i), t_i)]^* \Delta \hat{\psi}(t_i),$$
$$j = \overline{1, m}; \ p = \overline{1, r_j - 1}. \quad (10)$$

3.6. Relation between Coordinate Increments of Direct and Conjugate Systems.

After substituting expression (10) into system (8) we have the system of n linear algebraic equations with n unknowns

$$\Delta \tilde{x}(t_i) + \Delta \bar{x}(t_i) + 2 \sum_{j=1}^{m} \sum_{p=1}^{r_j-1} \Phi^{-1}(\nu_j^p(t_i), t_i) \times$$
$$\times B_j(\nu_j^p(t_i)) M_j S_j(p) \left\{ \left\{ [B_j(\nu_j^p(t_i))]^* A^*(\nu_j^p \right. \right.$$
$$(t_i)) - [\dot{B}_j(\nu_j^p(t_i))]^* \right\} [\Phi^{-1}(\nu_j^p(t_i), t_i)]^* \hat{\psi}(t_i) \right\}^{-1} \times$$
$$\times [B_j(\nu_j^p(t_i))]^* [\Phi^{-1}(\nu_j^p(t_i), t_i)]^* \Delta \hat{\psi}(t_i) +$$
$$+ \sum_{j=1}^{m} \Phi^{-1}(t_k^{(i)}, t_i) B_j(t_k^{(i)}) M_j S_j(r_j) \times$$
$$\times \Delta t_k^{(i)} = 0. \quad (11)$$

Unknowns of (11) are $(n - 1)$ increments $\Delta \hat{\psi}_\xi(t_i)$ and deviation $\Delta t_k^{(i)}$. On solving (11) we find $\Delta \hat{\psi}(t_i)$ and $\Delta t_k^{(i)}$ and, hence, the initial condition of the conjugate system $\hat{\psi}_n(t_i) = \hat{\psi}(t_i) + \Delta \hat{\psi}(t_i)$ corresponding to linear system (3) with new initial condition $x(t_i)$. The optimal control

$$u_j^o(t) = M_j \text{sign} [B_j(t)]^* [\Phi^{-1}(t, t_i)]^* (\psi(t_i) +$$
$$+ \Delta \psi(t_i)), \ j = \overline{1, m}, \quad (12)$$

moves (3) for time $T^{(i)} = t_k^{(i)} + \Delta t_k^{(i)} - t_i$ from point $x_\ast(t_i) = x(t_i)$ to the origin $x_\ast(t_k^{(i)} + \Delta t_k^{(i)}) = 0$. From formula (10) we compute $\Delta \nu_j^p(t_i)$, $j = \overline{1, m}; p = \overline{1, r_j - 1}$, which enable us to judge directly the extent every of the switching moments should be changed to, so that the

mal value of the increment $\max_{j,p}[|\Delta\nu_j^p(t_i)|, |\Delta t_k^{(i)}|]$ allows an evaluation to be made concerning the closeness of phase trajectories of systems (1) and (3) and the convergence of the computing procedure.

In time $\Delta t = t_{i+1} - t_i$ system (3) under control (12) will move from the initial point $x(t_i)$ to the final point

$$x_*(t_{i+1}) = \Phi(t_{i+1}, t_i)x(t_i) + \sum_{j=1}^{m} \int_{t_i}^{t_{i+1}} \Phi(t_{i+1}, \tau)B_j(\tau)u_j^\circ(\tau)d\tau.$$

Next we compute the value of the conjugate system at the instant t_{i+1} $\psi(t_{i+1}) = [\Phi^{-1}(t_{i+1}, t_i)]^*(\widehat{\psi}(t_i) + \Delta\widehat{\psi}(t_i))$ and carry out normalization $\widehat{\psi}(t_{i+1}) = \frac{\psi(t_{i+1})}{\psi_\alpha(t_{i+1})}$. The normalization of the conjugate system is carried out at every move $i = 1, 2, 3, \ldots$.

In the time required for computing optimal control (12) system (1) under the optimal control

$$u_j^\circ(t) = M_j \mathrm{sign}[B_j(t)]^*[\Phi^{-1}(t, t_i)]^*\psi(t_i),$$

$$t \in [t_i, t_{i+1}],\ j = \overline{1, m}$$

and an unknown disturbance F will move to some point $x(t_{i+1})$, coordinates of which are measured at the moment t_{i+1}. At the moment t_{i+1} the deviation $\Delta\tilde{x}(t_{i+1}) = x(t_{i+1}) - x_*(t_{i+1})$ is determined, the updated optimal control $u_j^0 = M_j\mathrm{sign}[B_j(t)]^*[\Phi^{-1}(t, t_{i+1})]^*\psi(t_{i+1})$, $j = \overline{1, m}$ is fed into system (1), and the next step of computations begins again. In this way, in following up the representative point of the optimal time control, sequential correction of the switching moments and the final control moment take place.

3.7. *The Features of Motion of the Controlled System about the Switching Varieties.*

If number of the switching moments over all the components of the control vector exceeds $(n - 1)$ then those elapsed are rest moments is equal to $(n - 1)$ then, on reaching an instant t_η of current switching moment $\nu_\beta^{\theta_\beta}(t_\eta)$, the further motion occurs in the neighbourhood of $(n - 1)$-dimensional switching variety $G^{(n-1)}$. In Eqs. (10), (11) we take $\nu_\beta^{\theta_\beta}(t_\eta) = t_\eta$ and solve (11). There are three possible cases: 1) $\Delta\nu_\beta^{\theta_\beta}(t_\eta) > 0$; 2) $\Delta\nu_\beta^{\theta_\beta}(t_\eta) = 0$; 3) $\Delta\nu_\beta^{\theta_\beta}(t_\eta) < 0$. The condition $\Delta\nu_\beta^{\theta_\beta}(t_\eta) > 0$ means that although the representative point of system (1) under F has deviated into the previous Y_1-subspace of X-space but under the action of the optimal control it is on the point of belonging to the variety $G^{(n-1)}$ again. Sliding mode begins. If $\Delta\nu_\beta^{\theta_\beta}(t_\eta) = 0$ then the representative point belongs to the variety $G^{(n-1)}$. If $\Delta\nu_\beta^{\theta_\beta}(t_\eta) < 0$ then the phase trajectory has gone through the variety $G^{(n-1)}$ and belongs to the Y_2-subspace. The switching moment $\nu_\beta^{\theta_\beta}(t_\eta)$ must be displaced an amount $\Delta\nu_\beta^{\theta_\beta}(t_\eta)$ left from the current moment t_η, which is actually unrealizable. In this case, the optimal control provides, for corresponding λ-th component (the switching serial number of which, in arranging switching moments in time, appears to be $(n - m)$), an additional short interval of sign constancy of control on the right end of the trajectory. For λ-th component, we set an additional switching at the instant $(t_k^{(\eta)} + \Delta t_k^{(\eta)}) = \nu_\lambda^{q_\lambda}(t_\eta)$. Simultaneously, we introduce a new final moment $(t_k^{(\eta)} + \Delta t_k^{(\eta)} + \Delta\tau)$ where $\Delta\tau = \mu T^{(\eta)}$, $0 < \mu \ll 1$. Here each component has an approximate initial value which is corrected in the process of computing. The moment $\nu_\beta^{\theta_\beta}(t_\eta)$ is canceled from the equations.

Thus, the sign of the deviation $\Delta\nu_\beta^{\theta_\beta}(t_\eta)$ indicates belonging of the representative point to one or another subspace while the magnitude $|\Delta\nu_\beta^{\theta_\beta}(t_\eta)|$ determines the value of the deviation from the variety $G^{(n-1)}$. Moving the representative point is retained in the neighbourhood of the variety $G^{(n-1)}$. On approaching the repre-

switching moments together and contracting the control time take place.

REFERENCES

Alexandrov, V.M. (1988). *Solution of optimal control problems on the basis of the method of quasioptimal control.* (in Russian) Proceedings of the Institute of Mathematics of SB USSR, vol. 10. Models and Methods of Optimization. Novosibirsk: Nauka. pp. 18–54.

ON QUASILINEAR POSITIONAL GAME PROBLEMS [1]

Al'brekht E.G. *

* *Institute of Mathematics and Mechanics, 16 Kovalevskaya St., 620219 Ekaterinburg, Russia, E-mail: ernst.albrekht@usu.ru*

Abstract: The aim of this paper is to present a way for constructing approximations of optimal positional strategies of two players counteracting in antagonistic games with a convex payoff function satisfying the Lipschitz condition and determining on terminal states of the controlled process. The behaviour of a controlled system is described by quasilinear differential inclusions which right-hand side is close to ellipsoids. The method of dynamic programming is substantiated as a useful tool for calculating the support functions of the attainability sets. A method for determining the value of the game and two types of optimal positional strategies of extremal aiming and extremal to maximal stable bridges with a prescribed accuracy in small parameters is obtained in the regular case. Sufficient conditions when the value and main elements of extremal constructions are analytic in small parameters are presented. *Copyright © 1998 IFAC*

Keywords: game theory, optimal control, nonlinear systems, feedback control, approximate analysis

1. INTRODUCTION

The paper deals with control problems by dynamic processes which behaviour is described by quasilinear differential inclusions under the presence of uncertain factors. These factors may be either unknown disturbances or controlling actions by the opposite side. The investigation is led within the framework of the conception of the positional differential game developed by (Krasovskii and Subbotin, 1974; Krasovskii and Subbotin, 1987; Subbotin and Chentsov, 1981; Krasovskii, 1985) and is a continuation of the work (Al'brekht, 1986), where a general approach to such problems has been presented for nonlinear processes described by differential inclusions. The aim of this work is to substantiate the use of the perturbation technique for calculating optimal strategies of players in antagonistic games with a prescribed accuracy in small parameters. The main attention is devoted to the construction of calculation algorithms which require a relatively small number of computations and may be realized on a computer with the aid of modern analytical manipulations combined with numerical calculations.

2. STATEMENT OF THE PROBLEM

Consider, on a prescribed time-interval $T = [t_0, \vartheta]$, two counteracting controlled objects with dynamics

$$\dot{y} = u \in P(t, y, \lambda), \quad \dot{z} = v \in Q(t, z, \lambda), \qquad (1)$$
$$P(t, y, \lambda) = \{u \in R_n : u'P_1(t)u + \lambda f_1(t, y, u, \lambda) \le \mu^2\},$$
$$Q(t, z, \lambda) = \{v \in R_n : v'Q_1(t)v + \lambda f_2(t, z, v, \lambda) \le \nu^2\},$$

where $y, z \in R^n$ are phase vectors, $u, v \in R^n$ are control actions, $P(t, y, \lambda)$ and $Q(t, z, \lambda)$ are control domains, λ is a small parameter. Vectors in the

[1] Supported by the Russian Fond of Fundamental Researches, project no. 97-01-00371.

space R^n are interpreted as column vectors, the prime denotes transpose.

Suppose that the payoff functional is of the form

$$J[u,v] = \sigma(y(\vartheta) - z(\vartheta)) = \sigma(x(\vartheta)), \qquad (2)$$

where $\sigma(x)$ is a given function of the vector $x = y - z$.

Assume that the first player strives to minimize the payoff while, on the contrary, the second player strives to maximize it. In accordance with the general definition of a positional differential game given in (Krasovskii and Subbotin, 1974; Krasovskii and Subbotin, 1987; Subbotin and Chentsov, 1981; Krasovskii, 1985) it is required to construct optimal positional strategies of the first player U^0 and the second player V^0 solving the minimax and the maximin problems. In these inverse problems optimal strategies U^0 and V^0 are defined for a fixed initial position $\{t_0, y_0, z_0\}$.

It is assumed throughout that:
(a). $P_1(t)$ and $Q_1(t)$ are continuous in t positively defined matrices.
(b). Functions $f_1(t,y,u,\lambda)$ and $f_2(t,z,v,\lambda)$ are continuous in t and analytic in all other variables.
(c). Function $\sigma(x) \geq 0$ is convex and satisfies the Lipschitz condition in R^n.

A way for approximating solutions of the minimax and the maximin problems by converging power series in the small parameter λ will be substantiated in the next two parts. First an approach for obtaining an analytical description of attainability sets of system (1) in terms of its support functions will be given. In the case investigated the most useful tool for calculating support functions is the method of dynamic programming. Further on, the programmed maximin (Krasovskii and Subbotin, 1974; Krasovskii and Subbotin, 1987) is obtained and sufficient regularity conditions for the programmed maximin ensuring its analytical dependence on parameters are considered.

3. ATTAINABILITY SET

Consider an auxiliary controlled system with the dynamic

$$\dot{x} = w \in R(t,x,\lambda), \qquad (3)$$
$$R(t,x,\lambda) = \{w \in R_n : w'R_1(t)w + \lambda f(t,x,w,\lambda) \leq \zeta^2\}.$$

Let the matrix $R_1(t)$ be continuous in t and positively defined. Differential inclusion (3) has a set $X[t_0, x_0, \lambda]$ of solutions $x(t; t_0, x_0, \lambda)$ starting from initial position $\{t_0, x_0\}$ and determined on the time-interval T under sufficiently small values of the parameter λ. The attainability set $G(t_0, x_0, \lambda)$ of system (3) from the initial position to the time ϑ consists of all the points $x(\vartheta; t_0, x_0, \lambda)$.

The attainability set $G(t,x,\lambda)$ from an arbitrary initial position $\{t,x\}$ to the time ϑ is convex, closed and bounded, its support function $\rho[l,t,x,\lambda]$ is a solution (Al'brekht, 1986) of the following Cauchy problem

$$\frac{\partial \rho}{\partial t} + \max_{w \in R(t,x,\lambda)} w' \frac{\partial \rho}{\partial x} = 0, \qquad (4)$$

$$\rho[l,\vartheta,x,\lambda] = l'x, \quad l \in R_n.$$

Let l be an arbitrary vector from the unit sphere $S = \{l \in R^n : l'l = 1\}$. At $\lambda = 0$ the equation (4) has the solution

$$\rho^{(0)}[l,t,x] = l'x + \zeta \int_t^{\vartheta} (l'R_1^{-1}(\tau)l)^{1/2} d\tau.$$

Matrix $R_1(t)$ being positively defined, therefore the maximum in (4) is attained on the unique vector

$$w^0(\psi,t,x,\lambda) = \zeta R_1^{-1}(t)\psi(\psi' R_1^{-1}(t)\psi)^{-1/2}$$
$$+\lambda w_1^0(\psi,t,x,\lambda), \psi = \partial\rho[l,t,x,\lambda]/\partial x,$$

where $w_1^0(\psi,t,x,\lambda)$ is continuous in t and analytic in ψ, t, x, λ under sufficiently small values of the parameter λ and values of ψ, x from every bounded domain. The solution of equation (4) under sufficiently small values of λ may be defined in the form of

$$\rho[l,t,x,\lambda] = \rho^{(0)}[l,t,x] + \lambda\kappa[l,t,x,\lambda].$$

The function $\kappa[l,t,x,\lambda]$ should be evaluated from the equation

$$\lambda\frac{\partial \kappa}{\partial t} - \zeta(l'R_1^{-1}(t)l)^{1/2}$$
$$+\zeta\left[\left(l + \lambda\frac{\partial \kappa}{\partial x}\right)' R_1^{-1}(t)\left(l + \lambda\frac{\partial \kappa}{\partial t}\right)\right]^{1/2}$$
$$+\lambda\left(l + \lambda\frac{\partial \kappa}{\partial x}\right)' w_1^0(l + \lambda\frac{\partial \kappa}{\partial x}, t, x, \lambda) = 0,$$

under the condition $\kappa[l,\vartheta,x,\lambda] = 0$.

Because $l'R_1^{-1}(t)l \neq 0$ for all $l \in S$ and $t \in T$ for an arbitrary bounded domain $\Gamma \subset R^n$ there exists a sufficiently small value of $\lambda^0(\Gamma) > 0$ such that the function $\kappa[l,t,x,\lambda]$ is analytic in $l \in S, x \in \Gamma, \lambda$ under sufficiently small values of $\lambda \in \Lambda = \{\lambda : |\lambda| \leq \lambda^0(\Gamma)\}$. Thus, the support function may be obtained on $S \times T \times \Gamma \times \Lambda$ in the form of convergence series

$$\rho[l,t,x,\lambda] = \rho_1^{(0)}[l,t,x] + \sum_{k=1}^{\infty} \lambda^k \kappa^{(k)}[l,t,x].$$

Theorem 1. Under conditions (a), (b) domains of attainability $G_1(t,y,\lambda)$ and $G_2(t,z,\lambda)$ of the objects (1) are convex, closed and bounded, their support functions $\rho_1[l,t,y,\lambda]$ and $\rho_2[l,t,z,\lambda]$ are continuous in $t \in T$ and analytic in all other variables on the set $S \times T \times \Gamma \times \Lambda$.

4. OPTIMAL STRATEGIES

Let $\Gamma^* \subset R^n$ be an arbitrary bounded domain to which belong the realizations $y[t,\lambda], z[t,\lambda]$ of motions of system (1) generated by strategies used by the players in the process of counteracting on the time-interval T, i.e.

$$y[t,\lambda] \in \Gamma^*, z[t,\lambda] \in \Gamma^*, t \in T.$$

Let $\{t,y,z\} \in T \times \Gamma^* \times \Gamma^*$ be the position realized at a time-instant $t \in T$. Here $y = y[t,\lambda], z = z[t,\lambda]$. Consider the programmed maximin (Krasovskii and Subbotin, 1987, p.191)

$$\varepsilon^0(t,y,z,\lambda) = \max_{z(\vartheta,\lambda)\in G_2} \min_{y(\vartheta,\lambda)\in G_1}$$

$$\max_{l\in L}\{l'z(\vartheta,\lambda) - l'y(\vartheta,\lambda) - \omega(l)\},$$

$$\omega(l) = \sup_{x\in R^n}\{l'x - \sigma(x)\},$$

$$L = dom\ \omega(l) = \{l \in R^n : \omega(l) < \infty\}.$$

With the aid of the general theorem on minimax and theorem 1 we may write

$$\varepsilon^0(t,y,z,\lambda) = \max_{l\in L}\{\rho_2[l,t,z,\lambda] - \rho_1[l,t,y,\lambda] - \omega(l)\}. \tag{5}$$

Suppose the following condition has taken place: (d).The function $\sigma(x)$ is such that there exists a sufficiently small value $\lambda^*(\Gamma^*)$ and there is an unique vector $l^0 = l^0(t,y,z,\lambda)$ which delivers maximum to the right-hand side of (5) in every position $\{t,y,z\}$ where $t < \vartheta, \varepsilon^0 > 0$ for $\lambda \in \Lambda^* = \{\lambda : |\lambda| \le \lambda^*(\Gamma^*)\}$. The vector - function $l^0(t,y,z,\lambda)$ is continuous in $t \in T$ and analytic in all other variables on the domain $\Gamma^* \times \Gamma^* \times \Lambda^*$.

In order to present examples when the condition (d) is fulfilled, let's consider games of approach and evasion when the payoff functional is the euclidean norm of the vector $x(\vartheta)$, i.e. $\sigma(x(\vartheta)) = ||x(\vartheta)||$. In this particular case of the differential game (1), (2) the programmed maximin (5) is determined by the equality

$$\varepsilon^0(t,y,z,\lambda) = \max_{l\in S}\{\rho_2[l,t,z,\lambda] - \rho_1[l,t,y,\lambda]\} = \max_{l\in S}\rho^*[l,t,y,z,\lambda]. \tag{6}$$

Introduce the condition
(e).The function $\rho^*[l,t,y,z,0]$ is strictly concave in $l \in S$ for every position $\{t,y,z\} \in T \times \Gamma^* \times \Gamma^*$.

Note that condition (e) will be fulfilled if $P_1(t) = Q_1(t)$ and $\mu > \nu$.

Theorem 2. Let conditions (a), (b), (e) be fulfilled and let $\sigma(x) = ||x||$. Then there exists a sufficiently small value $\lambda^*(\Gamma^*)$ such that the maximum in (6) is attained on an unique vector $l^0 = l^0(t,y,z,\lambda)$ on every position $\{t,y,z\} \in T \times \Gamma^* \times \Gamma^*$ where $t < \vartheta, \varepsilon^0 > 0$. The vector - function $l^0(t,y,z,\lambda)$ is continuous in $t \in T$ and analytic in all other variables and may be determined in the form of convergence series

$$l^0(t,y,z,\lambda) = l^{(0)}(t,y,z) + \sum_{k=1}^{\infty}\lambda^k l^{(k)}(t,y,z).$$

Here the vector $l^{(0)}(t,y,z)$ is a solution of extremal problem (6) at $\lambda = 0$.

Let condition (e) be fulfilled. Then the function $\rho^*[l,t,y,z,\lambda]$ is strictly concave in $l \in R^n$ for every position $\{t,y,z\} \in R^n$ at sufficiently small values of λ. Therefore, the maximum in (6) is actually attained (Krasovskii and Subbotin, 1987) on an unique vector $l^0 = l^0(t,y,z,\lambda)$. An analytic dependence of the vector $l^0(t,y,z,\lambda)$ on variables y,z,λ at sufficiently small values of the parameter λ follows from results of (Al'brekht and Loginov, 1977).

Definition 1. Strategies U_e and V_e are said to be (Krasovskii and Subbotin, 1987) of extremal aiming if on an every position $\{t,y,z\}$ where $\varepsilon^0(t,y,z,\lambda) > 0, t < \vartheta$ at sufficiently small values of λ they are defined by sets $U_e^*(t,y,z,\lambda)$, $V_e^*(t,y,z,\lambda)$ consisting of all vectors u_e and v_e which yield the maximum conditions

$$u_e'\partial\rho_1[l^0(t,y,z,\lambda),t,y,\lambda]/\partial y = \max_{u\in P(t,y,\lambda)} u'\partial\rho_1[l^0(t,y,z,\lambda),t,y,\lambda]/\partial y,$$

$$v_e'\partial\rho_2[l^0(t,y,z,\lambda),t,z,\lambda]/\partial z = \max_{v\in Q(t,z,\lambda)} v'\partial\rho_2[l^0(t,y,z,\lambda),t,z,\lambda]/\partial z.$$

In other cases the sets U_e^*, V_e^* can be selected arbitrarily.

Theorem 3. Under conditions (a), (b), (c), (d) the pair of strategies of extremal aiming $\{U_e, V_e\}$ delivers a saddle point to game (1), (2) . The value $c(t,y,z,\lambda)$ of the game coincides with the programmed maximin $\varepsilon^0(t,y,z,\lambda)$,(5). The function $c(t,y,z,\lambda) = \varepsilon^0(t,y,z,\lambda)$ is continuous in $t \in T$ and analytic in $\{y,z,\lambda\} \in \Gamma^* \times \Gamma^* \times \Lambda^*$ on domain $t < \vartheta, \varepsilon^0 > 0$ and satisfies the Isaacs-Bellman equation under condition $c(\vartheta,y,z,\lambda) = \sigma(y-z)$.

Consider arbitrary motions $\{y(t,\lambda), z(t,\lambda)\}$ of system (1) generated by some admissible strategies $\{U,V\}$ of the players and an absolutely continuous function $\varepsilon^0[t,\lambda] = \varepsilon^0(t,y(t,\lambda),z(t,\lambda),\lambda)$.

Under conditions of the theorem the right-hand side of (5) is continuously differentiable in $\{t,y,z\}$ and, moreover, while calculating the derivatives the dependence of extremal vector $l^0(t,y,z,\lambda)$ on the position $\{t,y,z\}$ should be ignored. Thus, it is not difficult to obtain the following equality

$$\frac{d\varepsilon^0[t,\lambda]}{dt} = \frac{\partial\rho_2[l^0[t,\lambda],t,z(t,\lambda),\lambda]}{\partial t} \tag{7}$$

$$+v'(t,\lambda)\frac{\partial\rho_2[l^0[t,\lambda],t,z(t,\lambda),\lambda]}{\partial z}$$
$$-\frac{\partial\rho_1[l^0[t,\lambda],t,y(t,\lambda),\lambda]}{\partial t}$$
$$-u'(t,\lambda)\frac{\partial\rho_1[l^0[t,\lambda],t,y(t,\lambda),\lambda]}{\partial y},$$

where $l^0[t,\lambda] = l^0(t, y(t,\lambda), z(t,\lambda), \lambda)$, $u(t,\lambda) = \dot{y}(t,\lambda)$, $v(t,\lambda) = \dot{z}(t,\lambda)$.

Let $J[t_0, y_0, z_0, \lambda; u, v]$ be a realization of the payoff (2) corresponding to an initial position $\{t_0, y_0, z_0\}$ and some control actions u and v. Suppose that the first player uses the strategy U_e of extremal aiming, while the second player uses an arbitrary strategy V. It then follows from definition 1 and (4), (7) that almost everywhere $d\varepsilon^0[t,\lambda]/dt \leq 0$. Thus, we will have that $J[t_0, y_0, z_0, \lambda; U_e, V] \leq \varepsilon^0(t_0, y_0, z_0, \lambda)$, whatever initial position $\{t_0, y_0, z_0\}$ from a given bounded domain for any strategy V at sufficiently small values of λ, i.e. the strategy U_e of extremal aiming is an optimal strategy U^0 of the first player solving the minimax problem. If the first player uses an arbitrary strategy U, while the second player uses the strategy V_e of extremal aiming, then almost everywhere $d\varepsilon^0[t,\lambda]/dt \geq 0$. Thus, we will have that $J[t_0, y_0, z_0, \lambda; U, V_e] \geq \varepsilon^0(t_0, y_0, z_0, \lambda)$ whatever initial position $\{t_0, y_0, z_0\}$ from a given bounded domain for any strategy U at sufficiently small values of λ, i.e. the strategy V_e of extremal aiming is an optimal strategy V^0 of the second player solving the maximin problem. Therefore,

$$J[t_0, y_0, z_0, \lambda; U_e, V] \leq J[t_0, y_0, z_0, \lambda; U_e, V_e] =$$
$$\varepsilon^0(t_0, y_0, z_0, \lambda) \leq J[t_0, y_0, z_0, \lambda; U, V_e].$$

It is easy to observe from (4), (7) that the value of the game satisfies the Isaacs-Bellman equation.

These results allow to use the perturbation technique for constructing stable bridges (Krasovskii and Subbotin, 1974; Krasovskii and Subbotin, 1987; Subbotin and Chentsov, 1981; Krasovskii, 1985) and optimal positional strategies extremal to stable bridges with any prescribed accuracy in λ. Let us introduce the following sets (Krasovskii and Subbotin, 1974; Subbotin and Chentsov, 1981):

$$W_\lambda^0 = \{\{t,y,z\} : c(t,y,z,\lambda) \leq c(t_0,y_0,z_0,\lambda)\},$$
$$W_0^\lambda = \{\{t,y,z\} : c(t,y,z,\lambda) \geq c(t_0,y_0,z_0,\lambda)\},$$
$$W_\lambda^0(t,z) = \{w \in R_n : \{t,w,z\} \in W_\lambda^0\},$$
$$W_0^\lambda(t,y) = \{w \in R_n : \{t,y,w\} \in W_0^\lambda\},$$
$$\overline{W}_\lambda^0(t,z) = \{w^0 \in W_\lambda^0(t,z) : \|y - w^0\| = \min_{w \in W_\lambda^0(t,z)} \|y - w\|\},$$
$$\overline{W}_0^\lambda(t,y) = \{w^0 \in W_0^\lambda(t,y) : \|z - w^0\| = \min_{w \in W_0^\lambda(t,y)} \|z - w\|\}.$$

Definition 2. The strategy U^e is called (Subbotin and Chentsov, 1981) extremal to the set W_λ^0 if in every position $\{t,y,z\}$ at sufficiently small values of λ it is determined by sets $U_*^e(t,y,z,\lambda)$ consisting of all vectors u^e which yield the maximum condition

$$(w^0 - y)'u^e = \max_{u \in P(t,y,\lambda)} (w^0 - y)'u$$

where $w^0 \in \overline{W}_\lambda^0(t,z)$ if the set $W_\lambda^0(t,z)$ is not empty and $w^0 = y$ if the set $W_\lambda^0(t,z)$ is empty.

Analogously, the strategy V^e is called extremal to the set W_0^λ if in every position $\{t,y,z\}$ at sufficiently small values of λ it is determined by sets $V_*^e(t,y,z,\lambda)$ consisting of all vectors v^e which yield the maximum condition

$$(w_0 - z)'v^e = \max_{v \in Q(t,z,\lambda)} (w_0 - z)'v$$

where $w_0 \in \overline{W}_0^\lambda(t,y)$ if the set $W_0^\lambda(t,y)$ is not empty and $w_0 = z$ if the set $W_0^\lambda(t,y)$ is empty.

From the general results of (Krasovskii and Subbotin, 1974; Krasovskii and Subbotin, 1987; Subbotin and Chentsov, 1981; Krasovskii, 1985) and theorems 1 - 3 follow.

Theorem 4. Under conditions (a), (b), (c), (d) the maximal stable bridges W_λ^0, W_0^λ may be constructed with any prescribed accuracy in the small parameter λ and the pair of strategies $\{U^e, V^e\}$ extremal to these stable bridges delivers a saddle point to game (1), (2) .

In the preceding sections a positional differential game with a fixed termination time and a convex payoff function was investigated within the framework of notions of near-optimal strategies of extremal aiming and strategies extremal to stable bridges. It may be shown that, under suitable conditions, this approach may be applied to various problems where the time of termination ϑ of a game is not fixed by introducing this parameter into the quantities considered above. For example, this situation arises in problems of pursuit and evasion.

5. CONCLUSION

Sufficient conditions when strategies of extremal aiming and strategies extremal to stable bridges may be calculated with a preassigned accuracy in small parameters are presented under the condition that the maximum in the equation for a programmed maximin is attained on an unique vector. Various types of sufficient and necessary, as well as sufficient conditions for the regularity of a programmed maximin ensuring the coincidence of its value with the value of a positional game are well known. Generally speaking, there may be several vectors yielding the maximum in the equation for a programmed maximin. If in a such case it is possible to verify a regularity condition for the programmed maximin then, the approach presented may be used for constructing near-optimal strategies of the players by using the notions of strategies extremal to stable bridges or in some cases (Loginov, 1982; Al'brekht and Samoilova, 1987) by a modification of strategies of extremal aiming.

6. REFERENCES

Al'brekht, E.G. (1986). On extremal strategies in nonlinear differential games. *Appl. Math. Mech.* **50**, 339–345.

Al'brekht, E.G. and M.I. Loginov (1977). On differentiability in a parameter of the value in a linear game of convergence. In: *Control problems with a guaranteed result* (A.I. Subbotin and S.A. Brykalov, Eds.). pp. 15–22. Ural Sci. Center, Inst. Math. Mech.. Ekaterinburg (in Russian).

Al'brekht, E.G. and T.I. Samoilova (1987). On extremal aiming in game problems of guidance when control domains depend on phase coordinates. In: *Liaypunov's function method in analysis of systems dynamics* (V.M. Matrosov and L.U. Anapol'skii, Eds.). pp. 302–308. Nauka, Sibirskoe Otdelenie. Novosibirsk (in Russian).

Krasovskii, N.N. (1985). *Control by dynamic system.* Nauka. Moskow (in Russian).

Krasovskii, N.N. and A.I. Subbotin (1974). *Positional differential games.* Nauka. Moskow (in Russian).

Krasovskii, N.N. and A.I. Subbotin (1987). *Game-theoretical control problems.* Springer-Verlag. New York, Berlin, Heidelberg, London, Paris, Tokyo.

Loginov, M.I. (1982). On one way of extremal control. *Appl. Math. Mech.* **46**, 893–899.

Subbotin, A.I. and A.G. Chentsov (1981). *Guaranteed optimization in control problems.* Nauka. Moskow (in Russian).

ON PROBLEMS OF GUARANTEED FILTERING FOR LINEAR FUNCTIONAL-DIFFERENTIAL SYSTEMS

B.I.Anan'ev *,[1]

* *Institute of Mathematics and Mechanics RAN*
16 Kovalevskaya Str., 620219, GSP-384, Yekaterinburg
Russia
abi@imm.uran.ru

Abstract: The problem of minimax guaranteed filtering is considered for some special classes of functional-differential systems with uncertain input and output disturbances belonging to a ball in a Hilbert space. As distinct from the problem investigated in previous papers , the case here is studied which, in general, cannot be treated as a problem for ordinary differential equations in Hilbert space with an appropriate semigroup approach. In the general case, informational sets in finite-dimensional space are constructed and the conditions for them to be bounded are given. In particular, the problem for differential systems with variable delay is considered in more detail and equations describing the dynamics of informational sets of current states are derived. *Copyright ©1998 IFAC*

Keywords: guaranteed filtering, informational sets, observability, functional-differential systems.

1. INTRODUCTION

This work is concerned with the state estimation for some special classes of functional-differential systems when there are uncertain input and output disturbances belonging to a ball in a Hilbert space. The uncertain quantities are not modelled as random variables or stochastic processes. The disturbances are only known to lie in the given convex compacts in appropriate vector spaces. The initial states of the system, as a rule, are unknown. The estimation problem consists in construction of an informational set containing all posible states of the system and determined by the system dynamics, the bounds on the given ball, and available observations. For linear systems this approach was taken for the first time in works (Krasovski and Kurzhanski, 1966; Kurzhanski, 1970; Kurzhanski, 1977) and independently in (Bertsekas and Rhodes, 1971; Schweppe, 1973). For systems with delays the state estimation problem was studied in (Anan'ev and Pishchulina, 1979) when initial states and disturbances were jointly constrained by an ellipsoid in a Hilbert space. In paper (Vodichev, 1987) the conditions of boundedness for the informational sets are given in terms of the restoration operator continuity and the the spectral observability of stationary systems with no information about initial states. In this work, the initial states for functional-differential systems are assumed to be also unknown and finite-dimensional. In general, the systems studied here, as distinct from previous papers (Anan'ev, 1995; Anan'ev, 1998), cannot be described as ordinary differential equations with infinitesimal generator and strongly continuous semigroup in appropriate Hilbert space. Nevertheless, the defining correlations for the informational sets are obtained as well as conditions for

[1] Supported by the Russian Foundation of Basic Research under Project 98-01-00117

them to be bounded are given. The evolutionary equations are derived for the parameters of the informational sets in some particular case of the differential system with variable delay. Other approaches based on stochastics can be found in papers (Kolmanovskii, 1974; Germani *et al.*, 1988).

2. PRELIMINARIES AND PROBLEM FORMULATION

Let a linear functional-differential system be described by the equation

$$(Fx)(t) = Bv(t), \quad t \geq 0, \tag{1}$$

and the corresponding equation of measurement has a form

$$y(t) = Cx(t) + Dv(t), \tag{2}$$

where $F : AC^n(0,\infty) \to L_2^n(0,\infty)$ is a linear operator; $B \in R^{n\times q}$, $C \in R^{m\times n}$, $D \in R^{m\times q}$, $rank\, D = m$, are constant matrices; $v(\cdot) \in L_2^q(0,\infty)$ is an uncertain disturbance constrained by the inequality

$$||v(\cdot)||_{L_2^q(0,\infty)} \leq \nu. \tag{3}$$

Here AC^n and L_2^n are the spaces of all n-vector absolutely continuous and square integrable functions respectively.

Vector $x(t)$ is the unmeasurable one, and it is necessary to estimate $x(t)$ according to the information $\{y(\tau),\ 0 \leq \tau \leq t\}$. This is a filtering problem. The solution of the problem consists in the construction of some informational sets in the state space R^n. Then the structure and the evaluation of the sets will be examined. To this end, it is supposed that equation (1) may be uniqually solved for any initial state $x_0 \in R^n$ and any function $v(\cdot) \in L_2^q(0,\infty)$ in the following way:

$$x(t) = U(t)x_0 + \int_0^t S(t,\tau)Bv(\tau)d\tau, \tag{4}$$

where $U(t) \in R^{n\times n}, t \geq 0, U(0) = I$, is a resolvent continuous matrix representing a solution $x(t) = U(t)x_0$ to equation (1) in homogeneous cases; $S(t,\tau)$ is a bounded piecewise continuous function defined in the infinite triangle $0 \leq \tau \leq t < \infty$ with the property $S(t,\tau) = 0$, when $\tau > t$.

Let us consider some examples of system (1).

Example 2.1. Operator F in (1) has a form

$$(Fx)(t) = \dot{x} - Ax(h(t)), \tag{5}$$

where variable delay $h(t)$ satisfies the conditions: $0 \leq h(t) \leq t$ and function $h(t)$ is piecewise continuous. Then

$$S(t,\tau) = I + A\overline{\chi}_1(t,\tau) + A^2\overline{\chi}_2(t,\tau) + \ldots,$$

$$\overline{\chi}_k(t,\tau) = \int_\tau^t \chi_k(s,\tau)ds, \quad \chi_{k+1}(s,\tau)$$

$$= \int_\tau^s \chi_1(s,\alpha)\chi_k(\alpha,\tau)d\alpha, \quad k = 1,2,\ldots,$$

$$\chi_1(s,\tau) = \begin{cases} 1, & 0 \leq \tau \leq h(s), \\ 0, & \text{otherwise.} \end{cases} \tag{6}$$

The series here converges uniformly with respect to t,τ on bounded intervals. It is well known that $S(t,\tau) = \exp(A(t-\tau))$, when $h(t) \equiv t$. In another particular case, $S(t,\tau) = I, \tau > 0$ and $S(t,0) = I + At$, when $h(t) \equiv 0$. Note that $0 \leq \overline{\chi}_k(s,\tau) \leq (s-\tau)^k/k!$; $U(t) = S(t,0)$, and the latter matrix, in general, is not a semigroup.

Example 2.2. Let operator F be written as $(Fx)(t) = \dot{x} - A_0x(t) - A_1x(t-h)$, where $x(t) = 0$ when $t < 0$. In this case, the constant variation formula (see (Hale, 1977)) may be used as representation (4). Besides, the infinitesimal generator and corresponding strongly continuous semigroup are easily constructed, (Krasovski and Kurzhanski, 1966).

Example 2.3. A more general case with a Lebesgue-Stieltjes integral when

$$(Fx)(t) = \dot{x} - \int_0^{\sigma(t)} dR(\tau)x(t-\tau),$$

$$0 \leq \sigma(t) \leq t,$$

was considered in book (Myschkis, 1972). However representation (4) for this case is very difficult if not impossible to realize.

In addition to Examples 2.1-2.3 where the retarded type systems are viewed it should be mentioned the systems of neutral type, (Bellman and Cooke, 1963). Following the terminology of book (Myschkis, 1972) the systems in Examples 2.1-2.3 have no "previous history" in contrast to the case of filtering problem in paper (Anan'ev, 1995) where the semigroup approach is used.

So, in what follows representation (4) will be the starting point. Let us present the main object of this investigation.

Definition 2.4. The informational set $X_t = X(t, y(\cdot))$ of states $x(t)$ of system (1) that are consistent with measurement data $y(\cdot) = \{y(\tau),\ 0 \leq \tau \leq t\}$, (2), and with inequality (3) is the collection of all those elements $x(t) \in R^n$ for each of which there exists a function $v(\cdot) \in L_2^q(0,t)$ satisfying (3) and an initial state $x_0 \in R^n$ that generate the above element $x(t)$ and the given signal $y(\cdot)$ in accordance with representation (4).

Along with set X_t it is useful to introduce one more set.

Definition 2.5. The compatible with measurements $\{y(\cdot)\}$ set $\Phi_t = \Phi(t, y(\cdot))$ of pairs $(x_0, v(\cdot))$ is the collection of all those elements which give rise to signal $y(\cdot)$ according to formulas (2)-(4).

The sets X_t and Φ_t are closely linked with each other. Namely, set X_t is the image of Φ_t according to the map

$$x(t) = U(t)x_0 + \mathcal{T}_t v(\cdot),$$
$$\mathcal{T}_t v(\cdot) = \int_0^t S(t,\tau)Bv(\tau)d\tau. \quad (7)$$

3. THE DEFINING CORRELATIONS

Introduce the operator

$$\mathcal{M}v(\tau) = Dv(\tau) + \int_0^\tau CS(\tau,\alpha)Bv(\alpha)d\alpha. \quad (8)$$

It follows from formulas (2), (4) that $y(\tau) = CU(\tau)x_0 + \mathcal{M}v(\tau)$. The general theory of paper (Anan'ev, 1998) gives the structure of set Φ_t.

Lemma 3.1 A pair $(x_0, v(\cdot)) \in \Phi_t$ iff

$$v(\cdot) \in \mathcal{M}_t^* S_t^{-1}(y(\cdot) - CU(\cdot)x_0) + ker\mathcal{M},$$
$$\|v(\cdot)\|_{L_2^q(0,t)} \le \nu, \quad (9)$$

where the operator $S_t = \mathcal{M}_t\mathcal{M}_t^*$.

Simple calculations yield the expression

$$S_t f(\tau) = DD'f(\tau) + \int_0^t K(\tau,\alpha)f(\alpha)d\alpha, \quad (10)$$

where the symbol $'$ means the transposition, $K(\tau,\alpha) = CS(\tau,\alpha)BD' + DB'S'(\alpha,\tau)C' + \int_0^t CS(\tau,\xi)BB'S'(\alpha,\xi)C'd\xi$; $K'(\tau,\alpha) = K(\alpha,\tau)$, and integral symmetric kernel $K(\cdot,\cdot)$ is a piecewise continuous matrix function. Operator S_t has a continuous inverse because $rankD$=m and $\mathcal{M}_t^* f(\tau) = D'f(\tau) + \int_\tau^t B'S'(\alpha,\tau)C'f(\alpha)d\alpha$ is a conjugate operator of Volterra type. For solving the equation $S_t f(\tau) = y(\tau)$ one needs to introduce the resolvent kernel $R(\tau,\alpha,t)$ as a solution to the matrix integral equations

$$R(\tau,\alpha,t) + \int_0^t K(\tau,\theta)(DD')^{-1}R(\theta,\alpha,t)d\theta$$
$$= K(\tau,\alpha) = R(\tau,\alpha,t)$$
$$+ \int_0^t R(\tau,\theta,t)(DD')^{-1}K(\theta,\alpha)d\theta.$$

As it follows from Fredholm theory the solution to this equation being unique has the symmetric property: $R'(\tau,\alpha,t) = R(\alpha,\tau,t)$. With the help of the resolvent kernel one can write

$$f(\tau,t) = S_t^{-1}y(\tau) = (DD')^{-1}(y(\tau)$$
$$- \int_0^t R(\tau,\theta,t)(DD')^{-1}y(\theta)d\theta). \quad (11)$$

Let

$$\Pi_t = id - \mathcal{M}_t^* S_t^{-1}\mathcal{M}_t. \quad (12)$$

Then from Lemma 3.1 and formulas (7) it follows that $x(t) \in X_t$ iff

$$x(t) = U(t)x_0 + \mathcal{T}_t(\mathcal{M}_t^* S_t^{-1}(y(\cdot) - CU(\cdot)x_0)$$
$$+\Pi_t v), \quad \int_0^t ((y(\tau) - CU(\tau)x_0)'S_t^{-1}(y(\tau)$$
$$-CU(\tau)x_0) + \|v(\tau)\|^2)d\tau \le \nu^2. \quad (13)$$

The support function $\rho(l|X_t) = \sup\{l'x : x \in X_t\}$ of the closed convex set X_t may be easily calculated if the following assumption takes place.

Assumption 3.2. The vector $l \in R^n$ is an observable direction, i.e. $U'(t)l \in imV(t)$, where the symbol imA means the image of matrix A and the matrix $V(t) = \int_0^t U'(\tau)C'S_t^{-1}CU(\tau)d\tau$.

In other words, Assumption 3.2 means that the condition $y(\tau) \equiv 0$ implies $l'x(t) = 0$. Consider a new inner product in $L_2^m(0,t)$, namely $[f,g] = \int_0^t f'(\tau)S_t^{-1}g(\tau)d\tau$, and let the operator Π_{st} be the orthogonal projection of $L_2^m(0,t)$ onto finite-dimensional subspace $\{CU(\cdot)x_0\}$ with respect to inner product $[\cdot,\cdot]$. Then

$$\rho(l|X_t) = l'\mathcal{T}_t\mathcal{M}_t^* S_t^{-1}y(\cdot) + \sup_{x_0}\{l'(U(t)$$
$$-\mathcal{T}_t\mathcal{M}_t^* S_t^{-1}CU(\cdot))x_0 + (\nu^2 - \|y^\perp\|_1^2$$
$$-\|CU(\cdot)(\bar{x}(t) - x_0)\|_1^2)^{1/2}(l'P(t)l)^{1/2}\}, \quad (14)$$

where $\|y\|_1 = [y,y]^{1/2}$, $\Pi_{st}y = CU(\cdot)\bar{x}(t)$, $y^\perp = (id - \Pi_{st})y$, $P(t) = \mathcal{T}_t\Pi_t\mathcal{T}_t^*$.

Calculating in (14) and using the elementary equality $\max_z\{(f,z) + (\nu^2 - \|z\|^2)^{1/2}a\} = \nu(\|f\|^2 + a^2)^{1/2}$ give the final assertion.

Theorem 3.3. The support function $\rho(l|X_t) < \infty$ iff l is an observable direction. In the observable case, the following formula is valid:

$$\rho(l|X_t) = l'\hat{x}(t) + (\nu^2 - \kappa(t))^{1/2}$$
$$\times(l'Q(t)l)^{1/2}, \quad (15)$$

where

$$\hat{x}(t) = \mathcal{T}_t\mathcal{M}_t^* S_t^{-1} y(\cdot) + (U(t) - \mathcal{T}_t\mathcal{M}_t^* S_t^{-1} \times CU(\cdot))\bar{x}(t), \qquad \bar{x}(t) = V^+(t) \int_0^t U'(\tau) \times C' S_t^{-1} y(\tau) d\tau; \qquad Q(t) = P(t) + (U(t) - \mathcal{T}_t\mathcal{M}_t^* S_t^{-1} CU(\cdot)) V^+(t) (U(t) - \mathcal{T}_t\mathcal{M}_t^* S_t^{-1} \times CU(\cdot))'; \qquad \kappa(t) = \|y(\cdot)\|_1^2 - (\int_0^t U'(\tau) C' \times S_t^{-1} y(\tau) d\tau)' V^+(t) (\int_0^t U'(\tau) C' S_t^{-1} y(\tau) d\tau) \tag{16}$$

Here symbol V^+ means the pseudoinverse matrix for V.

In order to approximate set X_t in some degenerate cases let us introduce

Definition 3.4. A set $X_t^{\epsilon k}$ is said to be the ϵk - informational one if it is given by formula (13) under the condition

$$\epsilon \|x_0\|^2 + \|y - CU(\cdot)x_0\|_1^2 + \|v(\cdot)\|_{L_2^q(0,t)}^2 \leq \nu^2 + k.$$

We have the formulas

$$\cup_{\epsilon > 0} X_t^{\epsilon k} = X_t^{0k} \supset X_t, \qquad \rho(l | X_t^{\epsilon k}) = l' \hat{x}^\epsilon(t) + (\nu^2 + k - \kappa^\epsilon(t))^{1/2} (l' Q^\epsilon(t) l)^{1/2},$$

where parameters $\hat{x}^\epsilon(t)$, $\kappa^\epsilon(t)$, and $Q^\epsilon(t)$ have the same structure as in (16), except that matrix $V^+(t)$ is replaced by $(V(t) + \epsilon I)^{-1}$.

Suppose that system (1), (2) is continuously observable, i.e

$$im U'(t) \subseteq im V(t) = im \int_0^t U'(\tau) C' C U(\tau) d\tau. \tag{17}$$

Then, as it follows from (15), set X_t is bounded.

Lemma 3.5. Under condition (17) the following convergences take place: $\hat{x}^\epsilon(t) \to \hat{x}(t)$, $\kappa^\epsilon(t) \to \kappa(t)$, $Q^\epsilon(t) \to Q(t)$, $X_t^{\epsilon k} \to X_t^{0k}$, as $\epsilon \to 0$. The latter convergence is meant in Hausdorff metric for sets in R^n.

Of course, condition (17) is fulfilled if

$$det \int_0^t U'(\tau) C' C U(\tau) d\tau \neq 0. \tag{18}$$

Condition (18) is more strong than (17) and it is called "the property of observability with respect to initial state", that means: $y(\tau) \equiv 0$ implies $x_0 = 0$, while condition (17) means: $y(\tau) \equiv 0$ implies $x(t) = 0$. Some criteria for (18) will be given below. Consider

Example 3.6. For system

$$\dot{x}_1 = -x_1(0), \quad \dot{x}_2 = -x_2(0), \quad y(t) = x_1(t)$$

condition (17) is valid for $t = 1$, while (18) is not fulfilled.

Now one needs to obtain the evolution equations for parameters of sets X_t. In general case, it is unlikely to do that. Nevertheless, in (Anan'ev, 1998) it was shown that

$$\partial S_t^{-1} y(\tau, t)/\partial t = -S_t^{-1} K(\tau, t) S_t^{-1} y(t, t) + S_t^{-1} \partial y(\tau, t)/\partial t.$$

Therefore, as $R(\tau, \alpha, t) = DD' S_t^{-1} K(\tau, \alpha)$ for all $\alpha \geq 0$, the above equality yields

$$R(\tau, \alpha, t)/\partial t = -R(\tau, t, t)(DD')^{-1} R(t, \alpha, t)$$

Further the special case as in Example 2.1 will be ivestigated in more detail.

4. THE EVOLUTIONARY EQUATIONS OF FILTERING FOR THE LINEAR SYSTEMS WITH THE VARIABLE DELAY

In this section, the system (5) is considered with equation (2) for measurements. First, note that $U(t) = S(t, 0)$ in our case, but in general terms matrix $S(t, \tau)$ is not a semigroup. It follows from (6) that matrix $S(t, \tau)$ may be differentiated and

$$\partial S(t, \tau)/\partial t = AS(h(t), \tau). \tag{19}$$

According to (12), (14) we have

$$P(t) = \mathcal{T}_t\mathcal{T}_t^* - \mathcal{T}_t\mathcal{M}_t^* S_t^{-1} \mathcal{M}_t\mathcal{T}_t^* = \int_0^t (S(t, \tau) BB' S'(t, \tau) - F(t, \tau) S_t^{-1} \times F'(t, \tau)) d\tau, \tag{20}$$

where

$$F(t, \tau) = S(t, \tau) BD' + \int_0^\tau S(t, \alpha) BB' \times S'(\tau, \alpha) d\alpha C'. \tag{21}$$

Matrix (21) satisfies the same equation (19) as matrix $S(t, \tau)$. With the help of matrix (21) one can rewrite

$$\mathcal{T}_t\mathcal{M}_t^* S_t^{-1} CU(\cdot) = \int_0^t F(t, \tau) S_t^{-1} CU(\tau) d\tau,$$

$$\mathcal{T}_t\mathcal{M}_t^* S_t^{-1} y(\cdot) = \int_0^t F(t, \tau) S_t^{-1} y(\tau) d\tau. \tag{22}$$

In order to differentiate the parameters (16) let us suppose that inequality (18) is fulfilled for any $t > 0$. In paper (Zuyev, 1996) the following results have been proved.

Lemma 4.1. Let $0 \leq h(t) \leq t$, $h(t) \not\equiv 0$, and function $h(t)$ is analitic in the neighborhood of zero. Then inequality (18) is fulfilled iff

$$\cap_{i=0}^{n-1} ker CA^i = \{0\}.$$

Lemma 4.2. Let function $h(t)$ be monotonically increasing and continuous, $h(t) < t$. Then inequality (18) is valid for given t iff

$$\cap_{i=0}^{r} ker CA^i = \{0\},$$

where $r = (n-1) \wedge k = \max\{n-1, k\}$,

$$k = \begin{cases} \min\{p : h_p(t) > 0\}, & h(t) > 0; \\ 0, & otherwise; \end{cases}$$

$$h_p(t) = \underbrace{h(h(\ldots h}_{p-times}(t)\ldots)).$$

Consider matrises

$$P_2(t,\alpha,\beta) = \int_0^{\alpha\wedge\beta} (S(\alpha,\tau)BB'S'(\beta,\tau) \\ -F(\alpha,\tau)S_t^{-1}F'(\beta,\tau))d\tau,$$
$$P_1(t,\alpha) = P_2(t,\alpha,t), \quad 0 \leq \alpha,\beta \leq t;$$
$$Q_2(t,\alpha,\beta) = P_2(t,\alpha,\beta) + (U(\alpha) \\ - \int_0^t F(\alpha,\tau)S_t^{-1}CU(\tau)d\tau)V^{-1}(t) \\ \times(U(\beta) - \int_0^t F(\beta,\tau)S_t^{-1}CU(\tau)d\tau)',$$
$$Q_1(t,\alpha) = Q_2(t,\alpha,t). \quad (23)$$

These matrices have the properties:

$$P_2'(t,\alpha,\beta) = P_2(t,\beta,\alpha), \quad P_1(t,t) = P(t);$$
$$Q_2'(t,\alpha,\beta) = Q_2(t,\beta,\alpha), \quad Q_1(t,t) = Q(t).$$

Differentiating yields

$$dQ(t)/dt = AQ_1(t,h(t)) + Q_1'(t,h(t))A' \\ +BB' - (BD' + Q(t)C')(DD')^{-1}(DB' \\ +CQ(t)); \quad \partial Q_1(t,\alpha)/\partial t = Q_2(t,\alpha,h(t))A' \\ -Q_1(t,\alpha)C'(DD')^{-1}(DB' + CQ(t)); \\ \partial Q_2(t,\alpha,\beta)/\partial t = -Q_1(t,\alpha)C'(DD')^{-1} \\ \times CQ_1'(t,\beta). \quad (24)$$

The set of equations (24) may be called the Riccati type system for our problem. Note that the initial condition $Q^\epsilon(0) = \epsilon^{-1}I$ there exists only for parameters of ϵk- informational set $X_t^{\epsilon k}$. Matrices $P(t)$, $P_1(t,\alpha)$, $P_2(t,\alpha,\beta)$ satisfy the same equations (24) with initial conditions $P(0) = 0$. Now it is necessary to differentiate the value $\hat{x}(t)$ or the center of informational set $X(t)$. In order to do that let us introduce the value

$$\hat{x}_1(t,\alpha) = \int_0^t F(\alpha,\tau)S_t^{-1}y(\tau)d\tau + (U(\alpha) \\ - \int_0^t F(\alpha,\tau)S_t^{-1}CU(\tau)d\tau)V^{-1}(t) \\ \times \int_0^t U'(\tau)CS_t^{-1}y(\tau)d\tau; \qquad \hat{x}_1(t,t) = \hat{x}(t).(25)$$

Doing as above gives the equations

$$d\hat{x}(t)/dt = A\hat{x}_1(t,h(t)) + (BD' + Q(t)C') \\ \times(DD')^{-1}(y(t) - C\hat{x}(t)); \\ \partial\hat{x}_1(t,\alpha)/\partial t = Q_1(t,\alpha)C'(DD')^{-1}(y(t) \\ -C\hat{x}(t)). \quad (26)$$

One needs to solve this system with initial condition $\hat{x}^\epsilon(0) = 0$ for set $X_t^{\epsilon k}$. At last, the equation for value $\kappa(t)$ has the form

$$d\kappa(t)/dt = (y(t) - C\hat{x}(t))'(DD')^{-1} \\ \times(y(t) - C\hat{x}(t)) \quad (27)$$

with initial condition $\kappa^\epsilon(0) = 0$ for set $X_t^{\epsilon k}$.

For equations (24), (26), (27) it was suposed that matrix $V^{-1}(t)$ exists. If $detV(t) = 0$ then pseudoinverse matrix $V^+(t)$ does exists for any t, and one can differentiate the support function $\rho(l|X_t)$ for any observable vector l, since

$$l'\dot{V}^+l = -l'V^+\dot{V}V^+l.$$

Summarising the results of differentiating leads us to conclusion.

Theorem 4.3. The parameters of informational sets $X(t)$ and $X^{\epsilon k}(t)$ satisfy the system of differential equations (24), (26), (27), if matrix $V^{-1}(t)$ exists for any $t > 0$. The solution to these equations is unique for given initial conditions $Q(t_0)$, $\hat{x}(t_0)$, and $\kappa(t_0)$, $t \geq t_0$.

In conclusion, it is necessary to note that the system of equations in Theorem 4.3 coincides when $h(t) \equiv t$ with known results on guaranteed filtering, (Kurzhanski, 1977). On the other hand, the equations for center $\hat{x}(t)$ and matrix $Q(t)$ coincide in that case with equations of Kalman filter.

5. REFERENCES

Anan'ev, B.I. (1995). A guaranteed filtering scheme for hereditary differential systems with no information on the initial states. In: *Proceedings of the ECC 95* (A. Isidori, Ed.). Vol. 2. pp. 966–971. The University of Roma. Roma, Italy.

Anan'ev, B.I. (1998). A guaranteed filtering scheme for hereditary linear systems. *Journ. of Math. Systems, Estimation, and Contr.* **8, No. 3**, 365–368.

Anan'ev, B.I. and I.Ya. Pishchulina (1979). Minimax quadratic filtering for systems with time lag. In: *Differential Systems of Control* (A.V. Kryazhimskii, Ed.). pp. 3–12. The Urals Science Center. Sverdlovsk.

Bellman, R. and K.L. Cooke (1963). *Differential-Difference Equations.* Academic Press. New York.

Bertsekas, D. and I.B. Rhodes (1971). Recursive state estimation for a setmembership description of uncertainty. *IEEE Trans. Automat. Control* **AC-16, No. 2**, 117–128.

Germani, A., L. Jetto and M. Piccioni (1988). Galerkin approximation for optimal linear filtering of infinite-dimensional linear systems. *SIAM Journ. on Control & Opt.* **26, No. 6**, 1287–1305.

Hale, J. (1977). *Theory of Functional Differential Equations.* Springer-Verlag. New York.

Kolmanovskii, V.B. (1974). On the filtering for some stochastic processes with time lag. *Automation and Remote Control, No. 1 (in Russian)* pp. 42–49.

Krasovski, N.N. and A.B. Kurzhanski (1966). On the observability problem for systems with delay. *Differential Equations (in Russian)* **2, No. 3**, 299–308.

Kurzhanski, A.B. (1970). On the duality of optimal control problems and observation. *Appl. Math. and Mech. (in Russian)* **34, No. 3**, 429–439.

Kurzhanski, A.B. (1977). *Control and Observation Under Conditions of Unsertainty.* Nauka. Moskow.

Myschkis, A.D. (1972). *Linear Differential Equations with Retarded Arguments.* Nauka. Moskow.

Schweppe, F.C. (1973). *Uncertain Dynamic Systems.* Prentice Hall.

Vodichev, A.V. (1987). On the continuous restoration of the current state for a time-delay system under observation. *Differential Equations (in Russian)* **24, No. 2**, 217–227.

Zuyev, M.A. (1996). On the observability of linear functional-differential systems. *Izvestiya Vuzov, Matematika (in Russian)* **8**, 49–53.

SPECIALIZED SCIENTIFIC VISUALIZATION SYSTEMS FOR OPTIMAL CONTROL APPLICATION

V.L. Averbukh[1], S.S. Kumkov[2], E.A. Shilov[3], D.A. Yurtaev[4], A.I. Zenkov[5]

[1,2,4] *Institute of Mathematics and Mechanics, Ural Branch, Russian Academy of Sciences, S.Kovalevskaya str., 16, Ekaterinburg, 620219, Russia*
[3,5] *Ural State University, Lenin str., 51, Ekaterinburg, 620083, Russia*
e-mail: [1] *averbukh@oso.imm.intec.ru,* [2] *2445@dialup.mplik.ru,*
[3] *2148@dialup.mplik.ru,* [4] *dmitry@channel4.mplik.ru,*
[5] *zenant@geocities.com*

Abstract: Software for visualization of three-dimensional sets appearing in differential game theory and in problems of control with incomplete information is discussed. Interactive interface for managing three-dimensional objects by two-dimensional pointing device (mouse) was created. *Copyright © 1998 IFAC*

Keywords: differential games, numerical simulation, computer graphics

1. INTRODUCTION

Developing an adequate displaying is very important in visualization of mathematics research results. Here "adequate" means that the image gives both the nature and properties of simulated objects and their internal mental formed in the researcher's mind. Specialized visualization systems have to provide the image conformation so that the user gets the maximally full information on geometry and topology of objects, which are under investigation.

In this paper, a method for working-out such specialized systems is proposed. First of all, the programmer, who develops the system, together with mathematicians (its further customers) considers the gist of the terms to be visualized and their geometrical sense. His duty is to understand how a mathematician sees and apprehends the explored phenomena, how he constructs imaginary the process of solving problem. After that, concrete visualization ways and interactive interface methods are chosen. Often, a number of objects (sometimes multidimensional), also varying in time, must be mapped simultaneously. This demands the usage of photorealistic graphics (for instance, such properties as transparency or illumination) and animation. For apparency of the problem solution process, direct and reverse time regimes are needful. Also sometimes step-by-step mode of the process browsing is useful. So, it is essential to create a dynamic visualization system, where "dynamic" means dynamic imaging of both the object evolving and the process of the mathematics problem solving. Thus, visualization ways depend on concrete problem and vary from one task to another.

On the basis of this methodology, a number of specialized systems of scientific visualization of some optimal control problems are realized. They are assigned to describe the results of numerical experiments, to illustrate them, to analyze the calculation model and (sometimes) to debug the computational program. When photorealistic drawing three-dimensional objects is implemented, some modern rendering algorithms (realized, particularly, in OpenGL language) are used.

In this paper, two specialized systems are considered:

1. The system for visualization of three-dimensional informational sets;
2. The system for visualization of maximal stable (Krasovskii) bridges in linear differential games.

2. VISUALIZATION OF THREE-DIMENSIONAL INFORMATIONAL SETS

The problem is to visualize informational sets, which appear when the differential game theory is applied to describe a spacecraft steering to a dangerous space object (Kumkov and Patsko, 1995, 1997). Informational set is a collection of all phase states compatible with the history of the observation-control process. The task for different visualization features was formulated, including:

- output of informational set dynamics;
- changing viewpoint of informational set observing;
- output of informational set in absolute and relative scales;
- different output regimes, including layer-by-layer one;
- changing the global scale;
- output of the true point of the observed dangerous object;
- color marking singularities and components of the informational set image.

This problem was successfully solved. In Fig. 1, three-dimensional informational sets are drawn where each layer is a plane convex set. The lower set shows evolution happened when new information (measurement) comes.

Additionally to developed system, a number of temptations of four-dimensional object visualization were made.

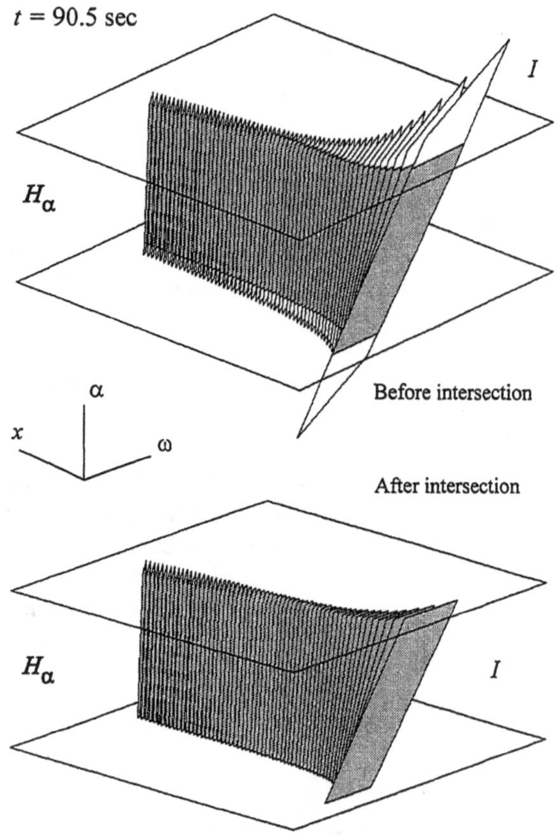

Fig. 1. Picture of informational sets.

3. VISUALIZATION OF MAXIMAL STABLE BRIDGES IN LINEAR DIFFERENTIAL GAMES

3.1 Software for visualization of maximal stable bridges

In linear differential games with fixed terminal time and terminal payoff function, each level set of the value function is a maximal stable bridge (Krasovskii and Subbotin, 1988). If the payoff function depends only on two coordinates of the phase vector at the terminal time, then transfer to equivalent game of the second order on phase variable can be applied. So, stable bridges for the equivalent game are built in the three-dimensional space where axes are time and two phase coordinates (Isakova, *et al.*, 1984). Thus, each bridge can be imagine as a "tube", which is determined by a collection of two-dimensional polygonal sections orthogonal to the time axis. Visualization system has to represent individual bridge or a number of them so that the value function structure and its peculiarities could be obvious.

Earlier, usual way to view these tubes was in drawing a number of its section projections in the plane of phase coordinates. But if the quantity of drawn contours grows, the image becomes unclear. Simultaneous view of a number of tubes is much more difficult. So, the problem was to show the object as a surface with its singularities. Here, "singularity" means a smoothness violation, which can born and disappear in the time on the surface of a tube. To get information about the value function, it is needful to view a number of tubes built for distinct meanings of value function. With that, separate tubes and the whole structure in general have to be recognizable.

Developed system allows to view interactively three-dimensional image of a tube or a collection of tubes. The picture can be seen in two regimes. The first is when tubes are represented in the traditional way as a number of contours. But now it is possible to project them not only to phase coordinates plane, but to an arbitrary one. This regime is useful when seeking an appropriate point of view. After it is found, the second view regime can be used. For it, a surface is built by triangulation on base of the collection of separate sections. Then, this surface is drawn using

Gouraud shading. The object is illuminated by a dot radiant, which position can be changed by user. Each tube has its own attributes: transparency/opaqueness and color. When a number of tubes are viewed each of them can be set visible/invisible independently.

The first version of the system was written using C language for UNIX-like operational systems with X-Window shell. As the base graphic tool, OpenGL library was used. User interface was implemented with help of Motif. Now, a new version of this system is developed. It is OS Windows 95 oriented. User interface is realized by Delphi interactive system.

Important demand is that algorithms built in such system have to find features interesting for user and to allow him concentrate attention on them. In general, considered system satisfies this demand. So, for more flexible searching non-smoothnesses on the surface of a tube, user can change the threshold angle, separating "smooth" and "non-smooth" vertices, in the range from 0 up to 180 degrees.

The central part of the system discussed is interactive tools for output managing and for changing view regimes. As it was mentioned above, it is possible to change position and orientation of examined object in three-dimensional space, scale of view, location of radiant and cut plane. Also, color and transparency of tubes can be changed.

Here, basic steps of the system work are enumerated:

1. Reading data files;
2. Reconstructing surface of each tube from separate sections;
3. Initializing window system and user interface;
4. Loop of processing window system messages.

Surface reconstruction from separate sections is made using the following algorithm. The aim is to build a system of triangles between each two neighbor sections. Firstly, corner points are detected accordingly to threshold angle set up by user. Starting value of the threshold angle is $\pi \cdot (1 - 1/N)$, where N is the average number of vertices of the tube sections. Further, a connection between these points from considered sections is established. In other words, corresponding points are connected. After that, points of arcs, laid between these corner points, are also connected so that a system of triangles appears.

If all viewed objects are opaque, then *z*-buffer algorithm is allowable and the order of drawing primitive elements is not important. Otherwise, if there are some transparent objects, then an additional problem appears. To get an adequate view, it is necessary to draw distant primitives before closer ones using the *alpha blending* procedure, for which the OpenGL tool permits to define necessary standard functions.

For output of a polygon, the OpenGL language allows to apply two regimes: *flat* and *smooth*. For the latter one, each vertex has its own normal. Then lightness of vertices is calculated on the base of the Lambert reflection law with actual properties of the reflecting surface, location of radiants and direction of the normal. Lightness of all other points of the polygon is computed using bilinear interpolation to two nearest vertices.

For changing the orientation of the object, so-called *arcball controller* algorithm (Shoemake, 1992) is used. This algorithm allows to rotate the object easily in three-dimensional space with help of two-dimensional coordinate pointer device (for example, by a mouse). The main idea of the algorithm is the following.

1. Choose a region of the screen as the projection of a sphere, which center $\vec{O} \in R^2$ is in the plane of the screen and radius is equal to ρ;
2. Let $\vec{S} \in R^2$ be the vector of cursor placement at the moment when the mouse button is pushed;
3. $\vec{s} = \overrightarrow{OS} = (s_x, s_y)$;
4. $r^2 = |s|^2 = s_x^2 + s_y^2$;
5. Map two-dimensional vector s onto a unit three-dimensional vector p: if $r > \rho$, then $\vec{p} = (\vec{s}/r, 0) = (s_x/r, s_y/r, 0)$; if $r \le \rho$, then $\vec{p} = \left(s_x/\rho,\, s_y/\rho, \sqrt{1 - (s_x/\rho)^2 - (s_y/\rho)^2}\right)$. That is, if the cursor is beyond the great circle of the chosen sphere, then the vector p is on the great circle in the plane of the screen, otherwise the vector p is on the chosen sphere somewhere in three-dimensional space;
6. For any new location T of the cursor, an unit vector $\vec{q} \in R^3$ is built by the above formula;
7. Parameters of the object rotation are computed by the following formulae: the axis $\vec{a} = \vec{p} \times \vec{q}$, the angle $\theta = 2 \cdot \arccos(\vec{p} \cdot \vec{q})$. Here "×" means vector production, "·" means scalar production.

So, user operates with an imaginary sphere. By pushing button, a point on it is fixed. Further, mouse motions rotate this sphere such that new location of the fixed point coincides with the cursor location. And the object rotates with the sphere.

Toolbar of the main window contains the following control elements:

a) for switching the rotation mode;

b) for switching the drag mode when user can move the object;

c) for switching the mode of cutting plane control;

d) for switching the mode of radiant control;

e) for choosing tubes attributes: color, transparency/opaqueness, visibility/invisibility;

f) for choosing the threshold angle;

g) for changing additional marking of corner points;

h) for switching between contour/solid regimes of view;

i) for choosing viewpoint:

- arbitrary;
- from Z axis (as projection on XY plane);
- from X axis (as projection on ZY plane);
- from Y axis (as projection on XZ plane).

As it was mentioned above, the first version of this software was written using C language for UNIX-like operational systems with X-Window graphic shell. Further, it was modified for OS Windows 95 using Delphi developing environment. A number of new features were added:

- capability to view contour skeleton in solid regime on each tube independently;
- two buttons for quick resetting original viewpoint and location of radiant;
- capability to view coordinate axes.

3.2 Description of work session

After loading the program, user has to read data files. Now, data files have no special marks about the problem and values of value function, they were calculated for. So, user must know which files he has to read. When all need data are transferred into computer memory and processed (tube surface are constructed from separate sections), contour view of all loaded surfaces appears (initially all tubes are set up as visible). Then user can switch off tubes, which are not need just at that moment.

One main stage is to find an acceptable viewpoint. For this stage, the first view regime – contour one – is used. All operations with the picture (rotation, zooming, etc.) can be done in the solid regime. But usually, it is carried out too slow when the computer does not equipped by a video card, which has hardware support of OpenGL commands. So, using the contour regime is the optimal way to seek for good viewpoint.

When the desirable aspect is found, the second regime (solid) can be chosen. Now, user has to look for a good placement of the radiant. As the moon craters can be seen only due to cast shadows, some interesting specialties of the tube surface are also visible when an irregularity of lightness exists. At this stage, views from coordinate axes are very convenient for global movements of the radiant.

If a number of tubes are viewed, the problem of visibility of their mutual collocation appears. The system suggests two ways for solving this problem. The first one is to make external tubes transparent. This way is acceptable when two or three tubes are drawn; otherwise, internal surfaces become unclear. The second way is to stay all tubes opaque, but to cut them by a plane. In actual version of the software, only a plane parallel to the axes of x_1 (X) and τ (Z) can be used for cutting. However, it is sufficient for majority of pictures.

When a transparent tube is examined, its space configuration is usually lost. It happens due to decreasing of lightness irregularity: the light is coming through the surface, not reflecting back to the observer. To recognize the third dimension of the surface, it is possible to draw the contour skeleton of the tube. The skeleton can be also drawn on opaque tubes if it is necessary for clearness of the view.

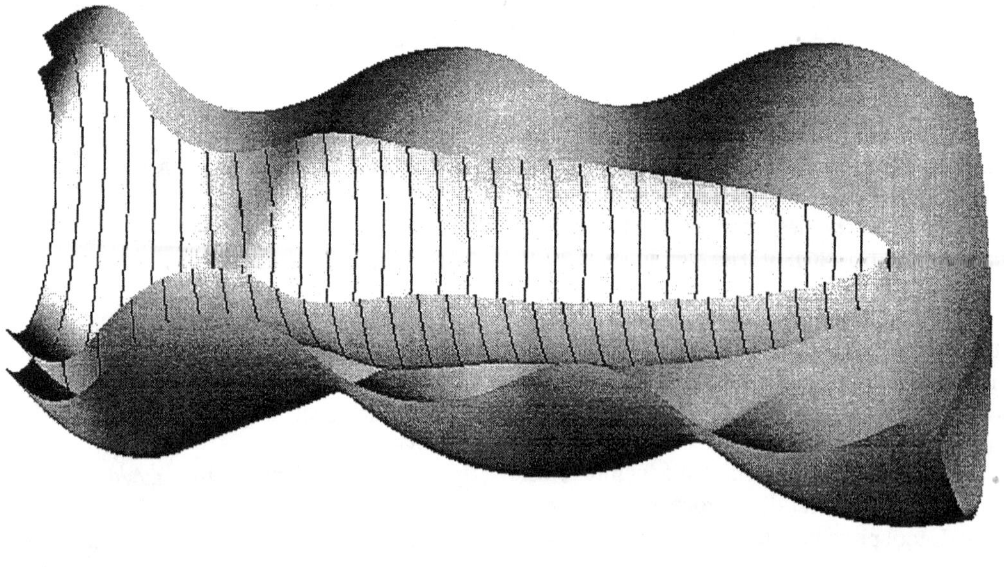

Fig. 2. Two level sets for "oscillator" system cut by a plane.

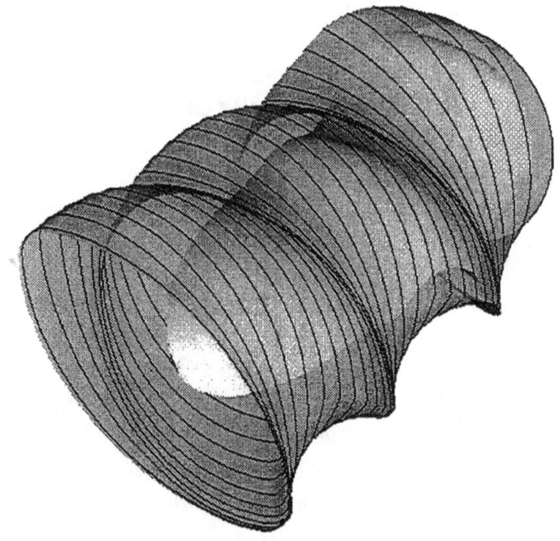

Fig. 3. Two level sets for "oscillator" system. The external one is transparent.

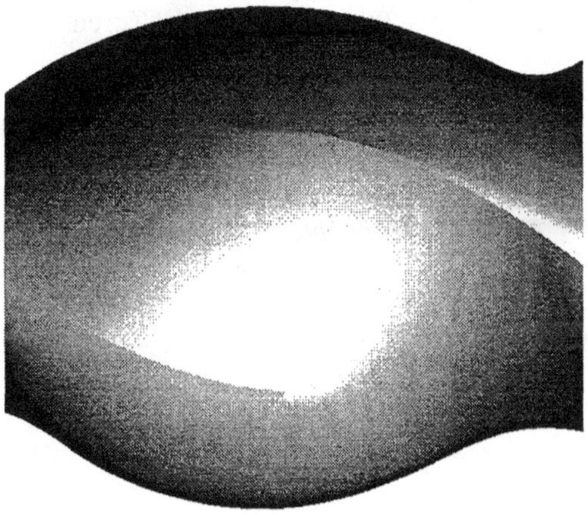

Fig. 4. Fragment of the external tube surface with discontinuous singular line.

Now, development of the system continues accordingly to users' suggestions. For example, it is planned to add such features as storing of good viewpoints, marking some lines (optimal motions of the system, lines of singularity of the value function, and so on) on tube surface, saving the drawn pictures in some popular graphics format, etc.

Development of such software for visualization of maximal stable bridges in differential games is extremely useful and allows to activate investigations of this subject. Also, this system can be used for visualization of objects, appearing in other branches of mathematics, if these objects are tube-like.

3.3 Examples

For computing example picture a number of model systems were taken. One of them is "controlled oscillator".

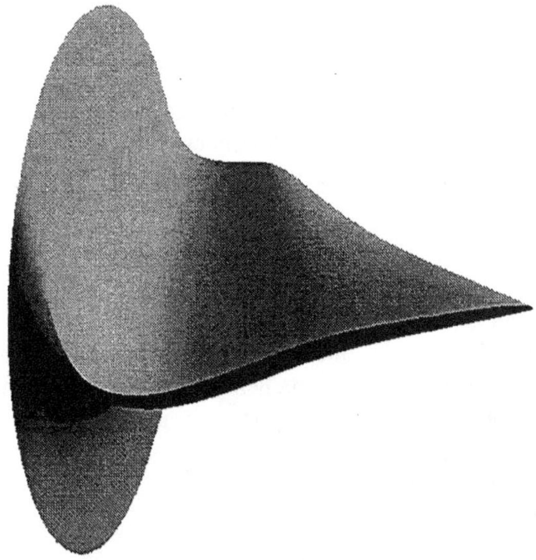

Fig. 5. One level set for "drawing-pin" system.

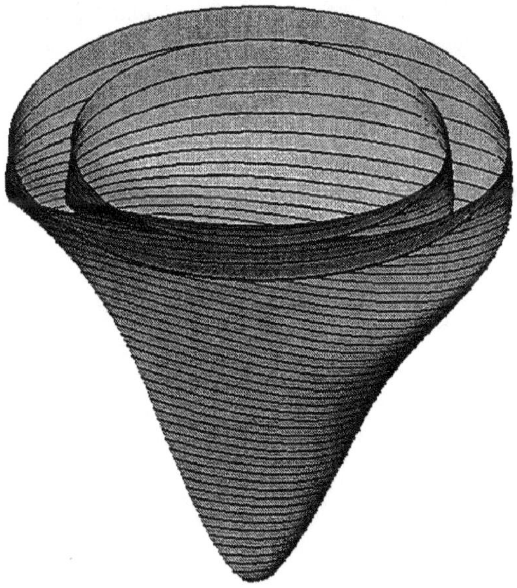

Fig. 6. Two level sets for "drawing-pin" system. Both are opaque with shown contour skeleton.

In Figure 2, two level sets are shown. They are cut by a plane and internal one has contour skeleton. The periodic nature of this system can be evidently seen.

The same sets are drawn in Fig. 3 from other viewpoint and using another mentioned way: without cutting plane, but the external tube is made transparent. For reproduction of its space structure, contour skeleton is represented.

In Figure 4, one of the subjects for investigation is shown. There is a fragment of the surface of external tube from Figs. 2. and 3. Location of radiant is such that singular line ("corner" line) is evident.

Figures 5 and 6 show level sets for so-called "drawing-pin" system. In Fig. 5, the singular line is easily seen.

CONCLUSION

The main goal of the research is to develop a convenient system of computer graphics for representation of multidimensional mathematical objects in differential games and control theory.

ACKNOWLEDGEMENTS

The statement of the problem was suggested by specialists from the Dynamic System Department of the Institute of Mathematics and Mechanics, Ekaterinburg. The system was developed in co-operation with V.S. Patsko, S.I. Kumkov, A.G. Ivanov.

REFERENCES

Kumkov, S.I. and V.S. Patsko (1995). Control of informational sets in a pursuit problem. In: *Annals of the International Society of Dynamic Games. New Trends in Dynamic Games and Applications* (G.J. Olsder (Ed)), pp. 191 – 206. Birkhauser, Boston.

Kumkov, S.I. and V.S. Patsko (1997). Informational sets in a problem of impulse control. *Automatics and Telemechanics*, No. 7, pp. 195 – 206 [in Russian].

Krasovskii, N.N. and Subbotin A.I. (1988). *Game-Theoretical Control Problems*. Springer-Verlag, New York.

Isakova, E.A., G.V.Logunova and V.S.Patsko (1984). Computation of stable bridges for linear differential games with fixed time of termination. In: *Algorithms and Programs for Solving Linear Differential Games* (Subbotin, A.I. and V.S. Patsko (Eds)), pp. 125 - 158. Inst. of Math. and Mech., Sverdlovsk [in Russian].

Shoemake, K. (1992). ARCBALL: A user interface for specifying three-dimensional orientation using a mouse. In: *Proceedings of Graphics Interface'92*, pp. 151 – 156. Canadian Information Processing Society, Toronto.

POSITIONAL OPTIMIZATION OF HIERARCHICAL CONTROL SYSTEM

N.V. Balashevich

Institute of Mathematics
National Academy of Sciences
Surganov str. 11, 220072 Minsk, Belarus
balash@im.bas-net.by

Abstract: A problem of optimal hierarchical control with two participants called a leader and a follower is considered. A leader constructs its control not paying attention to a follower while a follower knows a principle of acting a leader and uses it in deciding on its own behaviour. A method of constructing a realization of optimal feedback for a linear hierarchical control system is described. *Copyright © 1998 IFAC*

Keywords: optimal control, hierarchical system, feedback, algorithm

1. INTRODUCTION

Problems of optimal hierarchical control (Kleimenov, 1993; Petrosyan and Danilov, 1985) are intermediate between classical problems of optimal control and problems of the theory of differential games. Classical control problems are characterized by the choice of the common goal for all (possibly, different) participants of the control process.

In problems of the theory of antagonistic differential games participants of each of the two groups consider the other side as an enemy or a harmful obstacle (Krasovskii and Subbotin, 1974). Under hierarchical control as well as under classical control there is no counteraction of participants. However as in the theory of differential games in the problem of hierarchical control interests of participants do not coincide (Von Stackelberg, 1952).

In the paper two linear problems of optimal hierarchical control are investigated. Algorithms of operating optimal controllers able to realize positional solutions to the problems under consideration in real time are described. The work is based on the approach to the classical problem of optimal synthesis proposed in (Gabasov, *et al.*, 1992) and developed in (Gabasov, *et al.*, 1997).

2. PROBLEM STATEMENT

Consider a problem of hierarchical control with two participants. The first participant is said to be a leader and the second one is said to be a follower. The leader solves the problem

$$I_1(u) = c_1' x^1(t^*) \longrightarrow \max,$$
$$\dot{x}^1 = A_1 x^1 + b_1 u, \quad x^1(0) = x_{10}, \qquad (1)$$
$$H_1 x^1(t^*) = g_1, \quad |u(t)| \le 1, \quad t \in T,$$

in the class of piecewise continuous functions $u(t)$, $t \in T$, where $x^1(t)$ is an n_1-vector of state of the leader system at a moment t, $g_1 \in R^{m_1}$ is a given vector of output signals at the terminal moment t^*. It is assumed that

$$\mathrm{rank}(b_1, A_1 b_1, \ldots, A_1^{n_1-1} b_1) = n_1,$$
$$\mathrm{rank} H_1 = m_1 < n_1.$$

The follower organizes its behaviour by solving in the same class of piecewise continuous controls $v(t)$, $t \in T$, the extremal problem

$$I_2(u, v) = c_2' x^2(t^*) \longrightarrow \max, \qquad (2)$$
$$\dot{x}^2 = A_2 x^2 + A_{12} x^1 + b_2 v + b_{12} u, \quad x^2(0) = x_{20},$$
$$H_2 x^2(t^*) + H_{12} x^1(t^*) = g_2, \quad |v(t)| \le 1, \quad t \in T,$$

where $x^2(t)$ is an n_2-vector of state of the follower system, g_2 is an m_2-vector,

$$\mathrm{rank}(b_2, A_2 b_2, \ldots, A_2^{n_2-1} b_2) = n_2,$$

$$\mathrm{rank} H_2 = m_2 < n_2.$$

The main feature of hierarchical control is that the leader forms its behaviour not paying attention to the follower while the follower knows the principle of acting the leader and uses it in deciding on its own behaviour.

A piecewise continuous function $u(\cdot) = (u(t|\tau, z^1), t \in T_\tau = [\tau, t^*])$ is said to be an admissible open-loop control of the leader for the position (τ, z^1) if 1) it satisfies the constraint $|u(t|\tau, z^1)| \le 1, t \in T_\tau$, 2) the corresponding trajectory $x^1(t), t \in T_\tau$, of (1) with the initial state $x^1(\tau) = z^1$ reaches the terminal set $X_1^* = \{x \in R^{n_1} : H_1 x = g_1\}$ at the moment t^*.

A piecewise continuous function $v(t|\tau, z^1, z^2, u(\cdot))$, $t \in T_\tau$, is said to be an admissible open-loop control of the follower for the position $(\tau, z^1, z^2, u(\cdot))$ if 1) it satisfies the constraint $|v(t|\tau, z^1, z^2, u(\cdot))| \le 1, t \in T_\tau$, 2) the trajectory $x^2(t), t \in T_\tau$, of (2) with the initial state $x^2(\tau) = z^2$ corresponding to this function and the functions $u(t|\tau, z^1), x^1(t), t \in T_\tau$, reaches the terminal set $X_2^* = \{x \in R^{n_2} : H_2 x + H_{12} x^1(t^*) = g_2\}$ at the moment t^*.

Admissible controls $u_\tau^0(\cdot) = (u^0(t|\tau, z^1), t \in T_\tau)$, $v_\tau^0(\cdot) = (v^0(t|\tau, z^1, z^2, u_\tau^0(\cdot)), t \in T_\tau)$ for the position (τ, z^1, z^2) are said to be optimal open-loop controls of the leader and the follower if

$$I_1(u_\tau^0(\cdot)) = \max_u I_1(u),$$

$$I_2(u_\tau^0(\cdot), v_\tau^0(\cdot)) = \max_v I_2(u_\tau^0(\cdot), v).$$

Open-loop solutions $u^0(t), v^0(t), t \in T$, to (1) – (2) are natural for ideal situations when real control systems are not affected by disturbances and mathematical models reflect behaviour of these systems exactly. But, as a rule, mathematical models (1) – (2) reflect real behaviour of systems only approximately and do not take into account unavoidable and unknown beforehand disturbances. In this connection engineers prefer to use controls of feedback type which take into account states realized in a process of control.

Let $Z_{1\tau}$, $Z_{2\tau}$ be sets of states z^1, z^2 of the leader and the follower at a moment τ for which there exist optimal open-loop controls.

Following one of definitions of the classical optimal feedback, functions

$$u^0(\tau, z^1) = u^0(\tau|\tau, z^1),$$

$$v^0(\tau, z^1, z^2) = v^0(\tau|\tau, z^1, z^2, u_\tau^0(\cdot)), \quad (3)$$

$$z^1 \in Z_{1\tau}, \quad z^2 \in Z_{2\tau}, \quad \tau \in T,$$

will be called a positional solution to the problem of optimal hierarchical control (optimal hierarchical feedback).

Controls of feedback type are usually used for systems under unknown disturbances. Assume that the behaviour of systems (1), (2) closed by feedback (3) is described by the equations

$$\dot{x}^1 = A_1 x^1 + b_1 u^0(t, x^1) + w_1, \quad x^1(0) = x_{10},$$

$$\dot{x}^2 = A_2 x^2 + A_{12} x^1 + b_2 v^0(t, x^1, x^2) + \quad (4)$$
$$+ b_{12} u^0(t, x^1) + w_2, \quad x^2(0) = x_{20},$$

where $w_1(t), w_2(t), t \in T^0 = [0, t^0]$, are unknown piecewise continuous disturbances, $0 < t^0 < t^*$ is a moment of termination of affecting disturbances.

Thus the problem of effective control of a hierarchical system is reduced to constructing functions (3). By analogy with the classical control theory it can be named the problem of synthesis of optimal hierarchical systems.

In (Gabasov, *et al.*, 1992) a method of realization of optimal feedback without preliminary constructing the whole positional solution is proposed. Below this approach is extended to the problem of hierarchical control (1)–(2).

Suppose that functions (3) have been constructed and are used in real control process (4). Consider a concrete control process in which disturbances $w_1^*(t), w_2^*(t), t \in T^0$, are realized. From (4) it is clear that in the process under consideration functions (3) are used not for all the values $z^1 \in Z_{1\tau}, z^2 \in Z_{2\tau}, \tau \in T$. It is sufficient to use only the values $u^*(\tau) = u^0(\tau, x^{1*}(\tau))$, $v^*(\tau) = v^0(\tau, x^{1*}(\tau), x^{2*}))$ along the isolated trajectories $x^{1*}(\tau), x^{2*}(\tau), \tau \in T$, of equations (4) which are generated by the concrete realizations $w_1^*(t), w_2^*(t), t \in T^0$, and moreover it is enough to know the values $u^*(\tau), v^*(\tau)$ only at the moment τ.

Devices which in each concrete control process are able to generate the functions $u^*(\tau), v^*(\tau), \tau \in T$, in real time are called the optimal controllers of the leader and the follower.

The purpose of the paper is to describe the algorithm of operating optimal controllers.

3. OPTIMAL CONTROLLERS OF THE LEADER AND THE FOLLOWER

According to the definition for constructing values $u^*(\tau), v^*(\tau)$ at a current moment $\tau \in T$ optimal controllers use open-loop solutions $u_\tau(t) = u^0(t|\tau, x^{1*}(\tau))$, $v_\tau(t) = v^0(t|\tau, x^{1*}(\tau), x^{2*}(\tau), u^0(\cdot))$, $t \in T_\tau$, to the problems

$$c_1' x^1(t^*) \longrightarrow \max,$$

$$\dot{x}^1 = A_1 x^1 + b_1 u, \quad x^1(\tau) = x^{1*}(\tau),$$
$$H_1 x^1(t^*) = g_1, \quad |u(t)| \le 1, \quad t \in T_\tau,$$

$$c_2' x^2(t^*) \longrightarrow \max,$$
$$\dot{x}^2 = A_2 x^2 + A_{12} x^1 + b_2 v + b_{12} u_\tau(t),$$
$$x^2(\tau) = x^{2*}(\tau),$$
$$H_2 x^2(t^*) + H_{12} x^1(t^*) = g_2, \quad |v(t)| \le 1, \quad t \in T_\tau.$$

As is shown in (Lee and Markus, 1968) controls $u_\tau(t)$, $v_\tau(t)$, $t \in T_\tau$, have the form

$$u_\tau(t) = \mathrm{sgn}\Delta_\tau^1(t), \quad v_\tau(t) = \mathrm{sgn}\Delta_\tau^2(t), \quad t \in T_\tau,$$

where $\Delta_\tau^i(t) = \psi_\tau^{i\prime}(t) b_i$, $t \in T_\tau$, $i = 1, 2$, are optimal cocontrols of the leader and the follower, $\psi_\tau^i(t)$, $t \in T_\tau$, $i = 1, 2$, are solutions to the conjugate systems $\dot{\psi}_\tau^i = -A_i' \psi_\tau^i$, $\psi_\tau^i(t^*) = c_i - H_i' y^i(\tau)$, $i = 1, 2$; $y^i(\tau)$, $i = 1, 2$, are the Lagrange optimal vectors of the leader and the follower.

Thus at a moment $\tau \in T$ the controllers have to know the values of the elements

$$t_j^i(\tau), \; j = \overline{1, p_i(\tau)}; \quad y^i(\tau), \; i = 1, 2 \qquad (5)$$

where $t_1^i(\tau) < \ldots < t_{p_i(\tau)}^i(\tau)$, $i = 1, 2$, are switching points of optimal open-loop controls $u_\tau(t)$, $v_\tau(t)$, $t \in T_\tau$: $\Delta_\tau^i(t_j^i(\tau)) = 0$, $j = \overline{1, p_i(\tau)}$, $i = 1, 2$.

Denote

$$k_i(\tau) = \mathrm{sgn}\Delta_\tau^i(\tau) \quad \text{if } \Delta_\tau^i(\tau) \ne 0;$$
$$k_i(\tau) = -\mathrm{sgn}\partial\Delta_\tau^i(\tau)/\partial t\Big|_{t=\tau} \quad \text{if } \Delta_\tau^i(\tau) \ne 0;$$
$$l_{*i}(\tau) = 0 \quad \text{if } t_1^i(\tau) > \tau;$$
$$l_{*i}(\tau) = 1 \quad \text{if } t_1^i(\tau) = \tau;$$
$$l_i^*(\tau) = 0 \quad \text{if } t_{p_i(\tau)}^i(\tau) < t^*;$$
$$l_i^*(\tau) = 1 \quad \text{if } t_{p_i(\tau)}^i(\tau) = t^*;$$
$$L_i(\tau) = \{j \in \{1, \ldots, p_i(\tau)\} :$$
$$\partial\Delta_\tau^i(t)/\partial t\Big|_{t=t_j^i(\tau)} = 0\}, \quad i = 1, 2.$$

A set $S^i(\tau) = \{k_i(\tau), p_i(\tau), l_{*i}(\tau), l_i^*(\tau), L_i(\tau)\}$, $i = 1, 2$, is said to be a structure of optimal cocontrol $\Delta_\tau^i(t)$, $t \in T_\tau$. A structure $S^i(\tau)$ is nonsingular if

$$\beta_i(\tau) = l_{*i}(\tau) + l_i^*(\tau) + |L_i(\tau)| = 0.$$

Suppose that for any $\tau \in T^0$ and $i = 1, 2$ the following conditions are fulfilled
1) The equality

$$\mathrm{rank}(H_i F_i(t^* - t) b_i, \quad t \in T_i^*(\tau)) = m_i$$

can be violated only in isolated points $\tau \in T^0$.
2) The inequality $\beta_i(\tau) \le 1$ takes place and the equality $\beta_i(\tau) = 1$ is possible only in isolated points $\tau \in T^0$.
3) For any optimal cocontrol $\Delta_\tau^i(t)$, $t \in T_\tau$, such moments $\nu \in \mathrm{int}T_\tau$, $\mu \in \{\tau, t^*\}$ that

$$\Delta_\tau^i(\nu) = \partial\Delta_\tau^i(t)/\partial t\Big|_{t=\nu} = \partial^2\Delta_\tau^i(t)/\partial t^2\Big|_{t=\nu} = 0,$$
$$\Delta_\tau^i(\mu) = \partial\Delta_\tau^i(t)/\partial t\Big|_{t=\mu} = 0$$

do not exist.
Here

$$T_1^*(\tau) = \{t \in \mathrm{int}T_\tau : u_\tau(t-0) \ne u_\tau(t+0)\},$$
$$T_2^*(\tau) = \{t \in \mathrm{int}T_\tau : v_\tau(t-0) \ne v_\tau(t+0)\},$$
$$\dot{F}_i = A_i F_i, \quad F_i(0) = E.$$

It is not difficult to prove using the terminal constraints and the relations for switching points that elements (5) satisfy the equations

$$\Phi_1(\tau, x^{1*}(\tau), k_1(\tau), t_j^1(\tau), j = \overline{0, p_1(\tau)+1}) = 0,$$
$$q_1(y^1(\tau), t_j^1(\tau)) = 0, \quad j = \overline{1, p_1(\tau)}; \qquad (6)$$

$$\Phi_2(\tau, x^{i*}(\tau), k_i(\tau), t_j^i(\tau),$$
$$j = \overline{0, p_i(\tau)+1}, i = 1, 2) = 0,$$
$$q_2(y^2(\tau), t_j^2(\tau)) = 0, \quad j = \overline{1, p_2(\tau)}; \qquad (7)$$

Here

$$\Phi_1(\tau, x, k, t_j, j = \overline{0, p+1}) =$$
$$H_1 M_1(\tau, x, k, t_j, j = \overline{0, p+1}) - g_1;$$
$$M_i(\tau, x, k, t_j, j = \overline{0, p+1}) = F_i(t^* - \tau)x +$$
$$+\sum_{j=0}^{p} (-1)^j k \int_{t_j}^{t_{j+1}} F_i(t^* - t) b_i dt, \quad i = 1, 2;$$
$$\Phi_2(\tau, x^i, k_i, t_j^i, j = \overline{0, p_i+1}, i = 1, 2) =$$
$$= H_2\big[M_2(\tau, x^2, k_2, t_j^2, j = \overline{0, p_2+1}) +$$
$$+\int_\tau^{t^*} F_2(t^* - \tau) A_{12} F_1(t - \tau) dt x^1 + \sum_{j=0}^{p_1} (-1)^j k_1$$
$$\int_{t_j^1}^{t_{j+1}^1} F_2(t^* - t)(A_{12} \int_\tau^t F_1(t-s) b_1 ds + b_{12}) dt\big] +$$
$$+ H_{12} M_1(\tau, x^1, k_1, t_j^1, j = \overline{0, p_1+1}) - g_2;$$
$$q_i(y, t) = (c_i - H_i' y) F_i(t^* - t) b_i, \quad i = 1, 2;$$
$$t_0^i(\tau) = \tau, \quad t_{p_i(\tau)+1}^i = t^*, \quad i = 1, 2.$$

Thus we get $p_1(\tau) + p_2(\tau) + m_1 + m_2$ nonlinear equations with $p_1(\tau) + p_2(\tau) + m_1 + m_2$ variables.

Equations (6) are said to be the defining equations of the optimal controller of the leader and equations (7) are said to be the defining equations of the optimal controller of the follower.

To construct solution (5) for the moment $\tau \in T^0$ the solution for the moment $\tau - h$ is used as an initial approximation (h is the step of a grid while calculating). On intervals of nonsingularity of structures $S^i(\tau)$, $i = 1, 2$, $\tau \in T^0$, by virtue of the conditions 1)—3) the Jacobi matrix of system (6), (7) is nonsingular with respect to variables (5) and it allows to use the Newton method for solving system (6), (7) on these intervals. At a moment of singularity of either of two structures according to special rules structures for a following interval of nonsingularity are formed. A numerical method of solving the defining equations and rules of transfering between adjacent intervals with different structures are analogous to (Gabasov, *et al.*, 1992) where it is shown that such procedures can be realized in real time by microprocessors.

After solving the defining equations at a moment $\tau \in T^0$ on a regular interval we set $u^*(\tau) = k_1(\tau)$, $v^*(\tau) = k_2(\tau)$. The scheme of control on a sliding interval is similar to one described in (Gabasov, *et al.*, 1992). At the moment t^0 solving the defining equations is stopped and for $\tau \in [t^0, t^*]$ the controllers generate the values

$$u^*(\tau) = u^0(\tau|t^0, x^{1*}(t^0)),$$

$$v^*(\tau) = v^0(\tau|t^0, x^{1*}(t^0), x^{2*}(t^0), u^0(\cdot)).$$

4. THE CASE OF JOINT CONTROL

On interval $T = [0, t^*]$ consider the problem of optimization of the dynamic system

$$\dot{x} = Ax + b_1 u + b_2 v, \quad x(0) = x_0, \tag{8}$$

$$\left(x \in R^n, \quad u \in R, \quad v \in R\right),$$

with two participants. The leader uses control $u(t)$, $t \in T$, which it may choose among piecewise continuous functions satisfying the inequality

$$|u(t)| \leq 1, \quad t \in T. \tag{9}$$

The follower chooses the control $v(t)$, $t \in T$, in the same class of functions

$$|v(t)| \leq 1, \quad t \in T. \tag{10}$$

Each of participants is responsible for its own terminal constraints. The leader fulfils the constraint

$$h_i' x(t^*) = c_i, \quad i \in I_1 = 1, \ldots, m_1 , \tag{11}$$

$$H = \{h_i', \ i \in I_1\}, \quad \mathrm{rank} H = m_1 < n,$$

$$c = (c_i, \ i \in I_1),$$

the follower fulfils the constraint

$$g_i' x(t^*) = d_i, \quad i \in I_2 = \{1, \ldots, m_2\}, \tag{12}$$

$$G = \{g_i', \ i \in I_2\}, \quad \mathrm{rank} G = m_2,$$

$$m_1 + m_2 + 1 < n, \quad d = (d_i, \ i \in I_2).$$

The leader has to maximize the control criterion

$$J_1 = h_0' x(t^*) \longrightarrow \max. \tag{13}$$

The follower maximizes the control criterion

$$J_2 = g_0' x(t^*) \longrightarrow \max. \tag{14}$$

The interaction of participants is stipulated by the following principles of hierarchical control:
1) before the beginning of the control process the participants reach an agreement that the follower makes certain contribution in constraint (11) and in control criterion (13) of the leader;
2) during the control process the leader does not know the principle of behaviour of the follower;
3) during the control process the follower knows the principle of behaviour of the leader;
4) in every current moment the follower informs the leader about a control produced.

Let us write down terminal constraint (11) and control criterion (13) of the leader with the help of the Cauchy formula

$$HF(t^*)x_0 + H\int_0^{t^*} F(t^* - s)b_1 u(s)ds +$$

$$+ H\int_0^{t^*} F(t^* - s)b_2 v(s)ds = c,$$

$$h_0' F(t^*)x_0 + h_0' \int_0^{t^*} F(t^* - s)b_1 u(s)ds +$$

$$+ h_0' \int_0^{t^*} F(t^* - s)b_2 v(s)ds$$

where $F(t)$ is a fundamental matrix of solutions to the system $\dot{x} = Ax$. The agreement between the leader and the follower is that a number α and an m_1-vector c^v are chosen and the follower is obliged to use only such admissible controls for which the equalities

$$h_0' \int_0^{t^*} F(t^* - s)b_2 v(s)ds = \alpha, \tag{15}$$

$$H\int_0^{t^*} F(t^* - s)b_2 v(s)ds = c^v$$

are fulfilled.

This agreement allows to introduce the notion of admissible and optimal open-loop controls of the leader.

An accessible control $u(t), t \in T$, is said to be admissible if the equality

$$HF(t^*)x_0 + H\int_0^{t^*} F(t^* - s)b_1 u(s)ds = c - c^v$$

is fulfilled. An admissible control $u^0(t), t \in T$, is said to be an optimal open-loop control of the leader if the control criterion of the leader reaches the maximum value

$$J_1^0 = h_0' \int_0^{t^*} F(t^* - s) b_1 u^0(s) ds =$$

$$= \max_u h_0' \int_0^{t^*} F(t^* - s) b_1 u ds.$$

From the presented constraints it is seen that in deciding on its behaviour the leader is not interested in the principle of behaviour of the follower. The follower in deciding on its controls $v(t)$, $t \in T$, knows and takes into account the behaviour of the leader.

Let $u(\cdot) = (u(t),\ t \in T)$ be a control of the leader. An accessible control $v(t|u(\cdot))$, $t \in T$, of the follower is said to be admissible if

$$h_0' \int_0^{t^*} F(t^* - s) b_2 v(s|u(\cdot)) ds = \alpha,$$

$$H \int_0^{t^*} F(t^* - s) b_2 v(s|u(\cdot)) ds = c^v, \qquad (16)$$

$$GF(t^*)x_0 + G \int_0^{t^*} F(t^* - s) b_2 v(s|u(\cdot)) ds =$$

$$= d - G \int_0^{t^*} F(t^* - s) b_1 u(s) ds.$$

This definition takes into account the behaviour of the leader and the obligation (15) of the follower.

The first two equalities (16) may be presented with the help of the auxiliary control system

$$\dot{\chi} = A\chi + b_2 v, \quad \chi(0) = 0.$$

In terms of the new system the obligations of the follower take the form

$$h_0'\chi(t^*) = \alpha, \quad H\chi(t^*) = c^v.$$

An admissible control $v^0(t|u^0(\cdot)), t \in T$, is said to be an optimal open-loop control of the follower if on it the control criterion reaches the maximum value

$$J_2^0 = g_0' \int_0^{t^*} F(t^* - s) b_2 v^0(s|u^0(\cdot)) ds =$$

$$= \max_v g_0' \int_0^{t^*} F(t^* - s) b_2 v ds.$$

To introduce an optimal hierarchical control of feedback type (an optimal positional hierarchical control) let us imbed the problem (8)-(14) with the fixed initial state $x(0) = x_0$ into the family of problems depending on a scalar τ and n-vectors z, ξ:

$$h_0'x(t^*) \longrightarrow \max, \quad g_0'x(t^*) \longrightarrow \max,$$

$$\begin{gathered} \dot{x} = Ax + b_1 u + b_2 v, \quad x(\tau) = z, \\ Hx(t^*) = c, \quad Gx(t^*) = d, \\ \dot{\chi} = A\chi + b_2 v, \quad \chi(\tau) = \xi, \\ h_0'\chi(t^*) = \alpha, \quad H\chi(t^*) = c^v, \\ |u(t)| \le 1, \quad |v(t)| \le 1, \quad t \in T_\tau = [\tau, t^*]. \end{gathered} \qquad (17)$$

Let $Z_\tau \times \Xi_\tau$ be a set of couples $\{z, \xi\}$ for which problem (17) has optimal open-loop hierarchical controls $u_\tau^0(\cdot) = (u^0(t|\tau, z, \xi),\ t \in T_\tau)$, $v_\tau^0(\cdot) = (v^0(t|u_\tau^0(\cdot), \tau, z, \xi), t \in T_\tau)$.

Functions

$$\begin{gathered} u^0(\tau, z, \xi) = u^0(\tau|\tau, z, \xi), \\ v^0(\tau, z, \xi) = v^0(\tau|u_\tau^0(\cdot), \tau, z, \xi), \\ \{z, \xi\} \in Z_\tau \times \Xi_\tau, \quad \tau \in T, \end{gathered} \qquad (18)$$

are said to be optimal positional controls of the leader and the follower.

Assume that the behaviour of closed system (8) under unknown disturbances $w(t)$, $t \in T^0$, is described by the equations

$$\begin{gathered} \dot{x} = Ax + b_1 u^0(t, x, \chi) + b_2 v^0(t, x, \chi) + w(t), \\ x(0) = x_0, \\ \dot{\chi} = A\chi + b_2 v^0(t, x, \chi), \quad \chi(0) = 0. \end{gathered} \qquad (19)$$

In a concrete control process functions (18) are used not for all the values $z \in Z_\tau$, $\xi \in \Xi_\tau, \tau \in T$. It is sufficient to use only the values $u^*(\tau) = u^0(\tau, x^*(\tau), \chi^*(\tau))$, $v^*(\tau) = v^0(\tau, x^*(\tau), \chi^*(\tau))$ along the isolated curve $(x^*(\tau), \chi^*(\tau))$, $\tau \in T$, generated by a concrete realization $w^*(t)$, $t \in T^0$.

Devices which in each concrete control process are able to generate the functions $u^*(\tau), v^*(\tau), \tau \in T$, are called the optimal controllers of the leader and the follower.

The purpose of what follows is to describe algorithms of operating optimal controllers.

5. DEFINING EQUATIONS OF OPTIMAL CONTROLLERS

According to (Gabasov, 1994) optimal open-loop controls $u_\tau^0(t), v_\tau^0(t)$, $t \in T_\tau$, of problem (17) have the form

$$u_\tau^0(t) = \text{sign}\Delta_\tau^1(t), \quad v_\tau^0(t) = \text{sign}\Delta_\tau^2(t), \quad t \in T_\tau, \qquad (20)$$

where the functions $\Delta_\tau^i(t) = \psi_\tau^{i\prime}(t) b_i$, $t \in T_\tau$, $i = 1, 2$, are calculated along the solutions $\psi_\tau^i(t)$, $t \in T_\tau$, to the conjugate systems

$$\dot{\psi}_\tau^1 = -A'\psi_\tau^1, \quad \psi^1(t^*) = h_0 - H'y^1(\tau),$$

$$\dot{\psi}_\tau^2 = -A'\psi_\tau^2, \quad \psi^2(t^*) = g_0 - \sum_{k=0}^{m_1} y_{k+1}^2(\tau) h_k -$$

$$-\sum_{k=1}^{m_2} y^2_{k+m_1+1}(\tau)g_k.$$

Here $y^1(\tau) \in R^{m_1}$, $y^2(\tau) \in R^{m_1+m_2+1}$ are the optimal Lagrange vectors of the leader and the follower.

Assume that for any $\tau \in T^0$ and $i = 1,2$ the equalities

$$\text{rank}(HF(t^* - t)b_1, t \in T_1^*(\tau)) = m_1,$$

$$\text{rank}(HF(t^* - t)b_2, t \in T_2^*(\tau)) = m_1 + 1,$$

$$\text{rank}(GF(t^* - t)b_2, t \in T_2^*(\tau)) = m_2$$

can be violated only in isolated points $\tau \in T^0$ where

$$T_1^*(\tau) = \{t \in \text{int}T_\tau : u_\tau^0(t-0) \neq u_\tau^0(t+0)\},$$

$$T_2^*(\tau) = \{t \in \text{int}T_\tau : v_\tau^0(t-0) \neq v_\tau^0(t+0)\}.$$

The set (5) ($i = 1$) of the control $u_\tau^0(t)$, $t \in T_\tau$, satisfies the defining equations of the leader

$$\sum_{j=0}^{p_1(\tau)} \int_{t_j^1(\tau)}^{t_{j+1}^1(\tau)} HF(t^* - t)b_1 dt k_j^1(\tau) = c - c^v -$$

$$- HF(t^* - \tau)(x^*(\tau) - \chi^*(\tau)), \qquad (21)$$

$$(h_0' - y^{1\prime}(\tau)H)F(t^* - t_j^1(\tau))b_1 = 0, \quad j = \overline{1, p_1(\tau)},$$

For the follower ($i = 2$) the defining equations take the form

$$\sum_{j=0}^{p_2(\tau)} \int_{t_j^2(\tau)}^{t_{j+1}^2(\tau)} h_0' F(t^* - t)b_2 dt k_j^2(\tau) =$$

$$= \alpha - h_0' F(t^* - \tau)\chi^*(\tau),$$

$$\sum_{j=0}^{p_2(\tau)} \int_{t_j^2(\tau)}^{t_{j+1}^2(\tau)} HF(t^* - t)b_2 dt k_j^2(\tau) = \qquad (22)$$

$$= c^v - HF(t^* - \tau)\chi^*(\tau),$$

$$\sum_{j=0}^{p_2(\tau)} \int_{t_j^2(\tau)}^{t_{j+1}^2(\tau)} GF(t^* - t)b_2 dt k_j^2(\tau) = d-$$

$$\sum_{j=0}^{p_1(\tau)} \int_{t_j^1(\tau)}^{t_{j+1}^1(\tau)} HF(t^* - t)b_1 dt k_j^1(\tau) -$$

$$-GF(t^* - \tau)x^*(\tau),$$

$$(g_0' - \sum_{k=0}^{m_1} y^2_{k+1}(\tau)h_k' - \sum_{k=1}^{m_2} y^2_{k+m_1+1}(\tau)g_k')$$

$$F(t^* - t_j^2(\tau))b_2 = 0, \quad j = \overline{1, p_2(\tau)}.$$

Here $t_0^i(\tau) = \tau$, $t^i_{p_i(\tau)+1} = t^*$, $k_j^i(\tau) = \text{sign}\Delta_\tau^i(t_j^i(\tau) + 0)$, $j = \overline{0, p_i(\tau)}$, $i = 1,2$.

Thus we get $p_1(\tau) + p_2(\tau) + 2m_1 + m_2 + 1$ nonlinear equations with $p_1(\tau) + p_2(\tau) + 2m_1 + m_2 + 1$ variables.

After solving the defining equations at a moment $\tau \in T^0$ on a regular intervals the values $u^*(\tau) = k_0^1(\tau)$, $v^*(\tau) = k_0^2(\tau)$ are fed to the inputs of the systems. The values $u^*(\tau)$, $v^*(\tau)$ on sliding intervals are calculated according to the scheme (Gabasov, *et al.*, 1992). For $\tau \in [t^0, t^*]$ the open-loop values

$$u^*(\tau) = u^0(\tau | t^0, x^*(t^0), \chi^*(t^0)),$$

$$v^*(\tau) = v^0(\tau | u_\tau^0(\cdot), t^0, x^*(t^0), \chi^*(t^0))$$

are used.

6. CONCLUSION

In the paper a problem of optimal hierarchical control with two participants is investigated. By analogy with the classical optimal controls of feedback type positional solutions of the problem under consideration are introduced. A way of realizing such solutions with the help of optimal controllers is suggested.

The research has been done under support from the Byelorussian Foundation for Basic Research (grant Ф97M-139) and the State Programme for Basic Research of Belarus.

REFERENCES

Gabasov, R. (1994). Adaptive method of solving linear programming problems. *Preprint Series of the University of Karlsruhe*, Institute for Statistics and Mathematics, Germany, Karlsruhe.

Gabasov, R., F.M. Kirillova and O.I. Kostyukova (1992). Constructing of optimal controls of feedback type in a linear problem. *Soviet Math. Dokl.*, **44**, No. 2, 608–613.

Gabasov, R., F.M. Kirillova and O.I. Kostyukova (1997). Construction of a positional solution to a linear hierarchical control problem. *Doklady Mathematics*, **56**, No. 2, 805–808.

Kleimenov, A.F. (1993). *Nonantagonistic Differential Games*. Nauka, Ekaterinburg.

Krasovskii, N.N. and A.I. Subbotin (1974). *Positional Differential Games*. Nauka, Moscow.

Petrosyan L.A. and N.N. Danilov (1985). *Cooperative Differential Games and Its Applications*. TSU Publishing House, Tomsk.

Von Stackelberg, H. (1952). *The Theory of the Marketeconomy*. Hodge, London.

PROBLEM OF ADAPTIVE CONTROL FOR MANUFACTURING-AND-SELLING ENTERPRISE UNDER ALTERATING DEMAND*

Baev I.A., Rogova E.F., Shiryaev E.V.

*Applied Mathematics Department of Southern Ural State University,
Lenin av. 76, Chelyabinsk, 454080, Russia*

Abstract: One of the practical application of optimization algorithms is control for manufacturing-and-selling enterprise under the condition of inexact and partial information on market and the enterprise. The control is understood as adaptation to deviating demand by means of alteration of enterprise's parameters. *Copyright © 1998 IFAC*

Keywords: economic systems,dynamic modells, optimization problems, control theory, adaptive algorithms.

Adaptation of enterprises under the condition of instable economic relations is one of the most actual problems in Russia. Adaptive management of enterprises provides external changes and processes that may cause errors by means of classical feedback approach with prediction of market's state and a priori calculation of more effective techniques to compensate input disturbances by means of deviation of company model's parameters.

This work is based on approaches adopted at Department of Applied Mathematics, South Ural State University. Initially, we use well- known J. Forrester's method of economic modeling (Forrester, 1971). For real-time estimation, identification and control we use the following mathematical model (Rogova, et al., 1997)

$$\begin{bmatrix} \mathbf{X} \\ \mathbf{Y} \end{bmatrix}_{k+1} = \begin{bmatrix} \mathbf{A(p)} & \mathbf{B(p)} \\ \mathbf{C(p)} & \mathbf{D(p)} \end{bmatrix} \begin{bmatrix} \mathbf{X} \\ \mathbf{Y} \end{bmatrix}_k + \begin{bmatrix} \mathbf{G(p)} \\ \mathbf{H(p)} \end{bmatrix} \mathbf{F}[(\mathbf{X},\mathbf{Y})_k] + \begin{bmatrix} \mathbf{M(p)} \\ \mathbf{T(p)} \end{bmatrix} U_k ; \quad k = 1, \dots, N, \tag{1}$$

where $\mathbf{X}$ implies the vector of parameters for selling department; $\mathbf{Y}$ means vector of manufactoring parameters; $\mathbf{A}, \mathbf{B}, \mathbf{C}, \mathbf{D}, \mathbf{G}, \mathbf{H}, \mathbf{M}, \mathbf{T}$ are the given matrices depending on parameters of production and sales $\mathbf{p}$; F is a non-linear vector-function; we denote a demand rate as U_k that is an input for this system; k is a discrete time. Values of $\mathbf{X}_0$ is known.

*This work is supported by Russian Foundation of Basic Researches, grant 96-01-00460

Dimension of the system depends on character of concrete enterprise, model's completeness and chosen goals of control. In experimental model dimension of vectors $\mathbf{X}$ and $\mathbf{Y}$ are 13 and 46, respectively. Coordinates of these vectors include different parameters which mean the firm's state at the given moment of time. We consider such important characteristics of manufacturing- and-selling enterprise such as income and processing of orders, products and primary goods holding in manufacturing and selling departments, production cycle, purchase of primary goods, labour management, earnings and outgo of funds, pure profit and dividend calculation.

Demand $U(t)$ for company's production is an input signal in the system. System's reaction on to demand variation actually depends on parameters $\mathbf{p}$ of the model. These parameters are different delays of flows (for example, a firm needs a time to book orders, to manufacture and transport goods to customers), characteristics of production cycle and firm structure.

For analysis of model's behaviour we use abrupt, or, stepwise increase of demand. System's reaction to this input signal shows that the firm loses by inequality between demand and production offer. These losses are provided by deficit and overshoot of goods, overspending for holding, bad organization of manufacturing, low rate of funds return, etc.

We consider short-term cases when the structure of enterprise is constant. It means that we cannot increase firm facility or change its organization essentially. In this case technical and economic characteristics might be improved by setting and solution of respective optimization problem.

Any parametric changes are associated with investment of funds which are generally limited. Besides, we have to define ranges of these changes because parameters imply characteristics of real physical objects. As a result, we use the following restrictions

$$\begin{gathered} \mathbf{p}_i^L \le \mathbf{p}_i \le \mathbf{p}_i^H,\ i = 1,\ldots,m, \\ \sum_{i=1}^{m} \mathbf{c}_i(\mathbf{p}_i^0 - \mathbf{p}_i) \le R, \end{gathered} \tag{2}$$

where $\mathbf{p}_i^0, \mathbf{p}_i^L, \mathbf{p}_i^H$ are vectors of initial, minimal, maximal values of parameters $\mathbf{p}_i$; $\mathbf{c}_i$ is a vector of expenses for the parameters deviation; R means avaliable investments for the parameters changing and m is number of parameters.

Depending on control goals, we can choose a criterion of for efficiency of firm functioning efficiency. The most popular criterion for economic systems like this is summary pure profit at given moment of time. But the main difficulty in this case is that the reaction of profit rate to demand changes follows not immediately and we need to take a big interval of time to calculate this functional.

We offer the following criterion for manufacturing-and-selling efficiency. It shows how the system reacts to the input signal, i.e. how quickly the firm reacts to market conditions changes. The expression for the criterion is

$$L_{k+1} = L_k + T|U_k - F_k|,\ k = 0,\ldots, \tag{3}$$

where L_k is the value of total losses, $L_0 = 0$; sign U_k means the demand rate; F_k is the value of shipment and T implies the period between calculations.

To provide an adaptive control under alteraiting demand, we minimize total losses (3) for system (1) over some period N by means of deviation of parameters $\mathbf{p}$ under limited funds (2) provided for this aim.

Under changes of market conditions we can find optimal parameters for the firm repeatively. As a result, the enterprise control will adapt the firm to demand alteration in a real-time mode. For optimization we chose the random search algorithm with adaptive structure and the possible directions algorithm (on the base of Topkins-Veinott's method).

Results of application this optimization procedure are demonstrated on Fig. 1, where we can see different reactions of the system with initial and optimal parameters to deviation of demand. Experiments on adaptive enterprise management have shown the following results. The system with optimal parameters is characterized by both less length of transition period and less amplitudes of vectors coordinates variances that means better firm functioning.

Also we noted that starting at the certain moment, the increase of allocated funds produces less variation of functional value. This resource marginal utility principle is well-known for economists. It

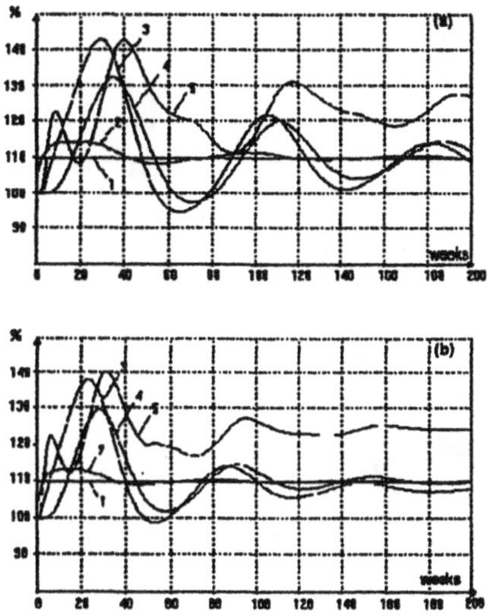

Fig. 1. Reaction of the system with initial (a) and optimal (b) parameters to 10% abrupt increase of demand. Here 1 - demand, 2 - outstanding orders in spelling department, 3 - outstanding orders in manufacturing department, 4 - work forcewhich is involved in production cycle, 5 - rate of pure profit.

means that there is a moment, after which the system is saturated and every new unit of resource produces less effect for enterprise's efficiency. So we have the problem to find the optimal value of investments R and to determine the expenses recoupment time. This problem can be solved using this model as well.

As a result of our research, we can conclude the following. We constructed the manufacturing-and-selling enterprise model described by the system of differential equations (1), developed algorithms and special software for research and adaptive enterprise management by means of optimization of production process, work force and financial state of the firm under alterating demands.

The offered method of enterprise modeling is flexible enough. Depending onf characters of the concrete firm it allows to decrease system dimension or, contrariwise, to make the model more detailed by way of addition of new parameters or, nay, new subsystems. Also we suggest to transfer the research results to the case when the firm operates under the condition of uncertainty, i.e. under the condition of inexact information on demand and firm's state vectors. Obtained results may be a basis for real-time adaptive control systems that use new information technologies.

REFERENCES

Forrester, J. (1971). *Industrial Dynamics.* Progress, Moscow.

Rogova, E.F. , E.V. Shiryaev, V.I. Shiryaev (1997). One optimization problem for industrial dynamics, labour and finances under alterating demand. In: *Proc. of International Conf. "Optimization Problems and Economic Applications".* 132. Omsk.

Shiryaev, V.I. , E.V. Shiryaev (1996). Estimation of firm's state and optimization under uncertain conditions. In: *Mulpile criteria and game problems under uncertainty. Abstract of the Fourth Internation Workshop.* 105. Moscow.

ON EXISTENCE OF PERIODIC SOLUTIONS FOR LINEAR DIFFERENCE SYSTEM WITH CONTINUOUS ARGUMENT UNDER PERIODIC CONTROL

Michael Blizorukov * Gab-Sang Yoo **

* *Institute of Engineering Scince of Russian Academy of Sciences, Ekaterinburg, Russia*
** *Information Technology Co., Seoul, Korea*

Abstract: This paper is devoted to consideration of linear difference equations with constants coefficients, continuous argument under periodic control. The general solution of a system without control is constructed. The conditions of existence of the periodic motion of systems under control and without control for periods both commensurable, and incommensurable with a step of of the system are obtained.

Keywords: difference equations, continuous time system, periodic motion

1. INTRODUCTION

The purpose of this paper is to obtain conditions for the system of difference equations with continuous argument under continuous and periodic control $u(t)$

$$x(t+1) = Ax(t) + u(t), \qquad (1)$$
$$x(\tau) = \varphi(\tau), \quad \tau \in [0,1),$$

that should be fulfilled for realization of a periodic motion by this system. Here control $u(t)$ is a continuous and T-periodic by t n-vector-function, $t \in R^+$, A is a real nonsingular $n \times n$ matrix, $x(t)$ is a real phase vector, $\varphi(t)$ is an initial n - vector-function. Solution of this problem is one of an initial steps in stabilization of a periodic motion for the system (1).

The extensive literature is devoted to development of the theory of a such kind of systems. The rather full bibliography may be found, for example, in (Sharkovskii *et al.*, 1986; Chermanescu, 1960). There has been much interest recently in the construction of general or special solutions of such systems, see e.g. (Harris and Sibuya, 1964; Tanaka, 1970; Blizorukov, 1996). In particular, the conditions of existence and uniqueness of the periodic solutions of difference systems with continuous time for a case of integer period, were established, for example, in (Pelukch, 1996).

This paper is devoted to obtaining of conditions for existence of the periodic solution of the system (1) with any period (both commensurable, and incommensurable with a step of the system). In the second section the general solution of the system without control (homogeneous system): $x(t+1) = Ax(t)$, is constructed. The theorem 1, presenting main result of the third section, establishes existence conditions of the periodic solutions of a system without control as a restrictions to eigenvalues of a matrix A and to initial function $\varphi(\tau)$. The fourth section is devoted to consideration of the system under control (1).

2. GENERAL SOLUTION OF A SYSTEM WITHOUT CONTROL

Let us consider a system without control, corresponding to (1):

$$x(t+1) = Ax(t), \quad (2)$$

$$x(\tau) = \varphi(\tau), \ \tau \in [0,1).$$

We assume that A there has normal Jordan form, and all eigenvalues of A are simple or semisimple. We denote eigenvalues of A as: $\rho_j = e^{\lambda_j}(j = 1,2,...n)$, and corresponding eigen vectors as $h_j(j = 1,2,...n)$. Let us search the solution of an equation (2) in the special form:

$$x(t) = e^{\lambda t} h c(t), \quad (3)$$

where λ is a number, h is an n-dimensional vector, $c(t)$ is a scalar 1-periodic function. If $\det(A - e^{\lambda}E) = 0$, then $x(t)$ defined by (3) sets a solution of (2) with arbitrary 1-periodic vector function $c(t)$. The system (2) has n independent solutions $x_j(t) = e^{\lambda_j t} h_j c_j(t), j = 1,2,...n$, difiniendums within to arbitrary 1-periodic functions $c_j(t+1) = c_j(t)(j = 1,2,...,n)$. The linear combination of such solutions

$$x(t) = \sum_{j=1}^{n} e^{\lambda_j t} h_j c_j(t), \quad (4)$$

is a general solution of (2).

In case of multiple eigenvalues the construction of the general solution is connected to bulky transformations, but it does not contain basic difficulties. It is created similarly, with the help of construction analogous to (Halanay and Wexler, 1968). The general solution, in this case, may be recorded in the form (Pelukch, 1994).

3. EXISTENCE PERIODIC MOTION FOR THE SYSTEM (2)

Let us consider the problem of existence of the T-periodic solution of the system (2). In case of integer T such problem was considered, for example, in (Pelukch, 1996). Here we shall not limit class of periods commensurable with a step of the difference equation. As above, let us assume, that all eigenvalues of a matrix A are simple or semisimple.

It is known (Malkin, 1956), that fulfillment of condition

$$x(t+T) = x(t), \quad t \in R^+, \quad (5)$$

is necessary and sufficient for T-periodicity of the solution $x(t)$. Substituting (4) into (5), and equating coefficients at independent vectors h_j, we receive $e^{\lambda_j T} c_j(t+T) = c_j(t), (j = 1,2,...,n)$. Thus, existence condition of the periodic solution for (2) is reduced to determination of λ and ψ satisfying the system

$$\begin{cases} e^{\lambda T}\psi(t+T) &= \psi(t), \\ \psi(t+1) &= \psi(t), \end{cases} \quad (6)$$

where $e^{\lambda T} = \rho_j^T \ (j = 1,2,...,n)$, are multiplicators of the system, and $\psi(t)$ is an initial function. First, let us consider the case of integer period $T = N$, $N \in Z$. The following lemma is true.

Lemma 1. Existence condition of the T-periodic solution for the system (2) in case $T = N, N \in Z$ are: $\rho = exp(2\phi ik/N)$, $\psi(t)$ is arbitrary 1-periodic function.

Proof. From the system (6) we obtain: $e^{\lambda N}\psi(t+N) = \psi(t)$ and $\psi(t+N) = \psi(t)$. Therefore, $\rho = exp(2\phi ik/N)$, and $\psi(t)$ there is any 1-periodic function.

For T, that is not necessary be multiple or rationally commensurable with a step of the system, the following results are true.

Lemma 2. In order that the system (2) has a T-periodic solution it is necessary that $e^{\lambda T} = 1$.

Proof. Hereinafter we shall denote as $[a]$ and $\{a\}$ the integral and fractional parts of number a respectively. From (6) and the fact that $\psi(t)$ is a 1-periodic function follows: $e^{s\lambda T}\psi(\{sT\}) = \psi(0)$. Under Dirichlet theorem (Shidlovskii, 1987), it is possible to find a sequences of integers $q_n \in [1,n]$ and p_n that $|Tq_n - p_n| < n^{-1}$. By passing in this inequality to a limit on $n \to \infty$, and using a continuity of a function $\psi(t)$ on $[0,1]$, we shall receive $|e^{\lambda T}| = 1$. The substitution of $\psi(t)$ as a Fourier series into (6) allows to receive $e^{\lambda T} = 1, (\rho^T = 1)$.

Lemma 3. Functions $\psi(t)$, supplying the solution to the system (2) depending on T is: an arbitrary $(1/m)$-periodic function, if T is rational $T = k/m$ (k, m are relatively prime natural numbers); and is an arbitrary constant, if T is irrational.

Proof. For rational period $T = k/m$ (k, m - relatively prime natural numbers) from (6) we have $\psi(t + k/m) = \psi(t)$, and $\psi(t+1) = \psi(t)$. Therefore, $\psi(t)$ is any $(1/m)$-periodic function. For irrational T, it is known (Shidlovskii, 1987), that $\{nT\}, n \in N$ is an everywhere dense point set on $[0,1]$. From the system (6) follows that $\psi(\{nT\}) = \psi(0) = \psi(1)$. Therefore, function $\psi(t)$ is equal to $\psi(0)$ almost everywhere in $[0,1]$ and, by virtue of a continuity, it is a constant on $[0,1]$.

Statements proved above, allow us to formulate the following theorem.

Theorem 4. The system (2) has the T-periodic solution if and only if, the matrix A has the eigenvalues in the form $\lambda = (2\pi ik/T)$. Thus the initial function ψ, depending on the period, should be one of the following

a. $\psi(t)$ is an arbitrary 1-periodic function, if $T = N$, $N \in Z$;

b. $\psi(t)$ is an arbitrary $(1/m)$-periodic function, if $T = k/m$ (k, m are relatively prime natural numbers);

c. $\psi(t)$ is an arbitrary constant, if T is irrational.

The obtained results may be widespread on the case of multiple eigenvalues. Moreover, the procedure of a construction of the periodic solutions will be similar to that was elaborated for differential equations in (Malkin, 1956), or for finite-difference equations in (Halanay and Wexler, 1968; Martinuk, 1972).

4. EXISTENCE CONDITIONS AND EVALUATION OF THE PERIODIC MOTION OF CONTROLLED SYSTEMS

Let us consider now controlled system

$$x(t+1) = Ax(t) + u(t), \quad (7)$$

$$x(\tau) = \psi(\tau), \ \tau \in [0,1);$$

where $u(t)$ is a control, continuous and T-periodic vector function. As above we assume, that A has the diagonal form. The following statements are true.

Theorem 5. If any eigenvalues of the matrix A are not equal modulo to unit, then the system (7) with T-periodic contro $u(t)$ has the T-periodic solution with the initial function $x(\tau) = \psi(\tau) = 0, \tau \in [0,1]$.

Proof. Let A is a diagonal matrix with eigenvalues not equal modulo to unit. Let us divide up A into two blocks: $A = diag(A_+, A_-)$, where A_+, and A_- are matrixes with eigenvalues by large and smaller unit respectively. Then the system (7) will break up to two independent subsystems. The general solution (7) has a form $x(t) = e^{\lambda_1 t}h_1c_1(t) + e^{\lambda_1 t}h_1c_1(t) + \cdots e^{\lambda_1 t}h_1c_1(t) + x_*(t)$. Here $\rho_j = e^{\lambda_j t}, h_j (j = 1, 2, \ldots, n)$, are eigenvalues and appropriate eigenvectors of a matrix respectively, $c_j(t)(j = 1, 2, \ldots, n)$, are 1-periodic complex-valued scalar functions. By a direct substitution it is possible to see, that the particular solution of (7), may be taken as $x_* = \sum_{k=1}^{+\infty} A_-^{k-1}u(t-k)$ for $\rho < 1$ and $x_* = -\sum_{k=-\infty}^{0} A_+^{k-1}u(t-k)$ for $\rho > 1$. The convergence of vectorial series is equivalent coordinate-wise convergence. Bounds $|u_j(t)| \le \max_{0\le t\le T} |u_j|, t \in R^+$, $\rho_j \le \alpha < 1$ and, $\rho_j > 1$ allow us to conclude, that majorizing series converge absolutely, and consequently also formal series converge absolutely and uniformly (Weierstrass criterion). Under the theorem 1, in order to the solution of an equati on (7) be T-periodic, it is necessary to put $c_j(t) = 0$. In this case the system (7) will have T-periodic solution

$$x(t) = \sum_{k=1}^{+\infty} A_-^{k-1}u(t-k)$$

and/or

$$x(t) = -\sum_{k=-\infty}^{0} A_+^{k-1}u(t-k).$$

The calculation of the periodic solution for (7) here implements as a Fourier series. It was shown, that under condition of absolute convergence of a series for periodic control $u(t) = \sum_{k=-\infty}^{+\infty} u_k e^{ik\nu t}$, where $\nu = 2\pi/T$, a series for $x(t)$ converges absolutely. Therefore, the periodic solution of system (7) in this case may be represented as $x(t) = \sum_{k=-\infty}^{+\infty} [e^{ik\nu t}E - A]^{-1}u_k e^{ik\nu t}$. The problem of existence of the periodic solutions in case of a diagonal matrix A has a critical eigenvalues now is reduced to investigation of a scalar equation

$$x(t+1) = \rho x(t) + u(t), \quad |\rho| = 1. \quad (8)$$

Let us consider separately cases of integer, rational and irrational period T. Concerning conditions of existence of the periodic motion of (8) the following statements are true.

Theorem 6. The N-periodic motion ($N \in Z$) of the system (8) under the control $u(t)$ with an integer period $T = N$ in resonant case ($\rho^N = 1$), exist if and only if $\sum_{j=-\infty}^{+\infty} \rho^{N-1}u(t+j-1) = 0$. Under this condition arbitrary N-periodic function is the solution of (8). If $\rho^N \neq 1$, then the system (8) has the unique N-periodic solution defined by an initial function $\psi = (1-\rho^N)^{-1} \sum_{j=-\infty}^{+\infty} \rho^{N-1}u(t+j-1)$, $t \in [0,1)$.

Proof. The condition of existence of the periodic solution for (8) has a form: $(1-\rho^N)x(t) = \sum_{j=1}^{N} \rho^{N-j}u(t+j-1)$ for anyone of t. If $1-\rho^N = 0$, and $\sum_{j=1}^{N} \rho^{N-j} u(t+j-1) \neq 0$ the equation has not the solutions for x. In this case there are no N-periodic solutions of the equation (8). If $1-\rho^N = 0$, and $\sum_{j=1}^{N} \rho^{N-j} u(t+j-1) = 0$ the solution of (8) is any N-periodic function. In case, when $1-\rho^N \neq 0$, the N-periodic solution of (8) is unequely determined by $(1-\rho^N)^{-1} \sum_{j=1}^{N} \rho^{N-j} u(t+j-1)$, i.e. specify by an initial function declared in the theorem.

Theorem 7. The equation (8) with a control $u(t)$ of rational period $T = k/m$, (k, m are relatively prime integers, $m \neq 0$) has a T-periodic solution if $\rho = 1$, and $\sum_{j=-\infty}^{+\infty} \rho^{k-1} u(t_1+j-1) = 0$, where $t_1 \in [0, T]$, in this case any p-periodic function is the solution (8). If $\rho \neq 1$, then the condition $\sum_{j=-\infty}^{+\infty} \rho^{k-j}[u(t_1+T+j-1) - u(t_1+j-1)] = 0$ is sufficient for existence of the T-periodic solution in this case the unique periodic solution will be defined by an initial function $\psi(t) = (1-\rho^k)^{-1} \sum_{j=-\infty}^{+\infty} \rho^{k-j} u(t+j-1)$, $t \in [0,1)$.

Proof. If conditions $1-\rho^k = 0$, is fulfilled then validity of this theorem follows from the theorem 3. If $1-\rho^k \neq 0$, the system (8) has the unique k-periodic ($T = k/m$) solution: $x(t) = (1-\rho^k)^{-1} \sum_{j=1}^{k} \rho^{k-j} u(t+j-1)$. Let us clarify when this solution will be also lT-periodic, where l is an integer, $0 < l < m$. For $t_1 \in [0,1)$, the solution (8) satisfy an equalities: $x(t_1+lT) - x(t_1) = x(t_1+lT+k) - x(t_1+k) = \sum_{j=1}^{k} \rho^{k-j}[u(t_1+lT+j-1) - u(t_1+j-1)]$. The solution will be lT-periodic if the last expression is equal to zero. Therefore, in case $1-\rho^k \neq 0$, and $\sum_{j=1}^{k} [\rho^{k-j} u(t_1+T+j-1) - u(t_1+j-1)]$ equation of (8) has unique T periodic solution, that is defined by initial function $\psi(t) = (1-\rho^k)^{-1} \sum_{j=1}^{k} \rho^{k-j} u(t_1+j-1)$.

Let us consider now case of irrational period T. We assume, that the control $u(t)$ may be represented as an absolutely converging series:

$$u(t) = \sum_{k=-\infty}^{+\infty} u_k e^{ik\nu t}, \quad \nu = \frac{2\pi}{T}. \qquad (9)$$

Theorem 8. The equation (8) with $\rho = e^{i\omega}$, $u(t)$ represented as series (9), and irrational period T for $\omega T = 2\pi(N+MT)$, M, $N \in Z$ has not the T-periodic solutions, if $u_N \neq 0$, and has a one-parameter T-periodic solution with parameter α: $x(t) = \sum_{k=-\infty, k\neq N}^{+\infty} (x_k e^{ik\nu t}) + \alpha e^{iN\nu t}$, if $u_N = 0$. For $\omega T \neq 2\pi(N+MT)$, M, $N \in Z$, the equation (8) has the unique T-periodic solution $x(t) = \sum_{k=-\infty}^{+\infty} x_k e^{ik\nu t}$, if only formal series determining all these solutions converge.

The proof of the theorem is carried out by a substitution of periodic control $u(t)$ as Fourier series into the system. The solution (8) is searched also as Fourier series. At the proof of convergence of series for solutions, indicated in the theorem, the problem of small devisors is arised. The conditions of convergence may be obtained by using of the Poincare and Sigel theorems (Arnold, 1978).

5. REFERENCES

Arnold, V.I. (1978). *The additional chapters of the theory of ordinary differential equations.* Nauka. Moscow.

Blizorukov, M.G. (1996). On construction of solutions for linear difference systems with continuous time. *Different. Uravnenia* **32**(1), 127–128.

Chermanescu, M. (1960). *Ecuatii functionale.* Bucuresti.

Halanay, A. and D. Wexler (1968). *Teoria calitativa a sistemelor cu impulsuri.* Editua Academiei. Republicii Romania.

Harris, W.A. and Y. Sibuya (1964). Note on linear difference equations. *Bulletin AMS* **70**(1), 123–127.

Malkin, I.G. (1956). *Some problems of the theory of non-linear oscillations.* Nauka. Moscow.

Martinuk, D.I. (1972). *The lectures on the qualitative theory of difference equations.* Naukova Dumka. Kiev.

Pelukch, G.P. (1994). The general solution for systems of linear difference equations with continuous argument. *Dokl. Ukraine Acad. Nauk* **1**, 16–21.

Pelukch, G.P. (1996). On the periodic solutions of difference equations with continuous argument. *Ukrainian Math. J.* **48**(1), 140–145.

Sharkovskii, A.N., Yu.L. Maistrenko and E.Yu. Romanenko (1986). *Difference equations and their applications.* Naukova Dumka. Kiev.

Shidlovskii, A.B. (1987). *Transcendental numbers.* Nauka. Moscow.

Tanaka, Sen-Isiro (1970). On asymptotic solutions of the functional difference equations associated with some nonlinear differential equations. *Publications of the Research Institute for Mathematical Sciences* **6**(2), 205–236.

DYNAMICAL RECONSTRUCTION OF UNBOUNDED CONTROLS THROUGH MEASUREMENTS OF A "PART" OF PHASE COORDINATES

Marina Blizorukova *,1 **Vyacheslav Maksimov** *,1

* *Institute of Mathematics and Mechanics, Ekaterinburg, Russia*

Abstract: The problems of dynamical reconstruction of a pair "control-trajectory" for the systems described by ordinary nonlinear differential equations and by equations with hereditary are considered. An algorithm which is stable with respect to informational noises and computational errors is suggested. This algorithm is based on the known in the theory of positional control principle of extremal aiming.
Copyright ©1998 IFAC.

Keywords: control, reconsrtuction, hereditary systems.

1. INTRODUCTION

Problems of reconstruction of input disturbances, determining motion of a dynamic system, through the measurements of a part of its phase coordinates are embedded in the theory of inverse problems of control systems dynamics, which is well developed at the present time. Inputs, as a rule, are the factors, which uniquely determine the system's motion. Output can be any the available information on the process. Usually, such information is some signal about a part of current system's state. One of approaches to solving such problems based on the methods of positional control theory (Krasovskii and Subbotin, 1988; Krasovskii, 1985) was suggested in (Kryazhimskii and Osipov, 1983) and developed in (Osipov and Kryazhimskii, 1995; Osipov *et al.*, 1991; Maksimov, 1990; Maksimov, 1995). In the present paper which is continuing a cycle of researches on this theme new algorithms are suggested. The algorithms differ from those of a similar type by that the set of admissible inputs is supposed to be unbounded. The algorithms are stable with respect to informational noises and computational errors. These algorithms are dynamical and work in "real time".

Briefly, the essence of the problems under consideration may be formulated in the following way. There is a dynamic system Σ functioning during time period $T = [t_0, \vartheta]$. Its trajectory $x(t) = x(t; x_0, u(\cdot))$, $t \in T$ depends on an unknown time-varying input $u(\cdot) \in L_2(T; R^N)$. On the interval T a uniform net $\Delta = \{t_i\}_{i=0}^m$ with a step δ is taken, $t_{i+1} = t_i + \delta$, $t_m = \vartheta$. Output

$$y(t) = Cx(t)$$

is measured at the moments t_i (C is an $r \times n$-dimensional matrix). The results of inaccurate measurements are vectors $\xi_i \in R^r$ which satisfy the inequalities

$$|\xi_i - y(t_i)|_r \le h, \quad i \in [0, m-1],$$

where h is the value of informational noise. It is required to indicate an algorithm which allows to reconstruct some pair $(u_*^h(\cdot), x_*^h(\cdot))$ synchro with the process. This pair must be "close" to the set of all pairs $(u(\cdot), x(\cdot))$ compatible with an output $y(\cdot)$, i. e. the set

[1] Partially supported by the Russian Foundation for Basic Research (grant # 98-01-00046) and the Program of support of leading scientific schools of Russia (project # 961596116)

$$Z(y(\cdot)) = \{(u(\cdot), x(\cdot)) : u(\cdot) \in L_2(T; R^N),$$
$$x(\cdot) = x(\cdot; \varphi^1, u(\cdot)),$$
$$Cx(t; x_0, u(\cdot)) = y(t)\ \forall t \in T\}.$$

In the present report, the problem of dynamic reconstruction of a pair "control-trajectory" is considered in three cases, when Σ is described by a system

a) of ordinary differential equations

$$\dot{x}(t) = A(x(t)) + Bu(t) + f(t), \qquad (1)$$
$$x(t_0) = x_0.$$

b) of differential equations with hereditary of the following form

$$A\dot{x}(t) = L(x_t(s)) + Bu(t) + f(t), \qquad (2)$$
$$L(x_t(s)) = \sum_{j=1}^{l} A_j x(t - \tau_j) +$$
$$+ \int_{-\tau}^{0} A_*(s) x(t + \zeta)\, d\zeta,$$
$$x_{t_0}(s) = \varphi^1(s), \quad s \in [-\tau, 0],$$
$$\varphi^1(s) \in C^{(1)}([-\tau, 0]; R^n).$$

c) of differential equations with hereditary of the form

$$\dot{x}(t) = L(x_t(s)) + Bu(t) + f(t), \qquad (3)$$
$$x_{t_0}(s) = \varphi^1(s), \quad s \in [-\tau, 0],$$
$$\varphi^1(s) \in C^{(1)}([-\tau, 0]; R^n).$$

Here $t \in T = [t_0, \vartheta]$, $x \in R^n$ is a phase vector of the system, $u(t) \in R^N$ is a control, $f(\cdot)$ is a given function, τ are hereditaries: $0 = \tau_1 < \tau_2 < \ldots < \tau_l \leq \tau$; A_j, $j \in [1 : l]$, B are constant $n \times N$ and $N \times n$ matrices. The elements of $n \times n$-dimensional matrix function, $s \to A_*(s)$, belong to space $L_\infty([-\tau; 0]; R)$. Let us suppose also, that the function $f(\cdot)$ is differentiable and $\dot{f}(\cdot) \in L_2(T; R^n)$. The solutions of all equations corresponding to controls $u(\cdot) \in L_2(T; R^N)$, are understood in the sense of Caratheodory. This means that they are absolutely continuous functions, at almost all $t \in T$ satisfying (1), (2) or (3) respectively. Denote them by $x(\cdot; \varphi^1, u(\cdot))$.

Let us pass to the description of the algorithms for solution of described above problems, staying in details, for example, on case **c)** and **b)**.

2. SOLVING ALGORITHMS

Let us apply the approach developed in (Kryazhimskii and Osipov, 1983; Osipov and Kryazhimskii, 1995; Osipov *et al.*, 1991; Maksimov, 1990; Maksimov, 1995). According to this approach, a problem of approximate calculation of some pair $(x_*(\cdot), u_*(\cdot)) \in Z(y(\cdot))$ substitute by a problem of control of an auxiliary system M (a model), which is formed according to the feedback principle. The phase trajectory of the model is $Y^h(\cdot)$ and control is $U^h(\cdot)$. The process of control of the model is organized in conformity to the rule identified with some map $U^h(\cdot) = U^h(\cdot; \xi(\cdot), Y^h(\cdot))$. The process is realized so that under appropriate conditions of the coordination of concordance of some parameters the pair $(U^h(\cdot), Y^h(\cdot))$ is "close" in some sense to the set $Z(y(\cdot))$. Hereinafter, for the simplicity assume that $h \in (0,1)$, $\delta \in (0,1)$, $h/\delta \in (0,1)$ and that the following conditions are fulfilled

Condition 1. $\operatorname{rank} G = n + r$, where

$$G = \begin{pmatrix} C' & I \\ 0 & B' \end{pmatrix}$$

Condition 2. $\operatorname{vrai\,sup}\{|\dot{x}_*(t)|_n : t \in T\} \leq K = \text{const} \in (0, +\infty)$.

At first, we select a family of partitions of the interval T

$$\Delta_h = \{t_{i,h}\}_{i=0}^{m}, \quad t_{0,h} = t_0, \qquad (4)$$
$$t_{m,h} = \vartheta, \quad t_{i+1,h} = t_{i,h} + \delta(h), \quad m = m_h.$$

The model equations M may be given in the form

$$\begin{cases} \dot{z}^h(t) = u^h(t) \\ \dot{y}^h(t) = L(z_t^h(s)) + Bv^h(t) + f(t), \end{cases} \qquad (5)$$
$$z_{t_1}^h(s) = \varphi^h(s), \quad y^h(t_1) = \varphi^h(0)$$
$$|\dot{\varphi}^h(s) - \dot{\varphi}^1(s)|_C \leq h, \ |\varphi^h(s) - \varphi^1(s)|_C \leq h$$

with phase trajectory $Y^h(t) = (z^h(t), y^h(t)) \in R^{2n}$, $t \in T$ and control $U^h(t) = (u^h(t), v^h(t)) \in R^{n+N}$. We identify the rule of formation of control in the model $U^h(\cdot; \xi^h(\cdot), Y^h(\cdot))$ with a rule which puts vector

$$U^h(q_*^{(i)}(\cdot)) = \qquad (6)$$
$$= (u_i^h, v_i^h) = \begin{cases} 0, & c_i \geq 0 \text{ or } d(s_{1i}, s_{2i}) = 0, \\ c_i \sigma_i |\sigma_i|_{n+N}^{-2}, & \text{otherwise}, \end{cases}$$

in correspondence to each element $q_*^{(i)}(\cdot) = \{t_i, y^h(t_i), z_{t_i}(s), \xi_i^h, \xi_{i-1}^h\}$, $i \in [1 : m-1]$. Here

$$\sigma_i = (C' s_{1i} + s_{2i}, -B' s_{2i}) \in R^{n+N},$$
$$c_i = (s_{1i}, (\xi_i^h - \xi_{i-1}^h)\delta^{-1})_r +$$
$$+ (s_{2i}, L(z_{t_i}^h(s)) + f(t_i))_n,$$
$$s_{1i} = Cz^h(t_i) - \xi_{i-1}^h \in R^r,$$

$$s_{2i} = z^h(t_i) - y^h(t_i) \in R^n.$$

For each fixed $h \in (0,1)$, this process is decomposed into a finite number of identical steps. At the i-th step carried out during time interval $\delta_i = [t_i, t_{i+1})$, $i \in [0 : m-1]$, $m = m_h$, $t_i = t_{i,h}$ the following actions are fulfilled. First, at time moment t_i, vector $U^h(q^{(i)}(\cdot))$ is calculated according to the chosen rules (6). Then the constant control

$$(u^h(t), v^h(t)) = (u_i^h, v_i^h), \quad t \in \delta_i,$$

is fed to the input of the model. Then the phase state of the model is transformed: instead of $(z_{t_i}^h(s), y^h(t_i))$ we calculate $(z_{t_{i+1}}^h(s), y^h(t_{i+1}))$. The procedure stops at time ϑ.

Let $X(y(\cdot))$ and $U(y(\cdot))$ be the projections of set $Z(y(\cdot))$ on the spaces $L_2(T; R^N)$ and $C(T; R^n)$, respectively, i. e.

$$U(y(\cdot)) = \{u(\cdot) \in L_2(T; R^N) :$$

$$Cx(t; x_0, U(\cdot)) = y(t) \ \forall t \in T\},$$

$$X(y(\cdot)) = \{x(\cdot) \in C(T; R^n) :$$

$$\exists u_*(\cdot) \in U(y(\cdot)), \ X(\cdot) = x(\cdot; x_0, u_*(\cdot))\}.$$

Definition 3. A family of algorithms $D_h = (\Delta_h, M, U^h)$ of the form (4)–(6) is called ***weakly regularizing*** if **a)** $v^h(\cdot) \to U(y(\cdot))$ weakly in $L_2(T; R^N)$, **b)** $z^h(\cdot) \to X(y(\cdot))$ in $C(T; R^n)$ as $h \to 0$.

If the value of information error is h, step of grid Δ is $\delta = \delta_h$ and the relation h/δ are small enough, then a pair $(v^h(\cdot), z^h(\cdot))$, where $v^h = 0$, $z^h(t) = \varphi^h(-\tau)$ at $t \in [t_0 - \tau, t_0 - \tau + \delta(h)]$, may be taken as an "approximation" to some element from the set $Z(y(\cdot))$. It follows from the next theorem.

Theorem 4. Let $h/\delta(h) \to 0$, $\delta(h) \to 0$ as $h \to 0$, $v^h(t) = 0$, $t \in [t_0, t_1]$. Then the family of algorithms D_h, $h \in (0,1)$ of the form (4)–(6) is weakly regularizing.

Remark 5. Theorem is also true if the model is described by discrete system

$$z^h(t_{i+1}) = z^h(t_i) + u_i^h \delta,$$

$$y^h(t_{i+1}) = y^h(t_i) + \{L(z_{t_i}^h(s)) +$$

$$+ Bv_i^h + f(t_i)\}\delta.$$

Remark 6. One can easily see, that the linearity of the operator $L(x_t(s))$ is not mandatory. The theorem is also correct if $L(x_t(s))$ is a nonlinear function, for example, $L(x_t(s)) = f_1(x(t - \tau_1), \ldots, x(t - \tau_n)) + \varphi(x_t(s))$, where functions $f_1(\cdot) : R^n \to R^n$ and $\varphi(\cdot) : C([-\tau; 0]; R^n) \to R^n$ satisfy to standard Lipschitz and growth conditions (Osipov, 1971).

Let us discuss now the case **b)**, when the system Σ is described by differential equations of the form (2). Let condition 2 as well as the following two conditions be fullfiled.

Condition 7. The representation $C = EA$ is valid.

Condition 8. There exists a number $c \in (0, \infty)$ such that

$$d(s_1, s_2) \equiv |E's_1 + s_2|_n^2 + |B's_2|_N^2 \geq$$

$$\geq c(|s_1|_r^2 + |s_2|_n^2) \quad \forall s_1, s_2 \in R^n.$$

In this case the model is given by the equations

$$\begin{aligned} \dot{z}^h(t) &= w^h(t) \\ \dot{p}^h(t) &= u^h(t) \\ \dot{y}^h(t) &= L(z_t^h(s)) + Bv^h(t) + f(t), \end{aligned}$$

$$t \in [t_1, \vartheta], \quad t_1 = t_{1,h}.$$

with the initial conditions

$$p^h(t_1) = y^h(t_1) = A\varphi^h(0),$$

$$z_{t_1}^h(s) = \varphi^h(s) \in C^{(1)}([-\tau, 0]; R^n),$$

$$|\dot{\varphi}^h(s) - \dot{\varphi}^1(s)|_c \leq h,$$

$$|\varphi^h(0) - \varphi^1(0)|_n \leq h,$$

with phase trajectory

$$Y^h(t) = (z^h(t), p^h(t), y^h(t)) \in R^{3n}, \ t \in T$$

and control

$$U^h(t) = (w^h(t), u^h(t), v^h(t)) \in R^{2n+N}.$$

We take the rule of formation of control in the model $U^h(\cdot; \xi^h(\cdot), Y^h(\cdot))$ according to rule which puts in correspondence to each element $q^{(i)}(\cdot) = \{t_i, y^h(t_i), z_{t_i}(s), p^h(t_i), \xi_i^h, \xi_{i-1}^h\}$, $i \in [1 : m-1]$, vector

$$U^h(q^{(i)}(\cdot)) = (w_i^h, u_i^h, v_i^h),$$

where

$$(u_i^h, v_i^h) = \begin{cases} 0, & c_i \geq 0 \text{ or } d(s_{1i}, s_{2i}) = 0, \\ c_i\sigma_i|\sigma_i|_{n+N}^{-2}, & \text{otherwise,} \end{cases}$$

$$w_i^h = \begin{cases} -\rho_i \cdot A's_{3i}\,|A's_{3i}|_n^{-1}, & \text{if } |A's_{3i}|_n \neq 0, \\ 0, & \text{otherwise,} \end{cases}$$

$$s_{1i} = Ep^h(t_i) - \xi_{i-1}^h \in R^r,$$

$$s_{2i} = p^h(t_i) - y^h(t_i) \in R^n,$$

$$s_{3i} = Az^h(t_i) - p^h(t_i),$$

$$\sigma_i = (E's_{1i} + s_{2i}, -B's_{2i}) \in R^{n+N},$$

$$c_i = (s_{1i}, (\xi_i^h - \xi_{i-1}^h)\delta^{-1})_r +$$

$$+ (s_{2i}, L(z_{t_i}^h(s)) + f(t_i))_n,$$

$$\rho_i = (2\,c^{-1})^{1/2}|A^{-1}|_{n\times n}(2 +$$

$$+ \sup_{t\in T}|f(t)|_n + K|C|_{n\times r} +$$

$$+ (la + a_*\tau)|z_{t_i}(s)|_c),$$

$$a = \max_{j\in[1:l]}|A_j|_{n\times n},$$

$$a_* = \{\text{vrai}\max|A_*(s)|_{n\times n}\ : s \in [-\tau, 0]\}.$$

Here the symbol $|A|_{n\times n}$ denotes norm of matrix A, l is number of time lags.
In this case, the procedure solving the problem is analogous to one described above.

3. EXAMPLE

The work of the algorithm was illustrated by a model example. The dynamical system Σ described by the differential equation (3) was considered under the assumption that

$$x = (x_1, x_2) \in R^2, \quad j = 2,$$

$$A_1 = 0, \quad A_2 = \begin{pmatrix} 1 & 1 \\ 1 & 0 \end{pmatrix},$$

$$B = (1, 0), \quad T = [0, 2], \quad \tau_1 = 0,$$

$$f(t) = \begin{pmatrix} t - 3 - (t - \tau_2)^2 - 5\sin t \\ -(t - \tau_2)^2 - (t - \tau_2) - 2 \end{pmatrix}.$$

The real trajectory $x(t)$ and the control $u(t)$ were supposed to be equal to $(1 + t^2, 2 + t)$ and $5\sin t$, respectively. The equations of the system and model were substituted by difference ones. It was introduced a uniform net with a time step δ. When computing the control $v^h(\cdot)$, only the values at nodes of the net were used. The results of measurements ξ_i^h were assumed to be equal to

$$\xi_i^h = x_1(t_i) + h\cos(10t).$$

The results of computer modelling of the real and model controls $u(t)$, $v^h(t)$ are presented in the following figures. For fig. 1 $\delta = 0.01$, $h = 0.01$ and for fig. 2 $\delta = 0.001$, $h = 0.001$. The real and model control are pictured by dashed and solid lines, respectively.

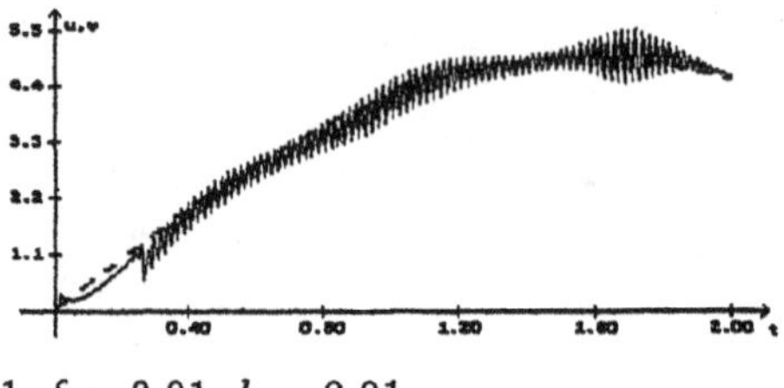

Fig. 1. $\delta = 0.01$, $h = 0.01$.

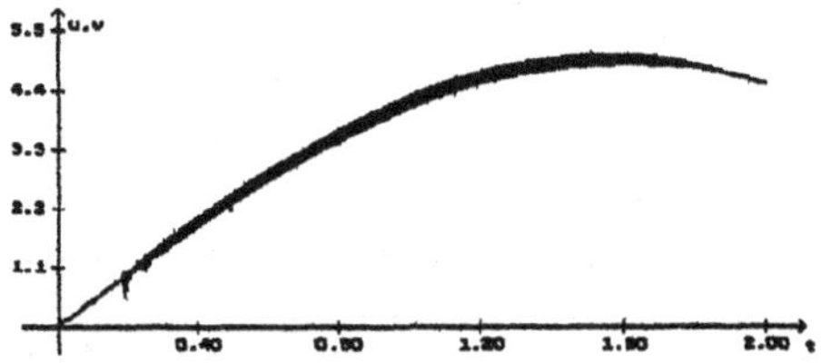

Fig. 2. $\delta = 0.001$, $h = 0.001$.

4. REFERENCES

Krasovskii, N.N. (1985). *Controlling of a dynamical system.* Nauka. Moscow.

Krasovskii, N.N. and A.I. Subbotin (1988). *Game—theoretical control problems.* Springer. Berlin.

Kryazhimskii, A.V. and Yu.S. Osipov (1983). On modelling of control in a dynamical system. *Izv. Akad. Nauk USSR, Tech. Cybern.* **2**, 51–60.

Maksimov, V.I. (1990). Stable reconstruction of controls on the basis of measurement results. *Avtomatika (Kiev)* **4**, 60–65.

Maksimov, V.I. (1995). On the reconstruction of a control through results of observations. In: *Proceedings of 3rd European Control Conference.* 3rd ed.. Vol. 3. pp. 3766–3771. Rome.

Osipov, Yu.S. (1971). Differential games for systems with hereditary. *Dokl. AN USSR* **196**, 779–782.

Osipov, Yu.S. and A.V. Kryazhimskii (1995). *Inverse problems of ordinary differential equations: dynamical solutions.* Gordon and Breach. London.

Osipov, Yu.S., A.V. Kryazhimskii and V.I. Maksimov (1991). *Dynamical regularization problems for distributed parameter systems.* Institute of Mathematics and Mechanics, Urals Branch, Russian Acad. Sci. Sverdlovsk.

A SCALAR CONFLICT CONTROL PROBLEM WITH NONLINEAR FUNCTIONAL CONDITION

S.A. Brykalov

Institute of Mathematics and Mechanics,
ul. Kovalevskoi, 16, Ekaterinburg, 620066, Russia
e-mail: brykalov@imm.uran.ru

Abstract: A scalar conflict control system with linear dynamics is considered with one of the players choosing a time moment. The payoff depends on the phase vector at this time moment. The efficiency of continuous and discontinuous feedback rules that appoint this moment is compared. It is closely related to the existence or nonexistence of solutions to some boundary value problem with solution dependent point in boundary condition. *Copyright ©1998 IFAC*

Keywords: differential games; nonlinear control systems; differential equations; boundary value problems.

1. INTRODUCTION

Positional differential games have been studied by many authors, see (Krasovskii and Subbotin, 1988; Krasovskii and Krasovskii, 1995) for the problem statements, basic results, and bibliography. Properties of continuous strategies were studied in (Barabanova and Subbotin, 1970). In (Krasovskii, 1978) one can find, in particular, properties of strategies given by Carathéodory functions. On pp 18-21 in (Subbotin and Chentsov, 1981) an example of a two-dimensional differential game was considered, for which the guaranteed result was found with respect to the classes of discontinuous and continuous strategies.

In (Brykalov, 1997) this control system was modified so that one of the players can choose the time moment when the payoff should be calculated. The payoff depends on the phase vector at this time moment. Thus, the quality index contains a variable point, which can specify the game's terminal moment assigned by feedback. It was shown in (Brykalov, 1997) that in this problem the methods of control on the basis of continuous feedback are of limited capability even in case of practically complete current information on the process, on the other hand, a discontinuous control law was constructed that can provide the desired result using but very limited current information and modest possibilities to influence the dynamics. The efficiency of a control law in the mentioned problem is directly connected with the existence or nonexistence of solutions to some boundary value problem with a solution dependent point in boundary condition. The corresponding boundary condition is described by a nonlinear functional of special form.

As for the proof in the planar case, (Subbotin and Chentsov, 1981) employed the Schauder fixed point theorem, and (Brykalov, 1997) used the fixed point theorem of Kakutani.

The present paper deals with a scalar example of conflict control system of this type with one of the players choosing the time moment. The efficiency of continuous and discontinuous control laws is compared. It will be shown that in this control system no rule for choosing the terminal time moment described by a continuous mapping can guarantee the corresponding player a nonzero result. However, simple discontinuous mappings can ensure the desired nonzero result. Two examples of such discontinuous rules will be given. One of them requires measuring the phase vector at one time moment only. The fact that the problem is scalar allows to simplify significantly the proofs and to do without fixed point theorems. Instead one can use here the simple fact that a scalar continuous function that takes values of different signs at the ends of an interval vanishes at some point.

2. CHOICE OF TERMINAL MOMENT IN A SCALAR CONFLICT CONTROL SYSTEM

Let the current state of an object be described by a scalar x. The evolution of x on a time interval $[0,2]$ is governed by a differential equation

$$\dot{x} = a(t)u + 1 - a(t), \quad 0 \le t \le 2, \tag{1}$$

with the coefficient given by

$$a(t) = \begin{cases} 1, & 0 \le t < 1, \\ 0, & 1 \le t \le 2. \end{cases} \tag{2}$$

Equation (1) should hold for almost all time moments t. The function $x(\cdot)$ is assumed to be absolutely continuous. The initial state is zero

$$x(0) = 0. \tag{3}$$

The payoff has the form

$$\gamma(x(\cdot)) = |x(\vartheta_0)|.$$

Note that $\gamma(x(\cdot))$ depends not only on $x(\cdot)$, but also on ϑ_0.

There are two players. One of them chooses the parameter $-1 \le u \le 1$ as a Lebesgue measurable function of time $u : [0,2] \to [-1,1]$. The aim of this player is to minimize the value of quality index $\gamma(x(\cdot))$. The other player chooses the number $0 \le \vartheta_0 \le 2$ with an aim to maximize $\gamma(x(\cdot))$.

For a fixed control $u(\cdot)$, the solution $x(\cdot)$ to the initial value problem (1),(3) has the form

$$x(t) = \begin{cases} \int_0^t u(\tau)d\tau, & 0 \le t \le 1, \\ x(1) + t - 1, & 1 \le t \le 2. \end{cases} \tag{4}$$

If the choice of control parameter ϑ_0 is known beforehand to the player that chooses u, then this player can make $\gamma(x(\cdot)) = 0$ putting $u(t) \equiv \min\{0, 1 - \vartheta_0\}$.

3. CONTINUOUS FEEDBACK RULES FOR CHOOSING THE TERMINAL MOMENT

Assume now that the player that chooses ϑ_0 employs some continuous mapping

$$\vartheta_0 : C^0 \to [0,2] \tag{5}$$

to appoint $\vartheta_0 = \vartheta_0(x(\cdot))$ on the basis of feedback. (Other possible restrictions on this mapping will be discussed below.) Here C^0 denotes the space of all continuous scalar functions endowed with the standard uniform norm. It turns out that whatever the continuous mapping (5) might be, it does not guarantee a nonzero result $\gamma(x(\cdot))$. For any continuous mapping $\vartheta_0 = \vartheta_0(x(\cdot))$ there exists a control u such that $\gamma(x(\cdot))$ vanishes. Here it suffices to employ constant controls $u \equiv$ const independent of t.

One can formulate this fact as an existence result for a boundary value problem as follows.

P r o p o s i t i o n. For an arbitrary continuous mapping (5) there exists a constant $u \in [-1,0]$ and an absolutely continuous scalar function $x(t)$, $t \in [0,2]$, that satisfy the boundary value problem (1),(3),(6) with

$$x(\vartheta_0(x(\cdot))) = 0 \tag{6}$$

and the coefficient $a(t)$ given by (2).

Proof of Proposition. For a number $u \in [-1,0]$ denote by $x_u(\cdot)$ the unique solution of the initial value problem (1),(3) that corresponds to the chosen value u. According to (4), one has

$$x_u(t) = \begin{cases} ut, & 0 \le t \le 1, \\ t + u - 1, & 1 \le t \le 2. \end{cases} \tag{7}$$

Thus,

$$x_{-1}(t) \le 0, \quad x_0(t) \ge 0 \tag{8}$$

for all $t \in [0,2]$. Consider a function

$$\Psi(u) = x_u(\vartheta_0(x_u(\cdot))).$$

Note that $\Psi(u)$ is a continuous scalar function defined for all $u \in [-1,0]$. It follows from (8) that $\Psi(-1) \le 0$, $\Psi(0) \ge 0$. So, the function Ψ vanishes at some point in $[-1,0]$. For this point u, the function $x_u(\cdot)$ satisfies condition (6). Proposition is proved.

In Proposition the continuous map (5) is arbitrary. This does not lead to any contradiction if one formally considers boundary value problem (1),(3),(6), or if one considers the corresponding control system assuming t to be some coordinate. (The case of coordinate in somewhat similar problems was dealt with in (Brykalov, 1990a; Brykalov, 1990b).) However, if t is treated as the time variable, it seems to make little sense to allow the player that chooses ϑ_0 to use information on the future evolution of the process. In this case the map (5) should be restricted by the following condition of non-anticipation.

(NA) For any $y(\cdot), z(\cdot) \in C^0$, if $y(s) \equiv z(s)$ for all $s \in [0, \vartheta_0(y(\cdot))]$, then $\vartheta_0(y(\cdot)) = \vartheta_0(z(\cdot))$.

Condition (NA) was not used in the proof of Proposition.

4. DISCONTINUOUS FEEDBACK RULES FOR CHOOSING THE TERMINAL MOMENT

Consider now the case when the map (5) is allowed to be discontinuous. To ensure a nonzero result, the corresponding player can choose ϑ_0 as follows

$$\vartheta_0(x(\cdot)) = \begin{cases} 1, & x(1) \leq -1/2, \\ 2, & x(1) > -1/2. \end{cases} \tag{9}$$

Relation (4) implies

$$x(2) = x(1) + 1 \tag{10}$$

for an arbitrary solution x to initial value problem (1),(3). Consequently, for either case in (9) one has $|x(\vartheta_0(x(\cdot)))| \geq 1/2$, and a result $\gamma(x(\cdot)) \geq 1/2$ is guaranteed.

Another rule for choosing the number ϑ_0 can be given by formula

$$\vartheta_0(x(\cdot)) = \min\{\tau : |x(\tau)| = 1/2\}. \tag{11}$$

From equalities (3),(10) and the continuity of function x it follows that the set of numbers τ in (11) is nonempty. As this set is closed, the minimum is attained. So, the number (11) is defined correctly, belongs to $[0, 2]$, and the described mapping provides the result $\gamma(x(\cdot)) = 1/2$.

On the other hand, whatever the map (5) is, a better result can not be ensured for the player that chooses ϑ_0 because for $u \equiv -1/2$ the unique solution (7) to initial value problem (1),(3) has the form

$$x(t) = \begin{cases} -t/2, & 0 \leq t \leq 1, \\ t - 3/2, & 1 \leq t \leq 2, \end{cases}$$

and satisfies $|x(t)| \leq 1/2$ for all $t \in [0, 2]$.

Note that both (9) and (11) are discontinuous mappings in the sense (5). Both of them satisfy condition (NA).

It is interesting to compare (9) and (11) as two ways of behaviour of the corresponding player. In the case (9), at the time moment $t = 1$ the player becomes aware of the value $x(1)$ and basing on this information a decision is taken either to stop the game immediately, or to wait until its end at $t = 2$. The rule (9) requires measuring the phase vector x only at one point $t = 1$, whereas (11) requires to measure x continuously. The time moment ϑ_0 calculated according to (11) is smaller than or equal to the one given by (9). By definition, for a fixed trajectory, the time moment given by (11) is the smallest possible one that achieves the result 1/2. The rule (11) always gives the guaranteed result 1/2, whereas (9) might provide a better result for the player that chooses ϑ_0 in case the opponent makes mistakes.

5. CONCLUSION

A scalar example of conflict control system can be given for which it is easy to show that discontinuous rules for assigning the terminal moment may be far more efficient than the continuous ones. Thus, the situation here is the same as for the differential game's strategies in traditional form, see e.g. pp 232–239 in (Krasovskii and Subbotin, 1988). This also resembles the case in (Brykalov, 1990a; Brykalov, 1990b), where the feedback assigned points were not time moments but coordinates of point heat sources on a rod being heated and a class of conflict control problems was considered.

6. ACKNOWLEDGEMENTS

This research was partially financed by Russian Foundation for Basic Research under grant 97-01-00160.

REFERENCES

Barabanova, N.N. and A.I. Subbotin (1970). On continuous strategies of evasion in game problems on encounter of motions. *Prikl. Mat. Mekh.*, **34,** 796–803. (in Russian)

Brykalov, S.A. (1990a). Stationary temperature distributions in systems with control and continuous feedback. *Izv. Akad. Nauk. Tekhnicheskaya Kibernetika,* no. 2, 162–165. (in Russian)

Brykalov, S.A. (1990b). The existence of temperature distributions close to a prescribed one in some control systems. *Probl. Control Inform. Theory,* **19,** 279–288.

Brykalov, S.A. (1997). Choice of the terminal moment in a differential game. *Izv. Akad. Nauk. Teoriya i Sistemy Upravleniya,* no. 1, 105–108. (in Russian)

Krasovskii, A.N. and N.N. Krasovskii (1995). *Control under lack of information,* Birkhäuser, Boston.

Krasovskii, N.N. (1978). Differential games. Approximative and formal models. *Mat. Sb.,* **107,** 541–571.

Krasovskii, N.N. and A.I. Subbotin (1988). *Game-theoretical control problems,* Springer-Verlag, New York.

Subbotin, A.I. and A.G. Chentsov (1981). *Optimization of guarantee in control problems,* Nauka, Moscow. (in Russian)

USE FINITE FAMILY OF MULTIVALUED MAPS FOR CONSTRUCTING STABLE ABSORPTION OPERATOR

S.V.Grigor'eva, V.N.Ushakov

Institute of Mathematics and Mechanics,
S.Kovalevskaya str., 16, Ekaterinburg, 620219, Russia
svtl@ods.imm.intec.ru, ushak@ods.imm.intec.ru

Abstract: The differential game of pursuit-evasion over a fixed time segment is considered (Krasovskii and Subbotin, 1970; Krasovskii and Subbotin, 1974). The problem of construction of the stable absorption operator of control system is investigated. The attainability sets is appointed with the help of the stable absorption operator. The partition of the conjugate space on the finite regions of convexity of Hamiltonian is used for constructing stable absorption operator.

Keywords: Differential equations, differential games, feedback control, stable bridges.

1. INTRODUCTION

We consider a conflict controlled system which dynamics over a time segment [0,T] is described by the equation

$$dx/dt = f(t,x,u,v),\ x[0] = x_0. \qquad (1)$$

Here $t \in [0,T]$, $x \in R^m$ is the phase vector of the system, u and v are control vectors of first and second players, restricted by constraints $u \in P \subset R^p$, $v \in Q \subset R^q$, P and Q are compacts.

The following conditions are satisfied:

A. The game takes place in bounded closed region D of variables $(t,x) \in [t_0,T] \times R^m$;

B. The function f(t,x,u,v) is continuous with respect to the aggregate of variables $(t,x,u,v) \in D \times P \times Q$ and Lipshitz continuous with respect to the variable x, namely:

$$| f(t, x^{(2)}, u, v) - f(t, x^{(1)}, u, v) | \leq L|x^{(2)} - x^{(1)}|$$

for all $(t, x^{(i)}, u, v) \in D \times P \times Q$, i=1,2;

C. Motions x(t) of the system (1) are continued on the segment $[t_0,T]$.

For the system (1) the pursuit-evasion problem with a target set $M \subset R^m$ to the fixed time T is regarded.

2. STABLE ABSORPTION OPERATOR

One of the effective approaches to the solution of the problem is that, determining the solution as strategies extremal to the stable bridges (Krasovskii et al., 1972). So the constructing stable bridges becomes the principal elements of the solution.

Stable bridges can be constructed by methods based on the notion of the stable absorption operator (Tarasiev et al., 1987; Grigor'eva et al., 1996) which is determined on the base of family of maps $\{F_\psi: D \rightarrow 2^{R^m}\}$ (where $D = R^m \times [t_0,T]$), satisfying conditions

A.1. for all $(t,x,\psi) \in D \times \Psi$ the set $F_\psi(t,x)$ is compact and uniform bounded;

A.2. for all $(t,x,l) \in D \times S$ the equality $\min_{\psi\in\Psi} \max_{f\in F_\psi(t,x)} <f,l>= H(t,x,l)$ is valid;

A.3. There exists a function $\omega^*(\delta)$ ($\omega^*(\delta) \downarrow 0$ when $\delta \downarrow 0$) such that for all for all (t_*,x_*), $(t^*,x^*) \in D$, for all for all $\psi\in\Psi$ the inequality

$$dist(F_\psi(t^*,x^*), F_\psi(t_*,x_*)) \le \omega^*(|t^*-t_*| + |x^*-x_*|) \quad (2)$$

is valid for the Hausdorf distance dist(F,G).

Let $Z^* \subset R^m$. Denote by $X_\psi(t^*; t_*,x_*)$ the attainability set of differential inclusion $\dot{x} \in F_\psi(t,x)$, $x[t_*]=x_*$, to the moment $t^* \in (t_*,T]$; by $X^{-1}_\psi(t_*; t^*, Z^*) = \{x_* \in R^m: X_\psi(t^*; t_*,x_*) \cap Z^* \ne \varnothing\}$.

D e f i n i t i o n 1. Call by stable absorption operator $\pi=\pi(t_*; t^*, Z^*)$ $(t_0 \le t_* < t^* \le T, Z^* \subset R^m)$ in the problem of pursuit of the target M to the moment T the map π, which is determined by equation

$$\pi(t_*; t^*, Z^*) = \bigcap_{\psi\in\Psi} X_\psi^{-1}(t_*; t^*,Z^*).$$

D e f i n i t i o n 2. Call closed set $W \subset D$ by u-stable bridge in the problem of pursuit of the target M to the moment T, if

1. $W_T \subset M$;
2. $W_{t_*} \subset \pi(t_*,t^*,W_{t^*})$ for all $t_*, t^* \in [t_0, T]$, $t_* < t^*$; $W_t = \{ x \in R^m: (t,x) \in W\}$

The assertion is true (Tarasiev and Ushakov 1985), that the set W^0 of positional absorption in the problem of pursuit of the target M to the fixed time T is the maximal u-stable bridge.

3. CONSTRUCTING FAMILY OF MULTIVALUED MAPS

Let us consider a partition of a unit sphere S into finite number of closed subsets $L_\psi(t,x)$ $(\psi \in \Psi)$, such that

B.1. For all $(t,x) \in D$ the cone $K(L_\psi(t,x))$ is a convex set;

B.2. For all $(t,x) \in D$ the Hamiltonian $H(t,x,l)$ is a convex function on $K(L_\psi(t,x))$ by l;

B.3. There exists a function $\tilde{\omega}(\delta)$ ($\tilde{\omega}(\delta) \downarrow 0$ when $\delta \downarrow 0$) such that for all (t_*,x_*) and $(t^*,x^*) \in D$, for all $\psi\in\Psi$ the following inequality takes place:

$$d(L_\psi(t^*,x^*), L_\psi(t_*,x_*)) \le \tilde{\omega}(|t^*-t_*| + |x^*-x_*|)$$

Here the partition of the sphere S will means the aggregate of sets $L_\psi(t,x)$, $\psi\in\Psi$, satisfying the equation $\bigcap_{\psi\in\Psi} L_\psi(t,x) = S$, $((t,x) \in D)$, at that some of them can intersect; The idea of partition of conjugate space on cones of convexity of Hamiltonian was used by Patsko and others (Isakova et al., 1975; Patsko, 1982; Patsko, 1989) while solving linear differential games.

Let $K(L_\psi(t,x))= \{l': l'=\lambda l, \lambda >0, l \in L_\psi(t,x) \}$ –be a cone pulled over the set $L_\psi(t,x)$. Let also that the following condition takes place:

B.4. $\operatorname{int}F_\psi(t,x) \ne \varnothing$ for all $(t,x,\psi) \in D \times \Psi$.
Let's denote

$$F(t,x) = co\{ f(t,x,u,v): u \in P, v \in Q\};$$

$$F_\psi(t,x)= \bigcap_{l\in L_\psi(t,x)} \{f\in R^m: <l,f>\le H(t,x,l)\} \cap F(t,x) \quad (3)$$

The following theorem was proved in (Grigor'eva and Ushakov, 1996) for control system with more strict conditions. That is, there was supposed, that the partition of conjugate space is not depended on phase state of the system.

T h e o r e m. The family of maps $\{F_\psi: D \to 2^{R^m} \}$ determined by equalities (3) satisfies conditions A.1, A.2, A.3.

P r o o f. Let's take into consideration

$$A_\psi(t,x)= \bigcap_{l\in L_\psi(t,x)} \Pi_l(t,x),$$

$$\Lambda_\psi^0(t,x) = \partial A_\psi(t,x) \cap \operatorname{int} F(t,x),$$

$$L_\psi^0(t,x)=cl \{l\in S: <l,z>= h_{A_\psi(t,x)}(l), z\in\Lambda_\psi^0(t,x)\}$$

Here ∂A and cl A – the boundary and the closure of the set A respectively.
Let's appoint the basic facts, which are used while proving.

U.1. $L_\psi^0(t,x) \subset L_\psi(t,x)$ for all $(t,x,\psi) \in D \times \Psi$;

U.2. For any point $(t,x) \in D$ and all vectors l,s $((l,s) \in S \times S)$ the inequality holds

$$h_{F_l(t,x)}(l) \le h_{F_s(t,x)}(l)$$

U.3. Let together with the set $F_\psi(t,x)$ $(\psi \in \Psi)$, there exists a convex compact Φ in F(t,x), such that $F_\psi(t,x) \cap \Phi = \varnothing$. Then there exists hyperplane $\Gamma_{l^*}(\alpha^0)= \{f\in R^m: <l^*,f>= \alpha^0\}$, $l^* \in L_\psi^0(t,x)$, $\alpha^0 \in (-\infty,+\infty)$, supporting to the set $F_\psi(t,x)$ and separating strictly sets $F_\psi(t,x)$ and Φ.

R e m a r k. U.2, formulated in (Krasovskii and Subbotin, 1974), is some extremal property of the system $\{ F_l(t,x): l \in S \}$, means that for every fixed $l \in S$ the minimum of the function $\chi(s) = h_{F_s(t,x)}(l)$ is attained for $s = l$. The essence of U.3 is that the presence of restriction on non-intersecting convex closed sets $F_\psi(t,x)$ and Φ allows to gather the addition information about position of some of separating them hyperplanes. U.2 and U.3 are useful facts (Subbotin, 1991; Tarasiev, et al., 1990; Tarasiev and Ushakov, 1985; Uspenskii, 1993) while analysis and synthesis of solutions of differential games, problems of optimal control theory, and generalised solutions of Hamilton-Jacobi equations.

Turn to the proving the theorem.

Proof A.1. Convex and closeness of the set $F_\psi(t,x)$ $((t,x,\psi) \in D \times \Psi)$ follows from definition of $F_\psi(t,x)$ as intersection of convex and closed sets. Inclusion $F_\psi(t,x) \subset G$ follows from inclusions $F_\psi(t,x) \subset F(t,x)$ and $F(t,x) \subset G$.

Proof A.2. Let (t,x,l) is arbitrary point in $D \times S$.

1) Let's prove the inequality

$$\min_{\psi\in\Psi} h_{F_\psi(t,x)}(l) \leq h(t,x,l) \qquad (4)$$

For some $\psi \in \Psi$ the inclusion $l \in L_\psi(t,x)$ holds. $A_\psi(t,x) \subset \Pi_l(t,x)$, and, hence, $F_\psi(t,x) \subset F_l(t,x)$. Then $h_{F_\psi(t,x)}(l) \leq h_{F_l(t,x)}(l) = h(t,x,l)$. The last inequality implies (4).

2) Let's prove, that

$$\min_{\psi\in\Psi} h_{F_\psi(t,x)}(l) \geq h(t,x,l) \qquad (5)$$

Assuming contrary to the (5), namely,

$$\min_{\psi\in\Psi} h_{F_\psi(t,x)}(l) \geq h(t,x,l) \qquad (6)$$

and taking into account, that the set Ψ is finite, we conclude, that $h_{F_{\psi^*}(t,x)}(l) < h(t,x,l)$ for some $\psi^* \in \Psi$. Then, supposing that $\Psi = \{f \in F(t,x): \langle l,f \rangle \geq h(t,x,l)\}$, we derive $\Phi \cap F_{\psi^*}(t,x) = \varnothing$.

In accordance with U.3, there exist $l^0 \in L^0_{\psi^*}(t,x)$ and $\alpha^0 \in (-\infty, +\infty)$ such that hyperplane $\Gamma_{l^0}(\alpha^0)$, supporting to $F_{\psi^*}(t,x)$, separating strictly sets $F_{\psi^*}(t,x)$ and Φ. Taking into account U.1, we conclude, that $l^0 \in L_{\psi^*}(t,x)$. Considering the last inclusion, and taking into account coincidence on the cone $K(L_{\psi^*}(t,x))$ values of functions $h_{F_{\psi^*}(t,x)}(l)$ and $h(t,x,l)$, we derive $\Gamma_{l^0}(\alpha^0) = \{f \in R^m: \langle l^0,f \rangle = h(t,x,l^0)\}$.

So, $\min_{z\in\Phi} \langle l^0,z \rangle > h(t,x,l^0) = h_{F_{l^0}(t,x)}(l^0)$. This inequality means, that $\Phi \cap F_{l^0}(t,x) = \varnothing$. Hence $h_{F_{l^0}(t,x)}(l) < h(t,x,l)$, that is, $h_{F_{l^0}(t,x)}(l) < h_{F_l(t,x)}(l)$. The last inequality contradicts U.2. Hence, (6) can't takes place, and (5) is proved. A.2 is proved.

Proof A.3. Let's take into account function $\omega^*(\delta) = \sup_{(\psi,t_*,x_*,t^*,x^*)\in Y(\delta)} d(F_\psi(t^*,x^*), F_\psi(t_*,x_*))$, where $Y(\delta) = \{ (\psi, t_*,x_*,t^*,x^*): \psi\in\Psi, (t_*,x_*) \in D, (t^*,x^*) \in D, |t^*-t_*|+|x^*-x_*| \leq \delta\}$, $\delta > 0$. The function $\omega^*(\delta)$ satisfies (2). Show that $\omega^*(\delta) \downarrow 0$ while $\delta \downarrow 0$. Really, supposing contradiction and taking into account finiteness of the set Ψ, we derive, that there exist such $\varepsilon > 0$, $\tilde\psi \in \Psi$ and sequences $\{(t^*_k,x^*_k)\}$, $\{(t^0_k,x^0_k)\}$ from D, that $|t^*_k - t^0_k| + |x^*_k-x^0_k| \downarrow 0$ while $k \to \infty$ and

$$d(F_{\tilde\psi}(t^*_k,x^*_k), F_{\tilde\psi}(t^0_k,x^0_k)) \geq \varepsilon > 0 \qquad (7)$$

Without less of community, we'll consider that there exists

$$\lim_{k\to\infty} (t^*_k,x^*_k) = \lim_{k\to\infty} (t^0_k,x^0_k) = (\tilde t, \tilde x) \in D. \qquad (8)$$

Show, that equalities are true

$$\lim_{k\to\infty} F_{\tilde\psi}(t^*_k,x^*_k) = F_{\tilde\psi}(\tilde t,\tilde x)$$
$$\lim_{k\to\infty} F_{\tilde\psi}(t^0_k,x^0_k) = F_{\tilde\psi}(\tilde t,\tilde x) \qquad (9)$$

Here convergence of the sets means the convergence *in the Hausdorff metric*.

Prove the first equality from (9). Let $f^* \in \lim_{k\to\infty} F_{\tilde\psi}(t^*_k,x^*_k)$. For any sequence $\{f_k \in F_{\tilde\psi}(t^*_k,x^*_k)\}$ the equality $f^* = \lim_{k\to\infty} f_k$ holds.

Vectors f_k satisfy relations
1) $\langle l, f_k \rangle \leq h(t^*_k,x^*_k,l)$ for all $l \in L_{\tilde\psi}(t^*_k,x^*_k)$,
2) $f_k \in F(t^*_k,x^*_k)$.

Take arbitrary vector $\tilde l \in L_{\tilde\psi}(\tilde t,\tilde x)$ and sequence $\{l_k \in L_{\tilde\psi}(t^*_k,x^*_k)\}$ such that $\lim_{k\to\infty} l_k = \tilde l$. Because of the supposition B.3 such sequence exists. The inequality $\langle l_k, f_k \rangle \leq h(t^*_k, x^*_k, l_k)$ is true. Because of continuity of function $h(t,x,l)$ on (t,x,l) and $F(t,x)$ on (t,x), derive
1) $\langle \tilde l, f^* \rangle \leq h(\tilde t,\tilde x,\tilde l)$
2) $f^* \in F(\tilde t,\tilde x)$,
that is, $f^* \in F_{\tilde\psi}(\tilde t,\tilde x)$.

The inclusion

$$\lim_{k\to\infty} F_{\tilde\psi}(t^*_k, x^*_k) \subset F_{\tilde\psi}(\tilde t, \tilde x) \qquad (10)$$

is proved. Let's prove the opposite inclusion

$$\lim_{k\to\infty} F_{\tilde\psi}(t^*_k, x^*_k) \supset F_{\tilde\psi}(\tilde t, \tilde x) \qquad (11)$$

Assume contrary to the (11), namely,

$$\text{int } F_{\tilde\psi}(\tilde t, \tilde x) \setminus \lim_{k\to\infty} F_{\tilde\psi}(t^*_k, x^*_k) \neq \varnothing. \qquad (12)$$

Let $f^* \in \text{int } F_{\tilde\psi}(\tilde t, \tilde x) \setminus \lim_{k\to\infty} F_{\tilde\psi}(t^*_k, x^*_k)$. Then for some small enough $\delta > 0$ the inclusion $O_\delta(f^*) \subset F_{\tilde\psi}(\tilde t, \tilde x) \setminus \lim_{k\to\infty} F_{\tilde\psi}(t^*_k, x^*_k)$ takes place, hence

$$\langle l, f^* \rangle \leq h(\tilde t, \tilde x, l) - \delta \qquad (13)$$

for all $l \in L_{\tilde\psi}(\tilde t, \tilde x)$;

$$O_\delta(f^*) \subset F(\tilde t, \tilde x). \qquad (14)$$

Also, because of uniform continuity of $h(t,x,l)$ on compact $D \times S$, and because of (8), the inequality

$$h(\tilde t, \tilde x, l) - \delta/2 \leq h(t^*_k, x^*_k, l^*) \qquad (15)$$

takes place for some k_0 and $\sigma > 0$ and for all $k \geq k_0$, $l \in S$, $l^* \in S$, $|l - l^*| \leq \sigma$. Taking into account (13) and (15), we derive, that for some k_0 and $\sigma > 0$ the inequality

$$\langle l, f^* \rangle \leq h(t^*_k, x^*_k, l^*) - \delta/2 \qquad (16)$$

takes place for all $k \geq k_0$, $l \in L_{\tilde\psi}(\tilde t, \tilde x)$, $l^* \in S$, $|l - l^*| \leq \sigma$. B.3 implies, that for some k_1 and all $k \geq k_1$ the relation

$$d(L_{\tilde\psi}(t^*_k, x^*_k), L_{\tilde\psi}(\tilde t, \tilde x)) \leq \min(\delta/2K, \sigma)$$

is true. Here $K = \max_{(t,x,u,v)\in D\times P\times Q} |f(t,x,u,v)|$.

Hence for any $k \geq k_1$ and $l^*_k \in L_{\tilde\psi}(t^*_k, x^*_k)$ there exists vector $l_k \in L_{\tilde\psi}(\tilde t, \tilde x)$, satisfies relation

$$\langle l^*_k, f^* \rangle - \langle l_k, f^* \rangle \leq |l^*_k - l_k|\,|f^*| \leq \delta/2, \quad |l^*_k - l_k| \leq \sigma \qquad (17)$$

Taking into consideration (16) and (17), we derive, that for any $k \geq \max\{k_0, k_1\}$, $l^*_k \in L_{\tilde\psi}(t^*_k, x^*_k)$ there exists such vector $l_k \in L_{\tilde\psi}(\tilde t, \tilde x)$, $(|l_k - l^*_k| \leq \sigma)$, that $\langle l_k, f^* \rangle + (\langle l^*_k, f^* \rangle - \langle l_k, f^* \rangle) \leq h(t^*_k, x^*_k, l^*_k) - \delta/2 + \delta/2$. That is, for any $k \geq \max\{k_0, k_1\}$, $l^*_k \in L_{\tilde\psi}(t^*_k, x^*_k)$ the inequality $\langle l^*_k, f^* \rangle \leq h(t^*_k, x^*_k, l^*_k)$ holds. The last inequality means, that for all $k \geq \max\{k_0, k_1\}$ the inclusion $f^* \in A_{\tilde\psi}(t^*_k, x^*_k)$ is true.

Taking into consideration continuity of $F(t,x)$ on (t,x), and inclusion (14), we derive, that for some k_2 and all $k \geq k_2$ the inclusion $f^* \in F(t^*_k, x^*_k)$ holds.

In the conclusion, for all $k \geq \max\{k_0, k_1, k_2\}$ the inclusion $f^* \in F_{\tilde\psi}(t^*_k, x^*_k)$ takes place. The last inclusion contradicts supposition (12).

(11) is proved. (10) and (11) together prove the first equality of (9). The second equality proves by analogous way. Equalities (9) contradict the inequality (8), that is why, that supposition (8) is not true. Hence $\omega^*(\delta) \downarrow 0$ while $\delta \downarrow 0$. A.3 is proved. The theorem is proved.

4. CONCLUSION

The constructing such family of maps, having final number of elements, facilitates the problem of constructing the stable absorption operator.

ACKNOWLEDGEMENTS

This research was supported by the Russian Foundation of Basic Researches under Grants № 96-01-00219 and № 96-15-96245.

REFERENCES

Grigor'eva S.V., Tarasiev A.M., Ushakov V.N., Uspenskii A.A. (1996) Constructions of nonsmooth analysis in numerical methods for solving Hamilton-Jacobi equations. *Nova Journal of Mathematics, Game Theory and Algebra*. Vol. 6. No. 1. P. 27-44.

Grigor'eva S.V., Ushakov V.N. (1996) Constructing stable absorption operator in differential games. *Izv. RAN. Theory systems of control*. No. 4. P. 69-76. (in Russian)

Isakova E.A., Logunova G.V., Patsko V.S. (1975) Constructing stable bridges in linear differential game with prescribed time-interval. Algorithms and programs of solving linear differential games. (1984) (Materials on software). Sverdlovsk. Ural Scientific Centre of Academy of Sciences of USSR.

Krasovskii N.N., Subbotin A.I. (1970) On structure of *differential* games. *Dokl. Akad Nauk SSSR*. Vol.190. No. 3. (in Russian)

Krasovskii N.N., Subbotin A.I. (1974) *Positional differential games*. Nauka. Moscow (in Russian)

Krasovskii N.N., Subbotin A.I., Ushakov V.N. (1972) Minimax differential game. *Dokl. Akad Nauk SSSR*. Vol. 206. No. 2. (in Russian)

Patsko V.S. (1982) Second order differential game of quality. *Prikl. matem. mekn.* Vol. 86. No 4.

Patsko V.S (1989) Quasilinear second order differential game of quality. I. *Problems of dynamic programming*, Sverdlovsk. Ural

Scientific Centre of Academy of Sciences of USSR
Subbotin A.I. (1991) *Minimax inequality and Hamilton-Jacobi equations*. 216 p. Moscow. Nauka. (in Russian)
Tarasiev A.M., Ushakov V.N., Khripunov A.P. (1987). On one computational algorithm for solving game control problems. *Prikl. matem. mekn.* Vol.51, No.2. P.216-222 (in Russian)
Tarasiev A.M., Uspenskii A.A., Ushakov V.N. (1990) On construction of solving positional procedures in a linear control problem. *IMACS, The Lyapunov Functions Method and Applications* P. 111-115 (in Russian)
Tarasiev A.M., Ushakov V.N. (1985) Algorithm of constructing stable bridge for linear differential game of pursuit with convex target. *Researching problems of minimax control.* Sverdlovsk. UrSC AS SSSR. P. 82-90 (in Russian)
Uspenskii A.A. (1993) Numerical procedures for constructing generalised solutions of Bellman-Isaacs equation. *PhD Dissertation*, Ekaterinburg, Institute of Mathematics and Mechanics UrB RAS. (in Russian)

DEGREE OF CONVERGENCE OF FINITE-DIFFERENCE OPERATOR WHILE SOLVING CAUCHY PROBLEM FOR HAMILTON-JACOBI EQUATIONS

S.V. Grigor'eva, A.A.Uspenskii

Institute of Mathematics and Mechanics,
S.Kovalevskaya str., 16, Ekaterinburg, 620219, Russia
svtl@cs.imm.intec.ru, uspen@cs.imm.intec.ru

Abstract: The Cauchy problem for the Hamilton-Jacobi equation is considered. The problem of approximate construction of the generalized soluton is investigated. Approximation scheme based on constructing local convex hulls is proposed.The estimation of degree of convergence of procedure is received. *Copyright © 1998 IFAC*

Keywords: algorithms; differential equations; differential games; stability properties; numerical methods.

1. INTRODUCTION

The Cauchy problem for the Hamilton-Jacobi equation

$$\frac{\partial w}{\partial t}(t,x) + H(t,x,\nabla w(t,x)) = 0, \qquad (1)$$
$$w(\vartheta,x) = \sigma(x),\ t \in [0,\vartheta],\ x \in R^n$$

is considered. Here $\nabla w(t, x) = (\partial w/\partial x_1\ (t, x) \ldots \partial w/\partial x_n\ (t, x))$ is the gradient of the function w; $H(t, x, s)$: $[0, \vartheta] \times R^n \times R^n \rightarrow R$ is the Hamiltonian; $\sigma(x)$: $R^n \rightarrow R$ is a function, satisfying local Lipschitz condition.

The solution of the problem is understood in the generalized (minimax or viscosity) sense (Subbotin, 1976; Subbotin, 1991; Crandall and Lions, 1983). The construction of converging computing procedures is nontrivial because of non-smoothness of the solution.

The proposed method uses subdifferential calculus and is based on constructing local convex hulls of functions, that approximate the solution by variable t (Grigor'eva, et al., 1996).

The methods for numerical approximate solution of the Cauchy problem (1) are developed with the help of the accompanying differential game (d.g.)

$$dx/dt = f(t, x, u, v) = f^{(1)}(t, x, u) + f^{(2)}(t, x, v). \qquad (2)$$

Here $t \in [0, \vartheta]$, x is n-dimentional phase vector of the system, u and v are control vectors of first and second players, obeying restrictions

$$u \in P \subset R^p, \quad v \in Q \subset R^q.$$

Here P and Q are compacta. The Hamiltonian H of the dynamic system (2) is determined by the formula

$$H(t,x,s) = \min_{u \in P} \langle s, f^{(1)}(t,x,u) \rangle + \max_{v \in Q} \langle s, f^{(2)}(t,x,v) \rangle$$

where $\langle s, f \rangle$ is the scalar product of vectors s and f.

Function $f(t, x, u, v)$ in the right-hand side of the system (2) satisfies the following conditions.

(f1) Function f is uniformly continuous by all variables and Lipschitz continuous with respect to variable x

$$\|f(t, x, u, v) - f(t, y, u, v)\| \leq \lambda_f(A) \|x-y\|$$

for all $(t, x) \in A$, $(t, y) \in A$. Here $A \subset [0, \vartheta] \times R^n$ is a compactum.

(f2) Function f satisfies the extendability condition for solutions of equation (2): there exists a constant $\kappa < +\infty$ such that

$$\|f(t, x, u, v)\| \leq \kappa(1+ \|x\|)$$

for all $(t, x, u, v) \in [0, \vartheta] \times R^n \times P \times Q$.

Conditions (f1), (f2) and the form of the right-hand side of the system (2) imply the following properties of the Hamiltonian H:

(H1) Uniform continuity of H on any compactum $A \subset [0, \vartheta] \times R^n \times P \times Q$.

(H2) The Lipschitz condition by variable x:

$$|H(t, x, s) - H(t, y, s)| \leq \lambda_H(A) \|s\| \|x-y\|,$$

for all $(t, x) \in A$, $(t, y) \in A$, $s \in R^n$. Here $A \subset R^n$ is a compactum.

(H3) The Lipschitz condition with respect to variable s:

$$|H(t, x, s_1) - H(t, x, s_2)| \leq K(A) \|s_1 - s_2\|,$$

where $(t, x) \in A$, $s_1 \in R^n$, $s_2 \in R^n$.

(H4) Positive homogeneity by variable s:

$$H(t, x, \lambda s) = \lambda H(t, x, s)$$

for all $(t, x, s) \in A \times R^n$, $\lambda \geq 0$, $A \subset [0, \vartheta] \times R^n$.

The payoff functional of d.g. is defined by the terminal function

$$\gamma(x(\cdot)) = \sigma(x(\vartheta)) \qquad (3)$$

Function $\sigma(\cdot)$: $R^n \to R$ satisfies the local Lipschitz condition:

$$|\sigma(x) - \sigma(y)| \leq \lambda_\sigma(A) \|x-y\|,$$

Here $A \subset R^n$ is a bounded set.

The functional γ estimates the quality of a trajectory $x(\vartheta)$ by the value $\sigma(x(\vartheta))$ of the payoff function σ.

Under conditions (f1), (f2) indicated for the right-hand side of the system (2) there exists the value function $(t, x) \to w(t, x)$: $[0, \vartheta] \times R^n \to R$ of d.g. (2)-(3).

The value function coincides with the generalized solution of the Cauchy problem (1) in which the Hamiltonian H satisfies conditions (H1) - (H4). Here the generalized solution of the problem (1) is determined uniquely by the boundary condition and the pair of differential inequalities

$$\inf_{s \in R^n} \sup_{h \in R^n} \left(< s,h > - \partial_- w(t,x) \,|\, (1,h) - H(t,x,s)\right) \geq 0$$

$$\sup_{s \in R^n} \inf_{h \in R^n} \left(< s,h > - \partial_+ w(t,x) \,|\, (1,h) - H(t,x,s)\right) \leq 0$$

Here symbols

$$\partial_- w(t,x) \,|\, (1,h) = \lim_{\delta \downarrow 0} \inf \delta^{-1}\left(w(t+\delta, x+\delta h) - w(t,x)\right)$$

$$\partial_+ w(t,x) \,|\, (1,h) = \lim_{\delta \downarrow 0} \sup \delta^{-1}\left(w(t+\delta, x+\delta h) - w(t,x)\right)$$

denote lower and upper derivatives of function w at point (t, x) in direction $(1, h)$.

The equivalence of the value function and the generalized solution gives the possibility to use methods of d.g. theory for constructing generalized solutions of Hamilton-Jacobi equations.

2. APPROXIMATION SCHEMES AND CONVERGENCE PROPERTIES

Two grid algorithms for constructing approximations w_1 and w_2 of the function w on the time step $\Delta > 0$ of the partition of $[0, \vartheta]$ is described below:

$$G_1(t, \Delta\ \varphi)(x) = \max_{s \in S_n} \min_{f \in F_1} \text{co}\,\varphi\,(t+\Delta, x+\Delta f),$$

$$G_2(t, \Delta\ \chi)(x) = \text{mim}_{s \in S_n} \max_{f \in F_2} \text{conc}\,\chi\,(t+\Delta, x+\Delta f)$$

Here $t \in [0, \vartheta - \Delta]$, $S_n = \{ s \in R^n : \|s\| = 1 \}$; $F_1(t, x, s) = \{ f \in F: <s, f> \geq H(t, x, s) \}$; $F_2(t, x, s) = \{ f \in F: <s, f> \leq H(t, x, s) \}$; $F = \{f \in R^n: \|f\| \leq K \}$; $K = \max_{(t,x) \in D} K(t,x)$, if $\sup (|H(t, x, s_1) - H(t, x, s_2)| - K(t, x) \|s_1 - s_2\| \leq 0$; $D \subset [0, \vartheta] \times R^n$; co $\varphi(t+\Delta, \cdot)$ is a convex hull of function φ on the ball $O(x, K\Delta) = \{ y \in R^n: \|y-x\| \leq K\Delta \}$:

$$\text{co}\,\varphi(y) = \inf\left\{ \sum_{i=1}^{n+1} \alpha_i \varphi(y^{(i)}) : \alpha_i \geq 0, i = 1, \ldots, n+1; \sum_{i=1}^{n+1} \alpha_i = 1; \sum_{i=1}^{n+1} \alpha_i y^{(i)} = y, y^{(i)} \in O(x, K\Delta) \right\}$$

conc $\chi(t+\Delta, \cdot) = -\text{co}(-\chi(t+\Delta, \cdot))$ is a concave hull of function χ on the ball $O(x, K\Delta)$; $<f_1, f_2>$ is a scalar product of vectors f_1 and f_2.

Note that under assumptions indicated on the right-hand side of equation (2) the family of set-valued maps

$$\{(t, x) \to F_j(t, x, s): s \in S_n\}, \quad j = 1, 2, \qquad (4)$$

generating the set D satisfies the following conditions

(F1) For any $(t, x, s) \in D \times S_n$ the set $F_j(t, x, s)$ is convex, closed and satisfies the inclusion $F_j(t, x, s) \subset F = \{f \in R^n: |f| \leq K\}$, j=1, 2.

(F2) For any $(t, x, s) \in D \times S_n$ the equalities

$$\max_{q \in S_n} \min_{f \in F_1(t,x,q)} < s, f >= H(t,x,s)$$

$$\min_{q \in S_n} \max_{f \in F_2(t,x,q)} < s, f >= H(t,x,s)$$

are valid.

(F3) For the maps $(t, x, s) \to F_j(t, x, s)$ there exists a function $\delta \to \hat{\omega}(\delta): R \to R$ ($\hat{\omega} \to 0$ when $\delta \to 0$) such that

$$dist\,(F_j(t_1, x, s), F_j(t_2, y, s)) \leq \hat{\omega}\,(|t_1 - t_2| + ||x - y||)$$

for all $(t_1, x) \in D$, $(t_2, y) \in D$ and $s \in S_n$, j=1, 2. Here $dist$(F*, F**) is the Hausdorf distance between sets F* and F**.

(F4) There exists a number $\lambda_F \in [0, +\infty)$ such that for any (t, x), $(t, y) \in D$ and any $s \in S_n$ the following relations take place

$$dist\,(F_j(t, x, s), F_j(t, y, s)) \leq \lambda_F\, ||x - y||, \quad j=1, 2.$$

Let's write another form of maxmin and minmax operators, which provide numerical realization.

$$G_1(t, \Delta\ \varphi)(x) = \max_{y \in O(x,K\Delta)} \max_{s \in \partial \text{co}\varphi(y)} \{\Delta H(t,x,s) + < s, x - y > + \text{co}\varphi\,(y)\},$$
$$G_2(t, \Delta\ \chi)(x) = \text{mim}_{y \in O(x,K\Delta)} \min_{s \in \partial \text{conc}\chi(y)} \{\Delta H(t,x,s) + < s, x - y > + \text{conc}\,\chi\,(y)\}, \qquad (5)$$

Here the symbol ∂ co $\varphi(y) = \{s \in R^n:$ co $\varphi(y^*)$ - co $\varphi(y) \geq <s, y^* - y>$ forall $y^* \in O(x, K\Delta)\}$ denotes the subdifferential of local convex hull of φ at a point $y \in O(x, K\Delta)$; the symbol ∂ conc $\chi(y) = \{s \in R^n$: conc $\varphi(y^*)$ - conc $\varphi(y) \leq <s, y^* - y>$ forall $y^* \in O(x, K\Delta)\}$ denotes the superdifferential of local concave hull of χ at a point $y \in O(x, K\Delta)$;

Proof of formula (5) can be obtained within methods of convex analysis. It is based on definition of maps (4).

Let $\Gamma = \{t_0, t_1, ..., t_n = \vartheta\}$ is the partition of segment $[0, \vartheta]$. Let's construct the set

$$\hat{D} = \bigcup_{i=0}^{N} (t_i, D(t_i)),$$

where $D(\tau) = \{x \in R^n: (x, \tau) \in D\}$ is the cutset of the set D at the moment $\tau \in [0, \vartheta]$. Let's denote functions w_1 and w_2 at the set $\hat{D}$ by using recurrent procedure:

$w_j(t_n, x) = \sigma(x)$,
$w_j(t_i, x) = G_j(t_i, \Delta_i, w_j)(x)$,
$\Delta_i = t_{i+1} - t_i$; $i = 0, 1, ..., N$-1; j=1, 2.

The estimation of degree of convergence of the procedure is received.

T h e o r e m. Let conditions (H1)-(H4) are satisfied, then there exist constants $C_1 > 0$ and $C_2 > 0$, such that for solution of the Cauchy problem (1) $w(t, x)$ for any $(t, x) \in \hat{D}$ the estimation

$$|w_j(t, x) - w(t, x)| \leq C_1\, \hat{\omega}(\Delta) + C_2 \sqrt{\Delta}, \quad (t, x) \in D,$$
j=1, 2.

takes place. Here $\Delta = \max\{\Delta_1, ..., \Delta_n\}$

R e m a r k. If, in addition to the above conditions, Hamiltonian satisfies Lipschitz condition with respect to variable t, then operators G_1 and G_2 satisfy conditions from (Souganidis, 1985), and, hence, provide the degree of convergence equals by $\sqrt{\Delta}$ (Tarasiev, 1994).

3. APPROXIMATION SCHEME FOR HAMILTONIAN PICEWISE-LINEAR WITH RESPECT TO IMPULSE VARIABLE

Let us consider control system in which we construct value functions w using approximation schemes for generalized solutions of the corresponding Cauchy problem (1). We use for this purpose approximation schemes with finite-difference operators $G_1(t, \Delta, w_1)$ and $G_2(t, \Delta, w_2)$. In the case when functions w_1 and w_2 is defined on the grid and Hamiltonian H is a piecewise linear and positive homogenious function with respect to variable s

$$G_1(t, \Delta\ w_1)(x) = \max_j \max_k \max_m \max_s \{\Delta(< s, f^{(1)}(t,x,u^{(k)}) > + < s, f^{(2)}(t,x,v^{(m)}) >) + \text{co}\ w_1(t+\Delta, y_j) - < s, y - x >\}$$

$$s \in L_{j,k,m}(t,x) = \partial \operatorname{co} w_1(t+\Delta, y_j) \bigcap L_k^P \bigcap L_m^Q$$

$$G_2(t, \Delta\, w_2)(x) = \min_j \min_k \min_m \min_s$$

$$\{\Delta(<s, f^{(1)}(t,x,u^{(k)})> + <s, f^{(2)}(t,x,v^{(m)})>) + \operatorname{conc} w_2(t+\Delta, y_j) - <s, y-x>\}$$

$$s \in L^*_{j,k,m}(t,x) = \partial \operatorname{conc} w_2(t+\Delta, y_j) \bigcap L_k^P \bigcap L_m^Q$$

In this formula symbols p_k denote vertexes of the polytope $f^{(1)}(t, x, P)$ and symbols L_k^P denote cones of linearity of the function $s \to \min_{u \in P} <s, f^{(1)}(t,x,P)>$, i.e. $L_k^P = \{s \in R^n: <s, p - p_k> \geq 0,\ p \in f^{(1)}(t, x, P)\}$, $k=1, ..., N_P$. Here N_P is a number of vertexes of the polytope $f^{(1)}(t, x, P)$.

Analogously symbols q_m denote vertexes of the polytope $f^{(2)}(t, x, Q)$ and symbols L_m^Q denote cones of linearity of the function $s \to \max_{v \in Q} <s, f^{(2)}(t,x,Q)>$, i.e. $L_m^Q = \{s \in R^n: <s, q - q_m> \leq 0,\ q \in f^{(2)}(t, x, Q)\}$, $m=1, ..., N_Q$. Here N_Q is a number of vertexes of the polytope $f^{(2)}(t, x, Q)$.

The set $L_{j,k,m}(t, x)$ is the convex intersection of the subdifferential $\partial \operatorname{co} w_1(y_j)$, $y_j \in O(x, K\Delta)$ with the cone of linearity of the Hamiltonian H. The maximized function here is linear by variable s. Therefore, computation of the value $G_1(t, \Delta, w_1)(x)$ is reduced to the series of linear programming problems.

The set $L^*_{j,k,m}(t, x)$ is the convex intersection of the superdifferential $\partial \operatorname{conc} w_2(y_j)$, $y_j \in O(x, K\Delta)$ with the cone of linearity of the Hamiltonian H. The minimized function here is linear by variable s. Analogously, computation of the value $G_2(t, \Delta, w_2)(x)$ is reduced to the series of linear programming problems.

4. CONCLUSION

In the paper difference approximate operators for Hamilton-Jacobi equation are considered. The estimate of degree of convergence is proposed.

ACKNOWLEDGEMENTS

This research was supported by the Russian Foundation of Basic Researches under Grants 96-01-00219 and 96-15-96245.

REFERENCES

Crandall M.G., Lions P.L. (1983) Viscosity solution{ESC} s of Hamilton-Jacobi equations. em *Trans. Amer. Math. Soc.* Vol. 277. No. 1. P. 1-42.

Grigor'eva S.V., Taras'ev A.M., Ushakov V.N., Uspenskii A.A. (1996) Constructions of nonsmooth analysis in numerical methods for solving Hamilton-Jacobi equations. *Nova Journal of Mathematics, Game Theory and Algebra.* Vol. 6. No. 1. P. 27-44.

Souganidis P.E. (1985) Approximation schemes for viscosity solutions of Hamilton-Jacobi equations. *J. Of Different. Equat.* Vol. 59. P. 1-43.

Subbotin A.I. (1976). Generalized of the basic equation of the theory of differential games. *Dokl. AN SSSR.* Vol. 226. No 6. P. 1260-1263. (in Russian)

Subbotin A.I. (1991). *Minimax inequations and Hamilton-Jacobi equations.* Moscow. Nauka. (in Russian)

Tarasiev (1994). Approximation schemes for constructions of minimax solutions of Hamilton-Jacobi equations. *Prikl. Math. Mekh.* Vol. 58. No. 2. P. 22-36.

SOFTWARE FOR THE COURSE OF CALCULUS OF VARIATIONS

A.G.Ivanov

Institute of Mathematics and Mechanics, Russian Academy of Sciences,
S.Kovalevskaya str., 16, Ekaterinburg, 620219, Russia
e-mail: u0121@cs.imm.intec.ru, WWW: http://home.ural.ru/~iagsoft/index.html

Abstract: In teaching the course of calculus of variations for students in mechanics at the Ural State University, a serious attention is paid to application of numerical methods in classical model problems. We consider the aerodynamic Newton problem, problems of the minimum rotation-surface and of brachistochrone. The last problem is examined also in non-classical formulation. The demonstration software developed with the participation of students is applied to these problems. *Copyright © 1998 IFAC*

Keywords: Variational analysis, education, computer software, numerical methods, multiprocessor systems

1. AERODYNAMIC NEWTON'S PROBLEM (ANP)

The problem consists of searching the generating line $y(x)$ of a rotation-body with the minimum resistance in a flow of rarefied ideal gas (Newton, 1916; Tikhomirov, 1990). The gas is represented as collection of infinitely small particles which do not mutually collide and are mirror-like reflected having collided with the body. At first, we solve the problem, following (Krasnov, *et al.*, 1973), under assumption that the value of y'^2 is small. It is shown, that this assumption isn't fulfilled on optimal solution. After that, the numerical search of the solution in the class of functions $y = x^p$ is demonstrated (Fig. 1). Further, in frames of the numerical approach, the program for finding the solution by the Euler method (as a piecewise linear approximation) is demonstrated. Here, the problem is reduced to searching the minimum of the function of many variables. To minimize this function, we use one technique of direct search. This technique is described in (Bunday, 1984) as the Hooke-Jeeves method. The optimal solution has a corner point (Newton, 1916) (Fig. 2).

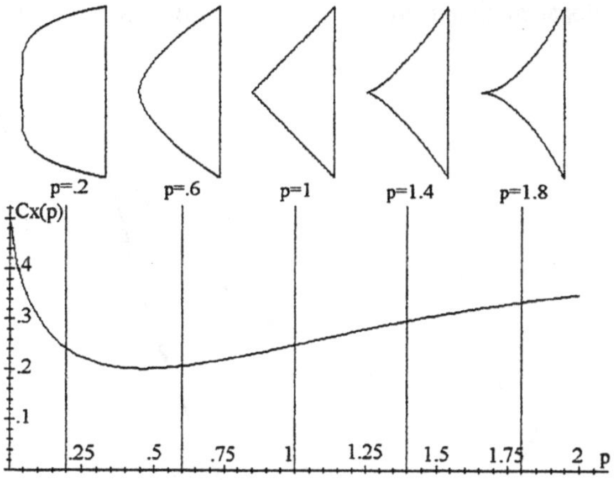

Fig. 1. Dependence of the drag coefficient upon the parameter p (i.e., the class of functions $y = x^p$). The sections of the rotation body for certain p are given in the upper part.

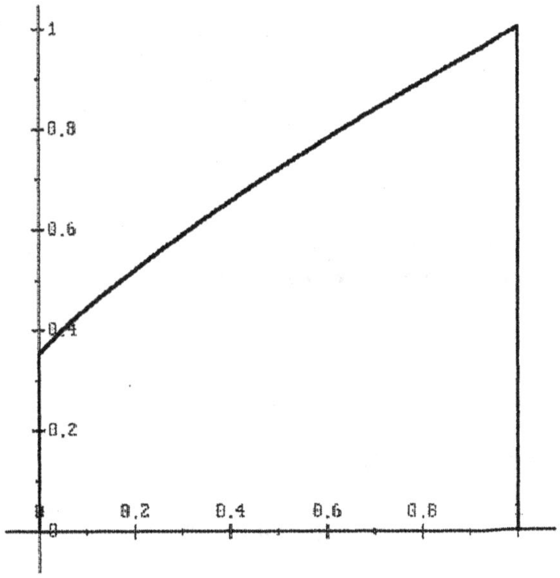

Fig. 2. Optimal solution for ANP

2. PROBLEM OF THE MINIMUM ROTATION-SURFACE (PMRS)

It is necessary to find a generating line of the rotation-surface with minimum area. It is known that the line to be found is a catenary. Depending on the concrete boundary conditions, there can exist either two solutions of the Euler equation or alone or none (Fig. 3). These solutions are catenaries. In traditional courses, one does not pay attention enough to the case of absence of the classic solution. In the numerical solution by the Euler method, the broken-wise solution is easily discovered. These are two disks for which the generating line consists of two vertical segments and a horizontal one lying on the axis of rotation. Moreover, the broken-wise solution can be a global minimum in the zone of existing the solution

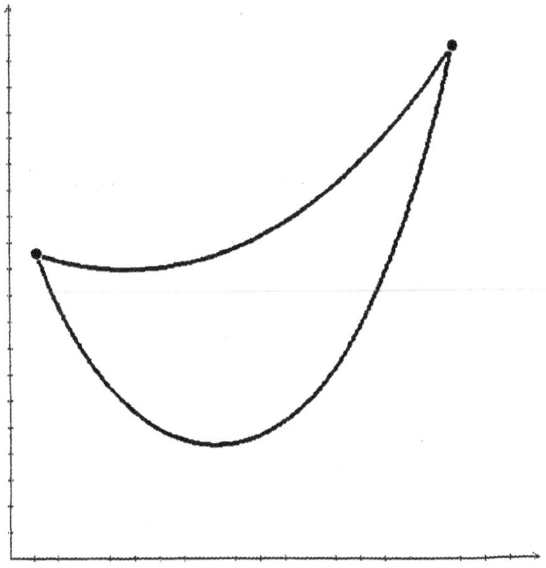

Fig. 3. Two solutions of the Euler equation for PMRS

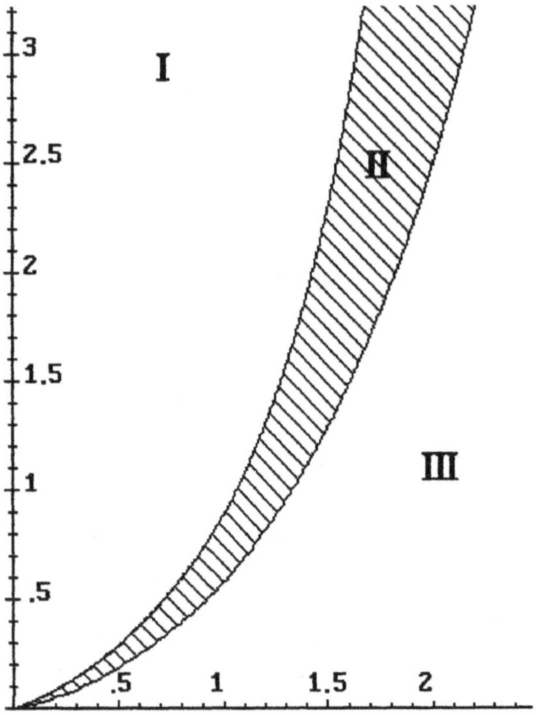

Fig. 4. Minimum existence depending on location of the right-end point:
I - catenary is a global minimum
II - catenatry is a local minimum only, disks are a global minimum
III - disks are the only global minimum

for the Euler equation (Ahiezer, 1955; Young, 1969; Cesari, 1983). If the left-end point is fixed at $(0,1)$, then one can construct a partition of the quadrant of feasible locations for the right-end point. If the right-end point is located in the area I (see Fig. 4), then the global minimum is a catenary. If this point is in the area II, then the global minimum is a broken-wise solution (disks) while the catenary gives only a local minimum. If the point is in the area III, then the only (global) minimum is the broken-wise solution, no solutions of the Euler equation exist. These facts can be found in (Anisimov, 1904; Ahiezer, 1955; Young, 1969; Cesari, 1983), however, they are not widely known.

3. BRACHISTOCHRONE PROBLEM (BP)

In this problem, it is necessary to find the line of fastest descent between two points in the vertical plane (Bernoulli, 1937; Tikhomirov, 1990). This line is an arc of the cycloid. The problem of finding the concrete cycloid for given boundary conditions is usually out of the frame of study. For arbitrary feasible boundary conditions, the program has been elaborated which finds and visually represents the optimal solution both in the form of the cycloid and in the class of two-link piecewise lines. The motion of a point along both the trajectories is animated. An Java-version of this program for Internet is also elaborated

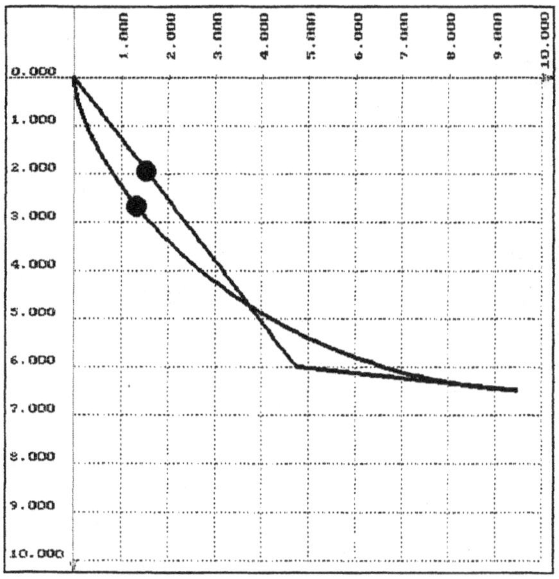

Fig. 5. Optimal solutions of BP for two class of lines

(see at http://home.ural.ru/~iagsoft/BrachJ.html).

For the case of motion in the central gravity field (BPCGF), the brachistochrone is calculated by the Euler method. For this case, the optimal solution (Fig. 6) and isochrones (Fig. 7) obtained on the multiprocessor system (Ivanov, 1995) are demonstrated. Here the "isochrone" means a line consisting of points with the same optimal time. In Fig. 6 the center of gravitation is at $(0,0)$, the starting point is $(1,0)$, optimal trajectories are given for certain endpoints having norm $\|(x_1, y_1)\| = 1$.

ACKNOWLEDGEMENT

This work was supported by the Russian Foundation of Basic Researches under Grant No. 97-01-00458.

REFERENCES

Ahiezer, N.I. (1955). *Lectures on the calculus of variations.* TehGIZ, Moscow [in Russian].

Anisimov, V. (1904). *Course on the calculus of variations.* V. I, Warshaw [in Russian].

Bernoulli, J. (1937). *Selected works on mechanics.* TehGIZ, Moscow, Leningrad [in Russian].

Bunday, B. (1984). *Basic optimisation methods.* Edward Arnold, London.

Cesari, L. (1983) *Optimization Theory and Applications Problems with Ordinary Differential Equations.* Springer-Verlag, N.Y., Heidelberg, Berlin.

Ivanov, A.G. (1995). An Experience of Level Line Constructions on Transputer Systems. In: *Algoritmy i Programmnye Sredstva Parallel'nykh Vychislenii (Algorithms and Program Means of Parallel Computations)* (Kukushkin A.P. and S.V. Sharf, (Eds.)), 69-78. Ural Branch of Russian Academy of Sciences,

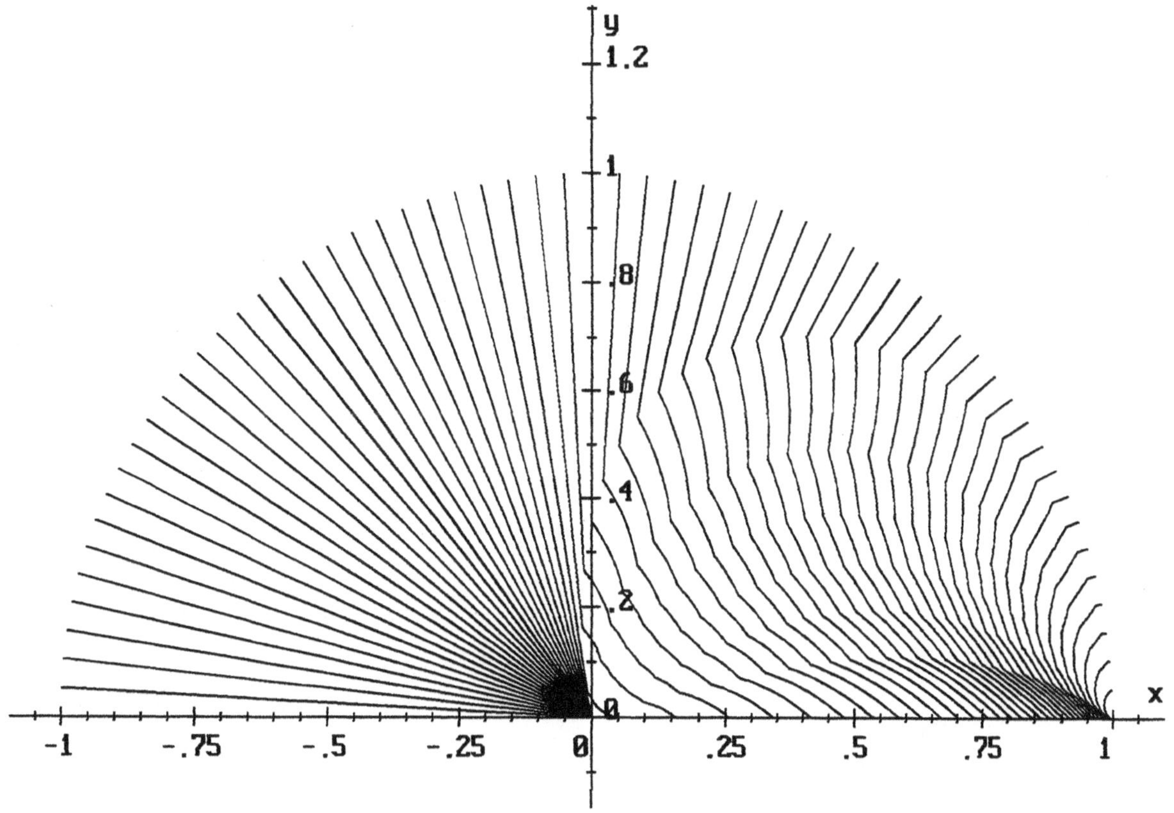

Fig. 6. Field of optimal trajectories for BPCGF

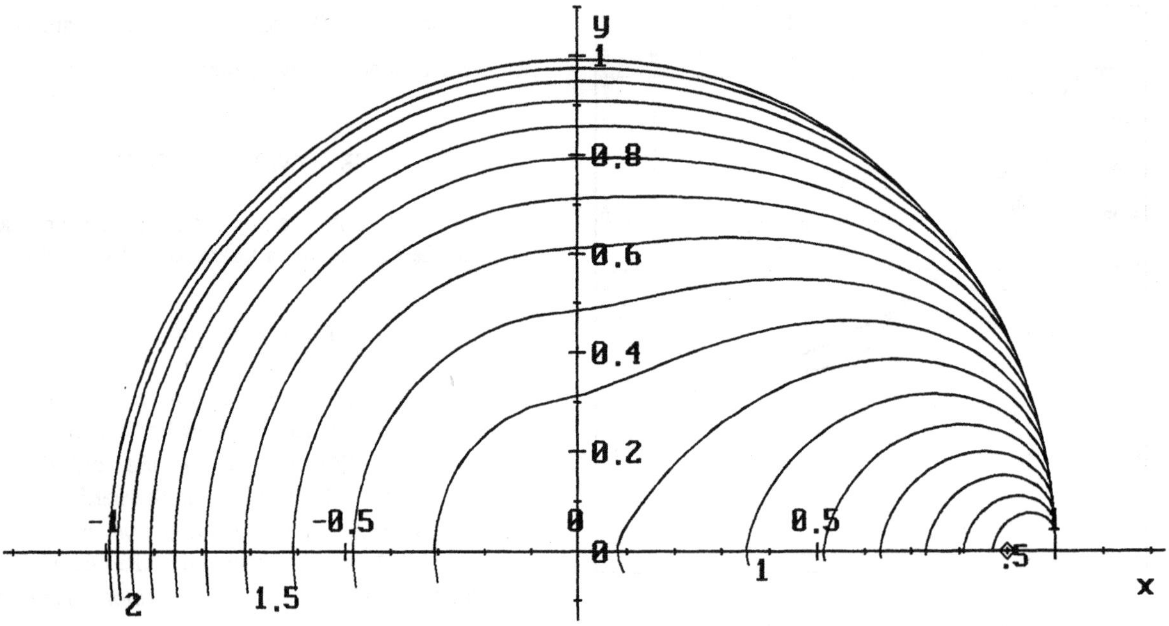

Fig. 7. Isochrones for BPCGF. The center of gravitation is at $(0, 0)$, the starting point is $(1, 0)$.

Ekaterinburg [in Russian].

Krasnov, M.A., G.I. Makarenko and A.I. Kiselev (1973). *Calculus of Variations.* Nauka, Moscow [in Russian].

Newton, I. (1916). *Philosophiae Naturalis Principia Mathematica.* S.-Petersburg [in Russian].

Tikhomirov, V.M. (1990). *Stories about Maxima and Minima. Translated from the Russian by Abe Shenitzer,* Amer. Math. Soc. and Math. Assoc. of America.

Young, L.C. (1969). *Lectures on the calculus of variations and optimal control theory.* W.B. Saunders Co., Philadelphia, London, Toronto.

THE PROBLEM OF OPTIMAL CONTROL OF ALMOST PERIODIC MOTIONS OF A LINEAR SYSTEM WITH QUADRATIC QUALITY FUNCTIONAL

Ivanov A.G.

Udmurt State University, Izhevsk, Russia

Abstract: In the present work the question of existence of a solution in the optimal control problem for almost-periodic motions of a system with quadratic quality functional is discussed and some examples of effective enlargement of periodic admissible processes to the a.p. ones are presented.

Keywords: speed control, controllability, linear systems.

Denote by $S_\infty(\mathbb{R},\mathbb{R}^n), B(\mathbb{R},\mathbb{R}^n)$ the collection of bounded functions $f:\mathbb{R}\to\mathbb{R}^n$ that are a.p. in the sense of Stepanov and Bohr, respectively. Recall that for every function $f\in S_\infty(\mathbb{R},\mathbb{R}^n)(f\in B(\mathbb{R},\mathbb{R}^n))$ there exists a mean

$$M\{f(t)\} \doteq \lim_{T\to\infty}\frac{1}{T}\int_0^T f(t)dt.$$

Consider further a linear stationary control system

$$\dot{x} = Ax + Bu, \quad x\in\mathbb{R}^n, \quad u\in\mathbb{R}^m, \qquad (1)$$

such that $Re(\lambda_j(A) \neq 0, j=1,...,n$. Note that in this case to each control $u(\cdot)\in\in S_\infty(\mathbb{R},\mathbb{R}^n)$, with $u=u(t), t\in\mathbb{R}$, there corresponds a unique Bohr a.p. solution $x(t) == x(t;u(\cdot), t\in\mathbb{R}$ of system (1). Introduce now the set

$$\mathbb{U} \doteq \{u(\cdot) = (u_l(\cdot))_{l=1}^m \in S_\infty(\mathbb{R},\mathbb{R}^m) : M\{u_l^2(t)\} \leqslant 1/2, \quad l=1,...,m\} \qquad (2)$$

of admissible controls of system (1) and select its subset

$$\mathbb{U}_{trig}^m \doteq \{u(\cdot)\in\mathbb{U} : u(\cdot) \text{ is a trigonometric polynomial of order } m\}.$$

We further fix matrices $Q\in \mathrm{Hom}(\mathbb{R}^n)$, $P\in \mathrm{Hom}(\mathbb{R}^m,\mathbb{R}^n)$, $D\in\mathrm{Hom}(\mathbb{R}^m)$ and consider the problem

$$I(x(\cdot),u(\cdot)) = M\{x^*(t)Qx(t) + x^*(t)Pu(t) + u^*(t)Du(t)\} \to \inf, \quad u(\cdot)\in\mathbb{U}, \qquad (3)$$

where $x(\cdot) = x(\cdot;u(\cdot)$ is a Bohr a.p. solution of system (1) corresponding to $u(\cdot)\in\mathbb{U}$.

Theorem 1. *We have the equality*

$$\inf_{u\in\mathbb{U}} I(x(\cdot),u(\cdot)) = \inf_{u\in\mathbb{U}_{trig}^m} I(x(\cdot),u(\cdot)).$$

Thus the greatest lower bounds of problem (3) and the following problem

$$J(x(\cdot),u(\cdot)) \to \inf, \quad u(\cdot)\in\mathbb{U}_{trig}^m \qquad (4)$$

do coincide.

Let us convexify problem (4). To this end, we consider the set

$$K_m \doteq \{(k_1,\ldots,k_m) : k_j \doteq (a_j,b_j)\in(\mathbb{R}^m\times\mathbb{R}^m), j=1,\ldots,m, \sum_{j=1}^m a_{jl}^2 + b_{jl}^2 \le 1, \quad l=1,\ldots,m,\},$$

as well as the set

$$\Xi_{trig}^m \doteq \bigcup_{q=0}^m \{(\mu(\cdot) \doteq \mu(\cdot;k_1,\ldots,k_m,\omega_{q+1},\ldots,\omega_m) : (k_1,\ldots,k_m)\in K_m, \omega_{q+1},\ldots,\omega_m > 0\},$$

in which

$$\mu(t;k_1,\ldots,k_m,\omega_{q+1},\ldots,\omega_m) \doteq$$
$$\doteq \begin{cases} \nu^{(q)} + \eta^{(q)}(t), & \text{if} \quad q \geq 1, \\ \eta^{(0)}(t), & \text{if} \quad q = 0, \end{cases} \tag{5}$$

where, in turn,

$$\nu^{(q)} \doteq \frac{1}{4}\sum_{j=1}^{q}\{\delta_{-a_j} + \delta_{a_j} + \delta_{-b_j} + \delta_{b_j}\},$$

$$\eta^{(q)} \doteq \sum_{j=q+1}^{m} \delta_{a_j \cos\omega_j t + b_j \sin\omega_j t}$$

(δ_u is the Dirac measure concentrated at the point $u \in \mathbb{R}^m$). Since for each function $c \in C(\mathbb{R}^m, \mathbb{R}^m)$ and any $\mu(\cdot) \in \Xi^m_{trig}$ the mapping

$$t \mapsto \langle \mu(t), c(u) \rangle \doteq$$
$$\doteq \frac{1}{4}\sum_{j=1}^{q}\{c(-a_j) + c(a_j) + c(-b_j) + c(b_j)\} +$$
$$+ \sum_{j=q+1}^{m} c(a_j \cos\omega_j t + b_j \sin\omega_j t)$$

belongs to $B(\mathbb{R}, \mathbb{R}^m)$, then [1] Ξ^m_{trig} are measure-valued a.p. functions. Moreover,

$$M\{\langle \mu(t;k_1,\ldots,k_m,\omega_{q+1},\ldots,\omega_m), u_l^2\rangle\} =$$
$$\frac{1}{2}\sum_{j=1}^{m} a_{jl}^2 + b_{jl}^2 \leq \frac{1}{2}, \quad l = 1,\ldots,m. \tag{6}$$

Next, on the set Ξ^m_{trig}, we define the mapping

$$\mu(\cdot) \mapsto \mathfrak{I}(x(\cdot), \mu(\cdot)) \doteq$$
$$\doteq M\{x^*(t)Qx(t) + x^*(t)P\langle\mu(t), u\rangle +$$
$$\langle \mu(t), u^*Du\rangle\},$$

where $x(\cdot) \doteq x(\cdot;\mu(\cdot))$ is a Bohr a.p. solution of the system

$$\dot{x} = Ax + B\langle\mu(t), u\rangle, \qquad (t,x) \in \mathbb{R} \times \mathbb{R}^n,$$

corresponding to $\mu(\cdot) \in \Xi^m_{trig}$ and we consider the problem

$$\mathfrak{I}(x(\cdot), \mu(\cdot)) \to \inf, \quad \mu(\cdot) \in \Xi^m_{trig}, \tag{7}$$

being (here see (5),(6) for $q = 0$) an enlargemant (convexification) of problem (4), in which the measure $\widehat{\mu}(\cdot) \in \Xi^m_{trig}$ is said to be the optimal control if $\mathfrak{I}(x(\cdot), \mu(\cdot)) \geqslant \mathfrak{I}(\widehat{x}(\cdot), \widehat{\mu}(\cdot))$ $(\widehat{x}(\cdot) \doteq x(\cdot;\widehat{\mu}(\cdot)))$ for all $\mu(\cdot) \in \Xi^m_{trig}$, and the pair $(\widehat{x}(\cdot), \widehat{\mu}(\cdot))$ is called the solution of this problem.

Lemma 1. *For every $\mu(\cdot) \in \Xi^m_{trig}$ there exists a sequence of controls $u_j(\cdot) \in \mathbb{U}^m_{trig} \in \mathbb{N}$ such that $I(x_j(\cdot), \mu_j(\cdot)) \to \mathfrak{I}(x(\cdot), \mu(\cdot))$ for $j \to \infty$.*

Theorem 2. *For any fixed matrices $Q \in \mathrm{Hom}(\mathbb{R}^n)$, $P \in \mathrm{Hom}(\mathbb{R}^m, \mathbb{R}^n)$, $D \in \mathrm{Hom}(\mathbb{R}^m)$ the solution of problem (7) exists.*

Corollary 1. *Let, in problem (7), the matrix Q be positive definite, and the matrix P be zero. Then for any fixed matrix D the optimal control will be represented by the measure*

$$\nu_0^{(m)} \doteq \frac{1}{4}\sum_{j=1}^{m}\{\delta_{-a_j^0} + \delta_{a_j^0} + \delta_{-b_j^0} + \delta_{b_j^0}\},$$

corresponding to some collection $(k_1^0, \ldots, k_m^0) \in K_m$.

Corollary 2. *Let the matrix $Q \in \mathrm{Hom}(\mathbb{R}^n)$ be negative definite. Then for any fixed matrix $D \in \mathrm{Hom}(\mathbb{R}^m)$ the problem*

$$J(x(\cdot), u(\cdot)) =$$
$$= M\{x^*(t)Qx(t) + u^*(t)Du(t)\} \to \inf,$$
$$u(\cdot) \in \mathbb{U},$$

has an optimal control $\widehat{u}(\cdot)$ belonging to the set $\mathbb{U}^m_{trig}$.

Next, we select in $\mathbb{U}$ a subset Π which consists of periodic controls, i.e., $u(\cdot) \in \Pi$ if, first, $u(\cdot) \in \mathbb{U}$, and, second, there exists an $\omega > 0$ such that $u(t+\omega) = u(t)$, $t \in \mathbb{R}$. It should be noted that the solution $x(\cdot)$ of system (1) corresponding to such a control will be ω–periodic too.

Thus, problem (3) is an enlargement of the following problem of periodic optimization (in general with a non-fixed period) $I(x(\cdot), u(\cdot)) \to inf$, $u(\cdot) \in \Pi$. This enlargement is said to be effective if the solution of problem (3) is an a.p. admissible process $\widehat{x}(\cdot), \widehat{u}(\cdot))$, such that for all $u(\cdot) \in \Pi$, the inequality $I(\widehat{x}(\cdot), \widehat{u}(\cdot)) \leq I(x(\cdot), u(\cdot))$ holds true.

We shall exemplify the above statements. Consider the matrix

$$C(\alpha,\beta) \doteq \begin{pmatrix} \alpha & \beta \\ -\beta & \alpha \end{pmatrix} \alpha \neq 0, \quad \beta > 0. \tag{8}$$

Using the definition of the spectral norm, we obtain

$$\begin{cases} |(C(\alpha,\beta) - i\omega E)^{-1}|^2 < \dfrac{1}{\alpha^2}, & \text{if} \quad \omega \neq \beta \\ |(C(\alpha,\beta) - i\omega E)^{-1}|^2 = \dfrac{1}{\alpha^2}, & \text{if} \quad \omega = \beta, \end{cases} \tag{9}$$

where $E \doteq id_{\mathrm{Hom}(\mathbb{R}^2)}$, $i^2 = -1$.

Example 1. Consider, for $m = 2$, the problem

$$J(x(\cdot), u(\cdot)) \doteq M\{u^*(t)Du(t) - |x(t)|^2\} \to inf, \tag{10}$$

where $u(\cdot) = (u_l(\cdot))_{l=1}^2 \in \mathbb{U}$, $x(\cdot) = x(\cdot;u(\cdot)) \in B(\mathbb{R}, \mathbb{R}^2)$ is the solution of the system

$$\dot{x} = C(\alpha,\beta)x - u(t), \quad x \in \mathbb{R}^2, \tag{11}$$

the nonnegative matrix $D = diag[d_1, d_2] \in Hom(\mathbb{R}^2)$ is such that $d_j < \frac{1}{\alpha^2}$, $j = 1,2$. By Corollary 2, the solution of problem (10) exists, and it will suffice to seek the optimal control among the controls of the form $u(t) = \sum_{j=1}^{2} a_j \cos\omega_j t + b_j \sin\omega_j t$, that belong to the set $\mathbb{U}^2_{trig} \subset \mathbb{U}$. For any such control $M\{u_l^2(t)\} = \frac{1}{2}(a_{jl}^2 + b_{jl}^2 \leq 12$, $l = 1,2$, and for the corresponding solution $x(\cdot)$ of system (11)

$$M\{|x(t)|^2\} \leqslant$$
$$\leqslant \frac{1}{2}\sum_{j=1}^{2} |(C(\alpha,\beta) - i\omega_j E)^{-1}|^2 \cdot (|a_j|^2 + |b_j|^2).$$

Whence it follows, by virtue of (9), that if $\omega_j \neq \beta$, $j = 1,2$, then

$$M\{|x(t)|^2\} < \frac{1}{2\alpha^2} \cdot \sum_{j=1}^{2}(|a_j|^2 + |b_j|^2)$$

and therefore, in this case, in view of the choice of d_1, d_2, we have

$$J(x(\cdot), u(\cdot)) > \frac{1}{2}\sum_{l=1}^{2}(d_l - \frac{1}{\alpha^2})\sum_{j=1}^{2}(a_{jl}^2 + b_{jl}^2) \geqslant$$
$$\geqslant \gamma \doteq \frac{1}{2}(d_1 + d_2) - \frac{1}{\alpha^2}. \quad (12)$$

On the other hand, it is easily seen that $\widehat{x}(t) = \frac{1}{\alpha} colon[\cos\beta t, -\sin\beta t]$ is the solution of system (11) corresponding to the admissible control $\widehat{u}(t) \doteq colon[\cos\beta t, -\sin\beta t]$, and moreover, $\mathfrak{I}(\widehat{x}(\cdot), \widehat{u}(\cdot)) = \gamma$. In view of (12), we conclude that the solution of problem (10) is a $2\pi/\beta$-periodic process $(\widehat{x}(\cdot), \widehat{u}(\cdot))$.

Example 2. Let us fix a block-diagonal matrix (here see (8)) $A = diag[C(\alpha_1, \beta_1), C(\alpha_2, \beta_2)] \in Hom(\mathbb{R}^4)$, $(\alpha_j \neq 0, \beta_j > 0, j = 1,2)$ and consider, for $m = 4$, a problem of the form (10), where $u(\cdot) = (u_l(\cdot))_{l=1}^4 \in \mathbb{U}$, the function $x(\cdot) = x(\cdot; u(\cdot)) \in B(\mathbb{R}, \mathbb{R}^4)$ is the solution of the system

$$\dot{x} = Ax - u(t), \quad x \in \mathbb{R}^4, \quad (13)$$

and the nonnegative matrix $D \doteq diag[d_1, \ldots, d_4] \in Hom(\mathbb{R}^4)$ is such that $d_j < \frac{1}{\alpha_1^2}$, $j = 1,2$, and $d_j < \frac{1}{\alpha_2^2}$, $j = 3,4$.

Let us demonstrate that if β_1 and β_2 are incommensurable, then, firstly, the solution of this problem is an a.p. process, and, secondly, the enlargement of the periodic admissible controls to the almost periodic ones is effective. In fact, by Corollary 2, we seek the optimal control in the form $u(t) = \sum_{j=1}^{4} a_j \cos\omega_j t + b_j \sin\omega_j t$ $(u(\cdot) \in \mathbb{U}^4_{trig} \subset \mathbb{U})$. From the structure of matrices A and D, using the same reasoning as in Example 1, we obtain that for any such control $J(x(\cdot), u(\cdot)) > \gamma \doteq \sum_{j=1}^{4} d_j - (\alpha_1^2 + \alpha_2^2)^{-1}$, if $\omega_1, \omega_2 \neq \beta_1$, $\omega_3, \omega_4 \neq \beta_2$, and for

$$\widehat{u}(t) \doteq colon[\cos\beta_1 t, -\sin\beta_1 t, \cos\beta_2 t, -\sin\beta_2 t],$$

the equality $J(\widehat{x}(\cdot), \widehat{u}(\cdot)) = \gamma$ holds true, where

$$\widehat{x}(t) = colon[\frac{1}{\alpha_1}\cos\beta_1 t, -\frac{1}{\alpha_1}\sin\beta_1 t,$$
$$\frac{1}{\alpha_2}\cos\beta_2 t, -\frac{1}{\alpha_2}\sin\beta_2 t]$$

is the solution of system (13) corresponding to $\widehat{u}(\cdot)$. Since β_1, β_2 are incommensurable, the solution $(\widehat{x}(\cdot), \widehat{u}(\cdot))$ of this problem is an a.p. process.

Consider further an arbitrary ω-periodic control $u(\cdot) \in \mathbb{U}$, to which we put in correspondence its Fourier series

$$u(t) \sim \sum_{j\in\mathbb{Z}} c_j \exp(i\frac{2\pi j}{\omega}t).$$

Because of their being incommensurable, β_1 and β_2 cannot belong simultaneously to the set $\{\frac{2\pi j}{\omega}, j \in \mathbb{Z}\}$. Hence, for some $l \in \{1,2\}$ and any $j \in \mathbb{Z}$ the inequality

$$|(C(\alpha_l, \beta_l) - i\frac{2\pi j}{\omega}E)^{-1}|^2 < \frac{1}{\alpha_l^2}$$

holds true, whence, by virtue of the equalities

$$M\{|u(t)|^2\} = \sum_{j\in Z} |c_j|^2,$$
$$M\{|x(t)|^2\} = \sum_{j\in Z} |(A - i\omega E)^{-1} c_j|^2$$

it follows that

$$J(x(\cdot), u(\cdot) \geqslant \sum_{l=1}^{4} d_l \sum_{j\in Z} |c_{jl}|^2 -$$
$$- |(C(\alpha_l, \beta_l) - i\frac{2\pi j}{\omega}E)^{-1}|^2 \cdot (|c_{j1}|^2 + |c_{j2}|^2) -$$
$$-|(C(\alpha_2, \beta_2) - i\frac{2\pi j}{\omega}E)^{-1}|^2 \cdot (|c_{j3}|^2 + |c_{j4}|^2) > \gamma.$$

Thus, for any periodic control $u(\cdot) \in U$, $J(x(\cdot), u(\cdot)) > \gamma$ which implies the effectiveness of enlargement of the periodic controls to the almost periodic ones.

The work is supported by Russian Foundation for Basis Reseach (grant 97 - 01 - 00413) and by Competition Centre of Fundamental Natural Science (grant 97 - 0 - 1.9).

REFERENCES.

Ivanov, A.G.(1991). *Measure-signed almost periodic functions.* Preprint Fiz.-Tekhn. Inst. UrO Akad. Nauk SSSR, Sverdlovsk.

THE PROGRAMMING ITERATIONS METHOD FOR DISCRETE-TIME CONTROL PROBLEMS ON L^*-SPACES

Vladimir M. Ivanov *

* *Ural State University of Economics, Ekaterinburg, Russia*

Abstract: A modification of the programming iterations method is described here for the discrete-time controlled dynamical systems in which spaces of controls and phase states are the spaces with a priori given notion of limit of a sequence, i. e. the L^*-spaces. For such systems described by nonlinear relations it is to be considered the game problem of keeping back of the system within given phase restrictions during an infinite time interval. The modification of the programming iterations method is used for solving the game problem for the first player on the classes of counterstrategies and quasistrategies. *Copyright ©1998 IFAC*

Keywords: approximate analysis, gradient methods, convex optimization, convex projections, convex programming

1. MAIN NOTIONS AND PROPERTIES

It is to be considered the following discrete-time controlled system described by the nonlinear relations

$$x_{m+1} = f(x_m, u),\ f \in F_m,\ m \in N_k,$$
$$u \in U,\ x_k = x,\ x \in X \qquad (1)$$

Here $k \in N_0$ – an initial time where $N_k = \{k, k+1, \ldots\}$ for any $k \in \{0, 1, 2, \ldots\}$; $F_m, m \in N_k$ – a nonempty subset of the space of sequentially continuous maps of the following kind $f : X \times U \to X$ where X and U are the L^*- spaces (Engelking, 1977).

An element of the Cartesian degree $\mathcal{U} = U^{N_0}$ is called the program, or the programmed control. An element of the Cartesian product $\mathbf{F}_k = F_k \times F_{k+1} \times \ldots$ is called the model of the system (1). The state of the system (1) is an arbitrary pair (k, x) where $k \in N_0, x \in X$. The motion of the system (1) from an initial state (k, x) corresponding to a model $f \in \mathbf{F}_k$ and a program $u \in \mathbf{U}$ is the element of the space X^{N_k} denoted by $\Phi_k(x, u, f)$ for which next relations are fulfilled for any $m \in N_k$:

$$(\Phi_k(x, u, f))(k) = x,$$
$$(\Phi_k(x, u, f))(m+1) =$$
$$= (f(m))((\Phi_k(x, u, f))(m), u(m))$$

The existence, the uniqueness and the sequential continuity of the motion $\Phi_k(x, u, f)$ with respect to the joint variable (x, u) are proved as in the work (Ivanov and Chentsov, 1987). The motion of the system (1) possesses the semigroup property. Note, that the functional notations are used only for the projections of the following elements: $u \in \mathbf{U}, f \in \mathbf{F}_k, \Phi_k(x, u, f) \in X^{N_k}$. Projections of other elements are indexed.

2. THE SETTING OF THE GAME PROBLEM OF KEEPING BACK

Let $\mathbf{X} = (2^X)^{N_0}$ and $M \in \mathbf{X}$, in addition the sets M_t for any $t \in N_0$ are not empty. Here and further, the notation M_t is used for the projection of any $M \in \mathbf{X}$ to the t-th factor of the Cartesian degree $\mathbf{X}$. It demands to find such a rule of constructing the program $u \in \mathbf{U}$ – a strategy – guaranteeing the inclusions

$$(\Phi_k(x,u,f))(t+1) \in M_{t+1}, t \in N_k \quad (2)$$

where $x \in M_k$. The opposite problem – the deviation problem – is formulated as follows: it demands to find such a rule of constructing the model $f \in \mathbf{F}_k$ – a strategy – guaranteeing the violation of the inclusion (2) for some $t \in N_{k+1}$ even though $x \in M_k$.

Both problems describe a conflict situation between two players if for any time $t \in N_k$ the first player is ordered by the selection of the control $u(t) \in U$, and the second player is ordered by the selection of $f(t) \in F_t$. Below it is to be given the solution of the problem for the first player when the second player is informationally discriminated, namely, the solution is given for two classes of strategies: quasistrategies and counterstrategies.

3. THE PROGRAMMING ITERATIONS METHOD FOR KEEPING BACK PROBLEM

The essence of the programming iterations method consists of the setting of an contraction operator $\mathbf{A} : \mathbf{X} \to \mathbf{X}$, the immovable point of which is used as the basis for definition of the first player resolving strategy at the keeping back problem.

Let us determine the operator $\mathbf{A}$ for an arbitrary $M \in \mathbf{X}$ and arbitrary $t \in N_0$ by the following ratio:

$$\mathbf{A}(M)_t = \{x \mid x \in M_t, \forall f \in \mathbf{F}_k \ \exists u \in \mathbf{U}$$
$$\forall s \in N_t \ (\Phi_t(x,u,f))(s) \in M_s\} \quad (3)$$

Later on it is to be assumed that for any $t \in N_0$ and any $M \in \mathbf{X}$ the relations $\mathbf{A}(M)_s \neq \emptyset$ are fulfilled for all $s \in N_t$ if the relation $M_s \neq \emptyset$ are fulfilled for all $s \in N_t$.

It is not difficult to verify that the operator $\mathbf{A}$ is contractive, that is, $\mathbf{A}(M)_t \subset M_t$ for all $t \in N_0$, and monotonic, i.e. for arbitrary $M, N \in \mathbf{X}$ and $t \in N_0$ if $M_t \subset N_t$ then $\mathbf{A}(M)_t \subset \mathbf{A}(N)_t$. The subset containing all elements $M \in \mathbf{X}$, for which M_t is sequentially closed in X for any $t \in N_0$, is denoted by $\tilde{\mathbf{X}}$. Using the sequential continuity of the motion $\Phi_t(x,u,f)$ with respect to the joint variable (x,u) and the sequential compactness of U it can be established the following statement: if $M \in \tilde{\mathbf{X}}$ then $\mathbf{A}(M) \in \tilde{\mathbf{X}}$.

Let degrees of the operator $\mathbf{A}$ are defined by the recurrence relations $\mathbf{A}^0(M) = M$, $\mathbf{A}^m(M) = \mathbf{A}^{m-1}(\mathbf{A}(M))$, for any $m \in N_1$ and any $M \in \mathbf{X}$, and let by definition

$$\mathbf{A}^\infty(M)_t = \bigcap_{n \in N_0} \mathbf{A}^n(M)_t,$$

$$\mathbf{W}(M) = \mathbf{A}^\infty(M) \quad (4)$$

for an arbitrary $t \in N_0$ and any $M \in \mathbf{X}$.

Theorem 1. For any $M \in \tilde{\mathbf{X}}$ the element $\mathbf{W}(M)$ is an immovable point of the operator $\mathbf{A}$, i.e. $\mathbf{A}(\mathbf{W}(M)) = \mathbf{W}(M)$.

Proof. At first it is to be proved $\mathbf{W}(M) \in \tilde{\mathbf{X}}$, and next - the equality $\mathbf{A}(\mathbf{W}(M)) = \mathbf{W}(M)$. Let $M \in \tilde{\mathbf{X}}$. Then for any $t \in N_0$ the sets $\mathbf{A}(M)_t$ are sequentially closed. Really, if $\{x_i\}_{i \in N_1}$ – is an arbitrary convergent sequence from $\mathbf{A}(M)_t$ and x is its limit, then in accordance with (3) there exists $u_i \in \mathcal{U}$ for any x_i and any $f \in \mathbf{F}_t$ which satisfies the inclusions $(\Phi_t(x_i,u_i,f))(s) \in M_s$ as $s \in N_t$. Due to the sequential compactness of $\mathcal{U}$, there is a convergent subsequence $\{u_{n_i}\}_{i \in N_1}$. Let u is its limit. Hence, as $(\Phi_t(x_{n_i},u_{n_i},f))(s) \in M_s$, $s \in N_t$ and M_s is sequentially closed, and the motion $\Phi_t(x,u,f)$ is sequentially continuous with respect to the joint variable (x,u) then it is fulfilled $(\Phi_t(x,u,f))(s) \in M_s$ for every $s \in N_t$. It means $x \in \mathbf{A}(M)_t$ i.e. the sequential closure of $\mathbf{A}(M)_t$ is proved. Consequently, taking into account (4), the set $\mathbf{W}(M)_t$ is sequentially closed for every $t \in N_0$.

To justify the equality $\mathbf{A}(\mathbf{W}(M)) = \mathbf{W}(M)$ it is necessary to prove only the inclusions $\mathbf{W}(M)_t \subset \mathbf{A}(\mathbf{W}(M))_t$, $t \in N_0$ because the inverse inclusions are obvious. Let $x \in \mathbf{A}(\mathbf{W}(M))_t$ for an arbitrary, but fixed, $t \in N_0$. Then, for every $k \in N_0$, it is true $x \in \mathbf{A}^{k+1}(M)_t = \mathbf{A}(\mathbf{A}^k(M))_t$ by definitions of the element $\mathbf{W}(M)$ and the operator $\mathbf{A}$. Since (3), for given $f \in \mathbf{F}_t$ and $k \in N_0$, there is the program u_k which satisfies for the relations $(\Phi_t(x,u_k,f))(s) \in \mathbf{A}^k(M)_s$ for each $s \in N_t$. So, due to the monotonicity of the operator $\mathbf{A}$, it is realized $(\Phi_t(x,u_k,f))(s) \in \mathbf{A}^n(M)_s$ for all $k \in N_n$ and each $n \in N_0$. As $\mathcal{U}$ is sequentially compact there is some convergent subsequence $\{u_{n_k}\}_{k \in N_1}$. If u is its limit then, because of the sequential closure of $\mathbf{A}^n(M)_s$, it follows $(\Phi_t(x,u,f))(s) \in \mathbf{A}^n(M)_s$. Thus, for the model $f \in \mathbf{F}_t$, it is found the program $u \in \mathcal{U}$ that last inclusions is fulfilled for each $n \in N_0$. That means $(\Phi_t(x,u,f))(s) \in \mathbf{W}(M)_s$, $s \in N_t$, this is, $x \in \mathbf{A}(\mathbf{W}(N))_t$. Theorem 1 has proved.

4. SOLUTION OF THE KEEPING BACK PROBLEM FOR THE CLASS OF QUASISTRATEGIES

A way (rule) of programmed control formation is called a strategy. Any map of the kind $\alpha : \mathbf{F}_t \to 2^{\mathcal{U}}$ satisfying for the restrictions

(a) $\alpha(f) \neq \emptyset$ for every $f \in \mathbf{F}_t$ (nonemptiness of image),

(b) for any $f_1, f_2 \in \mathbf{F}_t$, if $f_1 |_{\overline{t,m}} = f_2 |_{\overline{t,m}}$, then $\{u |_{\overline{t,m}} \mid u \in \alpha(f_1)\} = \{v |_{\overline{t,m}} \mid v \in \alpha(f_2)\}$ (physical realizability) is called a quasistrategy of the first player. Here the notation, for example, $f_1 \mid_{\overline{t,m}}$ denotes the limitation of the map f_1 on the integer set $\overline{t,m}$ where by definition $\overline{t,m} = \{t, t+1, \dots, m\}$.

Let, for any $t \in N_0$ and any $x \in \mathbf{W}(M)_t$, the map $\alpha_{t,x} : \mathbf{F}_t \to 2^{\mathcal{U}}$ is defined by the following relation for every $f \in \mathbf{F}_t$:

$$\alpha_{t,x}(f) = \{u \mid u \in \mathcal{U}\,,\, (\Phi_t(x,u,f))(s) \in \\ \in \mathbf{W}(M)_s \text{ for all } s \in N_t\}$$

Theorem 2. If $M \in \tilde{\mathbf{X}}$, $x \in \mathbf{W}(M)_t$ for $t \in N_0$, then the map $\alpha_{t,x}$ is the quasisfrategy of the first player solving the keeping back problem within given phase restrictions M from the initial state (t,x).

Proof. Obviously, it is sufficient to prove only that $\alpha_{t,x}$ is a quasistrategy, because this one solves the keeping back problem according to the definition itself of the sets $\alpha_{t,x}(f)\,,\, f \in \mathbf{F}_t$.
Nonemptiness of image. If $x \in \mathbf{W}(M)_t$, then $x \in \mathbf{A}(\mathbf{W}(M))_t$ according to the theorem 1. By the definition of the operator $\mathbf{A}$, it means that for any $f \in \mathbf{F}_t$ there is such a program $u \in \mathcal{U}$ that $(\Phi_t(x,u,f))(s) \in \mathbf{W}(M)_s\,,\, s \in N_t$, this is, $u \in \alpha_{t,x}(f)$.
Physical realizability. Let $f_1(\tau) = f_2(\tau)$ at $\tau \in \overline{t,m}$, where $f_1\,,\, f_2 \in \mathbf{F}_t$. It is to be proved only the inclusion

$$\{u \mid_{\overline{t,m}} \mid u \in \alpha_{t,x}(f_1)\} \subset \{v \mid_{\overline{t,m}} \mid v \in \alpha_{t,x}(f_2)\}$$

because the inverse inclusion can be justified in the same way, it remains only to interchange the roles f_1 and f_2, u and v, respectively.

If $u \in \alpha_{t,x}(f_1)$, then $(\Phi_t(x,u,f_1))(s) \in \mathbf{W}(M)_s$ for all $s \in N_t$. Let $f^* \in \mathbf{F}_{m+1}$ and $x^* \in \mathbf{W}(M)_{m+1}$ are determined by the formulas $x^* = (\Phi_t(x,u,f_1))(m+1)$ and $f_*(\tau) = f_2(\tau)$ for all $\tau \in N_{m+1}$ From here, due to the theorem 1, it is fulfilled $x^* \in \mathbf{A}(\mathbf{W}(M))_{m+1}$. So for the model f^* there exists such a program $v^* \in \mathcal{U}$ that $(\Phi_t(x^*,v^*,f^*))(s) \in \mathbf{W}(M)_s$ at $s \in N_{m+1}$. Consequently, the program, defined by the formula

$$v(\tau) = \begin{cases} u(\tau) & ,\ \tau \in \overline{0,m} \\ v^*(\tau) & ,\ \tau \in N_{m+1}, \end{cases}$$

satisfies for the inclusion $v \in \alpha_{t,x}(f_2)$. On the other hand, $u(\tau) = v(\tau)\,,\, \tau \in \overline{t,m}$ by the definition of the program v. That means $u \mid_{\overline{t,m}} \in \{v \mid_{\overline{t,m}} \mid v \in \alpha_{t,x}(f_2)\}$. Theorem 2 has proved.

5. SOLUTION OF THE KEEPING BACK PROBLEM FOR THE CLASS OF COUNTESTRATEGIES

Let us introduce the notation $\mathbf{U}_s = U^{X \times F_s}$ for any $s \in N_0$. The elements of the set $\mathcal{B} = \mathbf{U}_0 \times \mathbf{U}_1 \times \ldots \times \mathbf{U}_n \times \ldots$ are called counterstrategies. Further it is used the functional notation for an element $\Psi \in \mathcal{B}$, i.e. $\Psi(t)$ for any $t \in N_0$ is an element from $\mathbf{U}_t$. So that is some map of the kind $\Psi(t) : X \times F_t \to U$.

The set of all functions of the kind $y : N_t \to X$ possessing the properties

$$y(t) = x,$$

$$y(\tau{+}1) = (f(\tau))(y(\tau), (\Psi(\tau))(y(\tau), f(\tau)))\,,\ \tau \in N_t$$

for some $x \in X\,,\, f \in \mathbf{F}_t\,,\, \Psi \in \mathcal{B}$ is called the bundle of the motions of the system (1) from an initial state (t,x) according to a counterstrategy Ψ and denoted by $\mathcal{Y}(t,x,\Psi)$. It is not complicated to justify the relation $\mathcal{Y}(t,x,\Psi) \neq \emptyset$ at any $t \in N_0\,,\, x \in X\,,\, \Psi \in \mathcal{B}$, it remains only to use the method of the mathematical complete induction.

Lemma 1. If $M \in \tilde{\mathbf{X}}$, then for any $\tau \in N_0$ and any pair $(x,g) \in \mathbf{W}(M)_\tau \times F_\tau$ there exists such a control $u \in U$ at the point τ that $g(x,u) \in \mathbf{W}(M)_{\tau+1}$.

The proof of this lemma is based on the direct application of the theorem 1 and the definition of the operator $\mathbf{A}$.

Let formulate the keeping back problem for the system (1) within given phase restrictions $M \in \mathbf{X}$ for the class of counterstrategies as follows: for a given initial state (t,x), where $t \in N_0\,,\, x \in M_t$, it is necessary to find such a counterstrategy Ψ that $y(\tau) \in M_\tau$ for any motion $y \in \mathcal{Y}(t,x,\Psi)$ and any $\tau \in N_t$. The lemma allows to define the counterstrategy solving the keeping back problem.

Let us define the map $\psi_\tau : \mathbf{W}(M)_\tau \times F_\tau \to U$ at any $\tau \in N_t$ by the following way: for an arbitrary $x \in \mathbf{W}(M)_\tau$ and arbitrary $g \in F_\tau$ the element $\psi_\tau(x,g)$ is equal to some fixed control existing due to the lemma. Now let us define a map $\Psi_\tau : X \times F_\tau \to U$ as a fixed extension of the map ψ_τ. In addition, if $(x,g) \in \mathbf{W}(M)_t \times F_t$ then $g(x, \Psi_\tau(x,g)) \in \mathbf{W}(M)_{\tau+1}$. The solving counterstrategy is given by the formula $G(\tau) = \Psi_\tau$ for any $\tau \in N_0$. Said above allows to assert the following statement.

Theorem 3. If $M \in \tilde{\mathbf{X}}$, $x \in \mathbf{W}(M)_t$, for some $t \in N_0$, then the counterstrategy G solves the keeping back problem for the system (1) within given phase restrictions M from an initial state (t,x).

It remains to note when it is considered the class of so-called c-models ,i.e. $f(\tau) = f^* \in F^* = F_\tau$ for any $\tau \in N_t$, then it can be suggested a simplified procedure of the programming iterations method for solution the keeping back problem. The programming iterations method, coming from the extremal aiming idea (Krasovskii, 1970), has proposed for continuous controlled system in the book (Subbotin and Chentsov, 1981).

6. REFERENCES

Engelking, R. (1977). *General topology*. PWN. Warsaw.

Ivanov, V.M. and A.G. Chentsov (1987). On controlling by descrete-time systems for an infinite time interval. *J. of calculating mathematics and mathematical physics* **12**, 1780–1789.

Krasovskii, N.N. (1970). *Game problems on contact of motions*. Nauka. Moscow.

Subbotin, A.I. and A.G. Chentsov (1981). *Optimization of guarantee in problems of controlling*. Nauka. Moscow.

ON THE ESTIMATION PROBLEM FOR ONE CLASS OF NONLINEAR SYSTEMS

Kayumov R.I. *

* *Ural State University, Lenina st., 51, Ekaterinburg, 620083, Russia, E-mail: rashid.kayumov@usu.ru*

Abstract: This paper considers the estimation problem for the state of one class of nonlinear multistage and continuous systems under uncertainty and with the quantities observed described by linear equations. It is assumed that the system's initial state and the disturbances in measurement equations are inexact, and that the allowable *a priori* information on these two is limited to their range of variation. The solution of the estimation problem relies on describing the evolution of the system's state information sets which are consistent with both the measurement results and the *a priori* restrictions on uncertain quantities. An information set consists of all trajectories that are possible in the system and which, together with certain admissible disturbances in the observation, determine the given realization of the observed signal. A cross-section, at some time-instant, of information sets generated by the signal observation prior to that instant, is a multiple-valued analog of the system's current phase condition. Information sets provide guaranteed estimates of the system's state based on measurement results. *Copyright ©1998 IFAC*

Keywords: differential inclusions, dynamic system, state estimation, attainability set, nonlinear systems, uncertain system.

1. INTRODUCTION

In this paper, the solutions of determinate estimation problems for one class of nonlinear multistage systems with prior restrictions rest on approximation procedures that are based on optimizing the solutions of specially constructed problems with quadratic limitations. The same approach was used in works (Kurzhanski, 1980; Kurzhanski, 1981; Koscheev and Kurzhanski, 1983). Also, a transition from a nonlinear multistage system to a nonlinear continuous system is considered. For continuous nonlinear systems under uncertainty, a parameterized family of differential inclusions with no restrictions on the system's phase condition is introduced. The intersection of the attainability domains for the differential inclusions over the variety of all functional parameters yields the solution to the original problem (Kayumov, 1994).

2. A SINGLE-STAGE SYSTEM

Consider a single-stage process where the transition of state is described by equation

$$x_1 = x_0 + \delta f(t_0, x_0), \qquad (1)$$

where x is an n-dimensional system state vector, δ – a certain positive number, $f(t,x)$ – a given mapping from $R^1 \times R^n$ into R^n. The initial state x_0 is subject to constraint

$$x_0 \in X_0, \qquad (2)$$

where X_0 is a compact in R^n. The measurement device equation is given by

$$y_1 = G_1 x_1 + \xi_1, \qquad (3)$$

where y is an m-dimensional output signal, G_1 – a matrix of correspondent dimensions with a full rank, ξ_1 – a noise subject to constraint

$$\xi_1 \in \Xi, \qquad (4)$$

where Ξ is a convex compact in R^m. Let, by virtue of system (1), (3), $\hat{y}_1$ be the realized signal.

Definition. Informational domain (Kurzhanski, 1977) X_1 of the states of system (1) that are compatible with signal $\hat{y}_1$ under restrictions (2), (4) is defined as the set of such and only such vectors $x \in R^n$ that for each of them there exists a pair $\hat{x}_0 \in X_0$, $\hat{\xi}_1 \in \Xi$ such that the solution to the system (1), (3), found under $x_0 = \hat{x}_0$, $\xi_1 = \hat{\xi}_1$, satisfies the conditions $x_1 = x$, $y_1 = \hat{y}_1$.

Assumption 1. $f(t, x)$ is a continuous in (t, x) function satisfying the Lipschitz condition in x with constant $K > 0$ and the solution extendibility condition

$$|f(t,x)| \leq k_1(t)|x| + k_2(t),$$

where $k_1(t)$ and $k_2(t)$ are functions summable on any finite time interval.

Consider the following modified system. Let the pair $\{x_0, \xi_1\}$ be fixed, and the actual realization – generated not by the original system, but by the modified system

$$z_1 = z_0 + \delta f(t_0, z_0) + \delta h_0, \quad (5)$$

$$y_1 = G_1 z_1 + \xi_1 + w_1. \quad (6)$$

where $z_0 \in X_0$, h_0 and w_1 are additional fictitious noises subject to quadratic restriction

$$(h_0, N_0 h_0) + (w_1, L_1 w_1) \leq \mu^2, \quad (7)$$

where N_0 and L_1 are matrices of appropriate dimensions, and also $N_0 > 0$, $L_1 > 0$ (positively defined matrices). In restriction (7) the number μ can be chosen in such a way that the domain compatible with signal $\hat{y}_1$ by virtue of the modified system (5), (6) is not empty.

The informational domain of the modified system is an ellipsoid. Its center is given by

$$\hat{z}_1 = K_1^{-1}\frac{1}{\delta^2}N_0(z_0 + \delta f(t_0, z_0)) +$$
$$+ K_1^{-1}G_1^{\top}L_1(\hat{y}_1 - \xi_1), \quad (8)$$

where

$$K_1 = \frac{1}{\delta^2}N_0 + G_1^{\top}L_1G_1.$$

Denote D_1 to be the attainability domain of system (1), namely

$$D_1 = \{d | d = z_0 + \delta f(t_0, z_0), z_0 \in X_0\}. \quad (9)$$

Consider the sum of the centers of ellipsoids z_1^0 over all $z_0 \in X_0$ and $\xi_1 \in \Xi$:

$$Z_1^0(\Omega) = K_1^{-1}\frac{1}{\delta^2}N_0 D_1 +$$
$$+ K_1^{-1}G_1^{\top}L_1(\hat{y}_1 - \Xi), \quad (10)$$

where $\Omega = \{N_0, L_1\}$.

Theorem. The following equality is fulfilled

$$X_1 = \cap\{Z_1^0(\Omega)|\Omega\}, \quad (11)$$

where $\Omega = \{E_n, \varepsilon E_m\}, \varepsilon > 0$.

3. A MULTISTAGE SYSTEM

Consider the following multistage process described by equations

$$x_i = x_{i-1} + \delta f(t_{i-1}, x_{i-1}),$$
$$i = 1, 2, ..., 2^k, \quad (12)$$

where x is an n-dimensional system state vector, δ – partition step of segment $[t_0, \vartheta]$, $f(t,x)$ – a given mapping from $R^1 \times R^n$ into R^n. The initial state x_0 is not known in advance but is subject to geometrical restriction (2). Information on the system's state is provided by a measuring device described by equation

$$y_i = G_i x_i + \xi_i, \qquad i = 1, 2, ..., 2^k, \quad (13)$$

where y is an m-dimensional output signal, G_i – known matrices of appropriate dimension having a full rank. Disturbances ξ_i are subject to geometrical restrictions

$$\xi_i \in \Xi(t_i), \quad (14)$$

where $\Xi(t_i)$ are convex compacts in R^m.

In what follows, the symbol $y_{t,s} = \{y_t, y_{t+1}, ..., y_s\}$ will be used to denote the set of signals received by virtue of equation (13) for steps indexed $t, t+1, ..., s$.

Definition. Informational domain $X(s, t, M)$ is defined as the set of such and only such vectors $x \in R^n$ that, on step s, the trajectories of system (12), induced by the condition $x_t = \bar{x}$, constructed on interval $[t, s]$, and generating the realized signal $y_{t,s}$, under constraints

$$\bar{x} \in M, \quad \xi_i \in \Xi(t_i), \qquad i = t+1, ..., s,$$

– pass through these vectors.

From this definition one easily obtains

Lemma. For arbitrary s, t, r $(s \leq t \leq r)$ equality

$$X(s, t, M) = X(s, r, X(r, t, M)). \quad (15)$$

is fulfilled.

This equality means that the information sets possess semigroup properties. A description of the information sets' $X(s, 0, X_0)$ $(0 \leq s \leq 2^k)$ evolution is then required. Though unknown, the system's actual condition on step s by virtue of the information set's definition is in the set $X(s, 0, X_0)$. This means that the proposed estimation procedure for the unknown system state under uncertainty is "guaranteed".

Denote $X(s, 0, X_0) = X_s$, $s \in [0, 2^k]$. Consider a modified system

$$z_i = z_{i-1} + \delta f(t_{i-1}, z_{i-1}) + \delta h_{i-1} \quad (16)$$

$$y_i = G_i z_i + \xi_i + w_i, \quad i = 1, 2, ..., 2^k. \quad (17)$$

Let, in what follows, the pair $\{x_0, \xi\}$ be fixed. Here the following denotation is used $\xi = \{\xi_1, ..., \xi_{2^k}\}$.

Noises h_{i-1} and w_i are on every step subject to the following quadratic restriction

$$(h_{i-1}, N_{i-1}h_{i-1}) + (w_i, L_i w_i) \leq \mu_i^2,$$
$$i = 1, 2, ..., 2^k, \qquad (18)$$

where N_{i-1} and L_i are positively defined matrices of appropriate dimensions.

On every step, the domains that are compatible with the realized signal by virtue of system (16), (17) are ellipsoids. An ellipsoid's center satisfies relation

$$z_i^0 = K_i^{-1}\frac{1}{\delta^2}N_{i-1}(z_{i-1}^0 + \delta f(t_{i-1}, z_{i-1}^0)) +$$
$$+K_i^{-1}G_i^\top L_i(\hat{y}_i - \xi_i), \qquad (19)$$

where $K_i = \frac{1}{\delta^2}N_{i-1} + G_i^\top L_i G_i$.

Consider the following recurrent procedure. Denote Z_i^0 to be the following set

$$Z_i^0 = \{z_i^0 | z_i^0 = K_i^{-1}\frac{1}{\delta^2}N_{i-1}\times$$
$$\times(z_{i-1}^0 + \delta f(t_{i-1}, z_{i-1}^0)) +$$
$$+K_i^{-1}G_i^\top L_i(\hat{y}_i - \xi_i),$$
$$z_{i-1}^0 \in Z_{i-1}^0, \quad \xi_i \in \Xi(t_i)\}, \qquad (20)$$

where $Z_0^0 = X_0$. Denote

$$\tilde{Z}_i^0 = \{\cap Z_i^0 | (N_{j-1}, L_j) = (E_n, \varepsilon_j E_m), \varepsilon_j > 0,$$
$$j = 1, ..., i\}. \qquad (21)$$

Assumption 2. The mapping $f(t, x)$ is homeomorphic.

The following assertion is true:

Theorem. Under assumptions 1 and 2 the following equation is fulfilled:

$$\tilde{Z}_i^0 = X_i, \qquad i = 0, ..., 2^k. \qquad (22)$$

Let $t \in [t_0, \vartheta]$, G_i – values at instants $t_0 + i\delta$ of a continuous matrix $G(t)$, ξ_i – values at instants $t_0 + i\delta$ of function $\xi(t)$, which at every instant is subject to geometric restriction

$$\xi(t) \in \Xi(t), \qquad (23)$$

where $\Xi(t)$ is a convex compact in R^m, varying continuously in t. Let us presuppose $N_i = \delta N(t_0 + i\delta)$, $L_i = L(t_0 + i\delta)$, where $N(t)$ and $L(t)$ are matrices of correspondent dimensions continuous on the interval $[t_0, \vartheta]$.

Consider the following inclusion

$$z^0(t+\delta) \in z^0(t) + \{K^{-1}(t+\delta)\frac{1}{\delta}N(t)(z^0(t) +$$
$$+\delta f(t, z^0(t)) + K^{-1}(t+\delta)G^\top(t+\delta)L(t+\delta)\times$$
$$\times(\hat{y}(t+\delta) - \Xi(t+\delta)) - z^0(t)\}, \qquad (24)$$

$t = t_0, t_0 + \delta, ..., t_0 + (2^k - 1)\delta$. The assembly of all polygonal curves whose legs' ends satisfy inclusion (24) at instant $t+\delta$ defines a polygonal tube whose cross-section at instant t is set Z_i^0. A polygonal tube is a generalization of the notion of an Euler polygon for ordinary differential equations. Let us do the following conversion

$$K_i^{-1} = \delta^2(N_{i-1}^{-1} - \delta^2 N_{i-1}^{-1}G_i^\top L_i G_i N_{i-1}^{-1} +$$
$$+\delta^4 N_{i-1}^{-1}G_i^\top L_i G_i(N_{i-1} + \delta^2 G_i^\top L_i G_i)^{-1}\times$$
$$\times G_i^\top L_i G_i N_{i-1}^{-1}). \qquad (25)$$

The above conversion uses the equality

$$(A + \delta B)^{-1} = A^{-1} - \delta A^{-1}BA^{-1} +$$
$$+\delta^2 A^{-1}B(A + \delta B)^{-1}BA^{-1},$$

which is true for an invertible matrix A, an arbitrary matrix B, and sufficiently small δ. The right-hand side of inclusion (24) takes the following form

$$z^0(t) + \delta\{f(t, z^0(t)) - N^{-1}(t)G^\top(t+\delta)L(t+\delta)\times$$
$$\times(G(t+\delta)z^0(t) - \hat{y}(t+\delta) + \Xi(t+\delta))\} +$$
$$+\delta\varphi(\delta)B, \qquad (26)$$

where $\varphi(\delta) \to 0$ as $\delta \to 0$, B is a unit ball in R^n centered at the origin.

When assumptions 1, 2 and conditions (2) and (23) are true, the properties of R-solutions to differential inclusions (Panasyuk and Panasyuk, 1977; Tolstonogov, 1986) yield the following proposition:

Theorem. There exists an R-solution to differential inclusion

$$\dot{z}^0 \in f(t, z^0) - N^{-1}(t)G^\top(t)L(t)\times$$
$$\times(G(t)z^0 - \hat{y}(t) + \Xi(t)), \quad z(t_0) \in X^0. \qquad (27)$$

The solution is defined on the entire interval $[t_0, \vartheta]$. For every $t \in [t_0, \vartheta]$ an R-solution is an attainability set of inclusion (27) at instant t from the initial set X^0. In Hausdorff metric, an R-solution can be approximated by a polygon tube of inclusion (24).

4. A CONTINUOUS SYSTEM

Let the system's dynamics be expressed by equation

$$\dot{x} = f(t, x), \qquad t \in T = [t_0, \vartheta], \qquad (28)$$

where x is an n-dimensional system state vector. Assumption 1 is true of $f(t, x)$. The initial state $x(t_0) = x^0$ satisfies restriction (2). The measurement equation is

$$y = G(t)x + \xi, \qquad (29)$$

where y is an m-dimensional vector, $G(t)$ – a continuous matrix of appropriate dimensions. Functions $\xi(t)$ that represent disturbances are measurable. The noises are subject to geometric qualification (23). Let measurement $y^*(t)$ have realized on closed interval T by virtue of system (28) - (29).

Definition. Informational domain $X(\vartheta, y^*(\cdot)) = X(\vartheta, \cdot)$ of the states of system (28) at time-instant ϑ that are consistent with function $y^*(t)$ measured on closed interval T, is defined as the set of all vectors $x \in R^n$ such that for each of them, at instant ϑ, a trajectory of system (28) inducing, with noise $\xi(t)$ satisfying condition (23), by virtue of equation (29) a realization $y(t)$ such that $y(t) = y^*(t)$ almost everywhere on T, – passes through it.

A description of informational sets $X(\vartheta, \cdot)$ provides guaranteed estimates for the system's unknown actual phase states, and no other information on the actual state can possibly be derived. It is therefore only natural to accept informational sets as the system's current set-valued phase state.

Consider the following parameterized family of differential inclusions with no restrictions on the system's phase state

$$\dot{z} \in f(t,z) - N(t)G^\top(t)L(t)\times$$
$$\times(G(t)z - y(t) + \Xi(t)), \qquad (30)$$

where matrices $N(t)$ and $L(t)$ are positively defined for all t. Elements of matrices $N(t)$ and $L(t)$ are summable on T. The initial condition is $z(t_0) = z^0 \in X^0$. A Caratheodory solution to differential inclusion (30) is defined as an absolutely continuous function $z(t)$ defined on closed interval T and almost everywhere satisfying inclusion (30).

Requirement 1. Suppose there exist matrix functions $N(t)$ and $L(t)$ such that differential inclusion

$$dz/d\tau \in -N(t)G^\top(t)L(t)\times$$
$$\times(G(t)z - y(t) - \Xi(t)), \qquad (31)$$

where $\tau \geq 0$ and t is a parameter, has an asymptotically stable set $\Phi(t)$ (Filippov, 1979). Here multi-valued mapping $Z = \Phi(t)$ is a solution to inclusion

$$0 \in -N(t)G^\top(t)L(t)\times$$
$$\times(G(t)z - y(t) - \Xi(t)). \qquad (32)$$

Inclusion (31) is analogous to the so-called adjoined system for ordinary differential equations containing a small parameter multiplying the derivative, and inclusion (32) – to a confluent equation (Tikhonov, 1952; Vasilieva and Butuzov, 1978). Denote $\Omega = \{N(\cdot), L(\cdot)\}$.

Theorem. The following equality is fulfilled

$$X(\vartheta, \cdot) = \cap\{D(\vartheta, \Omega)|\Omega\}, \qquad (33)$$

where $D(\vartheta, \Omega)$ is the attainability set at instant $t = \vartheta$ of differential inclusion (30) with $y(t) = y^*(t)$.

Mention should be made of the fact that though, as demonstrated in (Donchev, 1987), the fundamental Tikhonov theorem (Tikhonov, 1952) can not be extended to differential inclusions in the general case, this extension becomes possible if requirement 1 is verified (Kayumov, 1994). Requirement 1 will be verified if, for example, the product of matrices $-N(t)G^\top(t)L(t)G(t)$ is a diagonal matrix with identical negative eigenvalues or a diagonal matrix such that some of its eigenvalues are negative and equal and the rest of them identically equal zero when $t \in T$. Because one is free to choose the form of matrices $N(t)$ and $L(t)$, this can be done, for instance, for coefficient matrices $G(t)$ invertible for every $t \in T$ or for a $m \times n$ constant matrix $G = G(t)$, $t \in T$, with rank m.

5. EXAMPLE

Consider a nonlinear system of the following type

$$\begin{cases} \dot{x}_1 = 0, \\ \dot{x}_2 = x_1^2, \ t \in T = [0,1]. \end{cases} \qquad (34)$$

The initial state $x(0) = x^0$ is subject to condition

$$x^0 \in X^0 = \{x : |x| \leq 1\}. \qquad (35)$$

The measurement equation is given by

$$y = x_2 + \xi. \qquad (36)$$

Noise in the measurement device is subject to geometric qualification

$$\xi(t) \in \Xi = \{\xi : |\xi| \leq 1\}. \qquad (37)$$

Let signal $y^*(t) \equiv 0$, $t \in T$ have realized (e.g. with the initial condition being $x^0 = (0,0)^\top$ and the noise being $\xi^*(t) \equiv 0$, $t \in T$).

The compatible domain is given by the inequality

$$x_1^2(1) - \sqrt{1 - x_1^2(1)} \leq x_2(1) \leq 1.$$

The attainability domain $Z(1, \varepsilon)$ for differential inclusion

$$\begin{pmatrix} \dot{z}_1 \\ \dot{z}_2 \end{pmatrix} \in$$
$$\in \left[\begin{pmatrix} 0 \\ z_1^2 - \varepsilon z_2 - \varepsilon \end{pmatrix} ; \begin{pmatrix} 0 \\ z_1^2 - \varepsilon z_2 + \varepsilon \end{pmatrix} \right] \qquad (38)$$

is given by

$$z_1(1) = z_1^0,$$
$$((z_1^0)^2 - \varepsilon)\frac{(e^{-\varepsilon} - 1)}{-\varepsilon} + e^{-\varepsilon} z_2^0 \leq z_2(1) \leq$$
$$\leq ((z_1^0)^2 + \varepsilon)\frac{(e^{-\varepsilon} - 1)}{-\varepsilon} + e^{-\varepsilon} z_2^0.$$

Letting $\varepsilon \to 0$, one gets set Z_1

$$z_1(1) = z_1^0, \quad z_2(1) = (z_1^0)^2 + z_2^0. \qquad (39)$$

Letting $\varepsilon \to \infty$, one gets set Z_2

$$z_1(1) = z_1^0, \quad -1 \leq z_2(1) \leq 1. \qquad (40)$$

In view of (39) and (40) one observes that

$$z_1(1) = z_1^0, \quad z_1^2(1) - \sqrt{1 - z_1^2(1)} \leq z_2(1) \leq 1,$$

i.e.

$$X(1, \cdot) = \cap\{Z(1, \varepsilon)|\varepsilon > 0\}.$$

6. CONCLUSION

In this paper, exact solutions of the initial nonlinear multistage and continuous systems are obtained. The employed analytical methods for the description of the said sets (proposed in (Kurzhanski and Filippova, 1986; Kurzhanski and Filippova, 1987) enable the development of computational procedure algorithms for the construction of these sets.

7. REFERENCES

Donchev, A. (1987). *Optimal control systems: Perturbations, approximations, and sensitivity analysis.* Mir. Moskow (in Russian).

Filippov, A.F. (1979). Stability for differential equations with discontinuous and multiple-valued right-hand sides. *Differentsialnye Uravneniya* **15**(6), 1018–1027.

Kayumov, R.I. (1994). Exact solution of the problem of estimation for one class of the nonlinear systems. *Izvestia RAN, Tehnicheskaya Kibernetika* (2), 177–182.

Koscheev, A.S. and A.B. Kurzhanski (1983). Adaptive estimation of multistage systems' evolution under uncertainty. *Izvestia AN SSSR, Tehnicheskaya Kibernetika* (3), 72–93.

Kurzhanski, A.B. (1977). *Control and observation under conditions of uncertainty.* Nauka. Moskow (in Russian).

Kurzhanski, A.B. (1980). Dynamic control system estimation under uncertainty conditions, part i. *Problems of Control and Information* **9**(6), 395–406.

Kurzhanski, A.B. (1981). Dynamic control system estimation under uncertainty conditions, part ii. *Problems of Control and Information* **10**(1), 33–42.

Kurzhanski, A.B. and T.F. Filippova (1986). On the description of the set of viable trajectories of a differential inclusion. *Doklady AN SSSR* **289**(1), 38–41.

Kurzhanski, A.B. and T.F. Filippova (1987). On the description of the set of viable trajectories of a control system. *Differentsialnye Uravneniya* **23**(1), 1303–1315.

Panasyuk, A.I. and V.I. Panasyuk (1977). *Asymptotic optimization of non-linear control systems.* Belorussian Univ. Press. Minsk (in Russian).

Tikhonov, A.N. (1952). Systems of differential equations containing a small parameter multiplying the derivatives. *Matematicheskiy Sbornik* **31**(3), 575–586.

Tolstonogov, A.A. (1986). *Differential inclusions in Banach space.* Nauka. Novosibirsk (in Russian).

Vasilieva, A.B. and V.F. Butuzov (1978). *Singularly disturbed equations in critical cases.* Moscow Univ. Press. Moscow (in Russian).

PROBLEMS OF CONTROL BY DYNAMICS FOR REPEATED BIMATRIX 2×2 GAMES WITH SWITCHING OF LOCAL CRITERIA FOR THE PLAYERS

Anatolii Kleimenov [*,1]

[*] *Institute of Mathematics and Mechanics,*
Ural Branch of Russian Academy of Sciences,
16, Kovalevskoi Str., Ekaterinburg, 620219, Russia
E-mail: kleimenov@imm.uran.ru

Elena Volegova [**,1]

[**] *Ural State University,*
51, Lenin Avenue, Ekaterinburg, 620083, Russia

Abstract: An approach to building dynamics in repeated 2×2 bimatrix games originating from theory of closed-loop differential games is presented. Natural formalism for viewing typical players' behaviors such as normal, altruistic, aggressive and paradoxical is discussed. A problem of finding a domain for which trajectory generating by this dynamics leads to COOPERATION for repeated Prisoner's Dilemma is solved. The problem of following trajectories of replicator dynamics of "prey - predator" type is discussed. *Copyright © 1998 IFAC*

Keywords: repeated bimatrix game, local and global goals, behavior types, auxiliary bimatrix game, Prisoner's Dilemma, dynamics of "prey - predator" type.

1. INTRODUCTION

There are various approaches to constructing dynamics in repeated games, see e.g. (Maynard Smith, 1982; Hofbauer and Sigmund, 1988). Here we outline an approach originating from theory of closed-loop differential games (Krasovskii, 1985; Krasovskii and Subbotin, 1987), and, more specifically, a nonzero-sum branch of this theory (Kleimenov, 1993; Kleimenov, 1997). The following points are characteristic in the approach:

(i) local displacements toward states nonimprovable for one or another players;

(ii) the selection of a local displacement through finding a Nash equilibrium in some "local" bimatrix game;

(iii) a classification of such typical behaviors as normal, altruistic, aggressive, and paradoxical.

In this paper the approach is applied to repeated Prisoner's Dilemma and to the replicator dynamics of "prey - predator" type.

2. REPEATED BIMATRIX 2×2 GAME

We deal with a mixed strategy bimatrix game with payoff matrixes

$$A = \begin{pmatrix} a_{11} & a_{12} \\ a_{21} & a_{22} \end{pmatrix}, \quad B = \begin{pmatrix} b_{11} & b_{12} \\ b_{21} & b_{22} \end{pmatrix} \quad (1)$$

[1] The work presented in this paper has been supported by the Russian Foundation of Basic Researches (grant 97-01-00161).

of players 1 and 2, respectively. Note that mixed strategies arise naturally when the game is assumed to be played out between agents from two large groups. If $(p, 1-p)$, $0 \le p \le 1$, is a mixed strategy of player 1 and $(q, 1-q)$, $0 \le q \le 1$, is a mixed strategy of player 2, then p and q are interpreted as shares of agents in the first and, respectively, second group who play the pure strategy 1. Payoffs to players 1 and 2 are given, respectively, by

$$\begin{aligned} f_1(p,q) &= Cpq - c_1 p - c_2 q + a_{22}, \\ f_2(p,q) &= Dpq - d_1 p - d_2 q + b_{22}, \end{aligned} \tag{2}$$

where

$$\begin{aligned} C &= a_{11} - a_{12} - a_{21} + a_{22}, \\ c_1 &= a_{22} - a_{12}, c_2 = a_{22} - a_{21}, \\ D &= b_{11} - b_{12} - b_{21} + b_{22}, \\ d_1 &= b_{22} - b_{21}, d_2 = b_{22} - b_{21}. \end{aligned} \tag{3}$$

Pair (p, q) in the unit square

$$E = \{(p,q): \quad 0 \le p \le 1; \quad 0 \le q \le 1\} \tag{4}$$

characterizes a *state* of the players (or the associated groups of agents).

Let the bimatrix game be repeated N times, where N is large. Let $(p_0, q_0) \in E$ be an initial state. To define a dynamics of states, or, equivalently, mixed strategies in the repeated bimatrix game, we introduce players' control and information areas and specify their objectives. Assume that in round $k+1$ $(k = 0, 1, ..., N-1)$ the players are allowed to pass from the current state $(p_k, q_k) \in E$ to any state $(p_{k+1}, q_{k+1}) \in M_{\alpha_1,\alpha_2}(p_k, q_k)$ where

$$\begin{aligned} M_{\alpha_1,\alpha_2}(p_k, q_k) = \{(p,q) \in E: \ |p - p_k| \le \\ \le \alpha_1, \ |q - q_k| \le \alpha_2\}. \end{aligned} \tag{5}$$

Here α_1 and α_2 are positive and sufficiently small. Small α_1 and α_2 indicate that the inner structure of the groups of interacting agents evolves slowly enough. Thus, $M_{\alpha_1,\alpha_2}(p_k, q_k)$ characterizes control abilities of players 1 and 2 at state (p_k, q_k). A game dynamics as a control process is described by

$$\begin{aligned} p_{k+1} &= p_k + u_k, \ |u_k| \le \alpha_1, \\ q_{k+1} &= q_k + v_k, \ |v_k| \le \alpha_2. \end{aligned} \tag{6}$$

Let us characterize information available to the players. At a current state (p_k, q_k) each player knows, first, (p_k, q_k) and, second, the values of the payoff functions f_1 and f_2 (2) in a rectangle $M_{\alpha_1,\alpha_2}(p_k, q_k)$.

Now we specify the objectives of the players. We assume two types of objectives, local and global. A local goal of player 1 (respectively, player 2) in round k is to maximize his (or her) current payoff $f_1(p_k, q_k)$ (respectively, $f_2(p_k, q_k)$). A global goal is given complementary.

To specify behavior patterns incorporating players' objectives, we invoke some ideas from theory of nonzero-sum closed-loop differential games. These are, first, the idea of "nondeterioration" of guaranteed results (in local or global understanding) for both players (Kleimenov, 1993) , second, the idea of maximum displacement (Krasovskii and Subbotin, 1987) toward a Pareto optimal state, and, third, the idea of finding local transitions via a search of Nash equilibria in appropriate "local" games (Kleimenov, 1997).

Another element of the approach discussed here is a formalizm for the analysis of such typical behaviors of the players as normal, aggressive, altruistic, and paradoxical.

Let (p_k, q_k) be a current state in the repeated game. We assume that the players find a transition to a next state, (p_{k+1}, q_{k+1}), via the analysis of their "local" extremal problems:

Problem 1. Find (p^1, q^1) such that

$$(p^1, q^1) = \underset{(p,q)\in M_{\alpha_1,\alpha_2}(p_k,q_k)}{\arg\max} f_1(p,q) \tag{7}$$

under the condition

$$f_2(p,q) \ge f_2(p_k, q_k). \tag{8}$$

Problem 2. Find (p^2, q^2) such that

$$(p^2, q^2) = \underset{(p,q)\in M_{\alpha_1,\alpha_2}(p_k,q_k)}{\arg\max} f_2(p,q) \tag{9}$$

under the condition

$$f_1(p,q) \ge f_1(p_k, q_k). \tag{10}$$

A decisionmaking pattern is as follows. The players check if one of the equalities

$$f_1(p^1, q^1) = f_1(p^2, q^2) \tag{11}$$

$$f_2(p^1, q^1) = f_2(p^2, q^2) \tag{12}$$

holds. If both (11) and (12) hold, then $(p_{k+1}, q_{k+1}) = (p^1, q^1)$ or $(p_{k+1}, q_{k+1}) = (p^2, q^2)$. If (11) holds and (12) does not hold, then $(p_{k+1}, q_{k+1}) = (p^2, q^2)$. Symmetrically, if (12) holds and (11) does not hold, then $(p_{k+1}, q_{k+1}) = (p^1, q^1)$. Finally, if neither (11), nor (12) hold, then the players play a pure strategy local bimatrix game in which the payoff matrixes of players 1 and 2 are, respectively,

$$\begin{aligned} A^c &= \begin{pmatrix} f_1(p^1,q^1) & f_1(p^1,q^2) \\ f_1(p^2,q^1) & f_1(p^2,q^2) \end{pmatrix}, \\ B^c &= \begin{pmatrix} f_2(p^1,q^1) & f_2(p^1,q^2) \\ f_2(p^2,q^1) & f_2(p^2,q^2) \end{pmatrix}. \end{aligned} \tag{13}$$

In this game strategy i ($i = 1, 2$) of player 1 is to switch from p_k to p^i, and strategy i of player 2 is to switch from q_k to q^i. To identify (p_{k+1}, q_{k+1}), the players find a pure strategy Nash equilibrium, NE. One can prove that this game always has a pure strategy NE; moreover, four situations for NE, and, respectively, (p_{k+1}, q_{k+1}), may occur. Namely, the next cases are only admissible:

1) $(1, 2)$ is a single NE, and $(p_{k+1}, q_{k+1}) = (p^1, q^2)$,

2) $(2, 1)$ is a single NE, and $(p_{k+1}, q_{k+1}) = (p^2, q^1)$,

3) $(2, 2)$ is a single NE, and $(p_{k+1}, q_{k+1}) = (p^2, q^2)$,

4) $(1, 2)$ and $(2, 1)$ are two NE, and (p_{k+1}, q_{k+1}) is either (p^1, q^2), or (p^2, q^1), with some (possibly equal) probabilities.

3. BEHAVIOR TYPES

Until now we assumed that each player is *normal* (nr) in the sense that his (or her) behavior is aimed at maximizing his (or her) own payoff. However, there might be other behavior types such as *altruistic (al)* ("the better to my rival, the better to me"), *aggressive (ag)* ("the worse to my rival, the better to me"), and *paradoxical (pr)* ("the worse to me, the better to me"). These three behavior types can be formalized in the following way.

Let us say that player 1 has

(i) the altruistic behavior whenever he (or she) identifies his (or her) payoff matrix with B,

(ii) the aggressive behavior whenever he (or she) identifies his (or her) payoff matrix with $-B$, and

(iii) the paradoxical behavior whenever he (or she) identifies his (or her) payoff matrix with $-A$.

For player 2 we assume symmetric definitions.

These definitions indicate extremes in players' behaviors. Real individuals behave partially normal, partially altruistic, partially aggressive, and partially paradoxical. In other words, mixtures of behavior types would better agree with real dynamics.

If we restrict each player to "pure" behavior types, we arrive at 16 admissible combinations shown in Table 1. In 4 combinations players' interests coincide, and they solve problems of optimal choice. In other 4 combinations the players have opposite interests and, thus, play zero-sum matrix games. The rest 8 pairs determine nonzero-sum bimatrix games.

Table 1.

	nr	**al**	**ag**	**pr**
nr	(A, B)	(A, A)	$(A, -A)$	$(A, -B)$
al	(B, B)	(B, A)	$(B, -A)$	$(B, -B)$
ag	$(-B, B)$	$(-B, A)$	$(-B, -A)$	$(-B, -B)$
pr	$(-A, B)$	$(-A, A)$	$(-A, -A)$	$(-A, -B)$

In the previous sections cooperative algorithms for building *normal-normal* game dynamics were defined. Replacing in these definitions the original matrix pair (A, B) by the matrix pairs given in Table 1, we obviously introduce 15 new types of game dynamics. If, moreover, the players moving along a game trajectory are allowed to switch their behavior types over time, "mixed" behavior game processes arise.

4. REPEATED PRISONER'S DILEMMA

Let a bimatrix 2×2 game of Prisoner's Dilemma type (A, B), $B = A^T$ be given. Without loss of generality we suppose that the elements of the matrix $A = (a_{ij}), i, j = 1, 2$ satisfy the conditions

$$a_{21} > a_{11} > a_{22} > a_{12}, \; 2a_{11} > a_{12} + a_{21}.$$

The payoff function of player 1 has the same form as (2). The payoff function of player 2 has the following form

$$f_2(p, q) = Cpq - c_2 p - c_1 q + a_{22}.$$

Let a global goal be to lead the initial state to the state COOPERATION, i.e. to the state $(p = 1, q = 1)$.

Problem is to construct a trajectory leading from a given initial state $(p_0, q_0) \in E$ to the state COOPERATION under the condition that dynamics of the game is as described above.

Let an initial state (p_0, q_0) be given. Now we describe a transition from (p_0, q_0) to the next state (p_1, q_1). Three cases can take place:

Case 1. The inequality

$$C(p_0 + q_0) - c_1 - c_2 > 0$$

is fulfilled. This case divides on three subcases.

1a) $\alpha_1 > \frac{(Cp_0 - c_2)\alpha_2}{c_1 - C(q_0 + \alpha_2)}$

First, we find the domain G in the plane $\{p, q\}$ where values of functions f_1 and f_2 do not decrease. Taking in account that grad$f_1 = (Cq_0 - c_1, Cp_0 - c_2)$, grad$f_2 = (Cq_0 - c_2, Cp_0 - c_1)$, where $Cq_0 - c_1 < 0, Cp_0 - c_2 > 0, Cq_0 - c_2 > 0$ and $Cp_0 - c_1 < 0$ for all $(p_0, q_0) \in E$ we obtain

$G = \{(p,q) : C(pq-p_0q_0)-c_2(p-p_0)-c_1(q-q_0) \leq 0, C(pq-p_0q_0)-c_1(p-p_0)-c_2(q-q_0) \geq 0, q \leq q_0+\alpha_2\}$. Solving problem 1 we find maximum of the function f_1 on the set G. It is not difficult to show that

$$(p^1,q^1) = (\frac{Cp_0q_0 - c_2p_0 + c_1\alpha_2}{C(q_0+\alpha_2) - c_2},\ q_0+\alpha_2).$$

Solving problem 2 we find maximum of the function f_2 on the set G. Then we get

$$(p^2,q^2) = (\frac{Cp_0q_0 - c_1p_0 + c_2\alpha_2}{C(q_0+\alpha_2) - c_1},\ q_0+\alpha_2).$$

Since neither (11), nor (12) hold, then we construct an auxiliary bimatrix game (A^c, B^c). The elements of the matrixes satisfy the conditions $f_1(p^1,q^1) > f_1(p^2,q^1), f_2(p^2,q^1) < f_2(p^2,q^2)$ because the point (p^1,q^1) $((p^2,q^2))$ is the point of maximum of function f_1 (f_2). The inequality $f_1(p^1,q^1) > f_1(p^2,q^2)$ holds because $q^1 = q^2$. The elements of the matrix B^c satisfy the equality $f_2(p^1,q^1) = f_2(p^1,q^2)$. We take the $NE = (p^1,q^2)$ as the next state (p_1,q_1), moreover $p_1 > p_0, q_1 > q_0$.

1b) The condition

$$\frac{(Cp_0 - c_1)\alpha_2}{c_2 - C(q_0+\alpha_2)} \leq \alpha_1 \leq \frac{(Cp_0 - c_2)\alpha_2}{c_1 - C(q_0+\alpha_2)}$$

is fulfilled. Reasoning in a similar way as in subcases 1a we get $G = \{(p,q) : C(pq-p_0q_0) - c_2(p-p_0)-c_1(q-q_0) \leq 0, C(pq-p_0q_0)-c_1(p-p_0)- -c_2(q-q_0) \geq 0, p-p_0 \leq \alpha_1, q-q_0 \leq \alpha_2\}$,

$$(p^1,q^1) = (\frac{Cp_0q_0 - c_2p_0 + c_1\alpha_2}{C(q_0+\alpha_2) - c_2},\ q_0+\alpha_2),$$

$$(p^2,q^2) = (p_0+\alpha_1,\ \frac{Cp_0q_0 - c_2q_0 + c_1\alpha_1}{C(p_0+\alpha_1) - c_2}).$$

Since neither (11), nor (12) hold, then we construct an auxiliary bimatrix game (A^c, B^c). It is not difficult to show that $NE = (p^1,q^2)$. And we take this NE as the next state (p_1,q_1), moreover $p_1 > p_0, q_1 > q_0$.

1c) $\alpha_1 < \frac{(Cp_0-c_1)\alpha_2}{c_2-C(q_0+\alpha_2)}$

Then the solutions of problems 1 and 2 are

$$(p^1,q^1) = (p_0+\alpha_1,\ \frac{Cp_0q_0 - c_1q_0 + c_2\alpha_1}{C(p_0+\alpha_1) - c_1})$$

and

$$(p^2,q^2) = (p_0+\alpha_1,\ \frac{Cp_0q_0 - c_2q_0 + c_1\alpha_1}{C(p_0+\alpha_1) - c_2}).$$

NE-solution of auxiliary bimatrix game (A^c, B^c) is the pair (p^1,q^2).

Case 2. The inequality

$$C(p_0+q_0) - c_1 - c_2 < 0$$

holds. This case divides on three subcases.

2a) $\alpha_1 \geq \frac{(Cp_0-c_1)\alpha_2}{c_2-C(q_0-\alpha_2)}$

We get $(p_1,q_1) = NE = (p^1,q^2)$,

$$(p^1,q^2) = (\frac{Cp_0q_0 - c_2p_0 + c_1\alpha_2}{C(q_0-\alpha_2) - c_2},\ q_0-\alpha_2).$$

2b)

$$\frac{(Cp_0 - c_2)\alpha_2}{c_1 - C(q_0-\alpha_2)} \leq \alpha_1 \leq \frac{(Cp_0 - c_1)\alpha_2}{c_2 - C(q_0-\alpha_2)},$$

$(p_1,q_1) = NE = (p_0-\alpha_1, q_0-\alpha_2)$.

2c) $\alpha_1 < \frac{(Cp_0-c_2)\alpha_2}{c_1-C(q_0-\alpha_2)}$

We get $(p_1,q_1) = NE = (p^1,q^2)$,

$$(p^1,q^2) = (p_0-\alpha_1,\ \frac{Cp_0q_0 - c_2q_0 + c_1\alpha_1}{C(p_0-\alpha_1) - c_2}).$$

Case 3. The equality

$$C(p_0+q_0) - c_1 - c_2 = 0$$

is fulfilled. This case divides on three subcases also. In all subcases problems 1 and 2 have two solutions because the level lines for functions f_1 and f_2 are symmetric with respect to line $q = p + q_0 - p_0$. The NE-solutions of corresponding auxiliary bimatrix games are chosen with equal probabilities.

Thus, the algorithm of finding the state (p_1,q_1) is described.

Note that, the equation of the line which separate the cases 1 and 2 is

$$q = -p + \frac{c_1+c_2}{C}. \tag{14}$$

If the initial state (p_0,q_0) lies below the line (14) then trajectory does not lead to the state COOPERATION.

Now we define another part of boundary of a domain for which the trajectory leads to the state COOPERATION.

Let an initial state (p_0,q_0) lie above the line (14) and $q_0 > p_0$. We want to find the sizes of information set M, i.e. values α_1 and α_2, such that the corresponding trajectory leads to the state COOPERATION. Suppose that the information set M has the form $\{|p_0-p| \leq \delta_1, |q_0-q| \leq \alpha_2\}$, where $\delta_1 = \delta_1(\alpha_2) = p^1 - p_0$, p^1 is the first component of the solution of problem 1 in square information set $(\alpha_1 = \alpha_2)$. First, we solve problems 1 and 2 on this information set. The pairs

$$(p^1,q^1) = (p_0+\delta_1, q_0+\alpha_2),$$

$$(p^2,q^2) = (p_0+\delta_1,\ \frac{Cp_0q_0 - c_2q_0 + c_1\delta_1}{C(p_0+\delta_1) - c_2})$$

give the solutions of these problems. Second, we construct the auxiliary bimatrix game (A^c, B^c) and find Nash equilibrium. The pair (p^1,q^2) is the NE. And $(p_1,q_1) = (p^1,q^2)$. The next state

(p_2, q_2) is found for (p_1, q_1) analogously and so on. As limit if $N \to \infty$ we obtain the trajectory which equation is $C(pq-p_0q_0)-c_1(p-p_0)-c_2(q-q_0)=0$. The equation of the hyperbola from this family passing through the state COOPERATION is

$$Cpq - c_1p - c_2q + a_{22} = a_{11}. \qquad (15)$$

If we take an initial state lying above the hyperbola (15), we can not select the sizes of information set so that the trajectory leads to the state COOPERATION.

The case when an initial state (p_0, q_0) lies above the line (14) and $q_0 < p_0$ is investigated analogously. We find the sizes of information set M such that the trajectory leads to the state COOPERATION. We get the following analog of the equation (15):

$$Cpq - c_2p - c_1q + a_{22} = a_{11}. \qquad (16)$$

If we take an initial state lying below the hyperbola (16), we can not select the sizes of information set so that the trajectory leads to the state COOPERATION.

Denote $S = \{(p,q) : q > -p + \frac{c_1+c_2}{C}, Cpq - c_1p - c_2q + a_{22} \le a_{11}, Cpq - c_2p - c_1q + a_{22} \ge a_{11}\}$.

So, we obtain the following result.

Theorem 1. For an initial state $(p_0, q_0) \in S$ the trajectory generated by suggested procedure leads to the state COOPERATION (i.e. to the state $(p = 1, q = 1)$).

Now we analyze the cases when an initial state (p_0, q_0) belongs to the set $E \setminus S$. Let an initial state (p_0, q_0) lie above the hyperbola (15). And let player 1 be altruistic. Altruistic behavior of players is often described in literature concerning evolutionary games. Suppose, a global goal is to lead an initial state into the boundary of the domain S. Then bimatrix game is of (A^T, A^T) type, i.e. we have maximization problem for the function f_2 on the set M. Its solution is

$$(p_1, q_1) = (p_0 + \alpha_1, q_0 - \alpha_2).$$

Let an initial state (p_0, q_0) lie below the hyperbola (16). Now player 2 is altruistic. Suppose a global goal is the same as above. Then bimatrix game is (A, A) type, i.e. we have maximization problem for the function f_1 on the set M. And its solution is

$$(p_1, q_1) = (p_0 - \alpha_1, q_0 + \alpha_2).$$

Theorem 2. If an initial state (p_0, q_0) lies above (below) the hyperbola (15) ((16)) and the first (second) player is altruistic then the trajectory generated by suggested procedure leads into the boundary of the domain S.

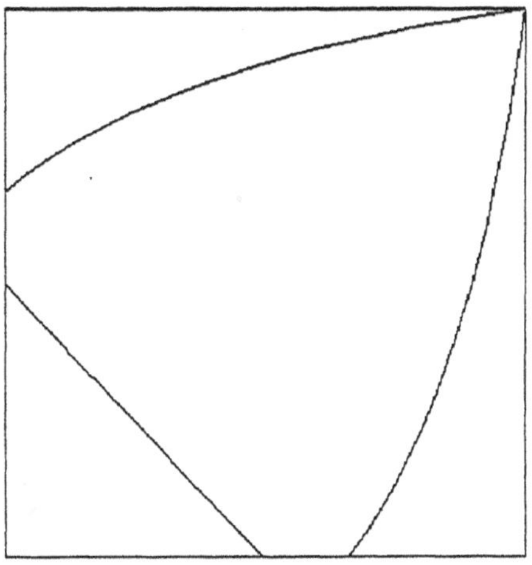

Fig. 1. The boundary of the domain S

Now we analyze the case when an initial state (p_0, q_0) lies below the line (14). Then all trajectories begining at these states lead to the Nash equilibrium if both players are normal. Suppose a global goal in this game is to lead the initial state into the boundary of the domain S. Let both players be agressive. Then the initial bimatrix game is reduced to another bimatrix game $(-A^T, -A)$. The payoff functions have the form

$$f_1^\alpha(p,q) = -Cpq + c_2p + c_1q - a_{22} = -f_2(p,q),$$

$$f_2^\alpha(p,q) = -Cpq + c_1p + c_2q - a_{22} = -f_1(p,q).$$

We suppose that the information set is square one $(\alpha_1 = \alpha_2)$. The solution of problem 1 is

$$(p^1, q^1) = (\frac{-Cp_0q_0 + c_2p_0 - c_1\alpha_2}{-C(q_0 + \alpha_2) + c_2}, q_0 + \alpha_2).$$

The solution of problem 2 is

$$(p^2, q^2) = (p_0 + \alpha_1, \frac{-Cp_0q_0 + c_2q_0 - c_1\alpha_1}{-C(p_0 + \alpha_1) + c_2}).$$

The Nash equilibrium in the auxiliary bimatrix game is the pair (p^1, q^2) moreover $p^1 > p_0, q^2 > q_0$. It is easy to verify, that among all 16 combinations of behavior types shown in Table 1 there exist only two combinations, namely *aggressive - aggressive* and *paradoxical - paradoxical*, for which trajectories lead into the boundary of the domain S.

Theorem 3. If an initial state (p_0, q_0) lies below the line (14) then in order to the trajectory leads into the boundary of the domain S it is necessary that players are aggressive or paradoxical, simultaneously.

5. REPLICATOR DYNAMICS

In this section we consider the game type of "prey - predator". Let the game (A, B) be given and the elements of these matrixes satisfy the conditions $a_{11} > a_{21}, a_{22} > a_{12}, b_{11} < b_{12}, b_{22} < b_{21}$. For this game the equations of replicator dynamics

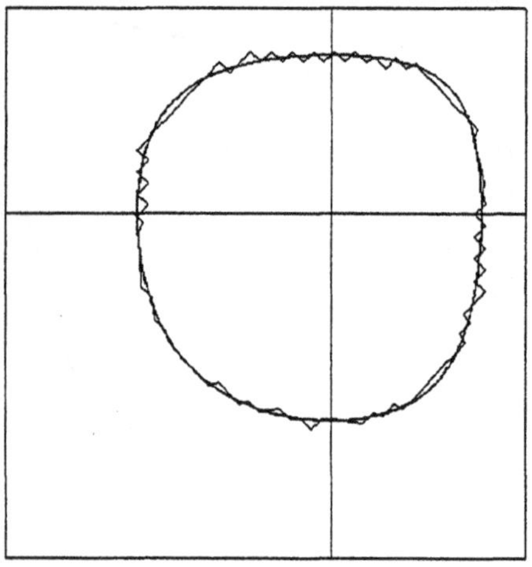

Fig. 2. Following trajectory of replicator dynamics

(Hofbauer and Sigmund, 1988) are resulted to the form

$$\begin{aligned} \dot{p} &= p(1-p)(Cq-c_1), \\ \dot{q} &= q(1-q)(Dp-d_2), \end{aligned} \tag{17}$$

where p, q are introduced in section 2, C, c_1, D, d_2 are introduced in (3). The suggested approach is applied to constructing dynamics for this game. Problem is to construct a sequence of states (p_k, q_k), $k = 0, 1, 2, ...N$ which follows trajectory generated by replicator dynamics. For the current state (p_k, q_k) one defines a point of aiming (p^*, q^*) lying on the trajectory of replicator dynamics and then finds the behavior types of players which provide the maximal displacement toward the point of aiming. The information set is given beforehand. The problem is solved in two variants: in "pure" behavior types and in "mixed" behavior types.

Let the game be considered in pure behavior types. Let (p_{k-1}, q_{k-1}) be a current state. According to the section 3 we have 16 admissible combinations of behavior types as is shown in Table 1. For each combination we can apply the procedure described in section 2 which allows to construct the pairs $(p_k^i, q_k^i), i = \overline{1, 16}$. Among these pairs (p_k^i, q_k^i) we select the pair which provides the maximal displacement toward the point of aiming and find the behavior type of players which generates this pair. And so on. Thus the behavior types of players is found on each step. The integral estimation of each behavior types of players, average for trajectory, is calculated.

Let the game be considered in mixed behavior types. For the current state (p_{k-1}, q_{k-1}) we find the pairs $(p_k^i, q_k^i), i = \overline{1, 16}$ again. Among these pairs (p_k^i, q_k^i) we select two pairs $(p_k^{i_1}, q_k^{i_1})$ and $(p_k^{i_2}, q_k^{i_2})$ which provide the maximal displacement toward the direction to the point of aiming and which lie on opposite sides of this direction. The vector from convex hull of vectors $(p_k^{i_1} - p_{k-1}, q_k^{i_1} - q_{k-1})$ and $(p_k^{i_2} - p_{k-1}, q_k^{i_2} - q_{k-1})$ whose direction coincides with the direction to the point of aiming determines the mixture of behavior types of players uniquely. Thus the behavior types of players are determined on each step. The integral estimation of behavior types of players is calculated too.

The numerical experiment was carried out. The following values of elements of matrixes A and B were chosen $a_{11} = 7, a_{12} = 9, a_{21} = 4, a_{22} = 14, b_{11} = 4, b_{12} = 7, b_{21} = 14, b_{22} = 9$. The trajectory of replicator dynamics passing through the point $(p_0 = 0.35, q_0 = 0.62)$ was chosen. This trajectory is represented on Figure 2. Both variants in pure and mixed behavior types were considered. A trajectory in pure behavior types is represented on this figure by the broken line. For this trajectory a share of normal behavior type equals 0.22, a share of altruistic behavior type is equal 0.345, a share of aggressive behavior type equals 0.0 and a share of paradoxical behavior type is equal 0.435. For trajectory in mixed behavior types (which coincide with trajectory of replicator dynamics) a share of normal behavior type equals 0.427, a share of altruistic behavior type is equals 0.235, a share of aggressive behavior type is 0.204 and a share of paradoxical behavior type is equal 0.134.

6. REFERENCES

Hofbauer, J. and K. Sigmund (1988). *The Theory of Evolution and Dynamic Systems.* Cambridge University Press. Cambridge.

Kleimenov, A.F. (1993). *Nonantogonistic Positional Differential Games.* Nauka. Ekaterinburg.

Kleimenov, A.F. (1997). Solutions in a nonatagonistic positional differential games. *Journal Applied Maths Mechs,* **61**, 717–723.

Krasovskii, N.N. (1985). *Control of a Dynamical System.* Nauka. Moskow.

Krasovskii, N.N and A.I. Subbotin (1987). *Game-theoretical control problems.* Springer-Verlag. New York.

Maynard Smith, J. (1982). *Evolution and Theory of Games.* Cambridge University Press. Cambridge.

DYNAMICAL RESTORATION OF CONTROLS FOR DISTRIBUTED PARAMETER SYSTEMS UNDER UNCERTAINTIES OF SYSTEM STATES

Alexander Korotkii *,[1]

* *Institute of Mathematics and Mechanics*
Ural Branch of Russian Academy of Sciences
16, Kovalevskoj Str., Ekaterinburg, 620219, Russia
E-mail: korotkii@imm.uran.ru

Abstract: An inverse problem of dynamics is considered, namely, it is necessary to reconstruct unknown controls (perturbations) applied to a nonlinear dynamical system and generating an observable motion of the system. The information for reconstruction is obtained by approximate measurements of varying informational sets of the system. The reconstruction is carried out in real time. This problem is known to be ill-posed. It is proposed to solve it by means of constructive, physical feasible, regularizing positional algorithms for the stable dynamical approximation of unknown controls. Copyright ©1998 IFAC.

Keywords: control, dynamic systems, real time, observability, inverse dynamic problem, uncertainty, identification algorithms, dynamic models, control algorithms, regularization.

1. INTRODUCTION AND STATEMENT OF THE PROBLEMS

Let us consider a controlled dynamical system which operates on a time interval $T = [t_0, \vartheta]$ $(-\infty < t_0 < \vartheta < +\infty)$ and is described by a nonlinear equation

$$\dot{x}(t) + A(t)x(t) = B(t, x(t))u(t) + f(t), \quad (1)$$
$$t \in T, \quad x(t_0) = x_0,$$

where A and B are given operators, f is a given function, $x(t)$ is a state of the system at the time moment $t \in T$, $u \in \Sigma$ is a control of the system, $u(t)$ is a realization of the control u at the time moment $t \in T$, $x(\cdot) = x(\cdot; t_0, x_0, u)$ is the motion (solution) on the time interval T of the dynamical system corresponding to the control $u \in \Sigma$ and the initial condition $x(t_0) = x_0 \in H$.

The control u in (1) is unknown. It is known only that its values belong to a set $P \subset U$. It is supposed that the exact state $x(t)$ of the system is not measured. However the observer knows a some set $G(t) \subset H$ such that $x(t) \in G(t)$, $t \in T$.The observer is assumed not to have enough information to estimate the set $G(t)$ more exactly or to find appropriate statistical description for distribution of the state $x(t)$ in the set $G(t)$. The informational sets $G(t)$ are sent to the observer in real time according to the movement of the system. It is supposed that the structure of the system is known.

It is necessary to construct an approximation u_h to the control u based on the approximations $Z(t)$ to observed sets $G(t)$, $t \in T$:

$$u_h = D(h, Z(\cdot), x_0) \in \Sigma, \quad h \geq 0,$$

$$\sigma(Z(t), G(t)) \leq h, \quad t \in T, \quad x_0 \in G(t_0),$$

$$\rho(u_h, u) \to +0, \quad h \to +0,$$

[1] Supported by the Russian Foundation of Basics Researches (grant 98-01-00046).

where ρ is a criterion of control approximation, σ is a criterion of set approximation. In the worst case it is necessary to construct a dynamical positional regularizing algorithm D satisfying the property:

$$u_h = D(h, Z(\cdot), x_0) \in \Sigma, \quad h \geq 0,$$

$$\sigma(Z(t), G(t)) \leq h, \quad t \in T, \quad x_0 \in G(t_0),$$

$$x(t; t_0, x_0, u_h) \in G(t)^{\varepsilon(h)}, \quad t \in T,$$

$$\varepsilon(h) \to +0, \quad h \to +0,$$

where $G(t)^{\varepsilon(h)}$ is a closed $\varepsilon(h)$-neighborhood of the set $G(t)$. The algorithms should possess the property of physical feasibility and should work in real time mode according to a feedback scheme.

The method for solving the problems described in (Osipov and Kryazhimskii, 1995) will be used. The method is based on the ideas of the theory of positional control (Krasovskii and Subbotin, 1987) and the theory of ill-posed problems (Tikhonov and Arsenin, 1977; Ivanov *et al.*, 1978). Similar approaches to determining the parameters of the dynamical systems based on guaranteed estimation methods were considered in (Kurzhanskii, 1977; Chernous'ko, 1988). This work continues the researches published in (Korotkii, 1997).

The problems is formulated more precisely. Let V be a separable reflexive Banach space, embedded continuously and densely in a Hilbert space H. Identifying H with its adjoins H^*, we obtain a triple of continuously and densely embedded spaces $V \subset H \subset V^*$. Let

$$X = \{x \in L^2(T; V) : \dot{x} \in L^2(T; V^*)\},$$

$$||x||_X = ||x||_{L^2(T;V)} + ||\dot{x}||_{L^2(T;V^*)},$$

X is reflexive Banach space that continuously embedded in $C(T; H)$. The definitions of functional spaces that we use as well as the main properties of such spaces may be found, for example, in (Gajewski *et al.*, 1974).

Let $\{A(t) : t \in T\}$ be a family of continuous monotone operators acting from V to V^* such that $< A(\cdot)v, w >$ is continuous function $T \to R$ for any fixed $v, w \in V$ and there exist constants $C_1 \in (0, \infty)$, $C_2 \in (-\infty, \infty)$, $C_3 \in (0, \infty)$ such that for any $t \in T$, $v \in V$

$$< A(t)v, v > \geq C_1 \cdot ||v||_V^2 - C_2,$$

$$||A(t)v||_{V^*} \leq C_3 \cdot (||v||_V + 1),$$

$< \cdot, \cdot >$ is canonical duality between V^* and V.

Let $\{B(t, z) : t \in T, z \in H\}$ be a family of continuous linear operators acting from Hilbert space U into H and there exists constant $C_4 \in (0, \infty)$ such that for any $t_1, t_2 \in T$, $z_1, z_2 \in H$

$$||B(t_1, z_1) - B(t_2, z_2)||_L \leq$$

$$\leq C_4 \cdot (||z_1 - z_2||_H + |t_1 - t_2|),$$

where $L = L(U, H)$ is the Banach space of continuous linear operators from U to H with the natural norm.

Let $f \in L^2(T; H)$, P be a convex bounded closed set of elements from U, let Σ be the set of all measurable mappings: $T \to P$.

The Cauchy problem (1) has a unique solution in the space X for any $x_0 \in H$, $u \in \Sigma$ (Gajewski *et al.*, 1974). The solution $x(\cdot) = x(\cdot; t_0, x_0, u)$ will also be treated as the motion of the dynamical system (1) over T in the phase space H from its initial state $x_0 \in H$ under the control $u \in \Sigma$.

Consider the set F of all set-valued maps $T \to \text{comp}(H)$, where $\text{comp}(H)$ is the space of all nonempty compact sets in H with the Hausdorff metric σ. A map $G(\cdot) \in F$ will be treated as ideal result of observation of the motion $x(\cdot) = x(\cdot; t_0, x_0, u)$ if $x(t) \in G(t)$, $t \in T$. Let us define the set

$$Z[h, G(\cdot)] = \{Z(\cdot) \in F :$$

$$\sigma(Z(t), G(t)) \leq h, t \in T\}, \quad h \geq 0.$$

A map $Z(\cdot) \in Z[h, G(\cdot)]$ will be treated as h-perturbation of the ideal result of the observation $G(\cdot)$.

Let W be the set of all maps $D : [0, \infty) \times F \times H \to \Sigma$ that possesses the property of physical realizability (feasibility): $u_1(t) = u_2(t)$ for almost all $t \in [t_0, \tau]$ if $u_1 = D(h, Z_1(\cdot), x_0)$, $u_2 = D(h, Z_2(\cdot), x_0)$, $h \geq 0$, $Z_1(\cdot) = Z_2(\cdot)$ for $t \in [t_0, \tau]$, $\tau \in T$, $x_0 \in H$. An element of W will be named an algorithm of reconstruction of a control.

Problem 1. It is necessary to construct an algorithm $D \in W$ with the property: for any $\varepsilon > 0$ there exists a number $h_0 > 0$ such that for any $h \in [0, h_0]$, $Z(\cdot) \in Z[h, G(\cdot)]$, $x_0 \in Z(t_0)$ the inclusion $x(t; t_0, x_0, u_h) \in G(t)^\varepsilon$, $t \in T$, takes place where $u_h = D(h, Z(\cdot), x_0)$.

Problem 2. It is necessary to construct an algorithm $D \in W$ with the properties:
1) for any $h \geq 0$, $Z(\cdot) \in Z[h, G(\cdot)]$, $x_0 \in Z(t_0)$ inclusion $x(t; t_0, x_0, u_h) \in G(t)^{\varepsilon(h)}$, $t \in T$, is true where $u_h = D(h, Z(\cdot), x_0)$, $\varepsilon(h) \geq 0$, $\varepsilon(h) \to +0$ as $h \to +0$;
2) for any realizations $Z(\cdot) \in Z[h, G(\cdot)]$, $h \geq 0$, $x_0 \in Z(t_0)$ the convergence $u_h = D(h, Z(\cdot), x_0) \to u_*$ as $h \to +0$ takes place where u_* is some control in Σ such that $x(t; t_0, x_0, u_*) \in G(t)$, $t \in T$.

2. SOLUTION OF THE PROBLEM 1

Let us introduce into consideration the controlled system-model

$$\dot{y}(t) + A(t)y(t) = B(t, y(t))v(t) + f(t), \quad (2)$$

$$t \in T, \quad y(t_0) = y_0; \quad v(t) \in P, \quad t \in T.$$

Define the strategy S of the control of the system-model as mapping $T \times H \times \text{comp}(H) \to P$ such that $S(t, y, Y)$ is an arbitrary fixed element of P if $y \in Y$ and $S(t, y, Y) \in \text{argmin}\{< y - z, B(t, y)v >_H \colon v \in P\}$ if $y \notin Y$ where $z \in \text{argmin}\{\|y - z\|_H : z \in Y\}$.

Let $\Delta = \Delta(h) = \{[t_i, t_{i+1}) : i = 0, ..., m\}$ is a partition of the segment $[t_0, \vartheta)$ by points t_i:

$$t_0 < t_1 < ... < t_{m+1} = \vartheta, \quad m = m(h),$$

$$d(\Delta) = \max\{t_{i+1} - t_i : i = 0, ..., m\} \leq C_5 \cdot h,$$

where $C_5 = const > 0$ is fixed, a function $m = m(h)$ is assumed to be fixed.

Let us construct an algorithm $D = D_S \in W$. Some $h \geq 0$ and elements $Y(\cdot) \in F$, $y_0 \in H$ are fixed. Let us fix the partition $\Delta = \Delta(h)$ and construct piecewise constant function $u_h : T \to P$ formed according to the rule

$$u_h(t) = u_i = S(t_i, y_i, Y(t_i)), \quad t_i \leq t < t_{i+1},$$

$$y_{i+1} = y(t_{i+1}; t_i, y_i, u_i), \quad i = 0, ..., m.$$

Now D_S can be defined by the rule

$$D_S(h, Y(\cdot), y_0) = u_h \in \Sigma. \quad (3)$$

Condition 1. The map $G(\cdot) \in F$ possesses the properties:
a) for any $t_1 \in [t_0, \vartheta)$, $t_2 \in (t_1, \vartheta]$, $x_1 \in G(t_1)$ there exists a control $u : [t_1, t_2] \to P$ such that $x(t_2; t_1, x_1, u) \in G(t_2)$;
b) there exists a number $C_6 > 0$ such that for any $t \in T$, $y \in H$, $h > 0$, $Z(\cdot) \in Z[h, G(\cdot)]$, $y_1 \in \text{argmin}\{\|y - z\|_H : z \in G(t)\}$, $y_2 \in \text{argmin}\{\|y - z\|_H : z \in Z(t)\}$ the unequality $\|y_1 - y_2\|_H \leq C_6 \cdot h$ holds.

Theorem 1. Let condition 1 be fulfilled. Then the algorithm D_S (3) solves the problem 1.

Proof. It is fixed an arbitrary $h \geq 0$, $Z(\cdot) \in Z[h, G(\cdot)]$, $x_0 \in Z(t_0)$. Taking into account the rule according to which the function $u_h = D_S(h, Z(\cdot), x_0)$ is constructed, one can obtain the following estimate

$$\varepsilon(h) = \max\{\varepsilon(h, t) : t \in T\} \leq C_7 \cdot h^{1/2}, \quad (4)$$

$$\varepsilon(h, t) = \min\{\|x(t) - z\|_H : z \in G(t)\},$$

$$x(t) = x(t; t_0, x_0, u_h).$$

From the obtained estimate it follows that

$$x(t; t_0, x_0, u_h) \in G(t)^{\varepsilon(h)}, \quad t \in T,$$

$$\varepsilon(h) \to +0 \text{ as } h \to +0.$$

The theorem is proved.

Let us describe dynamical realization of the algorithm (3). Before the initial time moment t_0 the partition $\Delta = \Delta(h)$ is defined and fixed in accordance with the error level h. Every point t_i, $i = 0, ..., m$, is considered as the begining point of the next step of calculation. At the time moment $t = t_0$ the observer gets the informational set $Z(t_0)$. He chooses and fixes any initial state $y_0 \in Z(t_0)$ for the system-model, calculates realization $u_0 = S(t_0, y_0, Z(t_0))$ for the segment $[t_0, t_1)$ and the state $y_1 = y(t_1; t_0, y_0, u_0)$ of the system-model for the time moment t_1. At the time moment t_1 the observer gets the informational set $Z(t_1)$ for calculation of realization $u_1 = S(t_1, y_1, Z(t_1))$ of the strategy for the segment $[t_1, t_2)$ and the new state $y_2 = y(t_2; t_1, y_1, u_1)$ of the system-model for the time moment t_2. And so on. At the final time moment ϑ the realization $u_h = D_S(h, Z(\cdot), y_0)$ will be constructed. It is clear that $D_S \in W$.

Let us consider another method for solution of the problem 1.

Condition 2. The map $G(\cdot) \in F$ possesses the property: for any $h \geq 0$, $Z(\cdot) \in Z[h, G(\cdot)]$, $t_1 \in [t_0, \vartheta)$, $t_2 \in (t_1, \vartheta]$, $x_1 \in Z(t_1)$ there exists a control $u : [t_1, t_2] \to P$ such that $x(t_2; t_1, x_1, u) \in Z(t_2)$.

Let Q be a rule which (according to the condition 2) for any $h \geq 0$, $Z(\cdot) \in Z[h, G(\cdot)]$, $t_1 \in [t_0, \vartheta)$, $t_2 \in (t_1, \vartheta]$, $x_1 \in G(t_1)$ defines some control $u = Q(h, Z(\cdot), t_1, x_1, t_2) : [t_1, t_2) \to P$ such that $x(t_2; t_1, x_1, u) \in Z(t_2)$.

Let us construct an algorithm $D = D_Q \in W$. Some $h \geq 0$ and elements $Z(\cdot) \in Z[h, G(\cdot)]$, $y_0 \in H$ are fixed. It is fixed the partition $\Delta = \Delta(h)$ of the segment $[t_0, \vartheta)$ and the control $u_h : T \to P$ is formed according to the rule

$$u_h(t) = u_i = Q(h, Z(\cdot), t_i, y_i, t_{i+1}),$$

$$t_i \leq t < t_{i+1},$$

$$y_{i+1} = y(t_{i+1}; t_i, y_i, u_i), \quad i = 0, ..., m.$$

The value $D(h, Z(\cdot), y_0)$ is an arbitrary fixed element of Σ if $Z(\cdot) \notin Z[h, G(\cdot)]$.

Now it is defined D_Q by the law

$$D_Q(h, Z(\cdot), y_0) = u_h \in \Sigma. \quad (5)$$

Theorem 2. Let condition 2 be fulfilled. Then the algorithm D_Q (5) solves the problem 1.

Proof. There exists a constant $C_8 > 0$ such that for any motion $x(\cdot; t_0, x_0, u)$ ($x_0 \in G(t_0)$, $u \in \Sigma$) the estimate

$$\|x(t+\tau; t_0, x_0, u) - x(t; t_0, x_0, u)\|_H \leq$$
$$\leq C_8 \cdot \tau^{1/2}, \quad t_0 \leq t \leq t+\tau \leq \vartheta.$$

is true. From this estimate it follows that (4) is valid and the problem 1 is solved. The theorem is proved.

The work of the algorithm D_Q and its dynamical realization are analogous to dynamical realization of the algorithm D_V.

3. SOLUTION OF THE PROBLEM 2

Let us consider the system-model (2) and a strategy Γ of control of it. Let Γ is a strategy (mapping) $T \times H \times \text{comp}(H) \to P$ such that $\Gamma(t, y, Y) = \text{argmin}\{< y - z, B(t, y)v >_H + \alpha(h) \cdot \|v\|_U^2 : v \in P\}$, where $z \in \text{argmin}\{\|y - z\|_H : z \in Y\}$ and $\alpha(\cdot)$ is a fixed function $[0, \infty) \to (0, \infty)$ with the property $h \cdot \alpha(h)^{-1} \to 0$ as $h \to 0$. Define the operator $D_\Gamma \in W$ similarly to (3).

Condition 3. The map $G(\cdot) \in F$ possesses the properties:
a) for any $t_1 \in [t_0, \vartheta)$, $t_2 \in (t_1, \vartheta]$ there exists a unique control $u_* : [t_1, t_2] \to P$ such that

$$G(t) = \{x(t; t_1, x_1, u_*) : x_1 \in G(t_1)\},$$
$$t_1 \leq t \leq t_2;$$

b) it is the property b) from the condition 1.

Theorem 3. Let condition 3 be fulfilled. Then the algorithm D_Γ solves the problem 2.

Proof. It is fixed an arbitrary $h \geq 0$, $Z(\cdot) \in Z[h, G(\cdot)]$, $x_0 \in Z(t_0)$. Taking into account the rule according to which the function u_h is constructed, one can obtain the following estimate for the functional Λ_h:

$$\max\{\Lambda_h(t) : t \in T\} \leq \gamma \cdot h,$$

where $\gamma > 0$ is some number which depends on the known data of the problem, but do not depend on the number h,

$$\Lambda_h(t) = \|x(t; t_0, x_0, u_h) - x(t; t_0, x_0, u_*)\|_H^2 +$$
$$+\alpha(h) \cdot \int_{t_0}^{t} (\|u_h(\tau)\|_U^2 - \|u_*(\tau)\|_U^2) d\tau,$$

u_* is the control satisfying the property a) from the condition 3 on the segment T ($u_* \in \Sigma$, $x(t; t_0, x_0^*, u_*) \in G(t)$, $t \in T$, $x_0^* \in G(t_0)$). In particular,

$$x(t; t_0, x_0, u_h) \in G(t)^{\varepsilon(h)},$$
$$0 \leq \varepsilon(h) \leq (\gamma \cdot h + 2 \cdot \alpha(h) \cdot (\vartheta - t_0) \cdot b)^{1/2},$$
$$b = \max\{\|v\|_U^2 : v \in P\},$$
$$\varepsilon(h) \to +0 \text{ as } h \to +0,$$
$$\|u_h\|_{L^2(T;U)}^2 \leq \|u_*\|_{L^2(T;U)}^2 + \gamma \cdot h \cdot \alpha(h)^{-1}.$$

From the obtained estimates it follows that

$$\|u_h\|_{L^2(T;U)} \to \|u_*\|_{L^2(T;U)},$$
$$u_h \to u_* \text{ weakly in } L^2(T; U),$$

consequently, $u_h \to u_*$ strongly in $L^2(T; U)$ and

$$\sup\{\|D_\Gamma(h, Z(\cdot), x_0) - u_*\|_{L^2(T;U)} :$$
$$Z(\cdot) \in Z[h, , G(\cdot)]\} \to 0 \text{ as } h \to 0.$$

The theorem is proved.

Remark. The operators D_S and D_Γ do not use controls which are described by conditions 1 and 3. The operator D_Q uses control which is described by condition 2.

4. REFERENCES

Chernous'ko, F.L. (1988). *Estimation of the phase state of dynamical systems.* Nauka. Moskow.

Gajewski, H., K. Groger and K. Zacharias (1974). *Nichtlineare operatorgleichungen und operatordifferentialgleichungen.* Akademie-Verlag. Berlin.

Ivanov, V.K., V.V. Vasin and V.P. Tanana (1978). *Theory of linear ill-posed problems and applications.* Nauka. Moscow.

Korotkii, A.I. (1997). On parameter reconstruction for dynamical systems under uncertainties. In: *Proceedings of the "15th World Congress on Scientific Computation, Modelling and Applied Mathematics"* (A.Sydow, Ed.). Vol. I. pp. 35–38. Wissenschaft and Technik Verlag. Berlin.

Krasovskii, N.N and A.I. Subbotin (1987). *Game-theoretical control problems.* Springer-Verlag. New York.

Kurzhanskii, A.B. (1977). *Control and observation under conditions of uncertainty.* Nauka. Moskow.

Osipov, Yu.S. and A.V. Kryazhimskii (1995). *Inverse problem of ordinary differential equations: dynamical solutions.* Gordon and Breach. London.

Tikhonov, A.N. and V.Ya. Arsenin (1977). *Solution of ill-posed problems.* Wiley. New York.

CONSTRAINT AGGREGATION PRINCIPLE IN THE PROBLEM OF OPTIMAL CONTROL OF DISTRIBUTED PARAMETER SYSTEMS

Franz Kappel *,[1] **Arkadii Kryazhimskii,** **,[1,2]
Vyacheslav Maksimov, ***,[1,2]

* *Institut für Mathematik Universität Graz, Austria*
** *V.A.Steklov Institute of Mathematics, Moscow, Russia*
*** *Institute of Mathematics and Mechanics, Ekaterinburg, Russia*

Abstract: A problem of optimal control for a linear dynamical system in a Hilbert space is discussed. Control variables are subject to mixed linear constraints. To construct a finite-step iteration procedure that stop at an approximate solution having a prescribed accuracy, a semigroup representation of trajectories is used. Some applications to dynamical systems governed by parabolic and functional-differential equations with time lags are presented. *Copyright ©1998 IFAC*

Keywords: optimal control, distributed parameter systems.

1. INTRODUCTION

In (Kryazhimskii and Osipov, 1987; Kryazhimskii, 1994; Ermoliev *et al.*, 1997), a general constraint aggregation method for convex optimization was proposed. At each iteration of the method, convex and linear constraints are replaced by a single inequality formed as a linear combination of the original ones. A modification of the method aggregates the original constraints in linear penalty terms added to the objective function in each iteration. After solving the simplified subproblem, new aggregation coefficients are calculated and the iterations are continued. The aggregation principle was implemented in particular optimization algorithms in finite- and infinite-dimensional spaces and adapted for systems of linear inequalities in Hilbert spaces (Kryazhimskii and Maksimov, 1998), dynamical inverse problems for parabolic equations (Kryazhimskii *et al.*, 1997*b*), problems of reconstruction of inputs in systems with hereditary (Blizorukova and Maksimov, 1997) and parabolic systems (Kryazhimskii *et al.*, 1997*a*), and inverse tomography problems of ray seismics (Digas *et al.*, 1998).

In the present report some applications of the constraint aggregation principle to optimal control problems for systems with distributed parameters are discussed.

2. PROBLEM STATEMENT

Let $(H, |\cdot|_H)$ and $(H_1, |\cdot|_{H_1})$ be real Hilbert spaces, $(U, \|\cdot\|_U)$ be a Banach space. We consider the problem of minimization of the functional

$$J(u(\cdot)) = \int_T \omega(t, u(t))\, dt \to \inf, \qquad (1)$$

$$T = [t_0, \vartheta], \quad t_0 \geq 0, \quad \vartheta < \infty$$

under the constraints

$$Gx(t) + Du(t) = g(t) \qquad (2)$$

$$\text{for a. a. } t \in T, \quad u(\cdot) \in P.$$

[1] Partially supported by INTAS (project 96-0816).
[2] Partially supported by the Russian Foundation for Basic Research (grant 97-01-01060).

Here $x(t) = x(t; x^0, u(\cdot))$ is a solution of the evolutionary equation

$$\dot{x}(t) = Ax(t) + Bu(t), \tag{3}$$

$$t \in T, \quad x(t_0) = x^0 \in H;$$

$P \subset L_2(T, U)$ is a convex, closed and bounded set; $A : H \to H$ is an infinitesimal generator of a strongly continuous semigroup of borded linear operators $T(t)$, $t \geq 0$;

$B : U \to H$, $G : H \to H_1$, $D : U \to H_1$ are linear operators;

$g(\cdot) \in L_2(T, H_1)$ is a given function;

$\omega(\cdot,\cdot) : T \times U \to \mathbf{R}$ satisfies the conditions: function $u \to \omega(t, u)$ is convex ($t \in T$), for every $u(\cdot) \in P$, function $t \to \omega(t, u(t))$ is integrable and

$$|J(u(\cdot))| \leq C_1 < \infty \quad \forall u(\cdot) \in P. \tag{4}$$

A solution of the equation (3) is understood in the weak sense: it is a continuous function $x(t)$, $t \in T$, defined by the equality

$$x(t) = x(t; x^0, u(\cdot)) = T(t - t_0)x^0 + $$

$$+ \int_{t_0}^{t} T(t - \tau)Bu(\tau)\, d\tau.$$

Define the operator $F : L_2(T, U) \to L_2(T, H_1)$ and element $b(\cdot) \in L_2(T, H_1)$ as follows:

$$(Fu(\cdot))(\xi) = \int_{t_0}^{\xi} GT(\xi - t)Bu(t)\, dt + Du(\xi)$$

$$(\xi \in T, \ u(\cdot) \in P)$$

$$b(\xi) = g(\xi) - GT(\xi)x^0 \quad (\xi \in T).$$

Under the above conditions the problem (1)–(3) is equivalent to the following extremal problem. *Problem 1.* Find

$$u^0 = \arg\min\{J(u) : u \in P, Fu = b\},$$

$$J^0 = \min\{J(u) : u \in P, Fu = b\}.$$

We assume that the set

$$\{u \in P : Fu = b\}$$

is nonempty. Let nonnegative numbers ν^F, ν^b, ν^J, linear continuous operators $F^\nu : L_2(T, U) \to L_2(T, H_1)$, elements $b^\nu \in L_2(T, H_1)$ and functionals $J^\nu(\cdot)$ on P be given and the next estimates hold:

$$|F^\nu u - Fu|_{L_2(T;H_1)} \leq \nu^F \quad \forall\, u \in P, \tag{5}$$

$$|b^\nu - b|_{L_2(T;H_1)} \leq \nu^b, \tag{6}$$

$$|J^\nu(u) - J(u)| \leq \nu^J \quad \forall\, u \in P. \tag{7}$$

Our aim is to design an approximate solution algorithm for the problem (1)–(3) which operates with inaccurate data on $\omega(\cdot,\cdot)$, A, B and x^0. At first, we give an approximate solution algorithm for Problem 1. Then, using the methodology of (Kryazhimskii and Osipov, 1987; Kryazhimskii, 1994; Ermoliev *et al.*, 1997), we return to the problem (1)–(3)

3. SOLUTION ALGORITHM FOR PROBLEM 1

We call $w_* \in W$ an *ε-solution* ($\varepsilon > 0$) of the extremal problem

$$\Phi(w) \to \inf, \ w \in W \neq \emptyset$$

$$\text{if } \Phi(w_*) \leq \inf\{\Phi(w) : w \in W\} + \varepsilon.$$

Denote by $Y_j^\nu(\delta, \alpha, \varepsilon)$, ν, δ, α, $\varepsilon \in [0, 1]$, $j \in \mathcal{N} = \{1, 2, 3, \ldots\}$ the finite sequence $\{y_i\}_{i=0}^{j}$ from U which is defined by the rule

$$y_0 = 0,$$

$$y_{i+1} = y_i + u_i\delta \quad (i = 0, \ldots, j - 1),$$

where u_i is a ε-solution of the problem

$$2\,\langle F^\nu y_i - i\delta b^\nu, F^\nu u\rangle_{L_2(T;H_1)} + \tag{8}$$

$$+ \alpha J^\nu(u) \to \inf, \quad u \in P.$$

Introduce the set

$$U(\gamma, \beta) = \{u \in P : |Fu - b|^2_{L_2(T;H_1)} \leq \gamma,$$

$$J(u) \leq J^0 + \beta\}$$

and numbers $K_1 \geq 0$, $K_2 \geq 0$ such that

$$K_1 = |b|_{L_2(T;H_1)},$$

$$|Fu|_{L_2(T;H_1)} \leq K_2 \ (u \in P).$$

Theorem 1. If $\{y_i\}_{i=0}^{j} = Y_j^\nu(\delta, \alpha, \varepsilon)$, $j \in \mathcal{N}$, then the relations

$$y_j/(j\delta) \in U(2\alpha C_1/(\delta j) + \delta_j^*, \delta_j^* j\delta/\alpha)$$

and

$$J^0(2\alpha C_1/(\delta j) + \delta_j^*) - J^0 \leq$$

$$\leq J(y_j/(\delta j)) - J^0 \leq \delta_j^* j\delta/\alpha$$

hold true.

Here

$$\delta_j^* = k_1\nu^F + k_2\nu^b + (k_3(\nu^F)^2 +$$

$$+ 2\alpha\nu^J)/(j\delta) + k_4/j + \varepsilon/(\delta j),$$

the constants k_i, $i \in [1:4]$ can be explicitly expressed through K_1 and K_2; number C_1 is defined in (4). Symbol $J^0(\gamma)$ denotes the optimal value of the γ-perturbed problem

$$J^0(\gamma) = \min\{J(u) : u \in P, |Fu - b|^2 \leq \gamma\}.$$

Theorem 1 is proved analogously (Kryazhimskii and Osipov, 1987; Kryazhimskii, 1994; Ermoliev *et al.*, 1997) and is based on the following lemma.

Lemma 2. If $\{y_i\}_{i=0}^j = Y_j^\nu(\delta, \alpha, \varepsilon)$, $j \geq 2$, then the inequality

$$|F(y_j/(\delta j)) - b|^2_{L_2(T;H_1)} +$$
$$+ \alpha\{J(y_j/(\delta j)) - J^0\}/(\delta j) \leq \delta_j^*,$$

is true, where $\delta_j^* = k_1\nu^F + k_2\nu^b + (k_3(\nu^F)^2 + 2\alpha\nu^J)/(j\delta) + k_4/j + \varepsilon/(\delta j)$.

Proof. Let us estimate the modification of the value

$$\Lambda_{i+1} = |F(y_i + \delta u_i) - t_{i+1}b|^2 +$$
$$+ \alpha \int_0^{t_{i+1}} J(\dot{y}(\tau))d\tau - \alpha J(u^0)t_{i+1} =$$
$$= \Lambda_i + \mu_i + |Fu_i - b|^2\delta^2, \quad i \geq 1,$$

where

$$\dot{y}(t) = u_i \quad \text{for} \quad t \in [t_i, t_{i+1}),$$
$$i \in \mathcal{N}, \quad t_i = i\delta, \quad y(0) = 0,$$
$$\mu_i = 2\left\langle Fy_i - t_ib, Fu_i - b\right\rangle\delta +$$
$$+ \alpha\delta\{J(u_i) - J(u^0)\}.$$

The symbols $|\cdot|$, $\langle\cdot,\cdot\rangle$ denote norm and scalar production in $L_2(T; H_1)$ respectively. As u^0 is a solution of the problem under cosideration, so the equality

$$Fu^0 = b.$$

is true. Therefore

$$\mu_i = 2\left\langle Fy_i - t_ib, Fu_i\right\rangle + \alpha\delta J(u_i) -$$
$$- 2\left\langle Fy_i - t_ib, Fu^0\right\rangle - \alpha\delta J(u^0).$$

We have

$$\lambda_i^* \equiv 2\left\langle Fy_i - t_ib, Fu_i\right\rangle\delta + \alpha J(u_i)\delta -$$
$$- 2\left\langle F^\nu y_i - t_ib^\nu, F^\nu u_i\right\rangle\delta +$$
$$+ \alpha J^\nu(u_i)\delta = \sum_{j=1}^{3} \lambda^{(j)},$$
$$\lambda^{(1)} = 2\left\langle Fy_i - t_ib, (F - F^\nu)u_i\right\rangle\delta,$$
$$\lambda^{(2)} = 2\left\langle (F - F^\nu)y_i - t_i(b - b^\nu), F^\nu u_i\right\rangle\delta,$$
$$\lambda^{(3)} = \alpha\{J(u_i) - J^\nu(u_i)\}\delta.$$

Besides

$$|Fy_i| = |\sum_{j=0}^{i-1} Fu_j\delta| \leq t_{i-1}K_2,$$
$$|F^\nu u| \leq |(F^\nu - F)u| + |Fu| \leq$$
$$\leq K_2 + \nu^F, \quad u \in P.$$

Note that due to the properties of the set P and the rule of choise of elements u_i the inclusion $y(t)/t \in P \;\; \forall \;\; t > 0$ is true. So (see (5) – (7))

$$\lambda^{(1)} \leq 2(K_2 + K_1)\nu^F\delta t_{i-1},$$
$$\lambda^{(2)} = 2(\nu^F + \nu^b)(K_2 + \nu^F)\delta t_{i-1},$$
$$\lambda_i^* \leq \delta\nu^F K_{1i} + \delta\nu^b K_{2i} + \alpha\nu^J\delta,$$

where

$$K_{1i} = 2((2K_2 + K_1)t_{i-1} + \nu^F),$$
$$K_{2i} = 2t_{i-1}(K_2 + \nu^F)\delta.$$

Analogously the estimation is true, if u_i is substituted by u^0. So

$$\mu_i \leq 2\left\langle F^\nu y_i - t_ib^\nu, F^\nu u_i\right\rangle\delta -$$
$$- 2\left\langle F^\nu u^0 - t_ib^\nu, F^\nu u^0\right\rangle\delta +$$
$$+ \alpha\delta\{J^\nu(u_i) - J^\nu(u^0)\} +$$
$$+ 2\{K_{1i}\nu^F + K_{2i}\nu^b + \alpha\nu^J\}\delta.$$

From (8) we deduce

$$\mu_i \leq 2\{K_{1i}\nu^F + K_{2i}\nu^b + \alpha\nu^J\}\delta + \varepsilon\delta.$$

Besides

$$J(y(t)/t) = J(\frac{1}{t}\int_0^t \dot{y}(t)dt) \leq \tag{9}$$
$$\leq \frac{1}{t}\int_0^t J(\dot{y}(t))dt.$$

Thus

$$\Lambda_{i+1} \le \Lambda_i + 2\{2((2K_2+K_1)(i-1)\delta + \\ + \nu^F)\nu^F\delta + 2(i-1)\delta(K_2+\nu^F)\nu^b\delta + \\ + \alpha\nu^J\delta\} + \{(K_2+K_1)\delta\}^2 + \varepsilon\delta.$$

So

$$\Lambda_{i+1} \le 2\{((2K_1+4K_2)(i-1)\delta + \\ + 2\nu^F)\nu^F\delta + 2(i-1)\delta(K_2+\nu^F)\nu^b\delta + \\ + \alpha\nu^J\delta\}i + \{(K_1+K_2)\delta\}^2 i + \varepsilon\delta i.$$

Dividing the right-hand part and the left-hand part by t_{i+1}^2 and using (9), we have

$$|F(y(t_{i+1})/t_{i+1}) - b|^2 + \\ + \alpha/t_{i+1}\{J(y(t_{i+1})/t_{i+1}) - J(u^0)\} \le \\ \le 4\{(K_1+2K_2)\nu^F + (K_2+\nu^F)\nu^b\} + \\ + 2\{2(\nu^F)^2 + \alpha\nu^J\}/((i+1)\delta) + \\ + (K_1+K_2)^2/(i+1) + \varepsilon/(\delta i).$$

Lemma is proved.

4. SOLUTION ALGORITHM FOR PROBLEM (1)–(3)

Now we return to the problem (1)–(3). Let the operators A, B and initial condition x^0 be given inaccurately by perturbed operators $A^\nu : H \to H$, $B^\nu : U \to H$, element $x^{0\nu} \in H$ and function $\omega^\nu(\cdot,\cdot) : T \times U \to \mathbf{R}$. We assume that

$$||B^\nu - B||_{L(U;H)} \le \nu^B, \quad |x^{0\nu} - x^0|_H \le \nu^{x^0},$$

the operator A^ν generates the semigroup of linear continuous operators $T^\nu(t)$, $t \ge 0$, the function $t \to \omega^\nu(t, u(t))$ is integrable for every $u(\cdot) \in P$ and $\sup_{t\in T} |T^\nu(t) - T(t)|_{L(H;H)} \le \nu^A$,

$$|J^\nu(u(\cdot)) - J(u(\cdot))| \le \nu^I,$$

$$J^\nu(u(\cdot)) = \int_T \omega^\nu(t, u(t))\,dt;$$

here ν^A, ν^B, ν^{x^0}, $\nu^I \in [0,1]$ are given numbers. Denote by

$$Z^\nu(j, \delta, \alpha, \beta_0, \beta_1, \beta_2)$$

$$(j \in \mathcal{N}, \nu, \delta, \alpha, \beta_0, \beta_1, \beta_2 \in [0,1])$$

the following finite sequence of elements

$$\{(y_i(\cdot), \varphi_i(\cdot), \psi_i(\cdot)\}_{i=0}^j$$

from $L_2(T,U) \times L_2(T;H_1) \times L_2(T,H)$:

$$y_0 = 0, \quad \varphi_0 = 0, \quad \psi_0 = 0,$$

$$y_{i+1}(\cdot) = y_i(\cdot) + \delta w_i(\cdot) \in U,$$

$$\varphi_{i+1}(\cdot) = \varphi_i(\cdot) + \delta\kappa_i(\cdot) \in H_1.$$

$$\psi_{i+1}(\cdot) = \psi_i(\cdot) + \delta\zeta_i(\cdot)\ (i = 0, 1, \dots, j) \in H;$$

$w_i(\cdot)$ is β_0-solution of the problem

$$\int_T \{2\langle B^{\nu*}\psi_i(t) + D^*\varphi_i(t),\, w(t)\rangle_H + \\ + \alpha\omega^\nu(t, w(t))\}\,dt \to \inf, \quad w(\cdot) \in P,$$

$$\gamma_i(\cdot) \in C(T, H), \quad |\gamma_i(\cdot) - \bar\gamma_i(\cdot)|_{C(T,H)} \le \beta_2;$$

$\bar\gamma_i(\cdot)$ is solution on T of the Cauchy problem

$$\dot\gamma = A^\nu\gamma + B^\nu w_i(t), \quad \gamma(t_0) = 0,$$

$$|\zeta_i(\cdot) - \bar\zeta_i(\cdot)|_{C(T,H)} \le \beta_1;$$

$$\kappa_i(t) = G\gamma_i(t) + Dw_i(t) - b^\nu(t) \quad (t \in T).$$

$\bar\zeta_i(\cdot)$ is solution on T of the Cauchy problem

$$\dot\zeta = -A^{\nu*}\zeta - G^*\kappa_i(t), \quad \zeta(\vartheta) = 0.$$

Here A^* is the operator adjoint to A.
Let

$$k_3 = \sup_{\nu\in[0,1]} ||T^\nu(\vartheta - t_0)||_{L(H;H)}.$$

Lemma 3. Let $\{y_i(\cdot), \varphi_i(\cdot), \psi_i(\cdot)\}_{i=0}^j = Z^\nu(j, \delta, \alpha, \beta_0, \beta_1, \beta_2)$. Then $\{y_i\}_{i=0}^j = Y_j^\nu(\delta, \alpha, \varepsilon)$, where $\varepsilon = j(C_2\beta_1 + C_3\beta_2) + \beta_0$.

Here C_2 and C_3 are constants depending on k_3, $||B||_{L(U;H)}$, C_1, $||G||_{L(H;H_1)}$, $||D||_{L(U;H_1)}$ which can be given explicitly. The next statement follows from Theorem 1 and Lemma 3.

Theorem 4. Let $\{y_i(\cdot), \varphi_i(\cdot), \psi_i(\cdot)\}_{i=0}^j = Z^\nu(j, \delta, \alpha, \beta_0, \beta_1, \beta_2)$. Then

$$y_j/(j\delta) \in U(2\alpha C_1/(\delta j) + \delta_j^*, \delta_j^* j\delta/\alpha)$$

and

$$J^0(2\alpha C_1/(\delta j) + \delta_j^*) - J^0 \le \\ \le J(y_j/(\delta j)) - J^0 \le \delta_j^* j\delta/\alpha.$$

5. EXAMPLES

The algorithm described above is applicable for the numerical solution of optimal control problems for systems governed by the parabolic equation

$$x_t(t,\eta) - \sum_{k,l=1}^{n} (a_{k,l}(t,\eta)x_{\eta_k}(t,\eta))_{\eta_l} +$$

$$+ b(\eta)x(t,\eta) = u(t,\eta) \text{ in } Q = T \times \Omega,$$

$$x(t,\sigma) = 0 \quad \text{for} \quad (t,\sigma) \in \Sigma = T \times \Gamma,$$

$$x(t_0,\eta) = x_0(\eta) \in L_2(\Omega),$$

$$0 < c_0 \le b(\eta) \le C^0 < +\infty \quad \text{for a. a. } \eta \in \Omega.$$

Here $u(\cdot) \in L_2(Q)$ is a control, coefficients $b(\cdot)$ and $a_{k,l}(\cdot) \in C_\infty(\Omega)$ satisfy the conditions

$$\sum_{k,l=1}^{n} a_{k,l}(\eta)\xi_k\xi_l \ge \omega|\xi|^2_{\mathbf{R}^n}$$

$$\text{for} \quad \xi \in \mathbf{R}^n \quad \text{and a. a.} \quad \eta \in \Omega, \quad \omega > 0,$$

$$a_{k,l}(\eta) = a_{l,k}(\eta) \quad \text{a. e. on } \Omega \quad \forall\, k,l \in [1:n].$$

A second class of optimal control problems, for which the solution algorithm described above can be used, is associated with control systems described by the differential-functional equations of the form

$$\dot{y}(t) = L(y_t) + Bu(t), \quad t \in T = [0,\vartheta],$$

$$L(y_t) = \sum_{i=0}^{l} A_i y(t-\tau_i) + \int_{\tau_l}^{0} A_*(s)y(t+s)ds$$

with the initial condition

$$y(t_0) = \phi^0, y(t_0+s) = \phi^l(s) \text{ for } s \in [-\tau, 0].$$

Here $y(t) \in \mathbf{R}^n$, $u(t) \in \mathbf{R}^r$, $\phi^0 \in \mathbf{R}^n$, $\phi^l(\cdot) \in L_2([-\tau_l,0] : \mathbf{R}^n)$, $0 = \tau_0 < \tau_1 < \ldots < \tau_l$, $y_t : s \to y(t+s)$, $-\tau_l \le s \le 0$, A_i and B are constant matrices of dimensions $n \times n$ and $n \times r$, respectively, the elements of the matrix function $s \to A_*(s)$, $s \in [-\tau_l, 0]$, are square integrable.
In the first case, $U = L_2(\Omega)$, and operator $A : H = L_2(\Omega) \to H$ is defined by the following way

$$Ax = A_D x \quad \forall \quad x \in D(A_H),$$

where $A_D \in L(V,V^*)$, $V = H_0^1(\Omega)$, $V^* = H^{-1}(\Omega)$,

$$\langle A_D x, y\rangle_{V\times V^*} = -\int_\Omega b(\eta)x(\eta)y(\eta)d\eta -$$

$$-\sum_{k,l=1}^{n}\int_\Omega a_{k,l}(\eta)x_{\eta_k}(\eta)y_{\eta_l}(\eta)d\eta \quad \forall\, x,y \in V,$$

$$D(A_H) = \{y \in H^2(\Omega) : y(\sigma) = 0 \text{ a. a. } \sigma \in \Gamma\}.$$

Here symbol $D(A)$ denotes the domain of definition of operator A, $H^2(\Omega)$ is the standard Sobolev space.
In the second case, $U = \mathbf{R}^n$ and operator A is given by the next form (Bernier and Manitius, 1978; Banks and Kappel, 1979)

$$D(A) = \{\phi = (\phi^0, \phi^l(s)) \in H :$$

$$\phi^l(\cdot) \in W^{l,2}([-\tau_l,0];\mathbf{R}^n), \phi^l(0) = \phi^0\},$$

$$A\phi = (L(\phi^l), \dot{\phi}^l), \quad \phi = (\phi^0, \phi^l(\cdot)) \in D(A).$$

6. REFERENCES

Banks, H.T. and F. Kappel (1979). Spline approximations for functional differential equations. *J. Differential Equations* **34**(3), 496–522.

Bernier, C. and A. Manitius (1978). On semigroups in $R^n \times L^p$ corresponding to differential equations with delays. *Canad. J. Math.* **30**(5), 897–914.

Blizorukova, M.S. and V.I. Maksimov (1997). On the reconstruction of an extremal input in a system with hereditary. *Vestnik PGTU. Functional-differential equations* **4**, 51–61.

Digas, B.V., V.I. Maksimov, B.G. Bukchin and A.V. Lander (1998). On an algorithm solving inverse problem of ray seismics. *Computational Seismology* **30**, to appear.

Ermoliev, Yu.M., A.V. Kryazhimskii and A. Ruszczynski (1997). Constraint aggregation principle in convex optimization. *Mathematical Programming* **76**, 353–372.

Kryazhimskii, A.V. (1994). Convex optimization via feedbacks. *Working Paper WP-94-109* pp. 1–25.

Kryazhimskii, A.V. and V.I. Maksimov (1998). On an iterative procedure for solving the control problem under phase constraints. *Computational Mathematics and Mathematical Physics* **9**, to appear.

Kryazhimskii, A.V. and Yu.S. Osipov (1987). To a regularization of a convex extremal problem with inaccurately given constraints. an application to an optimal control problem with state constraints. In: *Some Methods of Positional and Program Control.* pp. 34–54. Academic Press. Sverdlovsk.

Kryazhimskii, A.V., V.I. Maksimov and E.A. Samarskaia (1997*a*). On reconstruction of inputs in parabolic systems. *Mathematical Modelling* **9**(3), 51–72.

Kryazhimskii, A.V., V.I. Maksimov and Yu.S. Osipov (1997*b*). Reconstruction of extremal perturbations in parabolic equations. *Computational Mathematics and Mathematical Physics* **3**, 288–298.

LEVEL SETS OF VALUE FUNCTION AND SINGULAR SURFACES IN LINEAR DIFFERENTIAL GAMES

S.S. Kumkov*, V.S. Patsko**

Institute of Mathematics and Mechanics, Ural Branch, Russian Academy of Sciences,
S.Kovalevskaya str., 16, Ekaterinburg, 620219, Russia
*e-mail: * 2445@ dialup.mplik.ru   ** u0104@cs.imm.intec.ru*

Abstract: The paper deals with an algorithm of construction of level sets in linear differential games with fixed terminal time and convex payoff function depending on two components of the phase vector. Into this algorithm, the block for detection of singular points on the border of level set is included. Singular points give singular lines on the border of level set. Singular surfaces in the game space are collected using singular lines as skeleton. Examples of numerically calculated level sets and singular surfaces are represented. *Copyright © 1998 IFAC*

Keywords: differential games, terminal control, singularities, numerical methods.

1. INTRODUCTION

The concepts of the alternating integral (Pontryagin, 1967) and maximal stable bridge (Krasovskii and Subbotin, 1988) are the main ones in differential game theory. By means of these terms, the solvability set in a game of approach is usually described. For games with terminal payoff function, these terms define a level set of the value function. A system of level sets on a grid of values gives a representation of the value function in general. Information, obtained during construction of level sets, can be used for singularities analysis. Algorithmic description of this analysis is the main topic of the paper.

A linear antagonistic differential game

$$\begin{gathered}\dot{x} = A(t)x + B(t)u + C(t)v \\ x \in R^n,\ u \in P,\ v \in Q,\ T,\ \varphi(x(T))\end{gathered} \tag{1}$$

with fixed terminal time T and convex payoff function φ, which depends on two coordinates x_i, x_j of the phase vector, is considered. The first (second) player governs the control u (v) choosing it from the convex compact P (Q) and minimizes (maximizes) the value of the function φ.

It is known that the substitution $y(t) = X_{i,j}(T,t)x(t)$, where $X_{i,j}(T,t)$ is a matrix combined of two rows of the fundamental Cauchy matrix, provides the transformation to the equivalent differential game of the second order on phase variable.

At the beginning of the 80's, the backward constructions were elaborated (Subbotin and Patsko, Eds., 1984; Taras'ev and Ushakov, 1985) for building level sets of the value function in linear differential game (1). Software for interactive investigation of level sets was created recently in cooperation with V.L.Averbukh, D.A.Yurtaev, E.A.Shilov, A.I.Zenkov from the Department of System Support of the Institute of Mathematics and Mechanics.

As singular surfaces in theory of differential games, such sets in the game space are named where the optimal motions have some peculiarities (dispersion, refraction, junction, etc.). The classification of the singular surfaces was suggested by Isaacs (1965).

Necessary conditions, which characterize different types of singularity, were studied by Bernhard (1977) and Melikyan (1998).

In the paper, an attempt of elaboration of algorithm for global construction of complete system of singular surfaces is made. With that, level set of the value function is the base object. On the border of a level set, singular lines are detected. They are determined by various peculiarities of optimal motions coming along the border surface of the level set. Singular surfaces are built on the base of singular lines taken from a system of level sets. Similar idea of constructing singular surfaces was realized for a concrete problem by Shinar and Zarkh (1996).

Discussing algorithm for detection and classification of singularities is imbedded into the algorithm (Isakova, *et al.*, 1984) for backward construction of level sets. The level sets can be built for arbitrary polyhedra P and Q. The algorithm for detection of singularities has been elaborated now only for the case of scalar controls of the first and second players (i.e. the sets P and Q are segments).

2. CONSTRUCTING LEVEL SETS OF THE VALUE FUNCTION

Here, the algorithm (Isakova, *et al.*, 1984) for constructing level sets of the value function is described. It is needful for understanding further section.

2.1 Backward procedure

Assume that the transfer from the game (1) with the payoff function φ depending on two coordinates of the phase vector to the equivalent game

$$\begin{gathered} \dot{y} = D(t)u + E(t)v \\ y \in R^2,\ u \in P,\ v \in Q,\ T,\ \varphi(y_1(T), y_2(T)) \quad (2) \\ D(t) = X_{i,j}(T,t)B(t),\ E(t) = X_{i,j}(T,t)C(t) \end{gathered}$$

is already done.

Let on the interval $[0,T]$ the sequence of instants $t_i : t_N = T, \ldots, t_i = t_{i+1} - \Delta, \ldots, t_0 = 0$ dividing the interval with a step Δ is given. The interest is to find the time sections $W_c(t_i) = \{y \in R^2 : V(t_i, y) \le c\}$ of level set $W_c = \{(t,y) \in [0,T] \times R^2 : V(t,y) \le c\}$ of the value function V for the given value of parameter c.

Replace the dynamics (2) by the piecewise-constant dynamics

$$\begin{gathered} \dot{y} = \mathbf{D}(t)u + \mathbf{E}(t)v \\ \mathbf{D}(t) = D(t_i),\ \mathbf{E}(t) = E(t_i),\ t \in [t_i, t_{i+1}) \end{gathered} \qquad (3)$$

Instead of the sets P and Q, let us consider their polyhedral approximations $\mathbf{P}$, $\mathbf{Q}$. Let $\hat{\varphi}$ be the approximating payoff function. For any c, its level set $\mathbf{M}_c = \{y \in R^2 : \hat{\varphi}(y) \le c\}$ is a convex polygon.

The approximating game (3) is taken so that a game with simple motion dynamics (Isaacs, 1965), polyhedral convex control constraints and the convex polygonal target set appears for each step $[t_i, t_{i+1}]$ of the backward procedure. On the base of $W_c(t_N) = \mathbf{M}_c$ the game solvability set $W_c(t_{N-1})$ can be computed. Further, starting from $W_c(t_{N-1})$, set $W_c(t_{N-2})$ can be built, and so on. As a result, the collection of convex polygons is obtained, which approximate sections $W_c(t_i)$ of level set W_c of the value function in the game (2).

Let $\mathcal{P}(t_i) = -D(t_i)\mathbf{P}$, $\mathcal{Q}(t_i) = E(t_i)\mathbf{Q}$. The support function $l \to \rho(l, W_c(t_i))$ of the polygon $W_c(t_i)$ is the convex hull (Pschenichnyi and Sagaidak, 1970) of the function

$$\gamma(l,t_i) = \rho(l, W_c(t_i)) + \Delta\rho(l, \mathcal{P}(t_i)) - \Delta\rho(l, \mathcal{Q}(t_i)).$$

The function $\gamma(\cdot, t_i)$ is positively-homogeneous and piecewise-linear. The property of local convexity of this function can be violated only at the boundary of linearity cones of the function $\rho(\cdot, \mathcal{Q})$, i.e. at the boundary of cones generated by the normals to neighbor edges of the polygon $\mathcal{Q}$.

2.2 Algorithm of convex hull construction

Let us agree to omit the argument t_i in the notation of the function γ.

The linearity cones of γ are determined by the outer normals to the convex polygons $W_c(t_i)$, $\mathcal{P}(t_i)$, $\mathcal{Q}(t_i)$. Gathering the normals of these sets and ordering them clockwise, the collection L of the vectors is obtained. The collection of values $\gamma(l)$ on the vectors $l \in L$ is denoted by Φ. The collections L, Φ describe completely the function γ.

The collection of the normals to $\mathcal{Q}(t_i)$ ordered clockwise is denoted by S. Vectors from S are called "suspicious". This name is connected with the fact that the function γ is locally convex on the cones, which interior does not contain the normals of the set $\mathcal{Q}(t_i)$. The violation of the local convexity can appear only on the cones which interior contains at least one normal of the polygon $\mathcal{Q}(t_i)$.

Let $L^{(1)} = L$, $\Phi^{(1)} = \Phi$, $S^{(1)} = S$. The k+1 step of the multistep convexing process consists in replacing the collections $L^{(k)}$, $\Phi^{(k)}$ by the collections $L^{(k+1)} \subset L^{(k)}$, $\Phi^{(k)} \subset \Phi^{(k+1)}$. The collection $S^{(k)}$ is also replaced by the new one $S^{(k+1)}$.

Describe now one step of the convexing process. Suppose that the angle between two neighbor vectors from the collection $L^{(k)}$ counted clockwise is less than π. Let $l \to \gamma^{(k)}(l)$ be the piecewise-linear function determined by the collections $L^{(k)}$, $\Phi^{(k)}$. Since $L^{(k)} \subset L^{(k-1)} \subset \ldots \subset L^{(1)}$, $\Phi^{(k)} \subset \Phi^{(k-1)} \subset \ldots \ldots \subset \Phi^{(1)}$, then for any vector $\bar{l} \in L^{(k)}$ the value $\gamma^{(k)}(\bar{l})$ is equal to $\gamma(\bar{l})$.

Take some vector $l_* \in S^{(k)}$ and check the local convexity of the function $\gamma^{(k)}$ on the cone generated by the vector l_* and two its neighbor vectors l_- and l_+ selected counterclockwise and clockwise from the collection $L^{(k)}$. In other words, check whether the inequality $l_*'y \le \gamma(l_*)$ is active in the triple of the inequalities $l_-'y \le \gamma(l_-)$, $l_*'y \le \gamma(l_*)$, $l_+'y \le \gamma(l_+)$. If the system of three inequalities is compatible, then (by virtue of the ordering the vectors l_-, l_*, l_+) only the middle one can be inactive.

The algorithm of verification: find the intersection point y_* of the straight lines $l_-'y = \gamma(l_-)$, $l_*'y = \gamma(l_*)$, and then check the inequality $l_+'y_* < \gamma(l_+)$. If it holds, the local convexity takes place. Otherwise, the local convexity is absent.

In the first case, the vector l_* is taken away from the collection $S^{(k)}$, and the remained set is denoted by $S^{(k+1)}$. Let $L^{(k+1)} = L^{(k)}$, $\Phi^{(k+1)} = \Phi^{(k)}$.

In the second case, two situations are distinguished. Let α be the angle counted clockwise from l_- to l_+.

1. $\alpha < \pi$. The vector l_* is taken away from the collection $S^{(k)}$, and, simultaneously, the vectors l_- and l_+ are included into this collection (one of them or even both can be there already). Denote the new collection of the "suspicious" vectors by $S^{(k+1)}$. The difference of the new collection $L^{(k+1)}$ from the collection $L^{(k)}$ is that the vector l_* is absent in $L^{(k+1)}$. When processing $\Phi^{(k)}$ to $\Phi^{(k+1)}$, the value $\gamma^{(k)}(l_*) = \gamma(l_*)$ is taken away;

2. $\alpha \ge \pi$. The constructing is ceased.

One step of the convexing algorithm has been described.

The algorithm finishes at the step with the number j when for the first time $S^{(j)} = \varnothing$, i.e. when the collection of the "suspicious" vectors becomes empty. It means that the function $\gamma^{(j)}$, which corresponds to the collections $L^{(j)}$ and $\Phi^{(j)}$, is locally convex everywhere. Thus, the function $\gamma^{(j)}$ is the convex hull of the function γ. In this case, let us denote the final collections $L^{(j)}$ and $\Phi^{(j)}$ as $\overline{L}$ and $\overline{\Phi}$, respectively.

The second variant of the termination is following: the angle α between the vectors l_- and l_+ becomes greater or equal to π after elimination the checked vector l_* from the collection of the "suspicious" vectors at some step. It means that the convex hull of the function γ does not exist, i.e. $W_c(t_i) = \varnothing$. (If $\alpha = \pi$, it is possible that $W_c(t_i)$ is a degenerate polygon, i.e. $W_c(t_i)$ is a point or a segment. Further constructions are ceased in this case also.)

3. SINGULAR SURFACES

In this section, it is supposed that the sets P and Q are segments. So, $\mathcal{P}(t_i)$, $\mathcal{Q}(t_i)$ are also segments in the space y_1, y_2 for any time instant t_i.

Let assume in addition that the segments $\mathcal{P}(t_i)$, $\mathcal{Q}(t_i)$ are not parallel, and more than that, each of them is not parallel to neither one of edges of the polygon $W_c(t_{i+1})$. Let call this assumption "the non-parallelism condition".

3.1 Optimal motions and singular points

On each time interval $[t_i, t_{i+1}]$, the approximating game (3) is the game with simple motions. The first player tries to transfer the system from the polygon $W_c(t_i)$ onto the polygon $W_c(t_{i+1})$, the second player tries to prevent it. Entering the discrimination of the second player, determine the optimal motions.

Let us fix some arbitrary point y_0 on the boundary of the polygon $W_c(t_i)$. The second player control constant on the interval $[t_i, t_{i+1}]$ is called the optimal one if the first player can not direct the resulting motion from the point y_0 into the interior of the polygon $W_c(t_{i+1})$. For fixed optimal control of the second player, the first player control (also constant on the same interval) is called the optimal parrying one if the corresponding motion comes onto the boundary of the set $W_c(t_{i+1})$. Motion generated by optimal players' controls is called the optimal one.

The control v^* satisfying the condition of maximum $v^* = \arg\max\{l'v : v \in \mathcal{Q}(t_i)\}$ is called the extremal control of the second player on the vector l. Similarly, the extremal control of the first player on the vector l is the control u^*, which satisfies the condition of minimum $u^* = \arg\min\{l'u :\ u \in \mathcal{P}(t_i)\}$.

It is easy to see that if y_0 is an internal point of some edge of the polygon $W_c(t_i)$, then the constant control of the second player is optimal if and only if it is extremal on the normal vector to $W_c(t_i)$ at the given point. The optimal parrying control of the first player is extremal on the same vector.

If y_0 is a vertex of $W_c(t_i)$, then two normals correspond to the vertex. The following statements describe the structure of the optimal controls.

Proposition 1. For any vertex of the polygon $W_c(t_i)$, the second player control, being extremal at least on one of two normal vectors at this vertex, is optimal. With that, the optimal parrying control of the first player is extremal on the same vector.

A proof of this statement does not use a supposition of a scalar character of players' controls. Hence, the condition of non-parallelism is not used also. In the following statement, these assumptions are essential.

Proposition 2. Let the players' controls be scalar and the condition of non-parallelism is satisfied. Then for any vertex of the polygon $W_c(t_i)$, the totality of optimal controls of the second player consists only of the extremal controls on two normal vectors of the polygon $W_c(t_i)$ at this vertex.

The point y_0 is called regular if the optimal motion emanating from this point is unique and generated by the extreme players' controls. A point, which is not regular, is called the singular one. Here, "extreme" means that the constant control value is a boundary point of the segment $\mathcal{P}(t_i)$ (or $\mathcal{Q}(t_i)$).

3.2 Classification of singularities

Below, the classification of singular points of the polygon $W_c(t_i)$ is described. It is based on the analysis of character of the optimal motions emanating from these points. For the classification, two marks are used. These marks are attached to normals participated in the process of convex hull construction of the function $\gamma(\cdot,t_i)$. The mark *FS* ("former suspicious") is added to normals which, during the convexing process, were denoted as "suspicious" ones, but, after the process, were remained in the final collection $\overline{L}$ (this collection determines the polygon $W_c(t_i)$). The mark *NP* is added to normals, which were taken from the set $\mathcal{P}(t_i)$. The classification is represented in the Table.

The Table deals with the normals from the final collection $\overline{L}$. Individual normals or pairs of neighbor ones are analyzed. Quantity of considered normals is shown in the first column. In the second column, the marks are represented whose presence is checked out on the considered normals. With that, if only one mark is shown for a concrete vector, then the second mark is supposed to be absent. The third column contains the additional condition. Marks in the second column, together with the satisfaction of the additional condition, determine the singularity type (which is shown in the sixth column) of the object from the forth column. Such object can be the polygon $W_c(t_i)$ vertex, which is incident for edges determined by normals (row 1), either the edge correspondent to the considered normal (row 3) or one of two shown normals (row 2). Inside the case of the row 1, a subdivision exists which is determined by the condition from the fifth column. Type of singularity is determined by the character of the optimal motions emanating from the marked point or points of the marked edge. The names of singularities are coordinated with ones used in the R.Isaacs' book.

The term "*P*-normal" ("*Q*-normal") means a normal taken from the set $\mathcal{P}(t_i)$ ($\mathcal{Q}(t_i)$).

Table. Singularities classification

	Flags	Additional condition	Object for marking	Secondary condition	Type of singularity
1	*FS*, *FS*	*Q*-normal is strictly between	Vertex	*P*-normal is not between	Dispersal for 2nd
				P-normal is between	Dispersal
1	*NP* +*FS*	*Q*-normal is strictly between	*P*-normal edge		Equivocal
2	*NP*		Edge		Switching for 1st

The term "dispersal" corresponds to the situation when two optimal motions emanate from the given point. One of these motions is generated by players' controls, which are extremal on one of the regarded vectors with the mark *FS*. The second motion corresponds to the players' controls extremal on the second vector with the same mark. With that, all controls are extreme in its segments.

The term "equivocal" corresponds to the situation when from the vertex, determined by the normals with marks *FS* and *NP* +*FS*, two optimal motions go out. One motion is generated by players' controls, which are extremal on the vector with the mark *FS*. The second one is generated by controls extremal on the vector with the mark *NP* +*FS*. With that, on the first motion, both controls get the extreme values. But on the second one, the control of the first player is not extreme. (It is namely the difference of the equivocal case from the dispersal one.) The latter motion comes to the vertex of the polygon $W_c(t_{i+1})$ where the cone generated by two normals contains the considered normal with the mark *NP* +*FS*. Each other point of the marked edge emanates only one motion, which comes to the mentioned vertex.

The term "switching for the first player" corresponds to the case when only one optimal motion emanates from each internal point of the marked edge. An extreme control of the second player and non-extreme control of the first player generate it. The difference from the equivocal case is in the fact that only one motion emanates from both vertices of the edge. With that, the first player control gets one extreme value at one vertex of the edge, and the opposite value at another vertex.

On different sides of the equivocal edge, both players' controls change. In dispersal situation, either only the second player changes its control or both players change their controls together. So, a special subdivision appears. In the switching situation, only the first player changes its control.

It can be proved that any point on the boundary of the polygon $W_c(t_i)$ is regular if it is not a marked vertex or is not included to any marked edge. And vice versa, any point, which is a marked vertex or included to a marked edge, is singular, except maybe the case when it is an endpoint of a marked edge.

3.3 Constructing singular surfaces

The data for the algorithm of singularity classification are accumulated during the process of the convex hull construction, and after that the algorithm of classification begins to work. Really, its work is reduced to the verification of the presence of situations described by rows of the Table. As a result, the collection of the points and intervals with description of type of their singularity is obtained for the current backward time section of the level set.

For graphical presentation of the singular surfaces, a collection of level sets is calculated on the given grid of magnitudes of the value function, and, simultaneously, the singular points and edges are detected on each section of each level set. Constructing the singular surfaces on the base of the singular points and intervals is carried out in the program of visualization.

Validity of the algorithm elaborated was verified on the test example from (Patsko and Tarasova, 1985). In that work, the singular surfaces appeared had been investigated analytically in detail. Results of our calculations coincide well with the results of the mentioned paper.

4. EXAMPLE

In this section, the examples of constructing level sets of the value function and singular surfaces are represented. The Gouraud shading is used for visualization of level set surface ("tube") built from separate sections by triangulation. The surface is illuminated by dot radiant, which position can be changed by user. Now, visualization of singular surfaces is implemented by the simplest methods.

The system is a conflict-controlled oscillator

$$\begin{aligned} \dot{x}_1 &= x_2 + v \\ \dot{x}_2 &= -x_1 + u \end{aligned} \qquad |u| \le 1,\ |v| \le 0.9,\ \varphi(x_1, x_2) = x_1^2 + x_2^2.$$

Three level sets shown in Fig. 1 were calculated on the interval $[0, 8]$ of the backward time τ with the step $\Delta = 0.05$. The sets are shown for values $c = 1.05,\ 1.4,\ 2.7$. The internal tube terminates. The outer tubes are visualized non-transparent.

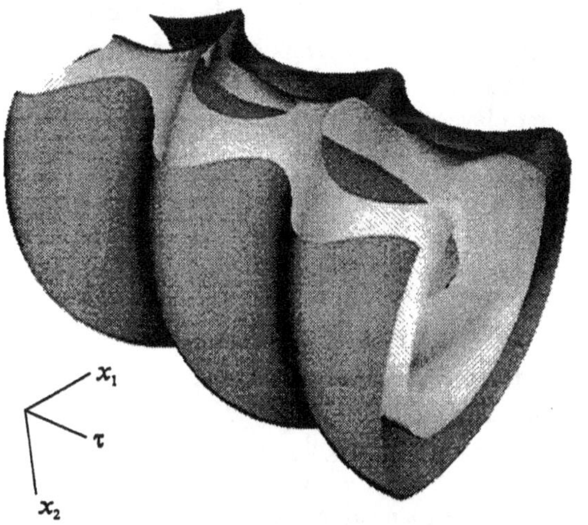

Fig. 1. Three level sets for "oscillator" system.

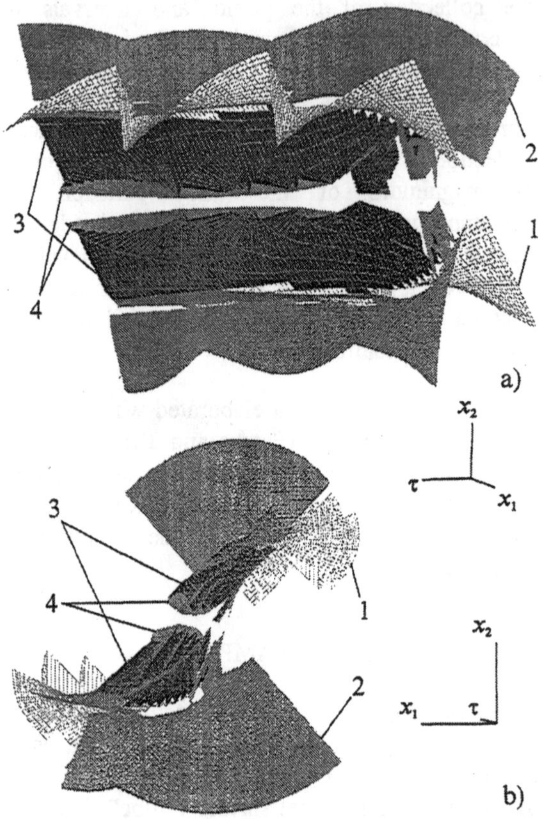

Fig. 2. Singular surfaces for "oscillator" system. a) View from the axis x_1. b) View from the axis τ. *Notations*: 1 – dispersal surface for the second player, 2 – switching surface for the first player, 3 – equivocal surface, 4 – dispersal surface.

To see the internal arrangement of the collection, two outer tubes are dissected by a plane parallel to the coordinate axes x_1, τ.

In Fig. 2, the singular surfaces are shown for two view-points: Fig. 2a corresponds to a view from the axis x_1, Fig. 2b is a view from the axis τ.

The surface closed to the axis τ is dispersal. On some distance from the axis τ, the equivocal surface is situated. It comes into the switching surface for the first player and dispersal one for the second player. The "empty" parts round the time axis can be filled by calculations with smaller time step Δ of the backward procedure.

CONCLUSION

In differential games, the analysis of singularities is important. In this paper, an algorithm for global construction of singular surfaces is suggested for linear differential games with scalar controls. In the future, it is planned to extend these ideas for the case of arbitrary geometric constraints on players' controls.

ACKNOWLEDGEMENTS

This research was supported by the Russian Foundation of Basic Researches under Grant No. 97-01-00458.

REFERENCES

Bernhard, P. (1977). Singular surfaces in differential games. In: *Lecture Notes in Control and Information Sciences*, Vol. 3 (P. Hagedorn, H.W. Knobloch and G.J. Olsder, Eds.), pp. 1 - 33. Springer-Verlag, Berlin.

Isaacs, R. (1965). *Differential games*, John Wiley and Sons, New York.

Isakova, E.A., G.V.Logunova and V.S.Patsko (1984). Computation of stable bridges for linear differential games with fixed time of termination. In: *Algorithms and Programs for Solving Linear Differential Games* (A.I. Subbotin and V.S. Patsko, Eds.), pp. 125 - 158, Inst. of Math. and Mech., Sverdlovsk [in Russian].

Krasovskii, N.N. and A.I. Subbotin (1988). *Game-Theoretical Control Problems*, Springer-Verlag, New York.

Melikyan, A.A. (1998). *Generalized Characteristics of First Order PDEs: Application in Optimal Control and Differential Games.* Birkhauser, Boston.

Patsko, V.S. and S.I. Tarasova (1985). Nonregular differential game of approaching. *Engenering. Cybernetics*, **22, No. 5**, pp. 59–67.

Pontryagin, L.S. (1967). Linear differential games, 2. *Soviet Math. Dokl.*, **8**, pp. 910–912.

Pschenichnyi, B.N. and M.I. Sagaidak (1970). Differential games of prescribed duration. *Cybernetics*, **6, No. 2**, pp. 72–83.

Shinar, J. and M. Zarkh (1996). Pursuit of a faster evader – a linear game with elliptical vectograms. In: *Proceedings of the Seventh International Symposium on Dynamic Games*, pp. 855–868, Yokosuka, Japan.

Subbotin, A.I. and V.S. Patsko, Eds. (1984). *Algorithms and Programs for Solving Linear Differential Games.* Inst. of Math. and Mech., Sverdlovsk [in Russian].

Taras'ev, A.M. and V.N. Ushakov (1985). Algorithm for constructing stable bridge in the linear problem of approach with convex target. In: *Investigations of Minimax Control Problems* (A.I.Subbotin and V.S.Patsko, Eds.), pp. 82–90, Inst. of Math. and Mech., Sverdlovsk [in Russian].

INFORMATIONAL SETS IN APPLIED PROBLEMS OF EVALUATION

S. I. Kumkov

Institute of Mathematics & Mechanics, Ural Branch, RAS, S.Kovalevskaya str.,16, Ekaterinburg, 620219, Russia, e-mail: kumkov@imm.uran.ru

Abstract: The paper deals with an application of informational sets in problems of evaluation and identification. The approach is based on theory of differential games and observation in uncertainty conditions. Originally, elaborated algorithms and numerical procedures were used in a problem of space vehicle control. Further, those were applied to a wide class of practical problems of evaluation and identification under uncertain and incomplete information. The approach is illustrated by examples where any probability characteristics of errors in the input information are absent.

Keywords: informational sets, perturbed measurements, uncertainty of information, evaluation algorithms, numerical methods.

1. INTRODUCTION

For a problem of a space vehicle control under uncertainty conditions, the methods and algorithms for informational sets building were elaborated (Kumkov and Patsko, 1995a,b and 1997). This approach is based on theory of differential games and observation in uncertainty conditions (Chernousko and Melikyan, 1978; Krasovskii and Subbotin, 1988; Kurzhanski, 1977).

Further investigations have shown that elaborated algorithms and numerical techniques have many similar features with ones in the bounding approaches (Milanese and Norton, 1996) and can be applied for solving a wide class of practical problems of evaluation and identification.

In the paper, an opportunity of application of elaborated algorithms founded on the informational sets is discussed in some applied problems. Those are identification of high temperature electrochemical process, information processing in high-precision metrological weighting, analysis of characteristics of high-strength steel fracture (Gladkovsky and Kumkov, 1997), evaluation of light radiation from laser-pumped solution of one or two chemicals.

In each of these problems, the input information is represented by a sampling of perturbed measurements of a true process. The measurement error has only the geometrical constraint, any other data about its properties are absent. The character of the measured process is determined by a function of some known type. The given type of function is described by a vector of its parameters.

Existing methods, such as the Least Square-Means Method (LSM), (*Direct Measurements*, 1986; *Metrological Characteristics*, 1984) for processing perturbed experimental data are mainly based on procedures of the mathematical statistics. In those, it is essential to know probability characteristics of the measurement error and to have a sampling of sufficient length (with sufficient number of measurements). Results of application of these methods are represented in the form of a point-wise estimate of the vector of parameters and its authentic interval with the given level of the authentic probability.

In discussed approach, any probability characteristics are absent, and it is hampered to validate an application of the LSM. As a result, it is difficult to show tolerances for parameters to be evaluated and limits for possible values of the given function.

The elaborated approach permits to evaluate a set of parameters (describing the given function) compatible with the measurements sampling and the given constraint on the error. It is customary to call such set the *informational set* (IS). In the essence, the IS is the analogue of the admissible domain of the parameters vector. Further, having the IS, it becomes possible to calculate the *tube* of admissible values of evaluated process. Any statistical information about errors and disturbances is not used.

2. PROBLEM OF EVALUATION WITH INFORMATIONAL SETS

A process is described by a function (dependence) of a given type $y = F(u,C)$, where y is a function value, u is an argument, C is a vector of parameters. The dependence is measured under the following conditions.

The argument u takes values $\{u_i,\ i=1,N\}$ known with an error, and the perturbed sampling of measurements $\{x_i,\ i=1,N\}$ is obtained. The structure of each measurement is

$$x_i = y(u_i^*, C) + \varepsilon_i, \quad (1a)$$

$$u_i = u_i^* + \delta_i. \quad (1b)$$

Here, $y(u_i^*, C)$ is an unknown true value of the function; ε_i is the error of the i-th measurement; u_i^* is an unknown true value of the argument; δ_i is its error.

As an example, the error in the argument and the error of measurement can be of the following type:

$$|\delta| \le \delta_{\max}, \quad (2)$$

$$|\varepsilon| \le \varepsilon_{\max}. \quad (3)$$

Here, $\delta_{\max}$ is the maximal (in modulus) value of error in the argument; $\varepsilon_{\max}$ is the maximal (in modulus) value of measuring error.

For the errors model (1)–(3), the following approximate *uncertainty set* can be shown for each measurement

$$G_i = [u_i - \delta_{\max},\ u_i + \delta_{\max}] \times [x_i - \varepsilon_{\max},\ x_i + \varepsilon_{\max}],\ i = 1,N. \quad (4)$$

For each reliable measurement, its uncertainty set contains true values of the argument and the function.

The measuring is carried out under uncertainty of information about the errors of measurements. Any statistical characteristics of the errors are absent. In each measurement, the errors can take arbitrary values from (2) and (3). Relations between these errors and relations between errors in neighbour measurements can also be arbitrary.

The problem is to evaluate both the set I_C of admissible values of the parameters vector C for the given dependence $y = F(u,C)$ and corresponding tube of its admissible values.

A dependence $y = F(u,C^*)$ and some concrete value C^* of the parameters vector are called the *admissible* ones if for each $i = 1,N$ there exists a point

$$(y,\ u) \in G_i, \quad (5)$$
$$\text{where } y = F(u,C^*).$$

Thus, the curve of the admissible dependence goes through all "windows" G_i (4).

In its essence, the problem of IS construction is a problem of the mathematical programming. To built the set I_C, it is necessary to find all C^* satisfying the system of insertions (5). Often, it is succeeded to construct the set by solving corresponding system of inequalities. For practical cases discussed below, the problem is solved in the form of intersection

$$I_C = \bigcap I_{i,j,\dots,m}. \quad (6)$$

Here, $I_{i,j,\dots,m}$ are partial informational sets corresponding to a group of uncertainty sets $\{G_i, G_j, \dots, G_m\}$ for $i,j,\dots,m = 1,N$ and $i \ne j \ne \dots \ne m$; each set $I_{i,j,\dots,m}$ determines the admissible set of parameters for which the corresponding curves pass through all windows $\{G_i, G_j, \dots, G_m\}$. A measurement is called the *inauthentic* one, if the set I_C, calculated with participation of this measurement, is empty, but becomes nonempty when the measurement eliminated.

Needed length of the group $\{G_i, G_j, \dots, G_m\}$ depends on dimension of the parameters vector C and properties of each uncertainty set. In cases where dimension of the parameters vector is equal to two, the set I_C is built by means of partial sets I_{ij}, which are constructed on pair combinations $\{G_i, G_j\}$, $i \ne j$ of the uncertainty sets.

From the structure of the procedure (6), it can be seen that for an authentic (compatible) sampling the greater dissipation of measurements the more IS decreases. In contrast to the LSM method, the result is obtained in the form of a set of admissible values of parameters determining the given function.

Thus, technically, the discussed approach to evaluation includes:

a) calculation of partial informational sets in the space of parameters;
b) implementation of operation of its intersection, and constructive description of the resultant IS;
c) calculation of the tube of the admissible values for a function of the given type.

3. EXAMPLES

For practically interesting cases, it was succeeded to construct fast algorithms for IS building and exact description of its frontiers. The algorithms were realised in computational programs for evaluation of parameters vector for functions of different types (constant, linear, polynomial, exponential, logarithmic, etc.) with a dimension of the parameters vector $1-5$. Direct grid methods were also applied for building IS.

In examples below, pithy physical notations of variables are used.

Example 1. Electrochemical process in a molten electrolyte. Relation of the electromotive force (EMF) in dependence of an electrolyte concentration is evaluated. Corresponding function is of a logarithmic type with dimension of the parameters vector equals to 2:

$$E = E_0 + \mathrm{Ln}(C) \cdot R \cdot T \,/\, (M \cdot F). \quad (7)$$

Here, E is a measured value of EMF, volt; E_0 , volt, M , dimensionless, are unknown parameters to be evaluated; R, F are given constant coefficients; C is an electrolyte concentration, the main argument, dimensionless; T is a fixed temperature of the experiment, degrees of Kelvin. Dimension of the multiplier $R \cdot T / (M \cdot F)$ is in volts.

In the experiment, the EMF E is measured with an error having maximal (in modulus) value $\varepsilon_{E\max}$; electrolyte concentration C takes values $\{C_i,\ i = 1,N\}$ with an error having maximal value $\delta_{C\max}$ (in modulus); the fixed temperature T is known with maximal (in modulus) error $\delta_{T\max}$. Because of errors both in the argument and in the function measurements, its approximate uncertainty sets are the rectangles in the plane $C \times E$:

$$[C_i - \delta_{C\max}, C_i + \delta_{C\max}] \times [E_i - \varepsilon_{E\max}, E_i + \varepsilon_{E\max}].$$

Figure 1 shows the sampling of EMF measurements (crosses) and their uncertainty sets (rectangles). The resultant tube of admissible values of the function (7) is marked with dotted lines.

Figure 2 represents the resultant informational set (solid line) in the plane $E_0 \times M$. The cross marks the point-wise estimate obtained formally by means of the LSM for demonstration. It is seen that the point lies out of the IS. Strictly saying, it means that the LSM estimate is not compatible with the given sampling so as the corresponding curve does not pass through uncertainty sets of some measurements.

Example 2. High-precision metrological weighting. A weight is measured. The given function to be evaluated is a constant. The maximal value of the measurement error is $\varepsilon_{\max} = 0.1\,\mathrm{gram\,(gr)}$. In simulation, the true weight W to be evaluated is 0.25 gr; the error is obtained by the random number generator with the uniform probability distribution law. The sampling length N is equal to 12. The sampling $\{W_i\}$, $i = 1,N$ has the following values: 4.1, 0.7, 1.6, -2.2, -3.9, 5.5, -9.5, 5.2, 6.3, 3.7, 8.9, -1.7 (in $\mathrm{gr} \times 10^{-2}$). The exact uncertainty set of each measurement is the interval

$$[W_i - \varepsilon_{\max}, W_i + \varepsilon_{\max}]. \quad (8)$$

As a result, the following informational set is obtained

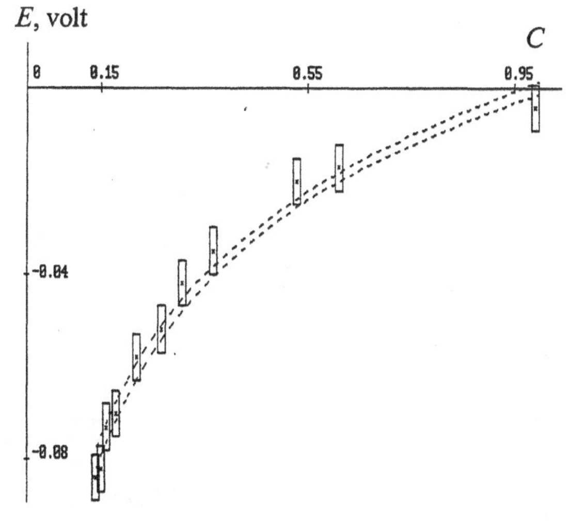

Fig. 1. EMF/Concentration dependence.

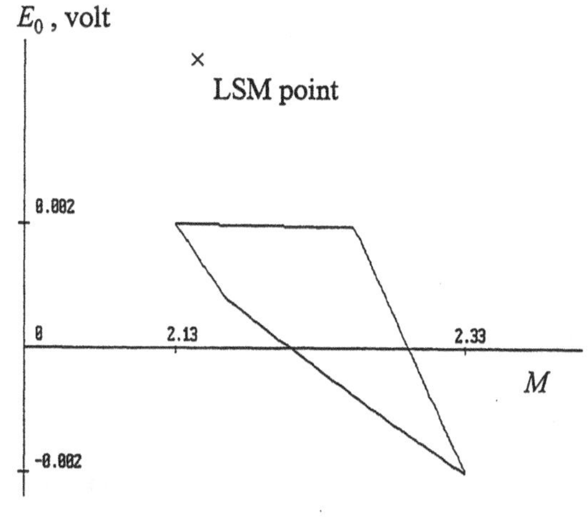

Fig. 2. Resultant informational set.

$I = [0.236\,\text{gr}, 0.252\,\text{gr}]$; its central point is $X_c = 0.247$ gr; the classic average mean is $X_m = 0.265$ gr. As it was noted above, for the given authentic sampling, the informational set is smaller than the origin uncertainty interval (8). The IS length is only 0.015 gr, maximal deviation is 0.0075 gr.

The properties of two the most important values were compared: the classic average mean X_m and the central point X_c of the IS. Let some "practical" tolerance μ ($\mu < \varepsilon_{\max}$) is given on the accuracy of the output evaluation. Statistical simulation confirms that the elaborated approach has the advantage in probability

$$P(|X_c - W| \le \mu) > P(|X_m - W| \le \mu) \qquad (9)$$

in interesting practical cases. The inequality takes place for all symmetrical probability distribution laws of the measurement error with essential dissipation in the interval (8). For example, it holds for essentially truncated Gaussian type and, always, for the probability uniform distribution law.

Example 3. Metrological verification of a weighting device. The measuring characteristic of a weighting device is verified, the time-drifting process is simulated and evaluated. The corresponding function is of the linear type, a dimension of parameters vector is equal to 2

$$D(t) = C + V \cdot t\,. \qquad (10)$$

Here, $D(t)$ is a measured drift, millivolt (mv); t is a time, the argument, hour (h); C is an unknown constant to be evaluated, volt; V is a velocity of the drift, an unknown constant to be evaluated, mv/h.

The true values of the drift parameters in (10) are V^*=1.0 (mv/h), C^*= 0.1 mv. The measurement error constraint $\varepsilon_{\max}$ is 0.05 mv. Time interval of measuring is [0, 0.7 h]; the sampling length is $N = 8$; the time net is uniform with the step 0.1 h. The measurement sampling $\{D_i\}$, $i = \overline{1, N}$ is the following: 0.055, 0.245, 0.260, 0.360, 0.545, 0.562, 0.651, 0.760 (in mv).

The error is present only in the function measurements, and the exact uncertainty sets are vertical intervals in the plane $t \times D$ (Figure 3):

$$[D_i - \varepsilon_{\max}, D_i + \varepsilon_{\max}]\,.$$

Figure 3 shows the results in the plane $V \times C$. There are:

– the initial partial informational set of the first and the second measurements, dotted line, with coordinates of its apexes (clock-wise from the right upper, abscissa in mv/h, ordinate in mv$\times 10^{-1}$) (1.9, 0.105), (2.9, 0.005), (1.9, 0.005), (0.9, 0.105);

– the resultant informational set I, solid line, with coordinates of its apexes (clock-wise from the right upper, abscissa in mv/h, ordinate in mv$\times 10^{-1}$) : (0,994, 0.105), (1.013, 0.094), (0.999, 0.095), (0.975, 0.105);

– the point of true parameters is marked by a cross;

– the point obtained by the LSM is marked by a circle, its value is 0.959 mv/h and 0.090 mv.

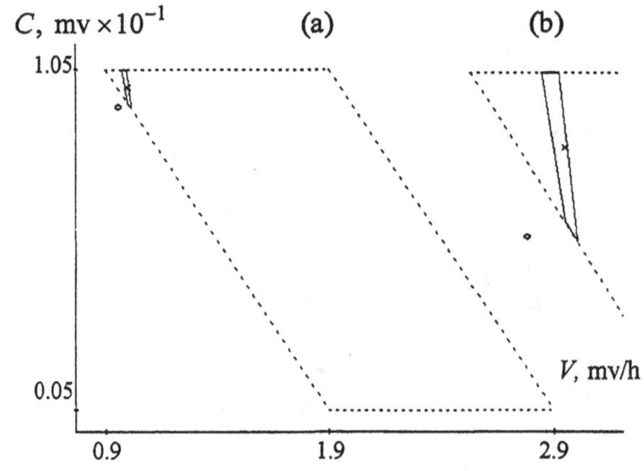

Fig. 3. Resultant informational set.

In Figure 3, the general view (a) and zoomed fragment (b) with the resultant informational set are shown.

Let underline the fact that the informational set contains the true point, but the LSM lies out of the set.

Figure 4 shows the results in the plane "time-drift". There are: the uncertainty intervals of the measurements (vertical segments and crosses); the frontiers of the original tube (dotted); the frontiers of the resultant tube calculated by means of the resultant informational set (solid lines); the line obtained by the LSM (dashed).

While the true process whole lies inside the resultant tube, the LSM line lies out of the tube and, for time instants 0.1 h and 0.4 h, is out of the uncertainty

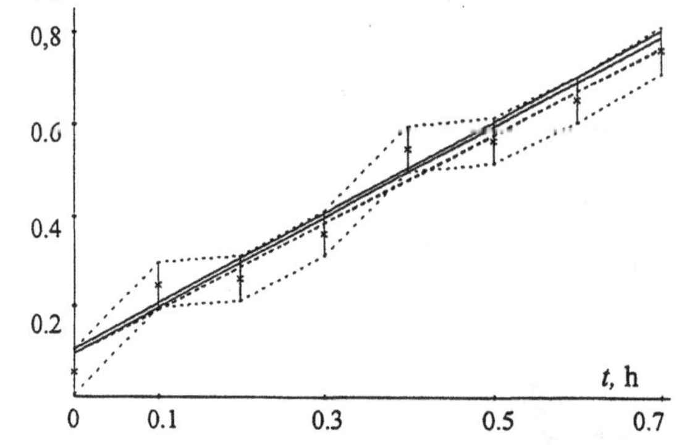

Fig. 4. Drift process, tube and the LSM line.

intervals of measurements No 2 and 5. It means that the LSM line is not consistent with the given sampling.

Example 4. Analysis of characteristics of high-strength steel fracture (Gladkovsky and Kumkov, 1997). A fracture toughness dependence as a function of impact toughness is investigated in different intervals of the latter variable. In the intervals, the type of the function can be parabolic, linear and of square-root character, the parameters vector dimension can be $2-3$. The given curves are

No 1 , power-type	$ST = A\cdot(IT)^{\alpha} + C$,
No 2 , linear	$ST = A\cdot IT + C$,
No 3 , semi-quadratic	$ST = A\cdot(IT)^{0.5} + C$.

Here, IT is an argument, the impact toughness, kg·m/cm^2; ST is a measured function, the static toughness, kg/mm$^{3/2}$; A, C, α are unknown parameters.

This practical example has one interesting feature. The values of the argument and the function in each measurement, in the fact, are not direct measurements.

They are obtained by complicated calculations over the original experimental data. (They can be regarded as "indirect" measurements.) So, only rough estimates of maximal errors ($\delta_{IT\max}$) in the argument and in the function ($\varepsilon_{ST\max}$) are known. Any statistical characteristics are absent.

Because of errors both in the argument and function measurements, its approximate uncertainty sets are the rectangles in the plane $IT \times \mathrm{ST}$

$$[IT_i - \delta_{IT\max},\ IT_i + \delta_{IT\max}] \times [ST_i - \varepsilon_{ST\max},\ ST_i + \varepsilon_{ST\max}].$$

The given sampling is represented in Figure 5. The totality of uncertainty sets (Figure 6) gives a picture of

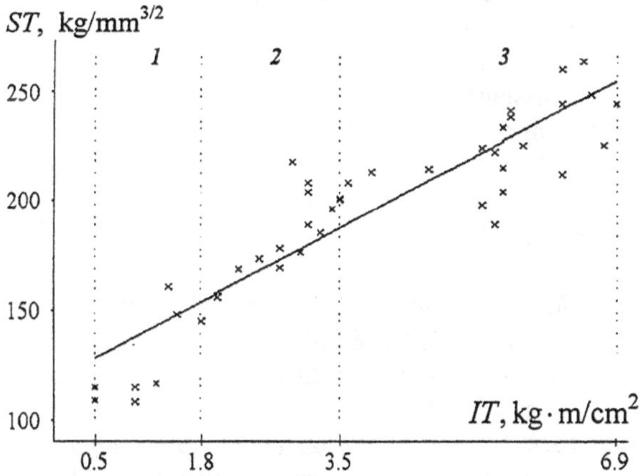

Fig. 5. The given sampling; measurements (crosses) LSM line (solid); 1, 2, 3 – argument intervals with different describing functions.

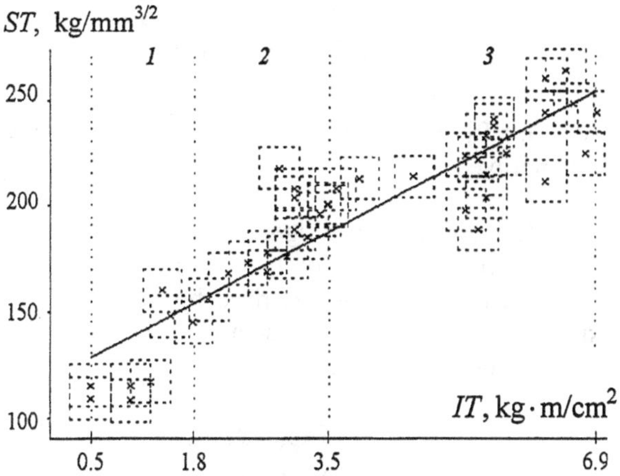

Fig. 6. Uncertainty sets.

possible dissipation of measurements. Under such conditions, an application of the LSM can give only demonstrative curves. As an example (Figures 5 and 6), the strait line (solid) as an approximating curve through the whole interval of the argument was calculated.

But it is too difficult to give any reasonable recommendations relatively the set of admissible values of the line parameters. And, also, the tube of lines that are consistent with the given sampling can not be shown. Only one useful conclusion can be done. Namely, the more delicate approximation is necessary on subintervals of the argument.

In Figure 6, measurements are marked by crosses, uncertainty sets by dotted rectangles, least-square line is denoted with the solid line; $1, 2, 3$ – argument intervals with different describing functions.

Results of processing the sampling on intervals with different functions are shown in Figure 7. The inauthentic measurements (circles) were found and eliminated out of the original sampling. The residual sampling (crosses) was processed.

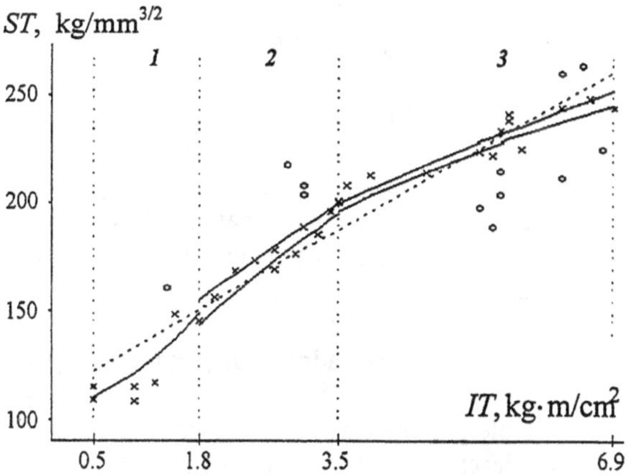

Fig. 7. Results; improved tubes (solid); consistent measurements (crosses); LSM line (dotted); inauthentic measurements (circles).

In the interval $IT \in [0.5, 1.8]$, the admissible values of the curve No 1 can lie in very narrow tube. In other intervals ($IT \in [1.8, 3.5]$, the curve No 2, and $IT \in [3.5, 6.9]$, the curve No 3), the resultant tubes are also improved.

Example 5. Evaluation of light radiation from laser-pumped solution of one or two chemicals. In time argument, the process is described by a descending exponential function or by a sum of two exponents, and a constant can also present

$$E(t) = E_0 \cdot \mathrm{Exp}(-\alpha t) + C,$$
$$E(t) = E_1 \cdot \mathrm{Exp}(-\alpha_1 t) + E_2 \cdot \mathrm{Exp}(-\alpha_2 t) + C.$$

The dimension of parameters vector can be two (E_0, α), three (E_0, α, C), four $(E_1, \alpha_1, E_2, \alpha_2)$ or five $(E_1, \alpha_1, E_2, \alpha_2, C)$. The constraints are given on maximal values of absolute or relative error in the radiation $E(t)$ measurements.

In different variants of the *Example 5*, the corresponding informational sets are constructed by various combinations of procedures with polygons building and its intersection, as well as with direct grid constructions in the space of parameters to be evaluated.

It is necessary to note once more that an application of the Least Square-Means Method (*Direct Measurements*, 1986; *Metrological Characteristics*, 1984) can not be validated in the problems involved. Besides, the numerical procedures of the LSM can uncontrollably stop at local minima, especially, in cases of multy dimensional space of parameters.

Feasible formulation of each mentioned evaluation problem was discussed with specialists in corresponding fields of experimental investigations. Simulation shows validity of suggested formulations, algorithms and its numerical realisations.

CONCLUSIONS

Discussed approach for evaluation of perturbed experimental data does not contradict to the procedures of now existing methods based on using procedures of theory of mathematical statistics, but can complement them when the error statistical characteristics of measurements are unknown.

The approach gives additional and practically meaning information: the full representation of the set of admissible parameters for the describing function; the full description of admissible values of the function to be evaluated (the tube). Elaborated algorithms for building informational sets and calculation of admissible tubes of the processes have rather simple numerical realisation.

ACKNOWLEDGEMENTS

The author would like to thank colleagues from the Institute of High-Temperature Electrochemistry (RAS), the State Institute of Metrology, the Laboratory of Metals Investigations (Ural State Technical University) and the Institute of Thermo-Physics (RAS). The work was partially supported by RFBR Grant No 97-01-00458.

REFERENCES

Chernousko, F.L. and A.A. Melikyan (1978). *Game Problems of Control and Search.* Nauka, Moscow, Russia [in Russian].

Direct Measurements with Multiple Observations. Methods of Processing Observation Results. GOST 8.207-86 (1986). The official edition. State Committee of Standards, Moscow, Russia [in Russian].

Gladkovsky, S.V. and S. I. Kumkov (1997). Application of approximation methods for fracture process peculiarities analysis and prediction of high-strength steel fracture toughness. In: *Collection of Scientific Works. Mathematical Modelling of Systems and Processes.* No 5, 26 – 34. Perm, Russia [in Russian].

Krasovskii, N.N. and A.I. Subbotin (1988). *Game-Theoretical Control Problems.* Springer-Verlag, New York.

Kumkov, S.I. and V.S. Patsko (1995a). Impulse corrections in a pursuit problem with incomplete information. *J. Comput. Systems Sci. Internat.*, **33**, No 5, 99 – 109.

Kumkov, S. I. and V. S. Patsko (1995b). Control of informational sets in a pursuit problem with incomplete information. In: *Annals of International Society of Dynamic Games. New Trends in Dynamic Games and Applications* (G.J.Olsder (Ed.)), 191 – 206. Birkhauser, Boston.

Kumkov, S.I. and V.S. Patsko (1997). Informational sets in a problem of impulse control. *Automatics and Telemechanics*, No 7, 195 – 206 [in Russian].

Kurzhanski, A.B. (1977). *The Control and Observation in Uncertainty Conditions.* Moscow, Nauka [in Russian].

Metrological Characteristics of Measurement Means and Precision Characteristics of Automation Means. General Methods of Estimate and Control. GOST 8.508-84 (1984). The official edition. State Committee of Standards, Moscow, Russia [in Russian].

Milanese, M. and J. Norton. (Eds) (1996). *Bounding Approaches to System Identification.* Plenum Press, London.

NUMERICAL MODELLING OF CONTROL TIME-DELAY SYSTEM

O.B.Kwon * A.V.Kim **,1,2 V.G.Pimenov *,1**

* *Kyung Sung College, Computer Science Dept.,*
Shin Prung-Ri 131, Pochun-Gun, Kyung-Kido, Seoul, Korea
e-mail: b1s2s3@chollian.net
** *Control Information Systems Laboratory*
School of Electrical Engineering
Seoul National University, Seoul, 151-742, Korea
e-mail: avkim@cisl.snu.ac.kr
*** *Ural State University*
Turgenev Str. 4, Ekaterinburg, 620083, Russia
e-mail: Vladimir.Pimenov@usu.ru

Abstract: Discrete–time control schemes are often used for realization of positional control strategies for continuous time control problems (Krasovskii and Subbotin, 1974; Osipov, 1971). However for general systems with delays (which are often called *functional differential equations* (FDE)) exact solutions in both continuous and discrete schemes can be found in exceptional cases. So, as rule, it is necessary to solve corresponding control problems using suitable *numerical algorithms*. In surveys (Cryer and Tavernini, 1972; Hall and Watt, 1976; Bellen, 1985) different numerical methods for FDE are discussed.

In this paper positional *implicit Runge–Kutta–like* numerical methods of solving general FDE are presented. The elaborated methods are direct analogs of the corresponding numerical methods of ODE case. Also we present new theorem (which generalizes the results of (Kim and Pimenov, 1997; Kim and Pimenov, 1998)) on convergence order.

The obtained results are based on distinguishing of finite dimensional and infinite dimensional components in the structure of FDE, on interpolation and extrapolation of the discrete prehistory of the model, on application of constructions of i–smooth analysis (Kim, 1996).

The notion of *an approximation order* for an optimal strategy in the discrete scheme of realization is introduced. The conditions, which guarantee required approximation order of the optimal strategy, are presented. Copyright ©1998 IFAC.

Keywords: Delay, positional control, discretization, Runge–Kutta method, implicit systems, accuracity.

[1] Supported in part by Russian Foundation for Fundamental Investigations (Grant N 98–01–00363) and in part by Korean Federation of Science and Technology Societies.

[2] On leave from the Institute of Mathematics & Mechanics Ural Branch Acad. Sci. of Russia; 16, Kovalevskoj Str., Ekaterinburg, 620219, Russia; e-mail: avkim@kim.imm.intec.ru.

1. CONTROL TIME–DELAY SYSTEMS

In the paper we consider control problems for systems with state delays

$$\dot{x}(t) = f(t, x(t), x_t(\cdot), u(t), v(t)), \qquad (1)$$

here $t \in [t_0, \theta]$, $x(t) \in R^l$ is the state vector, $x_t(\cdot) = \{x(t+s), -\tau \leq s < 0\}$ is the function–prehistory (delay), $u(t) \in R^q$ is the vector of control parameters, $v(t) \in R^r$ is a disturbance.

The set D_u (D_v) of *admissible* controls (disturbances) is a subset of the set of measurable on $[t_0, \theta]$ functions $u(t)$ ($v(t)$).

It is supposed that the mapping $f(t, x, y(\cdot), u, v) : [t_0, \theta] \times R^l \times C[-\tau, 0) \times R^q \times R^r \to R^l$

A1) is continuous with respect to $x, y(\cdot), u, v$ for every fixed t;
A2) is measurable in t for every fixed $x, y(\cdot), u, v$;
A3) satisfies the Caratheodory condition

$$\|f(t, x, y(\cdot), u, v)\| \leq \lambda(t),$$

where $\lambda(t)$ is an integrable function;
A4) satisfies the Lipschitz condition

$$\|f(t, x^1, y^1(\cdot), u, v) - f(t, x^2, y^2(\cdot), u, v)\| \leq$$
$$\leq L\|x^1 - x^2\| + M\|y^1(\cdot) - y^2(\cdot)\|_C$$

for some constants L and M.

Here $C[-\tau, 0)$ is the space of uniformly continuous on $[-\tau, 0)$ l–dimensional functions with the supremum norm $\|\cdot\|_C$.

If an initial conditions

$$x(t_0) = x^0, \quad x_{t_0}(s) = y^0(s), -\tau \leq s < 0, \; (2)$$

are fixed, then for every functions $u(\cdot) \in D_u$ and $v(\cdot) \in D_v$ there exists (Hale, 1977) unique absolute continuous solution $x(t) = x(t, t_0, y^0(\cdot), u(\cdot), v(\cdot))$ of problem (1)—(2).

Remark 1. Phase state of system (1) is a pair $\{x(t), x_t(\cdot)\} \in R^l \times C[-\tau, 0)$. However further we consider only continuous initial data, i.e. $x^0 = \lim_{s\to -0} y^0(s)$, so we will suppose that the initial function $y^0(s), s \in [-\tau, 0)$, is defined on whole interval $[-\tau, 0]$.

2. DISCRETE SCHEME OF A POSITIONAL CONTROL STRATEGY

In this section we consider (Krasovskii and Subbotin, 1974; Osipov, 1971) discrete scheme of realization of a positional control strategy $u[t]$ for system (1)—(2).

Let us fix a (uniform) partition of the time interval $[t_0, \theta]$ by points $t_n = t_0 + n\Delta, n = 0, ..., N$, $\Delta = (\theta - t_0)/N$. For the sake of simplicity we suppose that $\tau = m\Delta$, where m is an integer.

We define a *position* as a triplet

$$p_n = \{t_n, x(t_n), x_{t_n}(\cdot)\}.$$

Strategy is a mapping $U(p_n) = u[\cdot]$, where $u[\cdot] = u[t], t \in [t_n, t_{n+1})$, is a cut–off an admissible control on half–interval $[t_n, t_{n+1})$.

Movement $x^\Delta[t] = x^\Delta[t, t_0, y^0, U, v[t]]$, corresponding to a strategy U, is an absolute continuous solution of the system

$$\dot{x}^\Delta[t] = f(t, x^\Delta[t], x_t^\Delta[\cdot], u[t], v[t]), \qquad (3)$$
$$t \in [t_n, t_{n+1}),$$

with the initial condition

$$x_{t_0}^\Delta(s) = y^0(s), -\tau \leq s \leq 0. \qquad (4)$$

Here $u[t] = U(t_n, x^\Delta[t_n], x_{t_n}^\Delta[\cdot])$ is the realization of control strategy U, $v[t]$ is an admissible realization of disturbances.

As it was shown in (Osipov, 1971), this formalization of control strategy allows to solve an approach–evation problem, which is (Krasovskii and Subbotin, 1974) the basis for many different control problems.

However there are the following difficulties of modeling $x^\Delta[t]$:
1) exact solution of system (3)—(4) can be found only in some special cases, so it is necessary to solve this problem numerically in order to find an approximation of $z(t)$;
2) using numerical simulation we calculate a control $\hat{u}[t] = U(t_n, z[t_n], z_{t_n}(\cdot))$, but not the control $u[t] = U(t_n, x^\Delta[t_n], x_{t_n}^\Delta[\cdot])$;
3) exact calculation of a functional f is, as rule, impossible, especially if it contains distributed delays (of integral forms). In this case the value of the functional f can be calculated approximately by numerical methods.

Let us make the following assumptions:
A5) strategy U is weakly Lipschitz in the second and the third arguments, i.e. there exists a constant $L_U > 0$ such that for any positions $p_n^1 = \{t_n, x^1(t_n), x_{t_n}^1(\cdot)\}$, $p_n^2 = \{t_n, x^2(t_n), x_{t_n}^2(\cdot)\}$ and any $t \in [t_n, t_{n+1})$

$$\|f(t, x, y, u^1[t], v[t]) - f(t, x, y, u^2[t], v[t])\| \leq$$
$$\leq L_U\|x^1(\cdot) - x^2(\cdot)\|_C,$$

where $u^1[\cdot] = U(p_n^1)$, $u^2[\cdot] = U(p_n^2)$ (Note, that for programmed controls $u[t] = u(t)$ this condition is not necessary.) ;
A6) at any position it is possible to calculate approximate value f^Δ of the mapping f with an

approximation order p, i.e. there exists a constant $C_f > 0$, such that

$$\|f(t,x,y,u,v) - f^{\Delta}(t,x,y,u,v)\| \le C_f \Delta^p.$$

3. NUMERICAL MODEL OF IMPLICIT RUNGE–KUTTA TYPE

In this section we consider a problem of constructing numerical algorithms for solving the problem (3)—(4) on the basis of implicit Runge–Kutta–like (IRK) methods.

We define a *discrete numerical model* as the set of vectors $z_n \in R^l$, $n = -m, ..., N$.

The specific feature of systems (1) and (3) is that in order to construct at time moment t_n the adequate discrete model it is necessary to take into account *the prehistory of discrete model*, i.e. the set $\{z_i\}_n = \{u_i, \ \ i = n-m, ..., n\}$. The *initial prehistory* of the discrete model is the set

$$z_i = y^0(t_0 - t_i), \quad i = -m, ..., 0. \qquad (5)$$

In order to calculate the next value of the discrete model it is necessary at time moment t_n to make interpolation of values z_i, $i = n - m, ..., n$. Besides that, explicit (as well implicit) Runge–Kutta–like methods (Kim and Pimenov, 1997; Kim and Pimenov, 1998) require to make extrapolation of the model over the point t_n.

Let $c > 0$. We say, that on the set of discrete prehistories of a model $\{z_i\}_n$ an *interpolation-extrapolation operator* I is defined, if

$$I(\{z_i\}_n) = z(\cdot) \in C[t_n - \tau, t_n + c\Delta) \qquad (6)$$

Further we suppose that the operator I:
A7) is *consistent*, i.e. $z(t_i) = z_i$, $\ i = n-m, ..., n$;
A8) satisfies the Lipschitz condition, i.e. there exists a constant L_I such that for any two discrete prehistories $\{z_i^1\}_n$ and $\{z_i^2\}_n$, and for all $t \in [t_n - \tau, t_n + c\Delta)$

$$\|z^1(t) - z^2(t)\| \le L_I \max_{n-N_\tau \le i \le n} \|z_i^1 - z_i^2\|,$$

where $z^1(\cdot) = I(\{z_i^1\}_n), z^2(\cdot) = I(\{z_i^2\}_n)$.

We say, that *interpolation–extrapolation operator I has an order p* on a set of functions $X \subseteq C[t_n - \tau, t_n + c\Delta)$, if there exists a constant C_I such that for any $x(\cdot) \in X$ and $t \in [t_n - \tau, t_n + c\Delta)$

$$\|x(t) - \tilde{x}(t)\| \le C_I \Delta^p,$$

where $\tilde{x}(\cdot) = I(\{x_i\}_n)$, $x_i = x(t_i), i = n-m, ..., n$.

For example, the operator of piece-wise linear interpolation–extrapolation I: $\{z_i\}_n \to z(\cdot)$:

$$z(t) = \begin{cases} (z_{i+1}(t-t_i) + z_i(t_{i+1}-t))/\Delta, \\ \quad t \in [t_i, t_{i+1}), i = n-m, ..., n-1, \\ (z_{n+1}(t-t_n) + z_n(t_{n+1}-t))/\Delta, \\ \quad t \in [t_n, t_n + c\Delta) \end{cases}$$

is consistent, satisfies the Lipschitz condition with the constant $L_I = 1$, and on the set of continuous differentiable functions $x(t)$ has the second order. Examples of interpolation–extrapolation operators of high orders one can find in (Kim and Pimenov, 1998).

Implicit k–stage Runge–Kutta–like method (IRK) is the discrete model

$$z_{n+1} = z_n + \Delta \Psi^{\Delta}(t_n, z(\cdot)), \qquad (7)$$

$$n = 1, ..., N-1,$$

$$\Psi^{\Delta}(t_n, z(\cdot)) = \sum_{i=1}^{k} b_i h_i, , \qquad (8)$$

$$\hat{u}[t] = U(t_n, z[t_n], z_{t_n}(\cdot))$$

$$h_i = h_i(z(\cdot)) = f^{\Delta}(t_n + c_i\Delta, \qquad (9)$$

$$z_n + \Delta \sum_{j=1}^{k} a_{ij} h_j, z_{t_n + c_i\Delta}(\cdot),$$

$$\hat{u}[t_n + c_i\Delta], v[t_n + c_i\Delta]),$$

$$|c_i| \le c, \quad i = 1, ..., k.$$

Theorem 1. Let $\Delta < \dfrac{1}{kaL}$, $a = \max_{i,j} |a_{ij}|$. Then there exists a unique solution $H = (h_1, h_2, ..., h_k)'$ of system (9). Moreover, functionals $h_1, h_2, ..., h_k$ and Ψ^{Δ} satisfy the Lipschitz condition with respect to $z(\cdot)$.

The theorem can be proved using the iteration method of (Hairer *et al.*, 1987).

4. CONVERGENCE ORDER OF IRK–METHODS

Let $x_n = x^{\Delta}[t_n]$, $n = 0, 1, ..., N$, are values of the exact solution $x^{\Delta}[t]$ of problem (3)—(4) at points t_n. *Residual* of IRK–methods on $x^{\Delta}[t]$ at points t_n is the functional

$$g_n(x^{\Delta}[\cdot]) = \frac{x_{n+1} - x_n}{\Delta} - \Psi(t_n, x^{\Delta}[\cdot]),$$

where

$$\Psi(t_n, x^{\Delta}[\cdot]) = \sum_{i=1}^{k} b_i h_i,$$

$$h_i = f(t_n + c_i\Delta, x_n + \Delta \sum_{j=1}^{k} a_{ij} h_j, x^{\Delta}_{t_n + c_i\Delta}[\cdot],$$

$$u[t_n + c_i\Delta], \quad v[t_n + c_i\Delta]).$$

In case of ODE a residual order is defined, as rule, on some class of problems, for example, on a class of sufficiently smooth problems. The control problem (3)—(4) is defined by four parameters $(f, y^0, U, v[\cdot])$: a functional f in the right-hand side of system (1), an initial function (2), a control strategy U and disturbances v. Let F is a set (a class of problems) of $(f, y^0, U, v[\cdot])$ which satisfy the conditions A1—A8, and let $X(F)$ be a corresponding set of solutions $x^{\Delta}[\cdot]$ of problems (3)—(4).

We say, that *IRK-method has an order of approximation (residual) p on a class of problem F*, if for any $x^{\Delta}[\cdot] \in X(F)$ there exists a constant $C_{x^{\Delta}[\cdot]}$ such that

$$\|g_n(x^{\Delta}[\cdot])\| \leq C_{x^{\Delta}[\cdot]}\Delta^p, \quad n = 0, ..., N-1.$$

We say, that IRK-method (5)—(9) for problems (3)—(4) *converges*, if $(x_n - z_n) \to 0$ as $\Delta \to 0$ for every $n = 0, 1, ..., N$. The method *converges with the order p*, if there exists a constant C_z such that $\|x_n - z_n\| \leq C_z\Delta^p, \quad n = 0, 1, ..., N$.

For ODE a convergence order is defined only by approximation order. In case of FDE a convergence order of IRK-methods depends, besides that, on an order of interpolation-extrapolation operator. The following proposition is valid.

Theorem 2. Let on a class of problems F an IRK-method has an approximation order $p > 0$ and an interpolation-extrapolation operator I has an order p on the set $X(F)$. Then for $x^{\Delta}[\cdot] \in X(F)$ the IRK-method (5)—(9) converges with the order p.

Conditions, that define an approximation order, one can obtain by the expansion of a solution $x^{\Delta}[t]$ and a functional $\psi(t_n, x^{\Delta}[\cdot])$ into Taylor series at some neighborhood of a point t_n, and by equate the corresponding terms in Taylor series. In order to get Taylor series expansion of functionals one can use methods of i-smooth analysis (Kim, 1996).

It is important to note, that IRK-methods (5)—(9) are complete analoges of corresponding numerical methods of ODE, that is approximation orders are the same for smooth ODE and FDE with the same coefficients of numerical methods.

5. APPROXIMATION ORDER OF THE OPTIMAL CONTROL IN THE NUMERICAL MODEL

Let us consider a problem of minimization of a *cost functional* $J = J(x(\cdot))$ on trajectories $x(\cdot) \in C[t_0, \theta]$ of system (1)—(2). In the framework of discrete formalization of a movement (3)—(4) it means, that it is necessary to find a strategy U^0 with the following property: for every $\varepsilon > 0$ there exists $\Delta_0 > 0$ such that for any $\Delta < \Delta_0$ and any realization of disturbances $v[\cdot]$

$$J(x_0^{\Delta}[\cdot]) \leq J^0 + \varepsilon,$$

where $J^0 = \inf_U \sup_{\Delta, v[\cdot]} J(x^{\Delta}[\cdot, t_0, y^0, U, v[\cdot]])$ is the optimal value of the cost functional, $x_0^{\Delta}[t] = x^{\Delta}[t, t_0, y^0, U, v[\cdot]]$ is the optimal trajectory.

It was shown in (Krasovskii and Subbotin, 1974; Osipov, 1971), that under some additional conditions on the system (1)—(2), on sets of admissible controls (disturbances) D_u (D_v) and on the cost functional J, there exists an *optimal* strategy U^0 for this minimization problem.

Let us suppose that the following conditions are satisfied:
A9) the functional $J(x(\cdot))$ satisfies the Lipschitz condition, i.e. there exists a constant L_J such that for for every $x^1(\cdot)$ and $x^2(\cdot) \in C[t_0, \theta]$

$$\|J(x^1(\cdot)) - J(x^2(\cdot))\| \leq$$
$$\leq L_J\|x^1(\cdot) - x^2(\cdot)\|_{C[t_0,\theta]};$$

A10) optimal strategy U^0 exists and has *an approximation order p*, i.e. there exists a constant C_0 such that $J(x_0^{\Delta}[\cdot]) \leq J^0 + C_0\Delta^p$.

From this assumptions and from the theorem 2 follows

Theorem 3. Let the conditions A9 and A10 are satisfied. Then an optimal strategy U^0 guaranties an approximation order p for numerical model IRK-method (5)—(9), i.e. there exists a constant C such that $J(z(\cdot)) \leq J^0 + C\Delta^p$.

Remark 2. All results of the paper are valid for non-uniform partition of the time interval $[t_0, \theta]$. It allows to use variable steps for constructing numerical models. In the numerical experiments (simulations) the Runge-Kutta-Fehlberg-like 4(5) order numerical scheme with *automatic step size control* (Hairer *et al.*, 1987) and with the interpolation and extrapolation by cubic spline was realized.

Numerical simulation show effectivity of proposed algorithms for different classes of FDE with constant, time-varying and distributed delays. (The corresponding time-delay system toolbox for MATLAB was realized in cooperation with O.V.Onegova and A.B.Lozhnikov.)

Remark 3. The results of the paper can be spread on FDE with *control delays* (Osipov and Pimenov, 1981), and on case of a cost functional explicitly depending on control, for example, linear–quadratic problem (Andreyeva *et al.*, 1992).

6. REFERENCES

Andreyeva, E.A., V.B. Kolmanovskii and L.E. Shaiknet (1992). *Control of hereditary systems.* Nauka. Moscow.

Bellen, A. (1985). Consrained mesh methods for functional differential equations. In: *International Series of Numerical Mathematics* (Ch.Blanc, Ed.). Vol. 74. pp. 52–70. Verlag. Basel.

Cryer, C. and L. Tavernini (1972). The numerical solution of volterra functional differential equations by euler's method. *SIAM J. Numer. Anal.* **9**, 105–129.

Hairer, E., S. Norsett and G. Vanner (1987). *Solving ordinary differential equations. I. Nonstiff problem.* Springer-Verlag. Berlin.

Hale, J. (1977). *Theory of funtional differential equations.* Springer-Verlag. Berlin.

Hall, J. and J.M. Watt (1976). *Modern numerical methods for ordinary differential equations.* Clarendon Press. Oxford.

Kim, A.V. (1996). *i-Smooth analysis and functional differential equations.* IMM UrO RAN. Ekaterinburg.

Kim, A.V. and V.G. Pimenov (1997). Numerical methods for time-delay systems on the basis of *i*-smooth analysis. In: *Proceedings of the 15th World Congress on Scientific Computation, Modelling and Applied Mathematics* (Achim Sydow, Ed.). Vol. 1. pp. 193–196. Wissenschaft. Berlin.

Kim, A.V. and V.G. Pimenov (1998). On application of *i*-smooth analysis to elaboration of numerical methods for functional differential equations. In: *Trudy of the Inst. Math. Mekh. UrO RAN* (Achim Sydow, Ed.). Vol. 5. IMM UrO RAN. Ekaterinburg.

Krasovskii, N.N. and A.I. Subbotin (1974). *Positinal differential games.* Nauka. Moscow.

Osipov, Yu.S. (1971). Differential games for systems with delay. *Dokl. AN USSR* **196**, 779–782.

Osipov, Yu.S. and V.G. Pimenov (1981). Positional control in systems with control delay. *Prikl. Math. Mech.* **45**, 223–229.

ON THE PROBLEM OF CONTROL-OBSERVATION WITH A FINITE INFORMATIONAL SET

Loginov M.I. *

* *Urals State University, 51 Lenin St., 620080 Ekaterinburg, Russia, E-mail: mikhael.loginov@usu.ru*

Abstract:
In the problems of control-observation, as a rule, the indeterminacy is such that the corresponding informational sets are convex, which gives the possibility to use the powerful apparatus of convex analysis.
The paper deals with the problem of control constructing in a multistage system where the set of initial states consists of finite number of points, the proper initial state is unknown and on each step the signal is observed, which may differ from the "real" motion by the value of mistake. At each moment the mistakes are geometrically restricted. The peculiarity of the problem with finite informational set is that the control on each step must be constructed with the increasing of Chebyshev radius.
Such kinds of problems may appear in the process of "training" of the neural-like systems. During this process such tuning of the system parameters is making which gives the opportunity to recognize one of the finite numbers of possible images.

Keywords: control-observation, nonlinear systems, discrete informational set, image recognition, neural networks

1. INTRODUCTION

Nowadays the neural networks are widely used in different kinds of problems. These are, for example, the problems of combinatorial optimization (Hopfild, 1982; Hopfild, 1984; Melamed, 1994), the classification (pattern recognition) problems (Patrick, 1990), the generation of random variables sequence with the given first moments (Gorban and Mirkes, 1996) and approximation of the function of many variables (Umnov and Orlov, 1996). During the solution of such problems using neural networks we can point out the two following steps.

On the first step the network parameters are adjusted, this is the so-called training step. For the problems of combinatorial optimization an analog of energy function is constucted, its parameters are selected in such way that the minimum of our energy function gives the solution of optimization problem. The same approach is used in solution of other problems, for which the basis of the first step is the so-called training selection, i.e. the set of input-output pairs. For the problem of image recognition either Hebb method for Hopfield model, or the method of back error spreading for multilayed perceptrone (Hebb, 1957; Loskutov and Mikhailov, 1990; Rumelhart *et al.*, 1986) is used.

On the second step, after training, the dynamics of system state variation with known parameters is defined. As a result of system evolution after one, several or countable number of steps the system must come into one of the points of energy function minimum.

In this paper, using as an example the problem of pattern recognition, we suggest the method, in which both steps of problem solution, i.e. training and recognition of the "noisy" image, are solved at the same time.

The problem of image recognition in this approach is interpreted as a problem of control-observation with a special kind of dynamics, typical for neural-like networks.

The problems of control-observation in general form were solved by (Krasovskii and Subbotin, 1987; Kurzhanski, 1977; Shorikov, 1997),.

The singularity of the considered discrete problem of control-observation is the presence of non-linear functions like smoothed signum functions in the description of dynamics, and also discrete informational sets. This gives us some interesting effects.

2. STATEMENT OF THE PROBLEM FOR A SINGLE-STAGE SYSTEM

Consider a single-stage process where the transition of state is described by equation

$$x(1) = f(Ax(0) + b) \qquad (1)$$

where x is an n-dimensional system state vector, A is a square matrix of connections coefficients, b is an n-dimensional vector of displacement and $f(x)$ is a given mapping from R^n into R^n. All components of the vector-function f are of the same type and have one and the same activation function.

$$f_i = \frac{2}{\pi} \arctan \frac{x}{\alpha}$$

Note that $f_i(x) \to sign(x)$, where $\alpha \to 0, (\alpha > 0)$.

The initial state $x(0)$ is subject to constraint

$$x(0) \in X, \qquad (2)$$

where X is

$$X = \{x^{(i)} \in R^n : i = 1, ..., 2m\}.$$

Let us assume that the proper state $x(0)$ is unknown, and we know the set of possible initial states (standards) consisting of finite number of points $x^{(i)}, i = 1, ..., 2m$. For t=1 the current state of phase vector x is reported with a mistake ψ, i.e. we have a signal

$$y(1) = w(1) + \psi(1), \qquad (3)$$

where

$$||\psi|| \leq r, (0 \leq r < 2\sqrt{n}),$$

and $w(t)$ is the proper (unknown for us) state of the system at the moment t. $w(0) \in X$ and the dynamics of $w(t)$ is characterized by (1).

Let us introduce the following notations:

$$S_d(x) = \{z \in R^n : ||z - x|| \leq d\}.$$

If Z is a nonempty subset from R^n, then let $R(Z)$ be its Chebyshev radius, defined by the formula

$$R(Z) = \inf_{z_1 \in Z} \sup_{z_2 \in Z} ||z_1 - z_2||.$$

Definition. An informational domain $K(1)$ of the state of system (1) that is compatible with a signal $y(1)$ under restrictions (2), is the following set

$$K(1) = K(y(1), A, b) = S_r(y(1)) \cap f(AX + b)$$

Lemma. For each set $Z = \{z^{(i)} \in R^n : i = 1, 2, ..., 2m\}$ there exist such a matrix A and vector b, that the image W of the set Z, where $W = AZ + b$, consists of m elements with only positive coordinates and m elements with only negative ones.

As one of the variants of such a mapping we can suggest the following:

1) The expansion of Z along the x_1 with the transformation matrix $A_1 = \begin{pmatrix} k, 0, ..., 0 \\ 0, 1, ..., 0 \\ .... \\ 0, 0, ..., 1 \end{pmatrix}$

2) The rotation of the obtained set so that the basis vector $\mathbf{e} = (1, 0, ..., 0)^T$ will have the same direction as the vector $\mathbf{1} = (1, 1, ..., 1)^T/\sqrt{n}$, i.e. the transformation matrix is A_2.

So the required matrix $A = A_2 A_1$.

3) The displacement along the vector $\mathbf{1}$ until the number of points in the opposite orthants becomes equal to m.

It's easy to notice that the expansion coefficient k and the vector b are depending on location of the points from Z.

The following theorem takes place

Theorem. Let the set of initial points X consist of 2m elements. Then, choosing the control parameters A and b according to the Lemma (at Z=X), one can get the informational set K(1) consisting of m points, and its Chebyshev radius being more than r.

3. A MULTISTAGE SYSTEM

Consider a system, which dynamics is characterized by a discrete recurrent equation

$$x(t + 1) = f(Ax(t) + b), \qquad (1)$$

where $t = 0, ..., T$ (layer number); $x \in R^n$ is a phase vector, A is a square matrix of connections coefficients and b is an n-dimensional vector of displacement.

Let us assume that the proper state $x(0)$ is unknown, and we know the set of possible initial states (standards) consisting of finite number of points $x^{(i)}$, $i = 1, ..., N(N = 2^m)$. For each $t, (t = 1, 2, ..., T)$ the current state of phase vector x is reported with a mistake ψ, i.e. we have a signal

$$y(t) = w(t) + \psi(t), \qquad (2)$$

where

$$||\psi|| \le r, (r = 2),$$

is an admissible level of mistake, and $w(k)$ is the proper (unknown for us) state of the system.

The problem is to define, with the known signal $y(t)$, for which i the equality $w(0) = x^{(i)}$ is true, or, if it is impossible, to narrow down maximally the circle of "suspected".

As the control parameters we use matrix A and vector b, which will be corrected at each moment t, depending on realized signal y=y(t).

Theorem. Let the set of initial positions X consist of $N = 2^m$ points. Then, choosing A and b according to the Lemma, where $Z = K(t-1), K(0) = X, t = 1, 2, ..., m$, we shall get the informational set K(m), which consists of only one point.

So, to the moment T=m, using the signals y(t), t=1,2,...,m, one can know the initial state w(0) properly.

4. EXAMPLE

Let $n = 2, N = 4$,

$x^{(1)} = (1, 1)$; $x^{(2)} = (-1, 1)$; $x^{(3)} = (-1, -1)$; $x^{(4)} = (1, -1)$.

Let $A_1 = \begin{pmatrix} 2,0 \\ 0,1 \end{pmatrix}$,

$$A_2 = \begin{pmatrix} \sqrt{2}/2, -\sqrt{2}/2 \\ \sqrt{2}/2, \sqrt{2}/2 \end{pmatrix}.$$

Then $A = \begin{pmatrix} \sqrt{2}, -\sqrt{2}/2 \\ \sqrt{2}, \sqrt{2}/2 \end{pmatrix}$.

$b = (0, 0)^T$. Let, for example, $y(1) = (\sqrt{2}/2, 3/2\sqrt{2})$ (i.e. corresponding to the initial point $x^{(1)}$). Let $z^{(1)} = (\sqrt{2}/2, 3/2\sqrt{2})$ and $z^{(4)} = (3/2\sqrt{2}, 1/2\sqrt{2})$ - these are the images of $x^{(1)}$ and $x^{(4)}$ after the linear transformation with the chosen A and b. Then $K(1) = f(z^{(1)}) \cup f(z^{(4)})$.

At the next step we choose matrix A and vector b in such way that the points $f(z^{(1)})$ and $f(z^{(4)})$ should be located in the first and in the third quadrant respectively. After this for any y(2) the informational set K(2) will consist of the only point.

5. REFERENCES

Gorban, A.N. and Ye.N. Mirkes (1996). Estimations and reply interpretors for dual functioning neural nets. *Izvestiya vuzov, Priborostroyenie* **39**(1), 5–14.

Hebb, D.C. (1957). *The Organization of Bihavior: A Neuropsycological Theory.* Wiley. N.Y.

Hopfild, J.J. (1982). Neural networks and physical systems with emergent collective computational abilities. *Proc. of the Nat.Acad. of Sci.* **79**, 2554–2558.

Hopfild, J.J. (1984). Neurons with graded response have collective computational properties like those of two-state neurons. *Proc. of the Nat.Acad. of Sci.* **81**, 3088–3092.

Krasovskii, N.N. and A.I. Subbotin (1987). *Game-theoretical control problems.* Springer-Verlag. New York, Berlin, Heidelberg, London, Paris, Tokyo.

Kurzhanski, A.B. (1977). *Control and Observation Under Conditions of Uncertainty.* Nauka. Moscow (in Russian).

Loskutov, A.Yu. and A.S. Mikhailov (1990). *Introduction into Synergetics.* Nauka. Moscow (in Russian).

Melamed, I.I. (1994). Neural network and combinatorial optimization. *Automatics and Telemechanics* **4**, 3–40.

Patrick, K. (1990). *Simpson Artificial Neural Systems: Foundation, Paradigms, Application and Implementations.* Pergamon Press. NY.

Rumelhart, D.E., G.E. Hinton and R.J. Williams (1986). Learning internal representation by error propagation. In: *Parallel Distributed Processing.* Vol. 1. pp. 318–362. MIT Press.. Cambridge, MA.

Shorikov, A.F. (1997). *Minimax Estimation and Control in Discrete Dynamic Systems.* Ural Univ. Press.. Ekaterinburg (in Russian).

Umnov, N.A. and S.N. Orlov (1996). The comparison of rprop and scg algorithms for multilayer neural nets training. *Izvestiya vuzov, Priborostroyenie* **39**(1), 17–22.

ON APPROXIMATION OF ASYMPTOTIC ATTAINABILITY DOMAINS [1]

Morina S.I. *

* *Institute of Mathematics and Mechanics, 16 Kovalevskaya St.,
620219 Ekaterinburg, Russia
e-mail morina@imm.uran.ru*

Abstract: The asymptotics of attainability domains of linear control problems with integral constraints under infinitesimal perturbations of the latter is considered. This asymptotics is discribed by the attainability domain of a generalized problem in the class of finitely additive measures. The possibility of approximation of the generalized attainability domain by a finite number of step functions is discussed.

Keywords: attainability domain, finitely additive measure, extension, approximation.

1. INTRODUCTION

It is known that control problems with integral constraints are unstable. This means that small perturbations of the given constraints can lead to a jump of result; see, for example, (Warga, 1972, ch. III), (Chentsov, 1996; Morina, 1996). In particular, the attainability domain (AD) of the initial control system can not coincide with the asymptotic attainability domain. The latter is determined as a limit under $\varepsilon \to 0$ of AD of corresponding perturbed control problems (ε is a parameter characterizing the degree of perturbation of the given integral constraints). Since in practice constraints to control functions are given with some tolerance, it is natural instead of the given "severe" constraints to consider various relaxed variants of them and instead of AD of the initial problem to do an asymptotics of AD of perturbed problems. In the works of A.Chentsov (Chentsov, 1996; Chentsov, 1997) a regularization method for problems with integral constraints is supposed. This method is based on constructions of corresponding generalized problems in the class of finitely additive measures (FAM). In this case, the asymptotic attainability domain coincides with AD of the generalized problem. Since the space of solutions in the generalized problem, generally speaking, is not metrizable (it is a subset in the space of FAM), the problem of numerical construction of the asymptotic attainability domain was not touched so far.

The present work concerns questions of approximation of asymptotic attainability domains in the class of usual solutions (step functions).

2. SOME ELEMENTS OF TOPOLOGY AND FINITELY ADDITIVE MEASURE THEORY

Introduce some notions and designations required for the sequel constructions.

By $\mathbf{R}$ denote the real line, $\mathcal{N} \triangleq \{1, 2, \ldots\}$ is the integer, $\forall m \in \mathcal{N}\ \forall k \in \mathcal{N} : \overline{m,k} \triangleq \{i \in \mathcal{N} \mid m \le i \le k\}$, the symbol $\triangleq$ means the equality by definition. If (X, τ) is a topological space and A is a subset of X, by $cl(A, \tau)$ denote the closure of the set A in (X, τ). If A and B are sets, by B^A denote the set of all functions from A into B. (To avoid an ambiguity in designations, the integers are supposed to be not sets. Thus, if $m \in \mathcal{N}$ and H is a nonempty set, then $H^m \triangleq H^{\overline{1,m}}$). If $f \in B^A$ and C is a subset of A, suppose that

[1] Supported by the Russian Foundation of Fundamental Researches, project no. 97-01-00458.

$f^1(C) \triangleq \{f(x) : x \in C\}$. For any set A by $Fin(A)$ denote the set of all finite subsets of A.

Linear operations, product and ordering in spaces of functionals are understood point-wise.

Let E be an arbitrary nonempty set, $\mathcal{L}$ be a semi-algebra (Neveu, 1964), (Chentsov, 1996, p. 57) of subsets of E. By $(add)^+[\mathcal{L}]$ denote the cone of all nonnegative real-valued FAM on $\mathcal{L}$. Let $\mathbf{A}(\mathcal{L})$ be the linear subspace of $\mathbf{R}^{\mathcal{L}}$, generated by this cone. Following (Chentsov, 1985), by $B_0(E, \mathcal{L})$ denote the set of all step functionals on E in the sense of $(E, \mathcal{L})$. Let $B(E, \mathcal{L})$ stand for the closure of $B_0(E, \mathcal{L})$ in the space $\mathbf{B}(E)$ of all bounded functionals on E, equipped by the *sup*-norm $\|\cdot\|$ (Danford and Schwartz, 1958, p. 261). Elements of $B(E, \mathcal{L})$ are uniform limits of convergent sequences in $B_0(E, \mathcal{L})$ and only they. Then $B(E, \mathcal{L})$ as a closed subspace of $(\mathbf{B}(E), \|\cdot\|)$ is a Banach space whose topological dual $B^*(E, \mathcal{L})$ is isometrically isomorphic to $\mathbf{A}(\mathcal{L})$ (Chentsov, 1986, p. 75-76). In this case, a functional on $B(E, \mathcal{L})$ corresponds to each measure $\mu \in \mathbf{A}(\mathcal{L})$. Values of this functional are μ-integrals of functions from $B(E, \mathcal{L})$. The duality $(B(E, \mathcal{L}), \mathbf{A}(\mathcal{L}))$ induce the $*$-weak topology $\tau_*(\mathcal{L})$ in the space $\mathbf{A}(\mathcal{L})$.

Fix $\eta \in (add)^+[\mathcal{L}]$. If $f \in B(E, \mathcal{L})$, by $f * \eta$ denote the indefinite η-integral of f (Chentsov, 1986, p. 76). Suppose

$$(add)^+[\mathcal{L}, \eta] \triangleq \{\mu \in (add)^+[\mathcal{L}] \mid \forall L \in \mathcal{L} : (\eta(L) = 0) \Rightarrow (\mu(L) = 0)\}.$$

Let $\mathbf{A}_\eta[\mathcal{L}]$ be a subspace of $\mathbf{A}(\mathcal{L})$, generated by the positive cone $(add)^+[\mathcal{L}, \eta]$. Elements of $\mathbf{A}_\eta[\mathcal{L}]$ are called measures weakly absolutely continuous with respect to η. Further, $\forall b > 0$ introduce the set

$$\Xi[b] \triangleq \{\mu \in \mathbf{A}_\eta[\mathcal{L}] \mid V_\mu(E) \leq b\}, \qquad (1)$$

where $V_\mu(E)$ is variation of the measure $\mu \in \mathbf{A}(\mathcal{L})$. Besides, for any $b > 0$ along with $\Xi[b]$ introduce the set

$$M_b \triangleq \{f \in B_0(E, \mathcal{L}) \mid \int_E |f(x)| \, \eta(dx) \leq b\}.$$

In works of A.G.Chentsov the possibility of approximation of measures by indefinite integrals in the $*$-weak sense is considered. In particular, in (Chentsov, 1996) a theorem about every dense immersion of images of the spaces $B_0(E, \mathcal{L})$ and $B(E, \mathcal{L})$ into the set $\mathbf{A}_\eta[\mathcal{L}]$ under the integration operator is given (see theorem 4.3.2). For the sets (1) the corresponding density property is also true (Chentsov, 1996, p. 87):

$$\Xi[b] = cl(\{f * \eta : f \in M_b\}, \tau_*(\mathcal{L})). \qquad (2)$$

From this due to Alaoglue's theorem (Danford and Schwartz, 1958, ch. 5) the compactness of the set $\Xi[b]$ in the space $(\mathbf{A}(\mathcal{L}), \tau_*(\mathcal{L}))$ follows.

3. EXTENSION OF A CONTROL PROBLEM IN THE CLASS OF FINITELY ADDITIVE MEASURES

Consider the linear control system

$$\dot{x}(t) = A(t)x(t) + b(t)f(t), \quad t \in [t_0, \vartheta], \quad (3)$$

where $x(t_0) = x_0 \in \mathbf{R}^n$, $A(\cdot)$ is a $n \times n$- continuous matriciant, $b(\cdot)$ is a piecewise continuous, right continuous vector function from $[t_0, \vartheta[$ into $\mathbf{R}^n$. Control program $f(\cdot)$ is a piecewise constant and right continuous real-valued function on $[t_0, \vartheta[$. In the sequel, we will denote the set of such functions on $[t_0, \vartheta[$ (i.e. piecewise constant, right continuous and real-valued) by the symbol $\mathbf{F}([t_0, \vartheta[)$. Suppose that control programs $f \in \mathbf{F}([t_0, \vartheta[)$ satisfy the conditions:

$$\int_{t_0}^{\vartheta} |f(t)| \, dt \leq c, \qquad (4)$$

$$\left(\int_{t_0}^{\vartheta} s_1(t)f(t)\,dt, \ldots, \int_{t_0}^{\vartheta} s_m(t)f(t)\,dt\right) \in Y, \qquad (5)$$

where s_i, $i \in \overline{1,m}$, are piecewise continuous and right continuous real-valued functions on $[t_0, \vartheta[$, Y is a closed set in $\mathbf{R}^m$, $m \in \mathcal{N}$.

If $f \in \mathbf{F}([t_0, \vartheta[)$, by $x_f(\cdot) = (x_f(t), t_0 \leq t \leq \vartheta)$ denote the trajectory of the system (3), generated by the control program f from the initial position (t_0, x_0). By the Cauchy formula $\forall t \in [t_0, \vartheta]$ we have

$$x_f(t) = \Phi(t, t_0)x_0 + \int_{t_0}^{t} \Phi(t, \tau)b(\tau)f(\tau)\,d\tau, \qquad (6)$$

where $\Phi(\cdot, \cdot)$ is the fundamental matrix of solutions for the corresponding homogeneous system.

Let $\mathbf{F}_0$ be the set of all control programs $f \in \mathbf{F}([t_0, \vartheta[)$ satisfying the constraints (4), (5). As usualy (Krasovskii, 1970), the set

$$G \triangleq \{x_f(\vartheta) : f \in \mathbf{F}_0\}$$

is the AD of the system (3) under the given constraints to control program.

Along with the constraints (4)–(5) consider perturbed conditions on the choice of control. By Γ_0 denote the set of numbers $j \in \overline{1,m}$ for which the corresponding s_j function in (5) is a step one.

Let $\forall\alpha > 0$

$$Y_\alpha \triangleq \{y \in \mathbf{R}^m \mid \exists y^* \in Y \quad \forall i \in \overline{1,m}: |y_i - y_i^*| \leq \alpha\},$$

$$Y_\alpha^0 \triangleq \{y \in \mathbf{R}^m \mid \exists y^* \in Y : (\forall k \in \Gamma_0 : y_k = y_k^*) \;\&\; (\forall i \in \overline{1,m} \setminus \Gamma_0 : |y_i - y_i^*| \leq \alpha)\},$$

$$F_\alpha^{(1)} \triangleq \Big\{f \in \mathbf{F}([t_0, \vartheta[) \mid (\int_{t_0}^{\vartheta} |f(t)|\, dt \leq c + \alpha) \;\& \Big((\int_{t_0}^{\vartheta} s_1(t) f(t)\, dt, \ldots, \int_{t_0}^{\vartheta} s_m(t) f(t)\, dt) \in Y_\alpha \Big)\Big\},$$

$$F_\alpha^{(2)} \triangleq \Big\{f \in \mathbf{F}([t_0, \vartheta[) \mid (\int_{t_0}^{\vartheta} |f(t)|\, dt \leq c) \;\& \Big((\int_{t_0}^{\vartheta} s_1(t) f(t)\, dt, \ldots, \int_{t_0}^{\vartheta} s_m(t) f(t)\, dt) \in Y_\alpha^0 \Big)\Big\},$$

$$G_\alpha^{(1)} \triangleq \{x_f(\vartheta) : f \in F_\alpha^{(1)}\},$$

$$G_\alpha^{(2)} \triangleq \{x_f(\vartheta) : f \in F_\alpha^{(2)}\}.$$

Let ρ be the Euclidian metric in the space $\mathbf{R}^n$ and τ_ρ be the topology generated by this metric. Thus, in the sequal we will consider the topological space

$$(\mathbf{R}^n, \tau_\rho). \qquad (7)$$

According to (Chentsov, 1996, sect. 6.2) introduce asymptotic attainability domains:

$$Att^{(i)} \triangleq \bigcap_{\alpha > 0} cl(G_\alpha^{(i)}, \tau_\rho) \qquad i \in \overline{1,2}.$$

In this case, some point ω^* belongs to $Att^{(i)}$, $i \in \overline{1,2}$, iff (Morina and Chentsov, 1997) there exists a sequence of controls $(f^{(n)})_{n \in \mathcal{N}}$ suct that

1) $\forall \alpha > 0 \quad \exists n_\alpha \in \mathcal{N} \quad \forall n \in \overline{n_\alpha, \infty} : f^{(n)} \in F_\alpha^{(i)}$;

2) $(x_{f^{(n)}}(\vartheta))_{n \in \mathcal{N}} \to \omega^*$.

In other words, points of the sets $Att^{(i)}$, $i \in \overline{1,2}$, are realized as limits of terminal phase vector. In addition, for any $\alpha > 0$ the corresponding sequence of control functions satisfies α-weakened restriction from some instant.

For the description of $Att^{(1)}$, $Att^{(2)}$ consider the corresponding generalized problem in the space of FAM.

Fix $c \in [0, \infty[$. In the capacity of the triplet $(E, \mathcal{L}, \eta)$ we take the interval $T \triangleq [t_0, \vartheta[$ with the semialgebra $\mathcal{T}$ generated by the set of all intervals $[\alpha, \beta[$, $t_0 \leq \alpha \leq \beta \leq \vartheta$, and the measure l determined by the length of intervals. Then for the measure space $(T, \mathcal{T}, l)$ the set $B_0(T, \mathcal{T})$ coincides with $\mathbf{F}([t_0, \vartheta[)$ and the set of all piecewise continuous and right continuous real-valued functions on $[t_0, \vartheta[$ is contained in $B(T, \mathcal{T})$. Under this concretization for any peicewise continuous and right continuous function g on $[t_0, \vartheta[$ the integral determined by the scheme of (Chentsov, 1985; Chentsov, 1986) coincides with the Riemann integral (Chentsov, 1985, sect. 8):

$$\int_T g(t)\, l(dt) = \int_{t_0}^{\vartheta} g(t)\, dt. \qquad (8)$$

In the sequel constructions we mean the equality (8) when use the Riemann integral. Note that for the measure space $(T, \mathcal{T}, l)$ the space $\mathbf{A}_l(\mathcal{T})$ coincides with $\mathbf{A}(\mathcal{T})$; see (Chentsov, 1996, sect.4.4). We will also take into account this fact later on.

Introduce the sets

$$M_c \triangleq \{f \in \mathbf{F}([t_0, \vartheta[) \mid \int_{t_0}^{\vartheta} |f(t)|\, dt \leq c\},$$

$$\Xi[c] \triangleq \{\mu \in \mathbf{A}(\mathcal{T}) \mid V_\mu(T) \leq c\},$$

$$\Sigma_S \triangleq \Big\{\mu \in \mathbf{A}(\mathcal{T}) \mid (\int_T s_1(t)\, \mu(dt), \ldots, \int_T s_m(t)\, \mu(dt)) \in Y\Big\},$$

$$\Omega[c] = \Xi[c] \cap \Sigma_S.$$

By $\tilde{G} \triangleq \{\tilde{x}_\mu(\vartheta) \mid \mu \in \Omega[c]\}$ denote the AD of the generalized problem, where $\tilde{x}_\mu(\cdot)$ is a generalized solution of the system (3), determined $\forall t \in [t_0, \vartheta]$ by the extended Cauchy formula (Serov and Chentsov, 1990; Chentsov, 1992):

$$\tilde{x}_\mu(t) = \Phi(t, t_0) x_0 + \int_{[t_0, t[} \Phi(t, \tau) b(\tau)\, \mu(d\tau). \qquad (9)$$

As it was shown in (Chentsov, 1996, sect. 5.5, 6.4) the following equalities are true:

$$\Omega[c] = \bigcap_{\alpha > 0} cl(\{f * l : f \in F_\alpha^{(i)}\}, \tau_*(\mathcal{T})), \qquad (10)$$

$$Att^{(i)} = \tilde{G}, \quad i \in \overline{1,2}. \qquad (11)$$

4. THE DENSITY PROPERTY

Return to the measure space $(T, \mathcal{T}, l)$. Introduce the set of all uniform partitions of $T = [t_0, \vartheta[$, supposing $\forall s \in \mathcal{N}$

$$\tau_j^{(s)} \triangleq t_0 + j(\vartheta - t_0)/s, \; j \in \overline{0, s}; \qquad (12)$$

besides, let

$$\tilde{B}_0(T) \triangleq \Big\{f \in \mathbf{F}([t_0, \vartheta[) \mid \exists s \in \mathcal{N}\, \exists (\beta_i)_{i \in \overline{1,s}} \in \mathbf{R}^s :$$

$$f = \sum_{i=1}^{s} \beta_i \chi_{[\tau_{i-1}^{(s)}, \tau_i^{(s)}[} \Big\}$$

be the set of uniform step functions corresponding to uniform partitions of T. Here χ is the characteristic function.

Lemma 1. $\forall f \in B_0(T,\mathcal{T})$ $\forall g \in B(T,\mathcal{T})$ $\forall \varepsilon > 0$ $\exists \tilde{f} \in \tilde{B}_0(T)$:

$$\left| \int_T g(t) f(t)\, l(dt) - \int_T g(t) \tilde{f}(t)\, l(dt) \right| < \varepsilon.$$

P r o o f. Choose any $f \in B_0(T,\mathcal{T})$. Then by definition $\exists n_0 \in \mathcal{N}$ $\exists (\xi_i)_{i \in \overline{0,n_0}}$ ($\xi_0 = t_0$, $\xi_{n_0} = \vartheta$, $\xi_{i-1} < \xi_i$, $i \in \overline{1,n_0}$) $\exists (\alpha_i)_{i \in \overline{1,n_0}} \in \mathbf{R}^{n_0}$:

$$f = \sum_{i=1}^{n_0} \alpha_i \chi_{[\xi_{i-1},\xi_i[}. \tag{13}$$

Take any $g \in B(T,\mathcal{T})$ and $\varepsilon > 0$ and choose $k \in \mathcal{N}$ satisfying the condition

$$k > \frac{2\|g\|\,\|f\|\,n_0}{\varepsilon}. \tag{14}$$

To this k (14) the partition $\{[\tau_{i-1}^{(k)}, \tau_i^{(k)}[,\ i \in \overline{1,k}\}$, where $\tau_j^{(k)}$, $j \in \overline{0,k}$, are defined by (12), corresponds. Let $J : \overline{0,n_0} \to \overline{0,k}$ be the mapping such that

$$\xi_i \in [\tau_{J(i)}^{(k)}, \tau_{J(i)+1}^{(k)}[,\ i \in \overline{0,n_0-1},$$

$$J(n_0) = k. \tag{15}$$

Note that the mapping J satisfies the condition:

$$0 = J(0) \le J(1) \le \ldots \le J(n_0-1) \le J(n_0) = k,$$

and

$$\forall j \in \overline{0,k-1}\ \exists i \in \overline{1,n_0} :$$

$$J(i-1) \le j < J(i). \tag{16}$$

Further, suppose $\forall i \in \overline{1,n_0}$:

$$A_i \triangleq [\tau_{J(i-1)}^{(k)}, \tau_{J(i)}^{(k)}[. \tag{17}$$

Since $\tau_{J(i-1)}^{(k)} \le \xi_{i-1}$ and $\tau_{J(i)}^{(k)} \le \xi_i < \tau_{J(i)+1}^{(k)}$, we have

$$[\xi_{i-1},\xi_i[\setminus A_i = [\tau_{J(i)}^{(k)}, \xi_i[\subset [\tau_{J(i)}^{(k)}, \tau_{J(i)+1}^{(k)}[$$

and

$$l([\xi_{i-1},\xi_i[\setminus A_i) \le l([\tau_{J(i)}^{(k)}, \tau_{J(i)+1}^{(k)}[) = 1/k. \tag{18}$$

Introduce the function

$$\tilde{f} \triangleq \sum_{j=1}^{k} \beta_j \chi_{[\tau_{j-1}^{(k)}, \tau_j^{(k)}[},$$

where by (16)

$$\forall j \in \overline{0,k-1} : \beta_j = \alpha_i, \quad \beta_k = \alpha_{n_0}. \tag{19}$$

From (16), (17), (19) it follows that $\forall i \in \overline{1,n_0}$ $\forall t \in A_i \cap [\xi_{i-1},\xi_i[$:

$$f(t) = \tilde{f}(t). \tag{20}$$

By (18)–(20) we have the estimate

$$\left| \int_T g(t) f(t)\, l(dt) - \int_T g(t)\tilde{f}(t)\, l(dt) \right| \le$$

$$\|g\| \int_T |(f(t) - \tilde{f}(t))|\, l(dt) =$$

$$\|g\| \sum_{i=1}^{n_0} \int_{[\xi_{i-1},\xi_i[} |(f(t) - \tilde{f}(t))|\, l(dt) =$$

$$\|g\| \sum_{i=1}^{n_0} \int_{[\xi_{i-1},\xi_i[\cap A_i} |(f(t) - \tilde{f}(t))|\, l(dt) +$$

$$\|g\| \sum_{i=1}^{n_0} \int_{[\xi_{i-1},\xi_i[\setminus A_i} |(f(t) - \tilde{f}(t))|\, l(dt) =$$

$$2\|g\|\|f\| \sum_{i=1}^{n_0} l([\xi_{i-1},\xi_i[\setminus A_i) \le$$

$$\frac{2\|g\|\|f\| n_0}{k} < \varepsilon, \tag{21}$$

which completes the proof.

Note that from (21) and the estimate

$$|\ |f(t)| - |\tilde{f}(t)|\ | \le |\ f(t) - \tilde{f}(t)\ |, \quad t \in T,$$

it follows that

$$\left| \int_T |f(t)|\, l(dt) - \int_T |\tilde{f}(t)|\, l(dt) \right| \le \frac{2\|f\| n_0}{k}. \tag{22}$$

As a corollary of (22), we obtain that $\forall f \in M_c$ $\forall \varepsilon > 0$ $\exists \tilde{f} \in \tilde{B}_0(T)$:

$$\int_T |\tilde{f}(t)|\, l(dt) < c + \varepsilon. \tag{23}$$

Theorem 1. The following equality is true:

$$cl(\{f * l : f \in \tilde{B}_0(T)\}, \tau_*(\mathcal{T})) = \mathbf{A}[\mathcal{T}]. \tag{24}$$

P r o o f. Since $\tilde{B}_0(T) \subset B_0(T,\mathcal{T})$, we have (by theorem 4.3.2. (Chentsov, 1996))

$$cl(\{f * l : f \in \tilde{B}_0(T)\}, \tau_*(\mathcal{T})) \subset cl(\{f * l :$$

$$f \in B_0(T,\mathcal{T})\}, \tau_*(\mathcal{T})) = \mathbf{A}[\mathcal{T}].$$

We will prove that $\mathbf{A}[\mathcal{T}] \subset cl(\{f * l : f \in \tilde{B}_0(T)\}, \tau_*(\mathcal{T}))$. Choose any $\mu \in \mathbf{A}[\mathcal{T}]$. According to (Danford and Schwartz, 1958, ch.I), $\mu \in cl(\{f *$

$l : f \in \tilde{B}_0(T)\}, \tau_*(\mathcal{T}))$, if the intersection of the set $\{f * l : f \in \tilde{B}_0(T)\}$ with each element of the base at the point μ of the space

$$(\mathbf{A}(\mathcal{T}), \tau_*(\mathcal{T})) \quad (25)$$

is not empty. Let Z be an arbitrary element of the base of the space (25) at the point μ. Then according to (Chentsov, 1996, p. 71) $\exists p \in \mathcal{N}\ \exists (g_i)_{i\in\overline{1,p}} \in B(T,\mathcal{T})^p\ \exists \varepsilon > 0$:

$$Z = \Big\{\nu \in \mathbf{A}[\mathcal{T}] \mid \forall i \in \overline{1,p}:$$

$$|\int_T g_i(t)\nu(dt) - \int_T g_i(t)\,\mu(dt)| < \varepsilon\Big\}.$$

Since $\mu \in cl(\{f * l : f \in B_0(T,\mathcal{T}))\}, \tau_*(\mathcal{T}))$, we obtain that $\exists f \in B_0(T,\mathcal{T})\ \forall i \in \overline{1,p}$:

$$|\int_T g_i(t) f(t)\, l(dt) - \int_T g_i(t)\,\mu(dt)| < \varepsilon/2\}.(26)$$

Suppose f is represented in the form (13). Then from lemma 1 (in particular, from the estimate (21) it follows that if

$$k > \max_{i\in\overline{1,p}} \frac{4\|g_i\|\,\|f\|\,n_0}{\varepsilon},$$

then $\tilde{f} \in \tilde{B}_0(T)$ corresponding to the partition $\{[\tau_{j-1}^{(k)}, \tau_j^{(k)}[,\ j \in \overline{1,k}\}$, satisfies for $i \in \overline{1,p}$ the inequalities:

$$\Big|\int_T g_i(t)f(t)\,l(dt) - \int_T g_i(t)\tilde{f}(t)\,l(dt)\Big| <$$

$$\frac{2\|g_i\|\,\|f\|\,n_0}{k} < \varepsilon/2. \quad (27)$$

Thus, from (26), (27) it follows that $\forall i \in \overline{1,p}$:

$$|\int_T g_i(t)\tilde{f}(t)\,l(dt) - \int_T g_i(t)\,\mu(dt)| < \varepsilon,$$

i.e. the point $\tilde{f} * l$ belongs to Z. This means that $\mu \in cl(\{f * l : f \in \tilde{B}_0(T)\}, \tau_*(\mathcal{T}))$. Since $\mu \in \mathbf{A}[\mathcal{T}]$ was choosen arbitrary, we obtain the equality (24).

5. APPROXIMATION OF ASYMPTOTIC ATTAINABILITY DOMAIN

Let $\forall \varepsilon > 0\ \forall A \subset \mathbf{R}^n$ the set $\mathcal{O}_\varepsilon(A)$ be an ε-neighborhood of the set A in the metric space $(\mathbf{R}^n, \rho)$. Denote by $\mathcal{K}$ the class of compact sets in the space (7) and by $\rho_{\mathcal{K}}$ the Hausdorff metric on $\mathcal{K}$, i.e. $\forall A \in \mathcal{K}\ \forall B \in \mathcal{K}$:

$$\rho_{\mathcal{K}}(A,B) \triangleq \inf(\{\varepsilon > 0 : (A \subset \mathcal{O}_\varepsilon(B))\ \&$$

$$(B \subset \mathcal{O}_\varepsilon(A))\}).$$

Let $w : \mathbf{A}(\mathcal{T}) \to \mathbf{R}^n$ and $\psi : \mathbf{A}(\mathcal{T}) \to \mathbf{R}^m$ be the operators such that $\forall \mu \in \mathbf{A}(\mathcal{T})$

$$w(\mu) = \tilde{x}_\mu(\vartheta),$$

$$\psi(\mu) = \Big(\int_T s_1(t)\,\mu(dt), \ldots, \int_T s_m(t)\,\mu(dt)\Big).$$

Besides, introduce the mapping $W : B_0(T,\mathcal{T}) \to \mathbf{R}^n$, supposing $\forall f \in B_0(T,\mathcal{T})$:

$$W(f) \triangleq w(f * l) = x_f(\vartheta) \quad (28)$$

(see (6), (8), (9)). From continuity of the integral with respect to measure it follows that the operators w and ψ are continuous in the space (25). Then Σ_S is closed in (25) as inverse image of a closed set under continuous mapping. From this and the compactness of $\Xi[c]$ we obtain that $\Omega[c]$ is compact in (25). Therefore, $\tilde{G} = w^1(\Omega[c])$ is a compact set in the space (7) and $\forall \varepsilon > 0$ there exists a finite ε-net for the set $\tilde{G}$, i.e. $\exists A \in Fin(\mathbf{R}^n)\ \forall x \in \tilde{G}\ \exists a \in A : \rho(x,a) < \varepsilon$. Fix $\varepsilon > 0$ and take a set $A(\varepsilon)$ which is $\varepsilon/3$-net for the set $\tilde{G}$. Enumerate elements of the set $A(\varepsilon)$, supposing $A(\varepsilon) = (a_i)_{i\in\overline{1,k}} \in (\mathbf{R}^n)^{\overline{1,k}}$, where $k \in \mathcal{N}$. Let

$$\tilde{G}_i(\varepsilon) \triangleq \{x \in \tilde{G} : \rho(x, a_i) < \varepsilon/3\}. \quad (29)$$

Then

$$\tilde{G} = \bigcup_{i=1}^{k} \tilde{G}_i(\varepsilon) \subset \mathcal{O}_{\varepsilon/3}(A(\varepsilon)).$$

Take in each set $\tilde{G}_i(\varepsilon)$ a point $\tilde{x}_i = w(\mu_i)$, $\mu_i \in \Omega[c]$. In this case we have

$$\rho(\tilde{x}_i, a_i) < \varepsilon/3, \quad i \in \overline{1,k}. \quad (30)$$

By (10) and continuity of w we have $\forall \alpha > 0\ \forall j \in \overline{1,2}\ \ \forall i \in \overline{1,k}\ \ \exists f_i \in F_\alpha^{(j)}$:

$$\rho(w(\mu_i), w(f_i * l)) = \rho(w(\mu_i), W(f_i)) < \varepsilon/3.(31)$$

Moreover, from (29) -(31) it follows that $\forall x \in \tilde{G}_i(\varepsilon)$:

$$\rho(x, W(f_i)) \le \rho(x, a_i) + \rho(a_i, w(\mu_i)) +$$

$$\rho(w(\mu_i), W(f_i)) < \varepsilon, \quad i \in \overline{1,k}. \quad (32)$$

By (32) we obtain that $\forall \alpha > 0\ \forall j \in \overline{1,2}\ \exists F_\alpha(\varepsilon) \in Fin(F_\alpha^{(j)})$:

$$\tilde{G} \subset \mathcal{O}_\varepsilon(W^1(F_\alpha(\varepsilon))). \quad (33)$$

Note that $(W^1(F_\alpha(\varepsilon)) \in Fin(\mathbf{R}^n)$ and, therefore, $(W^1(F_\alpha(\varepsilon)) \in \mathcal{K}$. On the other hand, since $F_\alpha(\varepsilon) \subset F_\alpha^{(j)}$, $j \in \overline{1,2}$, then

$$W^1(F_\alpha(\varepsilon)) \subset G_\alpha^{(j)}, \quad (34)$$

and by theorem 6.4.3 (Chentsov, 1996) $\exists \alpha_* > 0\ \forall \alpha \leq \alpha_*$:

$$G_\alpha^{(j)} \subset \mathcal{O}_\varepsilon(\tilde{G}). \quad (35)$$

From this and (33)- (35) we obtain the following statement.

Theorem 2. $\forall \varepsilon > 0\ \exists \alpha_* > 0\ \forall \alpha \leq \alpha_*\ \forall j \in \overline{1,2}\ \exists F_\alpha(\varepsilon) \in Fin(F_\alpha^{(j)})$:

$$\rho_{\mathcal{K}}(\tilde{G}, W^1(F_\alpha(\varepsilon))) < \varepsilon.$$

Consider $\forall \alpha > 0$ the set

$$\tilde{F}_\alpha^{(1)} = F_\alpha^{(1)} \cap \tilde{B}_0(T).$$

From lemma 1 and the relations (23), (28) it follows that $\forall \varepsilon > 0\ \forall \alpha > 0\ \forall f \in F_\alpha^{(1)}\ \exists \tilde{f} \in \tilde{F}_{\alpha+\varepsilon}^{(1)}$:

$$\rho(w(f * l), w(\tilde{f} * l)) < \varepsilon.$$

Then by (11) $\forall \mu \in \Omega[c]\ \forall \varepsilon > 0\ \forall \alpha > 0\ \exists \tilde{f} \in \tilde{F}_\alpha^{(1)}$:

$$\rho(w(\mu), w(\tilde{f} * l)) = \rho(w(\mu), W(\tilde{f})) < \varepsilon. \quad (36)$$

By (36) we obtain the relation analogous to (33). Namely, $\forall \varepsilon > 0\ \forall \alpha > 0\ \exists \tilde{F}_\alpha(\varepsilon) \in Fin(\tilde{F}_\alpha^{(1)})$:

$$\tilde{G} \subset \mathcal{O}_\varepsilon(W^1(\tilde{F}_\alpha(\varepsilon)). \quad (37)$$

Besides, since $\forall \alpha > 0\ \tilde{F}_\alpha(\varepsilon) \subset \tilde{F}_\alpha^{(1)} \subset F_\alpha^{(1)}$, by (35) we obtain that $\forall \varepsilon > 0\ \exists \alpha_* > 0\ \forall \alpha \leq \alpha_*$:

$$W^1(\tilde{F}_\alpha(\varepsilon) \subset \mathcal{O}_\varepsilon(\tilde{G}). \quad (38)$$

From (37), (38) we have

Theorem 3. $\forall \varepsilon > 0\ \exists \alpha_* > 0\ \forall \alpha \leq \alpha_*\ \exists \tilde{F}_\alpha(\varepsilon) \in Fin(\tilde{F}_\alpha^{(1)})$:

$$\rho_{\mathcal{K}}(\tilde{G}, W^1(\tilde{F}_\alpha(\varepsilon))) < \varepsilon.$$

Thus, the asymptotic attainability domain can be approximated by a finite number of step functions and, moreover, of uniform step functions.

6. REFERENCES

Chentsov, A.G. (1985). *Applications of measure theory to control problems.* Sredne ural. Publishing House. Sverdlovsk (in Russian).

Chentsov, A.G. (1986). To the question on universal integrability of bounded functions. *Mat. sbornik* **131**, 73–93.

Chentsov, A.G. (1992). Stability of some nonlinear extremal problems with affects of impulse character. *Avtomatika i telemehanika* (5), pp. 30–41 (in Russian).

Chentsov, A.G. (1996). *Finitely additive measures and relaxation of extremal problems.* Plenum Publishing Corporation. New York, London and Moscow.

Chentsov, A.G. (1997). *Asymptotic Attainability.* Kluwer Academic Publishers. Dordrecht.

Danford, N. and J.T. Schwartz (1958). *Linear operators. Vol.1.* Interscience. New York.

Krasovskii, N.N. (1970). *Game problems on motion encounter.* Nauka. Moscow (in Russian).

Morina, S.I. (1996). On domains of asymptotic attainability in a control problem. *Izvestiya RAN. MTT* (3), 154–159 (in Russian).

Morina, S.I. and A.G. Chentsov (1997). Extension of control problems in the class of finitely additive measures. *Avtomatika i telemehanika* (7), pp. 207–216 (in Russian).

Neveu, J. (1964). *Bases mathematiques du calcul des probabilites.* Masson. Paris.

Serov, V.P. and A.G. Chentsov (1990). On one construction of extension of control problem under integral constraints. *Different. Eqns.* **26**, 442–450.

Warga, J. (1972). *Optimal control of differential and functional equations.* Academic Press. New York.

THE EXTENSION OF NAGUMO THEOREM

Nechaeva O.S.

Udmurt State University, Izhevsk, Russia

Abstract: In this paper the viability problem is interperted as a problem to construct the position control, which keeps trajectories of some dinamical system on the aim set. In viability theory the choice of the control is not made once and for all at some initial time, but the control is constructed at each point of trajectory so as to take into account the present state and the history of the dinamical system. Such control, which called the position control, is the function from phase space to some control set. The special attention is given to the systems with aftereffect because they have many applications in economic dynamics and others. The last part of the paper studies the discrete systems which arise from systems with time lag. *Copyright © 1998 IFAC*

Keywords: Differential equations, control systems, time-optimal control, available time, position control, time lag.

1. INTRODUCTION

Viability problems of control dinamical systems include a big number of actual problems like a construction of the control, which keeps the trajectories of the dinamical system on the preassigned manifold, a stabilizations, a speed control problems, the problems of optimal strategies in cooperative games. A big number of different models in economics, genetics, ecology, cognitive sciences can be consider like a viability problems.

The term is introduced by Aubin J.-P. The first viability theorem was proved by Nagumo (1942) for ordinary differential equations in order to deduce some results about global existence of the solution. The theorem is formulated in term of contingent Bouligand cone. The Haddad's extension of the Nagumo theorem was proved for ordinary differential inclusions. Sufficient conditions for ordinary differential inclusions are given in the work of Fazylov (1997) in term of the function without singularity. The viability theorem for differential inclusions with aftereffect was proved in Aubin monography (1991). This theorem is formulated in the case, then the aim set is given in phase space.

In this paper in first part the abstract viability problem is formulated and the main definitions are given. In second part the properties of the system connected with viability problem are formulated in term of directional derivatives, and these theorems are applied to a systems with aftereffect in the third part. It is shown that systems with aftereffect appear from ordinary system with integral constraints. The fourth part is devoted the extension of Nagumo theorem for systems with aftereffect in the case, when the aim set isn't given in the phase space. The theorem is formulated in term of contingent cone. Because by using the method of steps the system with time lag can be reduced to a discrete system in functional space, the fifth part is devoted the viability problem for discrete systems. These theorems are applied to the systems with time lag. In the last part the discrete systems in $\mathbb{R}^n$ are studied.

2. FORMULATION THE VIABILITY PROBLEM

Let $\mathbb{U}$ be a compact subset of Banach space

and X be a complete separable space (the phase space of dynamical system). For each $\mathrm{u} \in \mathbb{U}$ one-parameter semigroup $f_\mathrm{u}^t : X \to X$, $t \in \mathbb{R}_+$ is given. This semigroup continuously depends on (t, u, x). Let $\mathcal{U}$ be a set of admissible position controls $u\colon X \to \mathbb{U}$ which satisfy the following conditions:

1) if $u \in \mathcal{U}$, then for all $\mathrm{u} \in \mathbb{U}$ and for all bounded closed set $\mathcal{E} \subset X$ the variation of control u

$$x \to \mathrm{u}_\varepsilon(x) = \begin{cases} \mathrm{u}, & \text{if } x \in \mathcal{E}, \\ \mathrm{u}(x), & \text{if } x \notin \mathcal{E}. \end{cases}$$

belong to $\mathcal{U}$;

2) for each admissible control $u \in \mathcal{U}$ the semigroup f_u^t continuously depends on $(t, x) \in \mathbb{R} \times X$.

Let the aim set $K \subset X$ be given.

Definition 1. The set K is called *weakly invariant*, if for all initial state $x_0 \in K$ there exists the admissible control $u^* \in \mathcal{U}$ that the trajectory $\gamma_+(x_0, u^*) \doteq \{f_{u_*}^t x_0,\ 0 \geqslant t < \infty\}$ is contained in K.

Definition 2. The point $x \in \partial K$ is called *the exit point*, if for all $\mathrm{u} \in \mathbb{U}$ and for all $\varepsilon > 0$ there exists $t \in (0, \varepsilon)$ that $f_\mathrm{u}^t x \notin K$. Let $E \subseteq \partial K$ be an *exit point set.*

It is evidently that if $E = \varnothing$, then K is weakly invariant. In other case, if $E \neq \varnothing$, then $\vartheta(x, u)$ denotes the first time when the point $x \in K$ leaves the K under control u, i.e. $f_u^t x \in K$ if $0 \leqslant t < \vartheta(x, u)$ and for all $\varepsilon > 0$ there exists $t_\varepsilon \in [\vartheta(x, u), \vartheta(x, u) + \varepsilon)$ that $f_u^{t_\varepsilon} x \notin K$ (if $f_u^t x$ does not leave K after finite time then $\vartheta(x, u)$ is equal to infinity).

Definition 3. The value

$$\vartheta^*(x) = \sup_{u \in \mathcal{U}} \vartheta(x, u) \tag{1}$$

is called *the sojourn time* in the aim set and, if the sup in (1) is achieved under the control u^*, then the control u^* is called the *optimal sojourn control.*

For all $\vartheta \geqslant 0$ define the set

$$\mathfrak{D}^*(\vartheta) \doteq \{x \in K : \vartheta^*(x) \geqslant \vartheta\}$$

which is called the *ϑ-sojourn set.*

Definition 4. The set $\mathfrak{D}^* \doteq \bigcap_{\vartheta \geqslant 0} \mathfrak{D}^*(\vartheta)$ is called the *sojourn set.*

Definition 5. The set

$$\mathfrak{D}^\heartsuit \doteq \{x \in K:$$
$$\text{there exist } u \in \mathcal{U} \text{ that } f_u^t x \in K \forall t \in [0, \infty]\}$$

is called *the viability set.*

Proposition 1. *The viability set $\mathfrak{D}^\heartsuit$ is weakly invariant.*

Obviously that $\mathfrak{D}^\heartsuit \subseteq \mathfrak{D}^*$. But the contrary statement does not hold true. The viability problem is to describe the sets $\mathfrak{D}^*(\vartheta)$, $\mathfrak{D}^*$, $\mathfrak{D}^\heartsuit$ and to construct the function $\vartheta^* : K \to \mathbb{R}_+$ and the optimal sojourn control $u^* : K \to \mathbb{U}$, if it exists.

3. PROPERTIES OF THE SYSTEMS CONNECTED WITH VIABILITY

Let $a : X \to \mathbb{R}$ be a continuous function and f_u^t is the flow under control $u \in \mathcal{U}$. Then the directional right-hand derivative $\dot{a}(x, u)$ of the function a at x on the flow f_u^t is defined $\dot{a}(x, u) = \lim_{t \to +0} t^{-1}(a(f_u^t x) - a(x))$.

The aim set K is the subset of phase space. Suppose that $K = \{x \in X : a_i(x) \leqslant 0, i = 1, \ldots k\}$, where $a_i : X \to \mathbb{R}^1$, $i = 1, \ldots, k$, are continuous functions.

Theorem 1. *If for all $x \in \partial K$ and for all $i \in I_x \doteq \{i : a_i(x) = 0\}$, there exists $\mathrm{u}(x) \in \{\mathrm{u}_i(x) : i \in I_x\}$ that $\dot{a}_i(x, \mathrm{u}(x)) < 0$ and functions $\mathrm{u}_i : \partial K \to \mathbb{U}$ obey the minimum condition*

$$\dot{a}_i(x, \mathrm{u}_i(x)) = \min_{\mathrm{u} \in \mathbb{U}} \dot{a}_i(x, \mathrm{u}), \tag{2}$$

then $K = \mathfrak{D}^\heartsuit$.

If there exists the index i and the point $x \in \partial K$ that inequality $\min_{\mathrm{u} \in \mathbb{U}} \dot{a}_i(x, \mathrm{u}) > 0$ holds true, then $E \neq \varnothing$ and $x \in E$. For each $x \in E$ there exists the index $i(x)$ that $\min_{\mathrm{u} \in \mathbb{U}} \dot{a}_i(x, \mathrm{u}) \geqslant 0$.

The following theorem shows that the analog of Bellman equation is realized for sojourn time.

Theorem 2. *Let the optimal sojourn control u^* exist. Then the function $x \to \vartheta^*(x)$ obey equation*

$$min_{u \in \mathcal{U}} \dot{\vartheta}(x, u) = -1, \quad x \in \mathrm{int}(K \setminus \mathfrak{D}^\heartsuit),$$
$$\vartheta^*(x)|_{x \in E} = 0. \tag{3}$$

4. SYSTEMS WITH AFTEREFFECT

The linear control system of differential equation in $\mathbb{R}^2$

$$\begin{cases} \dot{x} = y \\ \dot{y} = u, \end{cases} \qquad \mathbb{U} = [0, 1] \tag{4}$$

with integral constraint

$$\int_{t-1}^{t} x^2(s) ds + y^2(t) \leqslant 1$$

is given.

A change the variables in order to get rid the integral constraint

$$z = \int_{t-1}^{t} x^2(s) ds$$

reduces the system (4)

$$\begin{cases} \dot{x}(t) = y(t) \\ \dot{y}(t) = u \\ \dot{z}(t) = x^2(t) - x^2(t-1) \end{cases}$$

to a new system with aftereffect and with geometric constraint

$$z(t) + y^2(t) \leqslant 1$$

in $\mathbb{R}^3$.

So, the viability problems for systems with aftereffect appear when one try to solve the ordinary differential systems with integral constraints.

Let $r \geqslant 0$ be a some real number. Consider Cauchy problem associated to the differential autonomy control system with aftereffect

$$\dot{x}(t) = F(x_t, u(x_t)), \quad u : \mathfrak{S} \to \mathbb{U} \qquad (5)$$

satysfying the initial condition

$$x_t(\cdot)|_{t=0} = \sigma, \qquad (6)$$

where $x_t(\tau) = x(t+\tau)$, $\mathfrak{S} \doteq C([-r,0],X)$, $F : \mathfrak{S} \times \mathbb{U} \to X$, $\sigma \in \mathfrak{S}$.

The phase space of differential system with aftereffect (5) is the functional space $\mathfrak{S}$. The aim set in viability problem associated with system (5) shoud be given in phase space $\mathfrak{S}$ or in Banach space X.

Consider the first case. Let $\alpha_i : \mathbb{R} \times \mathfrak{S} \to \mathbb{R}$, $\beta_i : X \to \mathbb{R}$, where $i \in \mathbb{I} \doteq \{1, \ldots, s\}$. The scalar functions $a_i(\sigma) = \beta_i(\sigma(0)) + \int_{-r}^{0} \alpha_i(s, \sigma(s))ds$ are defined and the aim set is defined

$$\Sigma_0 = \{\sigma \in \mathfrak{S} : \quad a_i(\sigma) \leqslant 0,\ i \in I\}.$$

Using the theorem 1 and 2 in this case, the following results are obtained.

Theorem 3. *Suppose that for all $k \in I$ there is a solution σ_k of the equation $a_k(\sigma) = 0$ that $a_i(\sigma_k) \leqslant 0$ for all $t \geqslant 0$, $i \in I$; S_k is the set of such solutions. Let $I_k(\sigma)$ be the set of index from I that each $\sigma \in S_k$ is the solution of the some equation $a_i(\sigma) = 0$, $i \in I_k(\sigma)$. For all $k \in I$, for all $\sigma \in S_k$ and for all $i \in I_k(\sigma)$ choose the control $\mathrm{u}_i^k(\sigma)$ satisfying the following condition*

$$\min_{\mathrm{u}\in\mathbb{U}} \dot{a}_i(\sigma, \mathrm{u}) = \dot{a}_i(\sigma, \mathrm{u}_i^k(\sigma)),$$

$$\dot{a}_i(\sigma, \mathrm{u}) = \langle \frac{\partial \beta_i(x)}{\partial x}|_{x=\sigma(0)}, F(\sigma, \mathrm{u})\rangle + \\ + \int_{-r}^{0} \langle \frac{\partial \alpha_i(t,x)}{\partial x}|_{x=\sigma(t)}, d\,\sigma(t)\rangle$$

and the control $\mathrm{u}^k(\sigma) \in \{\mathrm{u}_i^k(\sigma) : i \in I_k(\sigma)\}$ that for all $i \in I_k(\sigma)$ the inequality $\dot{a}_k(\sigma, \mathrm{u}^k(\sigma)) \geqslant \dot{a}_i(\sigma, \mathrm{u}_i^k(\sigma))$ holds true. The control $\mathrm{u} : \partial\Sigma_0 \to \mathbb{U}$ is defined by the equality $\mathrm{u}(\sigma) = \mathrm{u}^k(\sigma)$ for all $\sigma \in S_k$. Then if for all $\sigma \in \partial\Sigma_0$ and for all $i \in I(\sigma) = \{i \in I : a_i(\sigma) = 0\}$ the inequalities $\dot{a}_i(\sigma, \mathrm{u}(\sigma)) < 0$ hold true, then Σ_0 is not empty and is weakly invariant

If for some $k \in I$ there exists the solution σ_k of the equation $a_k(\sigma) = 0$ that the inequality $a_i(\sigma_k) \leqslant 0$ holds true for all $i \in I$ and the inequality

$$\min_{\mathrm{u}\in\mathbb{U}} \dot{a}(\sigma_k, \mathrm{u}) > 0,$$

iz realized, then the exit point set E is not empty.

Theorem 4. (estimate of sojourn time). *For all absolutely continuous solution $\sigma \in \mathfrak{S}$ of inequality*

$$\beta(\sigma(0)) + \int_{-r}^{0} \alpha(t, \sigma(t))\,dt \leqslant 0$$

choose the control $\sigma \to \mathrm{u}(\sigma)$ which satisfies the equality

$$\langle \partial\beta(\sigma(0))/\partial x, F(\sigma, \mathrm{u}(\sigma))\rangle = \\ = \min_{\mathrm{u}\in\mathrm{U}} \langle \partial\beta(\sigma(0))/\partial x, F(\sigma, \mathrm{u})\rangle.$$

Let $t \to x_t(\cdot, \sigma)$ be the solution of the Cauchy problem

$$\dot{x}(t) = F(x_t, \mathrm{u}(x_t)), \quad x_t|_{t=0} = \sigma(\cdot).$$

Let $\theta(\sigma)$ be a least of $t \geqslant 0$ which satisfies the equality

$$\beta(x_t(0,\sigma)) + \int_{-r}^{0} \alpha(s, x_t(s,\sigma)\,ds = 0,$$

then $\vartheta^*(\sigma) \geqslant \theta(\sigma)$ (if $\theta(\sigma) = \infty$, then $\vartheta^*(\sigma) = \infty$).

5. EXTENSION OF NAGUMO THEOREM

Consider the second case, when the aim set is the subset of X.

Definition 6. The aim set $K \subset X$ satisfies *the local viability property*, if for all initial function $\sigma \in \mathfrak{S}$ such that $\sigma(0) \in K$, there exists $T > 0$, the admissible control $\widehat{u} \in \mathcal{U}$ and the solution $x(\cdot)$ of system (5), under control $\widehat{u}$ starting at σ, that for all $t \in [0,T]$ $x(t) \in K$.

Let x belong to the aim set K.

Definition 7. *The contingent cone to K at x* is the set

$$T_K(x) = \{v \in X \mid \liminf_{h\to 0+} \frac{d_K(x+hv)}{h} = 0\},$$

where $d_K(y)$ denotes the distance of y to K.

Obviously that

$$\forall x \in \operatorname{int}(K) \quad T_K(x) = X.$$

The contingent cone is the extension of classic tangent cone. The following lemma describes the contingent cone in functional space in term of contingent cone in X.

Lemma 1. *Let σ' belong to $\mathfrak{S}_K \doteq \{\sigma \in \mathfrak{S} \mid \sigma(0) \in K\}$. Then*

$$T_{\mathfrak{S}_K}(\sigma') = \{\sigma \in \mathfrak{S} \mid \sigma(0) \in T_K(\sigma'(0))\}.$$

Theorem 5. (the extension of Nagumo theorem). *Let the aim set K be locally compact, and the function F be continuous from $\mathfrak{S} \times \mathbb{U}$ to X. Then K satisies the local viability property if and only if for all $\sigma \in \mathfrak{S}_K$ there exists $\mathrm{u}(\sigma) \in \mathbb{U}$, that $F(\sigma, \mathrm{u}) \in T_K(\sigma(o))$.*

The proof of this theorem is a modification of the Euler method of approximating the solution by piecewise linear functions in order to force the solution to remain in K. The main idea is concluded in following.

The system (5) defines the curve $\xi(\cdot)$ in the phase space $\mathfrak{S}$ by the rule $\xi(t) = x_t$. The system (5) is interpreted as an equation

$$\dot{\xi}(t)(0) = F(\xi(t), \mathrm{u}(\xi(t))). \tag{7}$$

Let $K_0 \doteq B_K(\sigma_0(0), \varepsilon)$, where K_0 is the compact neighborhood of $\sigma_0(0) \in K$. Let $\sigma_0 \in \mathfrak{S}$ be an initial state, $C \doteq B(F(\sigma_0), 1)$. C is bounded.

$$T \doteq \frac{\varepsilon}{||C||}, \quad ||C|| = \max_{x \in C} ||x||.$$

$$\mathfrak{S}_0 \doteq \{\sigma \in C^1([-r, 0], X) \colon ||\sigma - \sigma_0||_C \leqslant r, \; |\dot{\sigma}(t)| \leqslant M \text{ for some } M > 0 \quad \forall t \in [-r, 0]\}.$$

The set $\mathfrak{S}_0$ is compact in uniform topology in $\mathfrak{S}$.

Lemma 2. *For all $m \in \mathbb{N}$ there exists $\theta_m \in (0, m^{-1})$ that for all $\sigma \in \mathfrak{S}_K$ there exists $h > \theta_m$ and $\pi \in \mathfrak{S}$ that the conditions*

1) $\forall x \in [-r, 0]$ $\pi(x) \in C$;

2) $\sigma + h\pi \in \mathfrak{S}$;

3) $(\sigma, \pi(0)) \in B(Graph(F), \frac{1}{m})$

hold true.

Choose the initial state $\sigma_0 \in \mathfrak{S}_K$ and for all $m \in \mathbb{N}$ construct three sequences $\{h_j^m\}$, $\{\sigma_j^m\}$, $\{\pi_j^m\}$ by lemma 2, which obey the following conditions

1) $\sigma_{j+1}^m = \sigma_j^m + h_j^m \pi_j^m \in \mathfrak{S}_0$;

2) $(\sigma_j^m, \pi_j^m(0)) \in B(Graph(F), m^{-1})$ as long as $\sum_{i=0}^{j-1} h_i^m \leqslant T$. There exists J^m such that $h_1^m + \ldots + h_{J-1}^m \leqslant T \leqslant h_1^m + \ldots + h_J^m$. Let $r_j^m \doteq h_0^m + \ldots + h_{j-1}^m$, $j = 1, \ldots, J+1$. Define the piece-linear function $\xi \colon \mathbb{R} \to \mathfrak{S}$ by the rule

$$\forall t \in [r_j^m, r_{j+1}^m) \quad \xi^m(t) = \sigma_j^m(\cdot) + (t - r_j^m)\pi_j^m.$$

It is easy to prove these this piece-linear functions obey the conditions

1) $\forall t \in [0, T]$ $\xi^m(t) \in B(\mathfrak{S}_0, \varepsilon_m)$;

2) $\forall t \in [0, T]$ $(\xi^m(t), \xi^m(t)(0)) \in B(Graph(F), \varepsilon_m)$.

Because the sequence $\{\xi(\cdot)\}$ is equicontinuous and uniformly bounded, then there exists the subsequence which converges to some function $\xi(\cdot)$. By using condition 1) and 2) it is possible to prove that the limit function is the solution of the system (5).

6. VIABILITY PROBLEM FOR DISCRETE SYSTEMS

If the system (5) has a time lag (i.e. for functions $\sigma_1, \sigma_2 \in \mathfrak{S}$ such a $\sigma_1(\cdot)|_{[-r,-\delta]} \equiv \sigma_2(\cdot)|_{[-r,-\delta]}$) the equality $F(\sigma_1, \mathrm{u}) = F(\sigma_2, \mathrm{u})$ holds true for all $\mathrm{u} \in \mathbb{U}$) then it is possible by using the method of steps (Ehlsgolts, Norkin, 1971) to reduce the system (5) to a discrete system in functional space. Therefore it is important to study the discrete systems. Furthermore, the discrete systems are used in economic (especially bank) simulation (Andrianov, Polushkina, 1997).

Let X, W be a Banach space, U be a convex compact set in W. Consider the control discrete system with aftereffect

$$x_{n+1} = f(x_{n-\varrho}, \ldots, x_{n-1}, x_n, \mathrm{u}_n), \tag{8}$$

where $x_i \in X$, $\mathrm{u}_n \in \mathrm{U}$, $f \colon X^{\varrho+1} \times \mathrm{U} \to X$ is continuous. Let G be a given aim set.

Since the control in such system is the sequence of elements from U, then in capacity of the control set it is naturally to take the product $\mathcal{U} \doteq \mathrm{U}^{\aleph_0}$ of countable number of U. By the Tihonov theorem this set is compact in product topology.

When the control $\mathbf{u} = \{\mathrm{u}_n\}_{n=1}^{\infty} \in \mathcal{U}$ and the initial state $\mathbf{x}_0 = (x_{-\varrho}, \ldots, x_0) \in X^{\varrho+1}$ are given, then the sequence $\mathbf{x}(\mathbf{x}_0, \mathbf{u}) = \{x_n\}_{n=0}^{\infty}$, which called the trajectory of the system (8), is defined identically.

Definition 8. The trajectory $\mathbf{x}(\mathbf{x}_0, \mathbf{u})$ of the system (8) is called *viablity*, if $x_n \in G$ for all $n \geqslant 0$.

It is possible that for some initial state $\mathbf{x}_0$ the viability trajectory does not exist, but for all $n \in \mathbb{N}$ there exists the control $\mathbf{u}(n)$ such that the trajectory with initial state $\mathbf{x}_0$ under control $\mathbf{u}(n)$ is contained in the aim set at least n steps.

The sojourn set $\mathfrak{D}^*$, the viability set $\mathfrak{D}^{\heartsuit}$, the sojourn time and the optimal sojourn control are defined the same as in continuous case (definition (1),(4) and (5)). The main theorem is the following.

Theorem 6. *If in system (8) the function f is continuous and the aim set G is closed, then $D^{\heartsuit} = D^*$.*

Let $\mathbf{G} \doteq G^{\varrho+1}$. The next results follow from theorem 6.

Lemma 3. *Let the conditions of theorem 6 be realized. Let $\vartheta^*(\mathbf{x}_0, \mathbf{u})$ be an exit time of trajectory with initial state $\mathbf{x}_0$ under control $\mathbf{u} \in \mathcal{U}$ from* **G**. *Then there exists the neighborhood $\mathcal{O}(\mathbf{u})$ that for all $\hat{\mathbf{u}} \in \mathcal{O}(\mathbf{u})$ the inequality $\vartheta^*(\mathbf{x}_0, \hat{\mathbf{u}}) \leqslant \vartheta^*(\mathbf{x}_0, \mathbf{u})$ holds true.*

Theorem 7. *Let the conditions of theorem 6 be realized. If for initial point $\mathbf{x}_0 \in \mathbf{G}$ there does not exist the viability trajectory, then exists the number $N = N(\mathbf{x}_0)$ and optimal sojourn control $\hat{\mathbf{u}} \in \mathcal{U}$, that the corresponding trajectory leaves the aim set on step N, and others trajectories (with the same initial state and under other admissible control) leave the aim set not later then step N.*

Theorem 8. (the continuous dependence on the initial state and the control). *Let in system (8) the function f be continuous, $\mathbf{x}_0$ be a some initial state, $\mathbf{u}$ be a some admissible control. Then for each preassigned N and for all $\varepsilon > 0$ there exists $\delta_0 > 0$ and $\delta > 0$ that for all $\mathbf{x}'_0$, which satisfy the inequality $|\mathbf{x}'_0 - \mathbf{x}_0| < \delta_0$, and for all admissible controls $\mathbf{u}'$, which satisfy the condition $|\mathbf{u}' - \mathbf{u}| < \delta$, the inequality $|\mathbf{x}'_N - \mathbf{x}_N| < \varepsilon$ holds true, where $\mathbf{x}_N = \{x_k\}_{k=1}^N$ is the interval of trajectory with initial state $\mathbf{x}_0$ and under control* **u**.

If the system (5) has a time lag, it is possible to reduce the system to a discrete system in functional space. Using the theorem 6, the following result is obtained.

Theorem 9. *If the aim set $K \subset X$ is closed and the function F is continuous, then $\mathfrak{D}^* = \mathfrak{D}^\heartsuit$.*

7. DISCRETE SYSTEMS IN $\mathbb{R}^n$

Consider the discrete control system

$$x_{k+1} = x_k + \varepsilon v(x_k, \mathrm{u}(x_k)), \quad x_k \in \mathbb{R}^n, \quad v\colon \mathbb{R}^n \times \mathrm{U} \to \mathbb{R}^n, \quad \mathrm{u}\colon \mathbb{R}^n \to \mathrm{U}, \tag{9}$$

and the aim set $G = \{x \in \mathbb{R}^n : g(x) \leqslant 0\}$, where U is the compact set in $\mathbb{R}^m$, $g\colon \mathbb{R}^n \to \mathbb{R}$ is continuous, $\varepsilon > 0$.

Consider the boundary ε-layer

$$G_\varepsilon \doteq \{x \in \mathbb{R}^n : \varrho(x, \partial G) < \varepsilon\}.$$

The following propositions give some sufficient conditions which guarantee that the aim set be a weakly invariant.

Proposition 2. Let g be a convex function, $g \in C^2(\mathbb{R}^n)$, $|v(x, \mathrm{u})| \leqslant 1$ for all $(x, \mathrm{u}) \in G \times \mathrm{U}$ and there exist $\delta > 0$ that for x_1, x_2 from G, $|x_1 - x_2| < \varepsilon$, the inequality $|\operatorname{grad} g(x_1) - \operatorname{grad} g(x_2)| < \delta$ holds true. If inequality

$$\min_{\mathrm{u}} \langle v(x, \mathrm{u}), \operatorname{grad} g(x) \rangle < -\delta, \quad \mathrm{u} \in \mathrm{U},$$

is realized for all $x \in G_\varepsilon$, then the aim set G is weakly invariant.

Proposition 3. Let g be a convex function, $g \in C^2(\mathbb{R}^n)$ and for all $(x, \mathrm{u}) \in G \times \mathrm{U}$ the inequality $|v(x, \mathrm{u})| \leqslant 1$ holds true. Then if the inequality

$$\min_{\mathrm{u}} \langle v(x, \mathrm{u}), \operatorname{grad} g(x) \rangle \leqslant -\varepsilon M, \quad \mathrm{u} \in \mathrm{U},$$

$$M = \sqrt{m_1^2 + \cdots + m_n^2},$$

$$m_i = \max_{x \in G_\varepsilon} |\operatorname{grad} \frac{\partial g(x)}{\partial x_i}|$$

is realized for all $x \in G_\varepsilon$, then the aim set G is weakly invariant.

Proposition 4. Let $n = 2$ and for curvature $k(x)$ of the curve ∂G the inequality $k(x) \leqslant \varepsilon^{-1}$ holds true. If for all $x \in G_\varepsilon$ there exist $\mathrm{u}(x) \in \mathrm{U}$, that for each point couple $x_1, x_2 \in \partial G$, which satisfy the equality $|x_1 - x| = \varepsilon$, $|x_2 - x| = \varepsilon$, the inequality

$$\begin{aligned} &\langle x_1 + x_2, v(x, \mathrm{u}(x)) \rangle - 2\langle x, v(x, \mathrm{u}(x)) \rangle \leqslant \\ &\leqslant \frac{|v(x, \mathrm{u}(x))|}{\varepsilon} (\langle x_1, x_2 \rangle + |x_1|^2 + \\ &+ 2|x|^2 - 3\langle x, x_1 \rangle - \langle x, x_2 \rangle), \end{aligned} \tag{10}$$

holds true, then the aim set G is weakly invariant.

Remark 1. It is possible to show that in proposition 4 there exists precisely two point $x_1, x_2 \in \partial G$, which satisfy the equalities $|x_1 - x| = |x_2 - x| = \varepsilon$.

Remark 2. It is easy to check that the inequality (10) is symmetric on the point couple x_1 and x_2.

Theorem 10. *Consider the discrete system in* $\mathbb{R}^n$

$$x_{k+1} = x_k + v(x_k, \mathrm{u}(x_k)) \tag{11}$$

and the convex aim set $G \doteq \{x : g(x) \leqslant 0\}$, the boundary of this set is the smooth $(n-1)$-dimensional manifold $\partial G \doteq \{x : g(x) = 0\}$. Let $k(x)$ be a normal section curvature of the manifold ∂G at the point x. Suppose that $0 \leqslant k(x) \leqslant K$ for all $x \in \partial G$ and for some constant $K > 0$. Further, for all $(x, \mathrm{u}) \in G \times \mathrm{U}$ the inequality $|v(x, \mathrm{u})| \leqslant K^{-1}$ holds true and for all $x \in G_K \doteq \{x \in G : \varrho(x, \partial G) < K^{-1}\}$ there exists such $\mathrm{u}(x) \in \mathrm{U}$, that

$$\begin{aligned} &K|v(x, \mathrm{u}(x))|^2 + \\ &+ 2|v(x, \mathrm{u}(x))|(1 - \lambda(x)K) \cos \alpha(x) \leqslant \\ &\leqslant (2 - \lambda(x)K)\lambda(x), \end{aligned}$$

where $\lambda(x) = \varrho(x, \partial G)$, $\alpha(x)$ is the angle between $v(x, \mathrm{u}(x))$ and the normal to ∂G, which is costructed at the point x. Then the aim set G is weakly invariant.

8. ACKNOWLEDGEMENT

The work is supported by Russian Foundation of Basic Reseach (grant 97–01–00413) and by Competition Centre of Fundamental Natural Science (grant 97–0–1.9). The author thanks Professor Tonkov for formulation of the problem and essential help in the proving some results and the preparation of the paper.

REFERENCES

Andrianov, D.L. and G.L. Polushkina (1997). Forecast — analisys — solution. *Bankovskie tehnologii*, **8**, 54–57 (in Russian).

Andrianov, D.L. and G.L. Poluskina (1997). The solution of analitic center "Prognoz". The simulation of financail bank activity. *Banki i tehnologii*, **3**, 88–90 (in Russian).

Aubin, J.-P. (1990). A survey of viability theory. *SIAM J. Contr. and Optim.* **28**, **4**, 749–788.

Aubin, J.-P. (1991). *Viability theory.* Birkhauser, Boston.

Dubrovin, B.A., Novikov, S.P. and A.T. Fomenko (1979). *The modern geometry.* Nauka, Moscow (in Russian).

Ehlsgolts, L.Eh. and S.B. Norkin (1971). *Introduction in theory of differential equations with time lag.* Nauka, Moscow (in Russian).

Hale, G. (1984). *Theory of functional–differential equations.* Mir, Moscow (in Russian).

Nechaeva O.S. and E.L. Tonkov. (1998). Viability problems for dinamical systems. *Works of international conference "Mathematical Modelling Application for Problem in Science and Engineering".* Izhevsk (in Russian).

Tonkov, E.L. (1997). Dinamical viability problems. *Vestnik Permsk. gos. teh. un-t. Funct.-differ. uravnen. (spec. vip.),* **4**, 138–148 (in Russian).

Fillipov, A.F. (1985). *The differential equations with discontinuous right part.* Nauka, Moscow. (in Russian)

DIFFERENTIABILITY OF SPEED FUNCTION AND FEEDBACK CONTROL OF LINEAR NONSTATIONARY SYSTEM

Nickolayev S.F., Tonkov E.L.

Udmurt State University, Izhevsk, Russia

Abstract: This talk is dealing with the controllability sets structure, differentiability of the speed function and feedback control of linear nonstationary systems. *Copyright © 1998 IFAC*

Keywords: feedback control, speed control, controllability, non-stationary systems, linear systems.

1. INTRODUCTION

The problem of synthesis of positional speed control $u(t,x)$ (or feedback control) for dynamic system described by the equation

$$\dot{x} = v(t,x,u), \quad (t,x,u) \in \mathbb{R}^{1+n} \times U,$$

(U is a compactum in $\mathbb{R}^m$) remains one of difficult and insufficiently explored in the optimal control theory. The main reason obstructing to synthesize the positional speed control is absence of stability of the equation

$$\dot{x} = v(t,x,u(t,x)), \quad (t,x) \in \mathbb{R}^{1+n},$$

with respect to changing of the function $u(t,x)$ in sets of zero Lebeg measure in $\mathbb{R}^{1+n}$.

Apparently V.G. Boltyanskiy was the first who observes this anomalous behavior of the equation (1), but the simple example of such system (that is linear relative to phase coordinates and control function on the plane) was built in (Brunovski,1980a). Let us to give slightly modified example of P. Brunovski. Let examine the problem of end point control to the origin of coordinates with the system

$$\dot{x}_1 = -x_1 + u_1, \quad \dot{x}_2 = x_2 + u_2, \qquad (1)$$

where $u \in U \doteq \{u = (u_1, u_2)\colon\ |u_1 + u_2| \leqslant 1\}$. It's easy to make sure that the controllability set of the system (1) is the stripe

$$D = \{x = (x_1, x_2) \in \mathbb{R}^2\colon\ x_1 \in \mathbb{R},\ |x_2| < 1\},$$

and the optimal speed control is defined by the next equality

$$\widehat{u}(x_1,x_2) = \begin{cases} (0,0) \text{ if } & x_1 = x_2 = 0, \\ (+1,0) \text{ if } & x_1 < 0,\ x_2 = 0, \\ (-1,0) \text{ if } & 0 < x_1,\ x_2 = 0, \\ (0,-1) \text{ if } & 0 < x_2 < 1, \\ (0,+1) \text{ if } & -1 < x_2 < 0. \end{cases} \qquad (2)$$

The system (1) closed with the control function (2)

$$\begin{cases} \dot{x}_1 = -x_1 + \widehat{u}_1(x_1,x_2), \\ \dot{x}_2 = x_2 + \widehat{u}_1(x_1,x_2), \end{cases} \qquad (3)$$

has the next properties.

1°. Let $x(t, t_0, x^0)$ be the solution (defined by Caratheodory) of the system (3), $u(t, t_0, x^0) = \widehat{u}(x(t, t_0, x^0))$, then $u(t, t_0, x^0)$ is the optimal speed control of the system (1) for the point $(t_0, x^0) \in D$.

2°. Every non-trivial solution (defined in (Filippov, 1985)) of the system (3) started from D exponentially tends to zero when $t \to \infty$ but not achieves it for finite time.

This behavior takes place because the solutions of the system (3) defined by Caratheodory started from horizontal axis are solutions of the system $\dot{x}_1 = -x_1 - 1$, $\dot{x}_2 = x_2$ (if $x_1^0 > 0$), but the solutions defined by A.F. Filippov are solutions of the system $\dot{x}_1 = -x_1$, $\dot{x}_2 = x_2$ (if $x_1 > 0$). Therefore in this example there is no stability to perturbations of the positional speed control on zero measure sets. On figure 1 are shown velocity vectors v_C and v_F of the solutions defined by Caratheodory

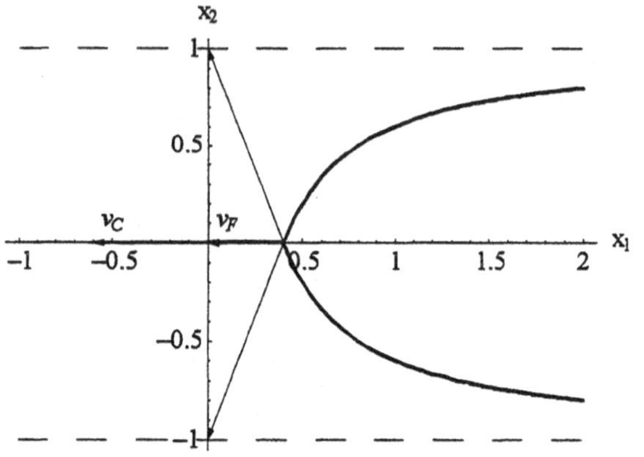

Fig. 1. Velocity vectors v_C and v_F of the solutions of the system (3) defined by Caratheodory and A.F. Filippov respectively

and A.F. Filippov respectively. It should be noted that for the system (1) for every $\varepsilon > 0$ there exists ε-optimal positional speed control $u_\varepsilon(x)$, that is stable to perturbations of closed system on zero measure sets.

This talk is devoted to description of class of linear non-stationary systems with one-dimensional input that have stable positional speed control. These systems are named *subcritical systems.* For the subcritical systems are established statements about extended controllability set structure, differentiability of speed function and switching surfaces of the positional speed control.

Form the last works concerned with this themes should be noted the works of F.L. Chernousko and his learners (Chernousko, Shmatkov, 1997), (Akoolenko, Shmatkov, 1998) and the works of E.G. Albrekht and his learners (Albrekht, Ermolenko, 1997).

2. SPEED FUNCTION AND CONTROLLABILITY SET

The subject of investigations is the controllability sets structure and differentiability of the speed function $(t, x) \to \tau_n(t, x)$ of linear nonstationary system

$$\dot{x} = A(t)x + b(t)u, \tag{4}$$

$x \in \mathbb{R}^n$, $n \geqslant 2$, $|u| \leqslant 1$, where u is scalar control function, and the function $t \to (A(t), b(t))$ belongs to the class C^r, $r \geqslant 0$.

Some common designations are used below. $\mathbb{R}^n$ is euclidean space of dimension n with norm operator $|x| = \sqrt{x^* x}$. By Greek letters are designated row vectors and by Latin letters are column ones. Operator $*$ is transposition. $\operatorname{int} D$ is interior of set D relatively to space $\mathbb{R}^n$ and $\operatorname{cl} D$ is closure of set D in $\mathbb{R}^n$. By *support function* $\xi \to \mathrm{c}(\xi, D)$ of set D is defined the function

$$\mathrm{c}(\xi, D) = \sup\{\xi x \colon x \in D\}.$$

The *speed function* $(t, x) \to \tau_n(t, x)$ of the system (4) is defined by the equality

$$\tau_n(t_0, x_0) = \min_{u(\cdot)\in\mathcal{U}} \{\vartheta \geqslant 0 \colon x((t_0 + \vartheta, t_0, x_0, u(\cdot)) = 0\},$$

where $\mathcal{U}$ is a join of measurable functions with values in $[-1, 1]$ and $x((t, t_0, x_0, u(\cdot))$ is the solution of the system (4) with the control function $u = u(t)$ and starting point $x(t_0) = x_0$. *Controllability set of the system* (4) *on interval* $[t_0, t_0 + \vartheta]$ is defined by the equality

$$D_\vartheta(t_0) \doteq \{x \in \mathbb{R}^n \colon \tau_n(t_0, x) \leqslant \vartheta\},$$

and *controllability set* of the system (4) is

$$D(t_0) \doteq \bigcup_{\vartheta \geqslant 0} D_\vartheta(t_0).$$

For the controllability sets the next equality is true (Rodionova, Tonkov, 1993):

$$D_\vartheta(t_0) = - \int_{t_0}^{t_0+\vartheta} X(t_0, t) b(t) U dt, \tag{5}$$

where $U = [-1, 1]$, $X(t, s)$ is the Caushi matrix of system $\dot{x} = A(t)x$ and integration is defined by Lyapunov (Ioffe, Tikhomirov, 1974). The system (4) is *differential controllable in point* t_0 if for every $\vartheta > 0$ the inclusion $0 \in \operatorname{int} D_\vartheta(t_0)$ holds. The system (4) is *differential controllable in interval* $J \subset \mathbb{R}$ if it is differential controllable in every point of J.

3. SUBCRITICALLITY

Let $\psi_1(t), \ldots, \psi_n(t)$ be a fundamental sequence of solutions of the conjugate system

$$\dot{\psi} = -\psi A(t), \tag{6}$$

and let $\sigma(t)$ be the least upper bound of those $\sigma > 0$ where on half-open interval $[t, t + \sigma)$ the system of functions

$$\xi_1(t) \doteq \psi_1(t) b(t), \ldots, \xi_n(t) \doteq \psi_n(t) b(t) \tag{7}$$

constitutes the Tchebyshev system (T-system). It means that every non-trivial linear combination of the functions (7) has on $[t_0, t_0+\sigma)$ not greater than $n - 1$ zeroes that are geometric distinguishable.

Definition 1. The system (4) is called *subcritical* on interval J if $\sigma(t) > 0$ for every $t \in J$.

All theorems are proved for subcritical systems (4).

Lemma 1. *If the system* (4) *is subcritical on interval* J *that it is differential controllable on* J.

Example 1. Let us consider the system

$$\begin{cases} \dot{x}_1 = a_1(t)x_1 + a_2(t)x_3 + u \\ \dot{x}_2 = x_1 \\ \dot{x}_3 = x_1 + a_3(t)x_3, \end{cases} \tag{8}$$

that describes airplane dynamics in linear approximation (Bodner, 1964). Here x_2 is the pitch angle, x_3 is the attack angle and u is the elevator angle. On figure (2) is shown parametric dependence of the function $\sigma(0)$ for the system (8) where $a_1 = 1,\ a_3 = 1$ and $-0.4 \leqslant a_2 \leqslant 0.1$.

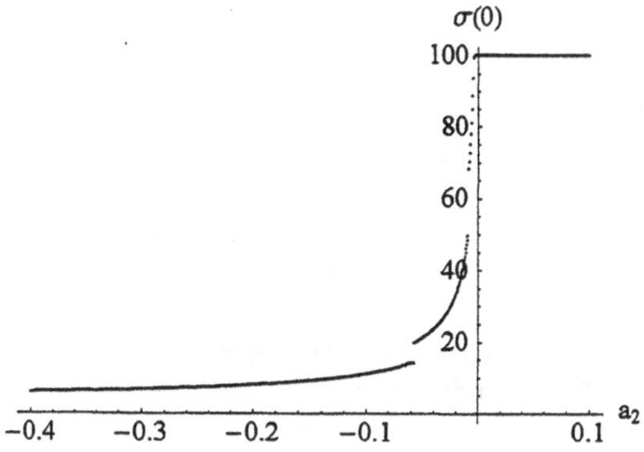

Fig. 2. $\sigma(0)$ for (8), $-0.4 \leqslant a_2 \leqslant 0.1$

On figure (3) is shown the function $t \to \sigma(t)$ for the system (8) where $a_1(t) = 1,\ a_2(t) = 1,\ a_3(t) = t$ and $-2 \leqslant t \leqslant 1$.

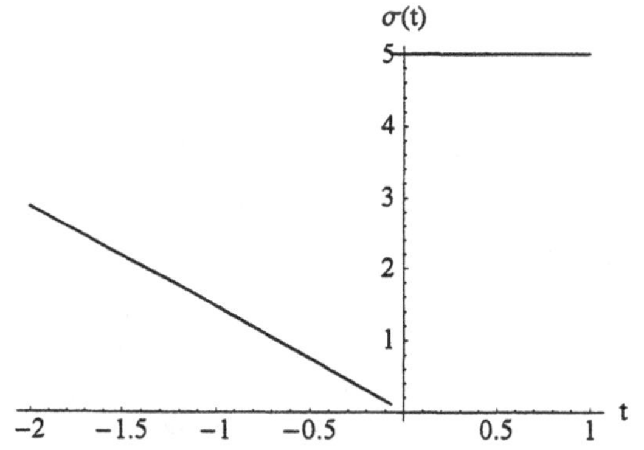

Fig. 3. $\sigma(t)$ for (8), $a_1 = a_2 = 1, a_3(t) = t$

On figure (4) is shown parametric dependence of the function $\sigma(0)$ for the system (8) where $a_1(t) = \cos t$ and $-1 \leqslant a_2 \leqslant 0,\ -1 \leqslant a_3 \leqslant 1$.

Note 1. Horizontal parts of the graphs (2)–(4) mean that on the test segments ($[0, 100]$, $[0, 5]$ and $[0, 10]$ respectively) there are no n-th zeroes of minimal linear combinations (see the section 4).

Theorem 1. *Every system of the form* (4) *that is reducible with non-singular transformation* $z(t) = L(t)x$ *(i.e.* $L(t)$ *is continuously differentiable and* $\det L(t) \neq 0,\ t \in J$*) to canonical system*

$$\dot{z} = F(t)z + g(t)u, \tag{9}$$

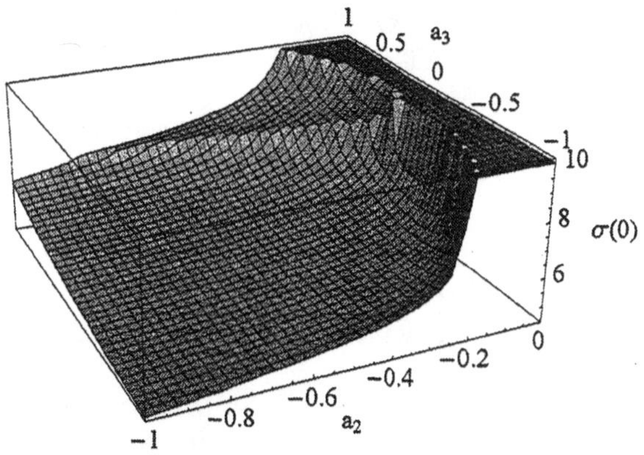

Fig. 4. $\sigma(0)$ for (8), $-1 \leqslant a_2 \leqslant 0,\ -1 \leqslant a_3 \leqslant 1$

is subcritical. Here is $g(t) = \operatorname{col}(\beta_1(t), 0, \ldots, 0)$,

$$F(t) = \begin{pmatrix} f_{11}(t) & f_{12}(t) & \ldots & f_{1n-1}(t) & f_{1n}(t) \\ -\beta_2(t) & f_{22}(t) & \ldots & f_{2n-1}(t) & f_{2n}(t) \\ 0 & -\beta_3(t) & \ldots & f_{3n-1}(t) & f_{3n}(t) \\ \ldots & \ldots & \ldots & \ldots & \ldots \\ 0 & 0 & \ldots & -\beta_n(t) & f_{nn}(t) \end{pmatrix},$$

where functions $f_{ik}(t)$ *and* $\beta_i(t)$ *are continuous,* $\beta_i(t) > 0$ *for all* $t \in J$ *and* $i = 1, \ldots, n$.

Let us suppose that system (4) satisfies the following two conditions.

Condition 1. For every $i = 1, \ldots, n{+}1$ functions defined by equalities

$$q_1(t) = b(t), \ldots, q_i(t) = \dot{q}_{i-1}(t) - A(t)q_{i-1}(t)$$

are continuous and bounded on $\mathbb{R}$ and $\det Q(t) \neq 0$ for every $t \in \mathbb{R}$ where

$$Q(t) \doteq (q_1(t), \ldots, q_n(t)).$$

Condition 2. There are numbers $\nu_1, \ldots, \nu_{n-1}$ that is $\nu_1 \leqslant \nu_2 \leqslant \cdots \leqslant \nu_{n-1}$ and for roots $\lambda_1(t), \ldots, \lambda_n(t)$ of equation $\det(\lambda Q(t) - H(t)) = 0$ where

$$H(t) \doteq (q_2(t), \ldots, q_{n+1}(t))$$

inequalities

$$\lambda_1(t) \leqslant \nu_1 \leqslant \lambda_2(t) \leqslant \cdots \leqslant \nu_{n-1} \leqslant \lambda_n(t) \tag{10}$$

hold for every t.

Theorem 2. *If the conditions* (1) *and* (2) *are true,* $\sigma(t) = \infty$ *for all* $t \in \mathbb{R}$. *Moreover if there are constants* $\varepsilon > 0$ *and* $\delta \geqslant 0$ *that in addition to* (10) *for all sufficiently big* t *inequalities* $\delta \leqslant \lambda_1(t)$,

$$\nu_{i-1} + \varepsilon \leqslant \lambda_i(t) \leqslant \nu_i - \varepsilon, \quad i = 2, \ldots, n-1,$$

are true then for all $t \in \mathbb{R}$, *controllability set* $D(t)$ *of the system* (4) *is coincide with* $\mathbb{R}^n$.

Suppose further the functions

$$A\colon \mathbb{R} \to \operatorname{End}(\mathbb{R}^n) \quad \text{and} \quad b\colon \mathbb{R} \to \mathbb{R}$$

of the system (4) be bounded on $\mathbb{R}$ and belong to class C^r (i.e. differentiable r times on $\mathbb{R}$), $r \geqslant 0$, and the system (4) is subcritical.

4. PROPERTIES AND NUMERICAL APPROXIMATION OF THE FUNCTION $\sigma(t)$

In this section for formal definition of numerical algorithm the functions

$$\xi_1(t), \ldots, \xi_n(t) \tag{11}$$

will be arbitrary scalar functions bounded and continuous on segment $[t_0, t_0+\sigma]$ unless otherwise stipulated.

There are simple examples that the function $t \to \sigma(t)$ (it has non-negative finite values or $+\infty$) can be discontinuous.

Lemma 2. *Let t_0 be discontinuity point of the function $\sigma(t)$, the next inequalities are true*

$$\sigma(t_0 - 0) \leqslant \sigma(t_0) \quad \textit{and} \quad \sigma(t_0) \leqslant \sigma(t_0 + 0).$$

Let for some system of functions (11) and some point t_0 inequality $\sigma(t_0) < \infty$ is true. Without losing generality it is possible to consider in every linear combination

$$\xi(t) \doteq c_1\xi_1(t) + \cdots + c_n\xi_n(t) \tag{12}$$

multipliers $\{c_1, \ldots, c_n\}$ satisfy the next property

$$|\operatorname{col}(c_1, \ldots, c_n)| = 1, \tag{13}$$

because normalization of the vector

$$c \doteq \operatorname{col}(c_1, \ldots, c_n)$$

does not affect on the zeroes of the linear combination (12). Thus $c \in S^{n-1}$ and by virtue of compactness of the set S^{n-1} and linearity $\xi(t)$ by c, can be constructed convergent sequence $\{c^i\}_{i=1}^{\infty}$, to which corresponds the functional sequence $\{\xi(t; c^i)\}_{i=1}^{\infty}$ that has the next property (here $\phi_n(\xi(t))$ is n-th zero of a function $\xi(t)$):

$$\lim_{i\to\infty} \phi_n(\xi(t; c^i)) = t_0 + \sigma(t_0).$$

Appropriate limit $\widehat{\xi}(t) \doteq \lim_{i\to\infty} \xi(t; c^i)$ will be call *minimal linear combination* of the functions (11).

Numerical algorithm described below is searching of linear combination (12) closest to $\widehat{\xi}(t)$.

Of course, behavior of linear combinations $\xi(t)$ of the functions (11) must be investigated on whole semi-axis $[t_0, +\infty)$, but it is impossible by virtue of limited computer resources. Therefore all computations are provided on *test segment* $[t_0, t_0 + T]$ where T is some fixed parameter of the algorithm. If there are no linear combinations of the functions (11) with n zeroes on the segment $[t_0, t_0 + T]$ then assumed $\sigma(t_0) \geqslant T$.

1°. Let $t_0, t_1, \ldots, t_{N-1}$ be the partitioning of the test segment by $N-1$ parts where N is fixed parameter of the algorithm. All values of the functions (11) and their linear combinations are computing in these N points.

According to (13) $c \in S^{n-1}$ but it is sufficient to choose multiplier sets $\{c_1, \ldots, c_n\}$ so that $c \in S_+^{n-1}$ where S_+^{n-1} can be any hemisphere of the S^{n-1}, because linear combinations (12) with $c \in S_-^{n-1}$ differ by sign only.

2°. Let $r_1, \ldots, r_n$ be an arbitrary normalized basis in $\mathbb{R}^n$. Let

$$s_2 = r_1 \sin\theta_1 + r_2 \cos\theta_1$$
$$s_3 = s_2 \sin\theta_2 + r_3 \cos\theta_2$$
$$\ldots$$
$$s_n = s_{n-1} \sin\theta_{n-1} + r_n \cos\theta_{n-1}$$
$$c = \frac{s_n}{|s_n|},$$

where $0 \leqslant \theta_i < \pi$, $i = 1, \ldots, n-1$. Angles $\theta_1, \ldots, \theta_{n-1}$ are vector c coordinates in spherical coordinate system on S_+^{n-1}. Orthogonalization of the basis $r_1, \ldots, r_n$ is not required because it is simpler to normalize the vector s_n. Let M be one more fixed parameter of the algorithm. Let us separate the segment $[0, \pi]$ on $M+1$ parts and make computations of the $\xi(t)$ for every $\theta_i = \frac{2k_i\pi}{M+1}$, $k_i = 0, \ldots, M-1$, $i = 1, \ldots, n-1$. In this way on the hemisphere S_+^{n-1} there are M^{n-1} distinguishable points c, and for all of these points linear combination $\xi(t) = c_1\xi_1(t) + \cdots + c_n\xi_n(t)$ must be computed in every point of the test segment partitioning. From produced functions $\xi(t)$ let us point the one that has n-th zero closest to t_0 (corresponding vector c will be designated $\bar{c}$). If not exist let us assume $\sigma(t_0) \geqslant T$ and stop this process.

Note 2. On the second stage of the algorithm computation of the function $\xi(t)$ is required $N \cdot M^{n-1}$ times. It is clear for big n described process will take a very long time but in research purposes this algorithm is applicable.

Before to continue of the algorithm description let us examine two examples of the computed minimal linear combinations.

Example 2. On figure 5 is shown the minimal linear combination $t \to \xi(t)$ of the functions $\xi_1(t) = 1$, $\xi_2(t) = t^2$ computed with parameters $t_0 = -1$, $T = 1.3$, $N = 500$, $M = 1000$.

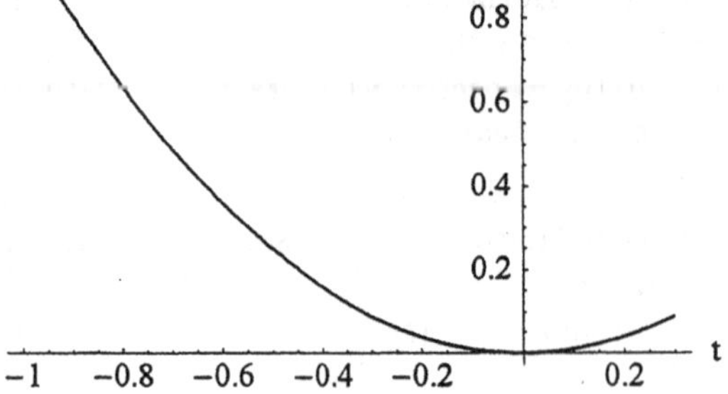

Fig. 5. $\xi_1(t) = 1$, $\xi_2(t) = t^2$

Example 3. On figure 6 is shown the minimal linear combination $t \to \xi(t)$ of the functions $\xi_1(t) = t$, $\xi_2(t) = \sin(t)$, $\xi_3(t) = \cos(t)$ computed with parameters $t_0 = -1$, $T = 1.8$, $N = 500$, $M = 1000$.

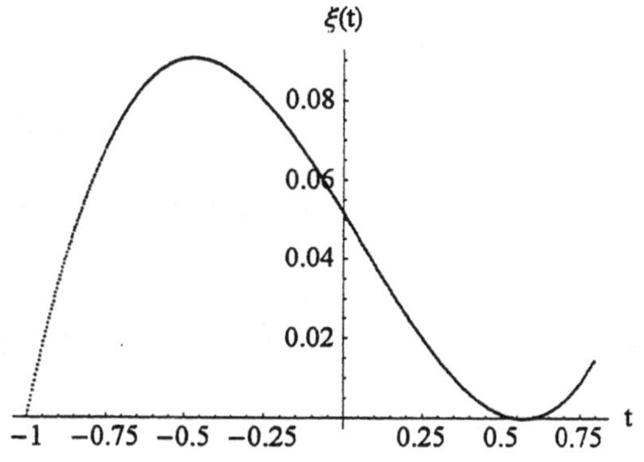

Fig. 6. $\xi_1(t) = t$, $\xi_2(t) = \sin(t)$, $\xi_3(t) = \cos(t)$

These examples are illustrating one important feature of minimal linear combinations. In some cases roots of computed minimal linear combination $\xi(t)$ are grouping together that in limit produces one root with multiplicity greater than 1. This event take place when the functions (11) are solutions of some differential equation or quasi-differential equation with order n. If present, it makes the next difficulty. A small changing of the multipliers $\{c_1, \ldots, c_n\}$ (in other words a small shift of the vector c on the hemisphere S_+^{n-1}) produces a big changing of the n-th zero of the function $\xi(t)$. The next two stages of the algorithm are intended to avoid this effect.

3°. Let $\overline{\theta}_1, \ldots, \overline{\theta}_{n-1}$ be the spherical coordinates of the vector $\overline{c}$. Let M_2 be one more parameter of the algorithm. This stage of the algorithm is repeating of the second stage with substitution of the hemisphere to the square (in spherical coordinates)

$$\left[\overline{\theta}_1 - \frac{\pi}{M-1}, \overline{\theta}_1 + \frac{\pi}{M-1}\right] \times \cdots \times \left[\overline{\theta}_{n-1} - \frac{\pi}{M-1}, \overline{\theta}_{n-1} + \frac{\pi}{M-1}\right]$$

and with substitution of the parameter M to the parameter M_2. By this way constructed more close to the minimal linear combination $\xi(t)$ with precision that can be achieve on the second stage with meaning of the parameter M equal to $M \cdot M_2$. As a matter of fact is produced $N \cdot (M^{n-1} + M_2^{n-1})$ computations of the function $\xi(t)$, that is significantly less than $N \cdot M^{n-1} \cdot M_2^{n-1}$ for big M and M_2.

4°. If after application of the algorithm's third stage the n-th and the $(n-1)$-th zeroes of the function $\xi(t)$ become more closely and distance between them become less than $\frac{2T}{N-1}$, it is proposed in limit these roots are coincide that produces one root with multiplicity greater than 1. In this case as the result of the algoritm application (or point $t_0+\sigma(t_0)$) takes the arithmetic mean of these roots. In other case as the result takes the n-th zero value. It is easy to make shure that this algorithm has an error not greater than $1/N$.

Described algorithm will be call *slow* because it can be improved in such cases when the first zero of the minimal linear combination $\widehat{\xi}(t)$ located in the point t_0. This event takes place when the functions (11) are defined by (7) and the system (4) satisfies the theorem 1. Let us designate

$$\xi_0 \doteq (\xi_1(t_0), \ldots, \xi_n(t_0))$$

and separate from the sphere S^{n-1} such points c' that satisfies $\xi_0 c' = 0$. The set of these points is the sphere S^{n-2}. Let us separate from this sphere the hemisphere S_+^{n-2} by arbitrary way. The *fast* algorithm is the modification of the described above with substitution of the hemisphere S_+^{n-1} to the hemisphere S_+^{n-2}. As the result of the fast algorithm application is the linear combination $\xi(t)$ that close to the minimal $\widehat{\xi}(t)$ and has the first zero close to the point t_0.

5. CONTROLLABILITY SET STRUCTURE

According to maximum principle of Pontryagin

$$\max_{u(\cdot)\in\mathcal{U}} \psi(t)b(t)u = \psi(t)b(t)u(t), \qquad (14)$$

$t_0 \leqslant t \leqslant t_0 + \vartheta$, $\vartheta \leqslant \sigma(t_0)$, for every $x_0 \in D_\vartheta(t_0)$ there exists integer k, $0 \leqslant k \leqslant n-1$ and vector $\tau \in M^k(\vartheta)$ where $M^0(\vartheta) \doteq \{0\}$,

$$M^k(\vartheta) \doteq \{\tau = (\tau_{n-k}, \ldots, \tau_{n-1}) \in \mathbb{R}^k : 0 < \tau_{n-k} < \cdots < \tau_{n-1} < \vartheta\},$$

$k = 1, \ldots, n-1$, such that control function transferring $x_0 = x(t_0)$ to the origin of coordinates by a minimal time has the values $+1$ and -1 with switching at points $t_0 + \tau_i$, $i = n-k, \ldots, n-1$. These points corresponds to the zeroes of the function

$$\xi(t) \doteq \psi(t)b(t),$$

where $\psi(t)$ is some non-trivial solution of the system (6) and according to $\vartheta < \sigma(t_0)$ the amount of these points is not greater than $n-1$. Later on the sets $M^k(\vartheta)$ are interprets as smooth manifolds with dimensions k imbedded in $\mathbb{R}^k$.

For every $k = 0, \ldots, n-1$ let us construct sets $N_+^k(t_0, \vartheta)$ and $N_-^k(t_0, \vartheta)$ as follows way. $N_+^k(t_0, \vartheta)$ is the set of points $x_0 \in D_\vartheta(t_0)$ such that is for every one there exists a point $\tau(t_0, x_0) \in M^k(\vartheta)$ such that optimal control function $u(t, x_0)$, $t_0 \leqslant t \leqslant t_0+\vartheta$ transfers point $x(t_0) = x_0$ to $x(t_0+\vartheta) = 0$ and switches at time moments $t = t_0 + \tau_i(t_0, x_0)$ only (before the first switching $u(t, x_0) = +1$). Note that the set $N_+^0(t_0, \vartheta)$ contents only the one point that can be found from the next equality

$$N_+^0(t_0,\vartheta)=\left\{-\int\limits_{t_0}^{t_0+\vartheta} X(t_0,t)b(t)\,dt\right\}.$$

The sets $N_-^k(t_0,\vartheta)$ are identical to the sets $N_+^k(t_0,\vartheta)$ with one exception: before the first switching $u(t,x_0)=-1$. The sets $N_+^k(t_0,\vartheta)$ (and analogously $N_-^k(t_0,\vartheta)$) have following properties.

Property 1. Let $\vartheta\leqslant\sigma(t_0)$. Then

$$N_+^k(t_0,\vartheta)\subset\partial D_\vartheta(t_0)$$

and to every point $x_0\in N_+^k(t_0,\vartheta)$ corresponds such single point $\tau(t_0,x_0)\in M^k(\vartheta)$ (the join of switching time moments) that control function $u(t,x_0)$ satisfies to the maximum principle (14) is transferring $x_0=x(t_0)$ to $x(t_0+\vartheta)=0$.

In accord to property (1) for every $\vartheta\leqslant\sigma(t_0)$ and any fixed $k=0,\dots,n-1$ is defined the function

$$f^{-1}\colon N_+^k(t_0,\vartheta)\to M^k(\vartheta)$$

that makes correspondence from the point $x\in N_+^k(t_0,\vartheta)$ to the point $\tau\in M^k(\vartheta)$. The function $f^{-1}=f_k^{-1}$ is depends on ϑ and index k that assumed below but not accented.

Property 2. The function f^{-1} is continuous and realizes homeomorphism of the sets $N_+^k(t_0,\vartheta)$ and $M^k(\vartheta)$. The inverse function $f:M^k(\vartheta)\to N_+^k(t_0,\vartheta)$ is defined by equality

$$f(\tau)=\sum_{i=n-k-1}^{n-1}(-1)^{i-n+k}\int\limits_{t_0+\tau_i}^{t_0+\tau_{i+1}} X(t_0,t)b(t)\,dt,$$

where $\tau_{n-k-1}=0,\ \tau_n=0$.

Property 3. Let $\vartheta\leqslant\sigma(t_0)$. For every $k=1,\dots,n-1$ and any point $\tau=(\tau_{n-k},\dots,\tau_{n-1})\in M^k(\vartheta)$ vectors

$$h(\tau_{n-k})\doteq X(t_0,t_0+\tau_{n-k})b(t_0+\tau_{n-k}),$$
$$\dots$$
$$h(\tau_{n-1})\doteq X(t_0,t_0+\tau_{n-1})b(t_0+\tau_{n-1})$$

are linearly independent.

In accord to properties (1)-(3) for any $\vartheta\leqslant\sigma(t_0)$ and every $k=1,\dots,n-1$ the set $N_+^k(t_0,\vartheta)$ is the smooth manifold of class C^1 with dimension k imbedded in $\mathbb{R}^n$. Moreover the next theorem is proved.

Theorem 3. *Let the system (4) be subcritical on $\mathbb{R}$. Then for every $\vartheta\leqslant\sigma(t_0)$ the controllability set $D_\vartheta(t_0)$ is strictly convex in $\mathbb{R}^n$ (i.e. $\operatorname{int}D_\vartheta(t_0)\neq\varnothing$ and for any $x,x_0\in\partial D_\vartheta(t_0)$ and any $\lambda\in(0,1)$ point $\lambda x+(1-\lambda)x_0\in\operatorname{int}D_\vartheta(t_0)$). The border $\partial D_\vartheta(t_0)$ of the set $D_\vartheta(t_0)$ is the union of nonintersecting smooth (of class C^{r+1}) manifolds $N_+^k(t_0,\vartheta)$ and $N_-^k(t_0,\vartheta)$, $k=0,1,\dots,n-1$ and the union*

$$\left(\bigcup_{i=0}^{k-1}N_-^i(t_0,\vartheta)\right)\bigcup\left(\bigcup_{i=0}^{k-1}N_+^i(t_0,\vartheta)\right)$$

is the common border of manifolds $\operatorname{cl}N_+^k(t_0,\vartheta)$ and $\operatorname{cl}N_-^k(t_0,\vartheta)$. In addition to every point $x\in N_+^k(t_0,\vartheta)$ corresponds the single control function that transfers $x_0=x(t_0)$ to $x(t_0+\vartheta)=0$ and has strictly k switching on $(t_0,t_0+\vartheta)$.

Example 4. On figures 7 and 8 are shown the set $D_\vartheta(t_0)$ and the manifold $N_+^2(t_0,\vartheta)$ of the system

$$\dot x_1=x_2,\quad \dot x_2=x_3,\quad \dot x_3=u, \qquad (15)$$

$|u|\leqslant 1$, with $\vartheta=3$, $t_0=0$.

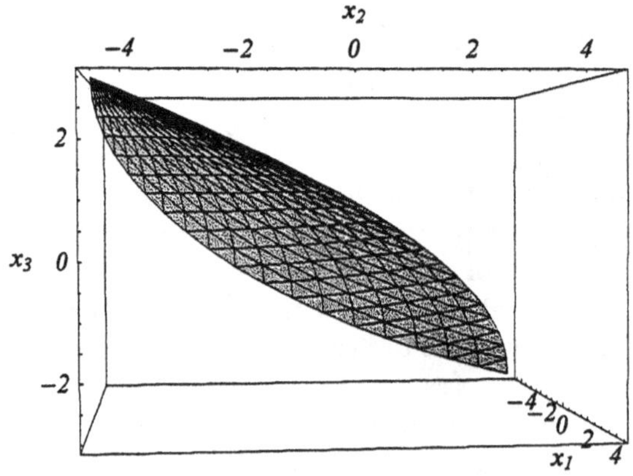

Fig. 7. Set $D_\vartheta(t_0)$ for (15), $t_0=0,\ \vartheta=3$

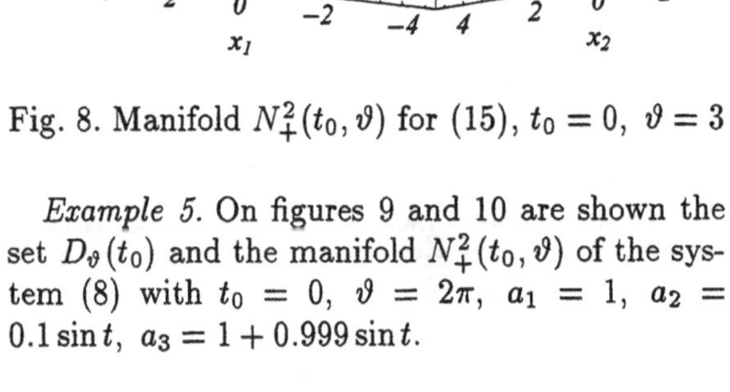

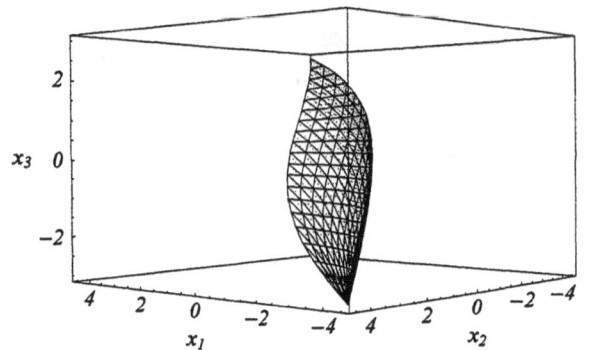

Fig. 8. Manifold $N_+^2(t_0,\vartheta)$ for (15), $t_0=0,\ \vartheta=3$

Example 5. On figures 9 and 10 are shown the set $D_\vartheta(t_0)$ and the manifold $N_+^2(t_0,\vartheta)$ of the system (8) with $t_0=0$, $\vartheta=2\pi$, $a_1=1$, $a_2=0.1\sin t$, $a_3=1+0.999\sin t$.

6. EXTENDED CONTROLLABILITY SET STRUCTURE

Let us introduce designation $\tau_n=\vartheta$ and for every $k=0,1,\dots,n$ and any $t\in\mathbb{R}$ let us define

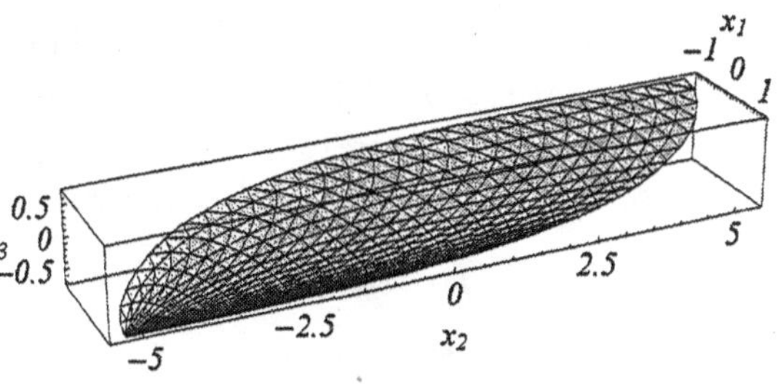

Fig. 9. Set $D_\vartheta(t_0)$ for (8), $t_0 = 0$, $\vartheta = 2\pi$

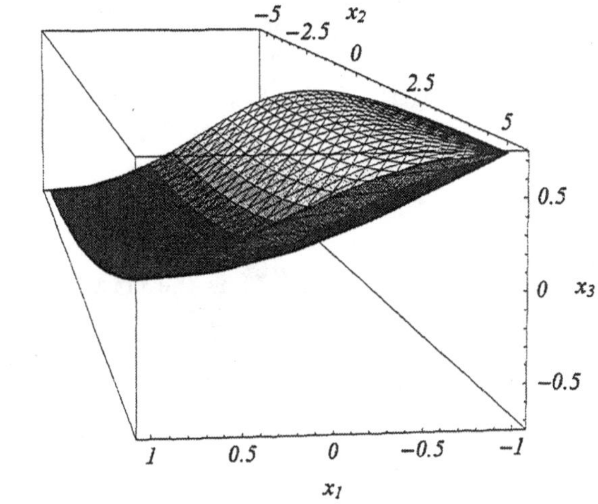

Fig. 10. Manifold $N_+^2(t_0, \vartheta)$ for (8), $t_0 = 0$, $\vartheta = 2\pi$

manifolds $\mathcal{M}^k(t)$ where $\mathcal{M}^0(t) \doteq \{0\}$,

$$\mathcal{M}^k(t) \doteq \{\tau = (\tau_{n-k+1}, \dots, \tau_n) : \\ 0 < \tau_{n-k+1} < \dots < \tau_n < \sigma(t)\},$$

$k = 1, \dots, n$, and manifold $\mathcal{M}^{1+k} \doteq \mathbb{R} \times \mathcal{M}^k(t)$. To every point $p = (t, \tau) \in \mathcal{M}^{1+k}$ let us correspond the point $q = (t, x)$ where $x = 0$ when $k = 0$ and

$$x = x(p) = \\ -\sum_{i=n-k}^{n-1} (-1)^{i-n+k} \int_{t+\tau_i}^{t+\tau_{i+1}} X(t,s)b(s)\,ds, \quad (16)$$

$\tau_{n-k} = 0$, when $k \geqslant 1$.

By the equality (16) for every k is defined the function $p \to F(p) = q$ with domain of definition $\mathcal{M}^{k+1}$ and range of values

$$\mathcal{N}_+^{1+k} \doteq F(\mathcal{M}^{1+k})$$

(lower index at $\mathcal{N}_+^1$ will be excluded below). Since $\mathcal{N}_+^{1+k} = \mathbb{R} \times \mathcal{N}_+^k(t)$ where $\mathcal{N}^0(t) = 0$ and for $k \geqslant 1$ the set $\mathcal{N}_+^k(t)$ is consist of points (16) then $\mathcal{N}_+^k(t) \subset D_{\sigma(t)}(t)$. It is proved that F is diffeomorphism of class C^{r+1} and therefore for every $k = 0, 1, \dots, n$ the set $\mathcal{N}_+^{1+k}$ is the smooth manifold of class C^{r+1}.

Theorem 4. *Let the system (4) is subcritical. Then extended controllability set $\mathfrak{D} \doteq \mathbb{R} \times D_{\sigma(t)}(t)$ can be represent as $\mathfrak{D} = \mathrm{cl}\,(\mathfrak{N}_+^{1+n} \bigcup \mathfrak{N}_-^{1+n})$ where*

$$\mathfrak{N}_+^{1+k} = \mathcal{N}_+^{1+k} \bigcup \mathcal{N}_-^k \bigcup \mathcal{N}_+^{k-1} \bigcup \dots \bigcup \mathcal{N}^1,$$
$$\mathfrak{N}_-^{1+k} = \mathcal{N}_-^{1+k} \bigcup \mathcal{N}_+^k \bigcup \mathcal{N}_-^{k-1} \bigcup \dots \bigcup \mathcal{N}^1,$$

$k = 0, \dots, n$. Manifolds $\mathfrak{N}_+^{1+k}$, $\mathfrak{N}_-^{1+k}$ are weakly invariant and for every $k = 0, \dots, n$ manifold $\mathfrak{N}_+^k \bigcup \mathfrak{N}_-^k$ is common border of the manifolds $\mathrm{cl}\,\mathfrak{N}_+^{1+k}$ and $\mathrm{cl}\,\mathfrak{N}_-^{1+k}$.

Example 6. On figures 11, 12 and 13 respectively are shown the function $t \to \sigma(t)$, fragment of the union of the manifolds $\mathcal{N}_+^n \bigcup \mathcal{N}_-^n$ and fragment of the extended controllability set $\mathfrak{D}$ of the system (this system describes pendulum behavior)

$$\dot{x}_1 = x_2, \quad \dot{x}_2 = -1 - 0.5\sin(2t)x_1 + u, \quad (17)$$

$|u| \leqslant 1$, with $0 \leqslant t \leqslant 6$.

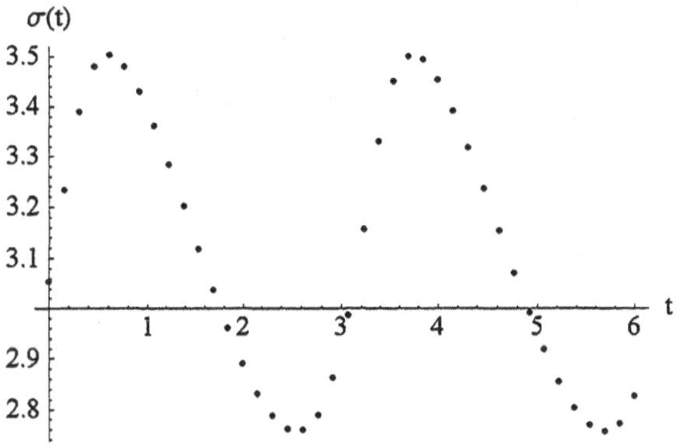

Fig. 11. The function $t \to \sigma(t)$ for the system (17)

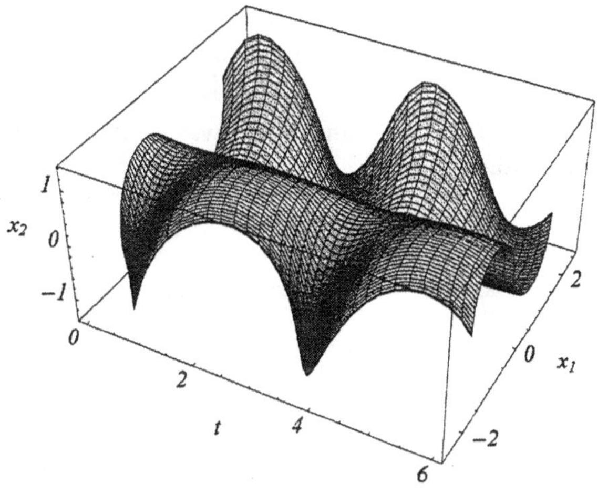

Fig. 12. Fragment of the union $\mathcal{N}_+^n \bigcup \mathcal{N}_-^n$ for the system (17)

7. DIFFERENTIABILITY OF SPEED VECTOR

Let point $q_0 = (t_0, x_0) \in \mathcal{N}_+^{1+n}$ then function

$$F^{-1} \colon \mathcal{N}_+^{1+n} \to \mathcal{M}^{1+n}$$

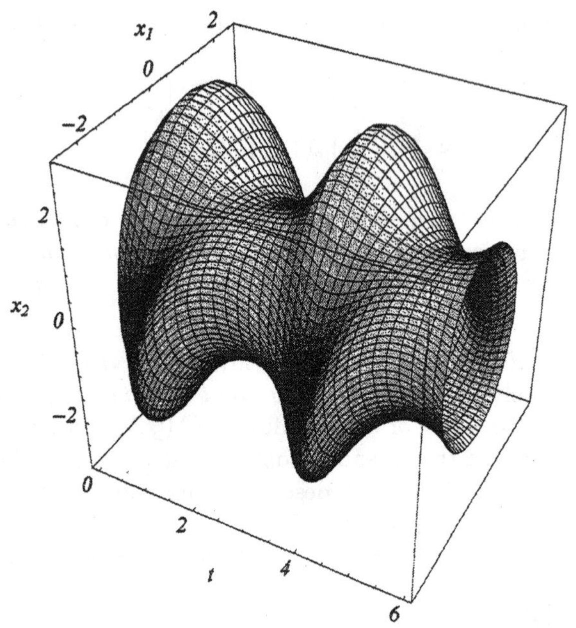

Fig. 13. Fragment of the set $\mathfrak{D}$ for (17)

that is reverse to F for $k = n$ give us the point

$$p_0 = (t_0, \tau_1(q_0), \ldots, \tau_n(q_0)) \in \mathcal{M}^{1+n}, \quad 0 < \tau_1(q_0) < \cdots < \tau_n(q_0).$$

Below it is shown that in the next problem for every $i = 1, \ldots, n$ the number $\tau_i(q_0)$ is the minimal time of transferring a point $x_0 \in \mathcal{N}_+^n(t_0)$ to manifold $\mathcal{N}_{\nu(i)}^{n-i}(t_0 + \vartheta)$:

$$\vartheta(u(\cdot)) \to \min_{u(\cdot)}, \quad u(\cdot) \in \mathcal{U}, \tag{18}$$

$$\dot{x} = A(t)x + b(t)u(t), \quad t_0 \leqslant t \leqslant t_0 + \vartheta, \tag{19}$$

$$x(t_0) = x_0, \quad x(t_0 + \vartheta) \in \mathcal{N}_{\nu(i)}^{n-i}(t_0 + \vartheta), \tag{20}$$

where $\nu(i)$ is sign "plus" if i is even and sign "minus" otherwise. In this way the vector

$$\tau(q) \doteq ((\tau_1(q), \ldots, \tau_n(q))$$

is naturally to call *speed vector*. Here is $\tau_i(q) = 0$ if $q \in \mathcal{N}_{\nu(i)}^{n-i}$.

Note that the set $D_{\sigma(t)}(t)$ is centrally symmetric and therefore the vector $\tau(q)$ actually is defined on $\mathcal{N}_+^{1+n} \bigcup \mathcal{N}_-^{1+n}$.

Theorem 5. *Let the system* (4) *is subcritical. Let us designate by* $(u^0(\cdot), \vartheta^0, x^0(\cdot))$ *the optimal process of the problem* (18)-(20) *with some fixed* $i \in \{1, \ldots, n\}$. *Then* $\vartheta^0 = \tau_i(q_0)$ *and on interval* $(t_0, t_0 + \tau_i(q_0))$ *the optimal control function* $u^0(t)$ *and corresponding to it optimal solution* $x^0(t)$ *of the system* (19) *are coinciding with the optimal control function and solution of the problem for transferring to the origin of coordinates (i.e. of the problem* (18)-(20) *with* $i = n$).

Let $\tau(q) = (\tau_1(q), \ldots, \tau_n(q))$ is the speed vector of the system (4). Note that by derivative $d\tau_i(q_0)$ of function $\tau_i : \mathcal{N}_+^{1+k} \to \mathbb{R}$ in point q_0 in direction of vector $w \in T_{q_0}\mathcal{N}_+^{1+k}$ (here and below the $T_{q_0}\mathcal{N}_+^{1+k}$ is tangent space for manifold $\mathcal{N}_+^{1+k}$ in point q_0) we call the linear transformation

$$d\tau_i(q_0) \colon T_{q_0}\mathcal{N}_+^{1+k} \to \mathbb{R}$$

that is defined by the equality

$$d\tau_i(q_0)w \doteq \left.\frac{d\tau_i(q(\varepsilon))}{d\varepsilon}\right|_{\varepsilon=0}$$

where $q(\varepsilon)$ is the class of equivalence of smooth curves of kind $q\colon (-1,1) \to \mathcal{N}_+^{1+k}$ with the following properties: $q(0) = q_0$, $dq(\varepsilon)/d\varepsilon|_{\varepsilon=0} = w$. Similarly are defined derivatives $d^s\tau_i$, $s \geqslant 2$:

$$d^s\tau_i(q_0)(w_1, \ldots, w_s) \doteq \left.\frac{d^s\tau_i(q(\varepsilon))}{d\varepsilon^s}\right|_{\varepsilon=0}, \tag{21}$$

where $q : (-1,1) \to \mathcal{N}_+^{1+k}$ is the class of equivalence of smooth curves of kind

$$q(\varepsilon) = q_0 + \varepsilon w_1 + \varepsilon^2 w_2/2! + \cdots + \varepsilon^s w_s/s! + o(\varepsilon^s).$$

A function $q \to \tau_i(q)$ belongs to class C^s on manifold $\mathcal{N}_+^{1+k} \bigcup \mathcal{N}_-^{1+k}$ if for any C^s-curve

$$q\colon (-1,1) \to \mathcal{N}_+^{1+k} \bigcup \mathcal{N}_-^{1+k}$$

the function $\varepsilon \to \tau_i(q(\varepsilon))$ is in the class C^s.

Theorem 6. *Let the system* (4) *is subcritical and the functions* $A\colon \mathbb{R} \to \mathrm{End}(\mathbb{R}^n)$ *and* $b\colon \mathbb{R} \to \mathbb{R}^n$ *are belong to the class* C^r. *Then for every* $k = 0, \ldots, n$ *the functions*

$$\tau_i\colon \mathcal{N}_+^{1+k} \bigcup \mathcal{N}_-^{1+k} \to \mathbb{R}, \quad i = 1, \ldots, n$$

are belong to the class C^{r+1}. *In particular the function* τ_i *is continuously differentiable* $r+1$ *times on* $\mathcal{N}_+^{1+n} \bigcup \mathcal{N}_-^{1+n}$.

8. BELLMAN EQUATIONS

By virtue of the theorem 6 speed vector $\tau(q) = (\tau_1(q), \ldots, \tau_n(q))$ of the system (4) where $q = (t, x)$ is differentiable along a directions tangent to corresponding manifolds. This fact permits to write Bellman equations for coordinates $\tau_i(q)$ of the speed vector in the extended controllability set $\mathfrak{D} = \mathbb{R} \times D_{\sigma(t)}(t)$.

In the first place let us to note that in all points q of the set $\mathcal{N}_+^{1+n} \bigcup \mathcal{N}_-^{1+n}$ all coordinates of the speed vector are differentiable along all directions, therefore these coordinates appear classic solutions of the equation

$$\frac{\partial\theta}{\partial t} + \frac{\partial\theta}{\partial x}(A(t)x + b(t)) = -1, \tag{22}$$

in the set $(t, x) \in \mathcal{N}_+^{1+n}$, and of the equation

$$\frac{\partial\theta}{\partial t} + \frac{\partial\theta}{\partial x}(A(t)x - b(t)) = -1, \tag{23}$$

in the set $(t,x) \in \mathcal{N}_-^{1+n}$. In addition by virtue of definition of the speed vector the function $\tau_i(t,x)$ turns into zero in all points

$$(t,x) \in \mathfrak{N}_+^{1+n-i} \bigcup \mathfrak{N}_-^{1+n-i}, \quad i = 1,\dots,n$$

(see (18)–(20)). Therefore the function $\tau_1(t,x)$ satisfies the equations (22), (23) and boundary condition

$$\tau_1(t,x) = 0 \quad \text{for all} \quad (t,x) \in \mathfrak{N}_+^n \bigcup \mathfrak{N}_-^n.$$

Further let the point $(t,x) \in \mathcal{N}_-^n \bigcup \mathcal{N}_+^n$ then the functions $\tau_2(t,x),\dots,\tau_n(t,x)$ are satisfy the equation

$$d\theta(t,x)(1, A(t)x - b(t)) = -1$$

in the set $(t,x) \in \mathcal{N}_-^n$, and the equation

$$d\theta(t,x)(1, A(t)x + b(t)) = -1$$

in the set $(t,x) \in \mathcal{N}_+^n$.

It is easy to write similar equations on other manifolds $\mathcal{N}_-^k \bigcup \mathcal{N}_+^k$, $k = n-1,\dots,1$. And the next statement is true.

Theorem 7. *In the domain $\mathbb{R} \times D_{\sigma(t)}(t)$ the speed function $(t,x) \to \tau_n(t,x)$ of the subcritical system (4) is continuous solution of the problem*

$$d\theta(t,x)w(t,x) = -1, \quad \theta(t,x)|_{x=0} = 0,$$

where $d\theta(t,x)w(t,x)$ is derivative of the function $\theta(t,x)$ in the point (t,x) along the direction of the vector $w(t,x) = (1, A(t)x + u(t,x)b(t))$,

$$u(t,x) = \begin{cases} 1, & \text{if} \quad (t,x) \in \bigcup\limits_{k=1}^{n} \mathcal{N}_+^{1+k}, \\ -1, & \text{if} \quad (t,x) \in \bigcup\limits_{k=1}^{n} \mathcal{N}_-^{1+k}. \end{cases}$$

9. POSITIONAL CONTROL OF NONSTATIONARY SYSTEM

Let function $q \to u(q)$ where $q = (t,x)$ defined on the interior of the extended controllability set $\mathfrak{D}$ has values on $U = [-1,1]$ and superpositionally measurable. By *C-solution* (solution defined by Caratheodory) of system

$$\dot{x} = A(t)x + b(t)u(t,x) \tag{24}$$

will be called any absolutely continuous function $t \to x(t)$ that satisfies for all t the equality

$$x(t) = X(t,t_0)x(t_0) + \int_{t_0}^{t} X(t,s)b(s)u(s,x(s))\,ds,$$

where t_0 is arbitrary fixed time moment. The main lack of the C-solutions is strong sensitivity to changes of function $u(q)$ on sets with zero measure. This lack is not peculiar to *F-solutions* (solutions defined by A.F. Filippov in (Filippov, 1985)) that are described below. Moreover F-solutions are preferable for applied problems that can be model by differential equations with discontinuities on phase coordinates.

To define the solutions in Filippov sense let us construct multiform function

$$q \to \mathbb{F}(q) \doteq \bigcap_{\varepsilon>0} \bigcap_{\text{mes}\,\mu=0} \overline{\text{conv}}\, u(O_\varepsilon(q) \setminus \mu), \tag{25}$$

$q \in \text{int}\,\mathfrak{D}$, where $O_\varepsilon(q)$ is ε-neighborhood of the point q; μ is any set in $\mathbb{R}^{1+n}$ with zero Lebeg measure and $\overline{\text{conv}}Q$ is closure of convex hull of set Q. F-solution of the system (24) is any absolutely continuous function $t \to x(t)$ that satisfies for almost all t differential inclusion.

$$\dot{x} \in A(t)x + b(t)\mathbb{F}(t,x).$$

The function $u_C\colon \mathfrak{D} \to U$ that is superpositionally measurable will be called *speed optimal positional C-control* (or optimal C-control in short) if for any point $q_0 \in \text{int}\,\mathfrak{D}$ the C-solution $x(t,q_0)$ of the problem

$$\dot{x} = A(t)x + b(t)u, \quad x(t_0) = x_0 \tag{26}$$

with $u = u_C(q)$ exists on semi-axis $[t_0,\infty)$, single, equal to zero when $t = t_0 + \tau_n(q_0)$ and $x(t,q_0) \equiv 0$ for $t > t_0 + \tau_n(q_0)$.

Similarly is defined *speed optimal positional F-control* (or optimal F-control in short). In this case the function $q \to u_F(q)$ must be defined for almost all (according to Lebeg measure in $\mathbb{R}^{1+n}$) points $q \in \text{int}\,\mathfrak{D}$ and provide the next property: for every $q_0 \in \text{int}\,\mathfrak{D}$ corresponds a single F-solution $x(t,q_0)$ of problem (26) with control function $u = u_F(q)$ and $x(t,q_0) \equiv 0$ for $t \geqslant t_0 + \tau_n(q_0)$. Note once again that by virtue of definition of F-solutions to construct the optimal F-control there are no necessity to define $u_F(q)$ in every point of interior of the extended controllability set $\mathfrak{D}$. It is sufficient to construct $u_F(q)$ on a set of full measure.

As shown in the introduction there are examples of such abnormal behavior of linear controllable systems (Brunovski P., 1980a, 1980b): the optimal C-control exists and is unique but optimal F-control does not exist. This occurence takes place (even for linear stationary systems) in case when optimal C-control that strictly defined from maximum principle of Pontryagin is defines on surfaces of discontinuity (this surfaces has zero Lebeg measure) velocity vector that is not co-directed with velocity vector from Filippov's construction (25).

Theorem 8. *Let the system (4) be subcritical. Then function*

$$u_C(q) = \begin{cases} 1, & \text{if} \quad q \in \mathcal{N}_+^{1+k} \text{ for any } k \\ 0, & \text{if} \quad k = 0 \\ -1, & \text{if} \quad q \in \mathcal{N}_-^{1+k} \text{ for any } k \end{cases}$$

where $k \in \{1, \ldots, n\}$, is the optimal C-control, and the function

$$u_F(q) = \begin{cases} 1, & \text{if } q \in \mathcal{N}_+^{1+n} \\ -1, & \text{if } q \in \mathcal{N}_-^{1+n} \end{cases} \qquad q \in \operatorname{int} \mathfrak{D}$$

is the optimal F-control.

10. ACKNOWLEDGEMENT

The work is supported by Russian Foundation of Basic Research (grant 97–01–00413) and by Competition Center of Fundamental Natural Science (grant 97–0–1.9).

REFERENCES

Akoolenko L.D, Shmatkov A.M. (1998) *Optimal speed control synthesis of material point reduction to the given position with zero velocity //* Prikladnaya matematatika i mehanika, Vol.62, No. 1, pp. 129 — 138 (in Russian).

Albrekht E.G., Ermolenko E.A. (1997) *Optimal speed control synthesis of linear systems //* Differents. uravneniya Vol.33, No.11. pp. 1443 — 1450 (in Russian).

Bodner V.A. (1964) *Theory of automatic control flight.* Moskva: "Nauka" (in Russian).

Brunovski P. (1980a) *Regular synthesis and singular extremas //* Lect. Contr. and Inform. Sci. 1980, V.22, P. 280-284.

Brunovski P. (1980b) *Existence of regular synthesis for general control //* J. Different. Equat., 1980, V.38, No. 3, P. 317-343.

Chernousko F.L., Shmatkov A.M. (1997) *Optimal speed control synthesis of one three-dimensional system //* Prikladnaya matematatika i mehanika, Vol.61, No. 5, pp. 723 — 731 (in Russian).

Filippov A.F. (1985) *Differential equations with discontinuos right hand.* Moskva: "Nauka", 223 P. (in Russian).

Ioffe A.D., Tikhomirov V.M. (1974) *Theory of extremal problems.* Moskva: "Nauka", 479 P. (in Russian).

Rodionova A.G., Tonkov E.L. (1993) *On Continuity of Function of Linear System in Critical Case //* Russian Mathematics (Iz. VUZ) Vol.37, No. 5, PP. 101 — 111.

RELAXATIONAL POLYHEDRON OF VEBER PROBLEM

A. V. Panyukov

Southern Ural State University
76, Lenina Ave., Chelyabinsk, 454080, Russia

Abstract: A location problem known as Veber problem is considered. It is proved $\mathcal{N}P$-hardness of Veber problem and representation of this problem as a integer linear programming problem is given. It is stated the relaxed Veber problem has a quasi-integer relaxational polyhedron. The integer version of simplex-method to solve Veber problem is developed. In this version the sequence of adjecent integer tops of Veber problem polyhedron is constructed by special linear programming problems.

Keywords: Algorithms, desision making, design, graph theoretic models, integer programing, linear programing, networks, optimization problems.

1. INTRODUCTION

Mathematical model of many problems of control, design and decision making is represented task $\Theta(G, V, b, c, \Phi)$

$$F(\varphi) = \sum_{\{i,j\}\in E} b(\{i,j\}, \varphi(i), \varphi(j)) + \sum_{j\in J} c(j, \varphi(j)) \to \min_{\varphi\in\Phi}$$

for preassigned graph $G = (J, E)$, *finite* set V, map $b: E \times V^2 \to \mathbf{Z}$, map $c: J \times V \to \mathbf{Z}$ and set Φ of the admissible arrangements of the elements of a set J in points of set V. In a case

$$\Phi = \Phi_V = \{\varphi : J \to V\}$$

(i.e. represents all univalent maps) the task $\Theta = \Theta_V$ is known as *Veber problem* (Francis and White, 1974). If set

$$\Phi = \Phi_W = \{\varphi : J \to V \mid (i \neq j) \Rightarrow \varphi(i) \neq \varphi(j)\}$$

(i.e. represents all injective univalent maps) the task $\Theta = \Theta_W$ is *the square-law assignment problem* (Horst and Tuy, 1993).

For many special cases task Θ_V is solvable for polynomial time, see (Francis and White, 1974; Panyukov and Pelzwerger, 1991). In the present work is proved that a common case of task Θ_V is $\mathcal{N}P$-hard.

Task Θ_V is considered as the integer linear programming problem. Its relaxational polyhedron is quasi-integer, i.e. in the polyhedron graph there is a path between any pair of integer tops taking place only through integer tops. For the first time polyhedrons with a similar property were remarked for the elementary task of the arrangement (Trubin, 1969) and for set covering problem (Balas and Padberg, 1972).

To solve problems with quasi-integer relaxational polyhedron is possible with the help of integer simplex-method (Trubin, 1969; Balas and Padberg, 1975). But highly degeneracy of problem Θ_V excludes the chance to use the known algorithms.

The reachings in the field of a linear programming (Khachiyan, 1979; Grotschel, *et al.*, 1981) allow in the integer simplex-method for polynomial time to define an extreme base of degenerate top. The found extreme base defines the all adjacent integer tops with the best value of the goal function. This approach to solve Veber problem develops in the work.

2. COMPLEXITY OF VEBER PROBLEM

For a common case of problem Θ_V we have

Theorem 1. The task $\Theta(G, V, b, c, \Phi_V)$ is $\mathcal{NP}$-hard ■

PROOF. Let's assume

$$C = \max_{(j,m)\in J\times V} c(j,m);$$

$$B = \max_{(\{i,j\},m,n)\in E\times V^2} b(\{i,j\},m,n);$$

$$\tilde{c}(j,m) = C - c(j,m), \quad (j,m) \in J \times V;$$

$$\tilde{b}(\{i,j\},m,n) = B - b(\{i,j\},m,n), \quad ([i,j],m,n) \in E \times V^2.$$

Uniqueness of maps $\varphi \in \Phi_V$ implies

$$F(\varphi) = |J|\,C + |E|\,B - \sum_{j\in J} (\tilde{c}(j,\varphi(j))) - \sum_{\{i,j\}\in E} \left(\tilde{b}(\{i,j\},\varphi(i),\varphi(j))\right) = |J|\,C + |E|\,B - \tilde{F}(\varphi).$$

Therefore Veber problem $\Theta(G, V, b, c, \Phi_V)$ on a minimum is equivalent to task $\tilde{\Theta}(G, V, \tilde{b}, \tilde{c}, \Phi_V)$:

$$\tilde{F}(\varphi) \to \max_{\varphi\in\Phi_V}.$$

Let's consider the cut maximum problem for graph (S, T) and weight function $w : T \to \mathbf{Z}^+$:

$$\sum_{t\in T\cap[S_1\times S_2]} w(t) \to \max_{\{S_1,S_2\}},$$

where $\{S_1, S_2\}$ - partition of a set S on two subsets. It is known (Garey and Jonson, 1979) this problem is $\mathcal{NP}$-complete. On the other hand it is a special case of the task $\tilde{\Theta}(G, V, \tilde{b}, \tilde{c}, \Phi_V)$, in which

$$G = (S,T), \; V = \{v_1, v_2\},$$

$$\tilde{c}(j,v_1) = \tilde{c}(j,v_2) = 0, \; j \in J;$$

$$\tilde{b}(\{i,j\},v_1,v_1) = \tilde{b}(\{i,j\},v_2,v_2) = 0, \; \{i,j\} \in E;$$

$$\tilde{b}(\{i,j\},v_1,v_2) = \tilde{b}(\{i,j\},v_2,v_1) = w([i,j]), \; \{i,j\} \in E.$$

Theorem 1 is proved ■

3. ILP REPRESENTATION OF VEBER PROBLEM

Let's consider the integer linear programming problem

$$\sum_{\{i,j\}\in E,\; m,n\in V} v^{i\,j}_{mn} b(\{i,j\},m,n) + \sum_{j\in J,\; m\in V} y^j_m c(j,m) \to \min_{(y,v)\in M_V}, \tag{1}$$

with admissible set M_V defined by restrictions

$$\sum_{m\in V} y^i_m = 1, \quad i \in J; \tag{2}$$

$$\sum_{n\in V} v^{i\,j}_{mn} = y^i_m, \quad \sum_{n\in V} v^{j\,i}_{mn} = y^j_m, \quad (\{i,j\},m) \in E \times V; \tag{3}$$

$$y^i_m \in \{0,1\}, \quad (i,m) \in J \times V; \tag{4}$$

$$v^{i\,j}_{mn} \in \{0,1\}, \quad (\{i,j\},m,n) \in E \times V^2. \tag{5}$$

Theorem 2. The single-valued function φ_* is an optimum solution of Veber problem $\Theta(G, V, b, c, \Phi_V)$ if and only if it exists an optimum solution $(\tilde{y}, \tilde{v})$ the tasks (1) - (5) such, that

$$\varphi_*(i) = m : \quad \tilde{y}^i_m = 1, \qquad i \in J \; ■ \tag{6}$$

PROOF. Let $\varphi \in \Phi_V$. Let's define (y, v) as follows

$$y^i_m = \chi_{\{m\}}(\;(i)), \quad (i,m) \in J \times V; \tag{7}$$

$$v^{ij}_{mn} = \chi_{\{(m,n)\}}(\varphi(i),\varphi(j)), \quad (\{i,j\},m,n) \in E \times V^2, \tag{8}$$

where $\chi_X(\cdot)$ is characteristic function of set X. The equality $F(\varphi) = F(y,v)$ in this case is obvious. The inclusion $(y, v) \in M_V$ follows from a uniqueness of map φ.

Back, if $(y, v) \in M_V$ than, in the correspondence with (2) and (4), function $\varphi : J \to V : \varphi(i) = m : y^i_m = 1$ is univalent. Besides it is follows from (3) and (5) that $v^{ij}_{mn} = 1$ if and only if $y^i_m = y^j_n = 1$. Therefore, has a place equality $F(\varphi) = F(y, v)$.

Thus, the correspondence between a set of admissible solutions of the task (1) - (5) and task $\Theta(G, V, b, c, \Phi_V)$ saving value of the goal function is established one-to-one.

Theorem 2 is proved ■

So the statement $\Theta(G, V, b, c, \Phi_V)$ as the tasks of an integer linear programming (1) - (5) is possible. Therefore in the further task (1) - (5) we shall identify with the Veber problem. Let's designate through M_V a set of admissible solutions of the task (1) - (5). A convex hull $\mathbf{conv}M_V$ we shall name as a polyhedron of Veber problem. Let's mark that each point from M_V is top of $(0-1)$-cube, and consequently, top of polyhedron $\mathbf{conv}M_V$.

4. RELAXED VEBER PROBLEM AND ITS POLYHEDRON

Let's name the task of a linear programming distinguished from the Veber problem by a replacement of integer conditions (4) - (5) with a nonnegativity variable y and v conditions

$$y^i_m \geq 0 \quad (i,m) \in J \times V; \qquad (9)$$

$$v^{i\,j}_{mn} \geq 0, \quad (\{i,j\}, m, n) \in E \times V^2. \qquad (10)$$

as *relaxed Veber problem* $\tilde{\Theta}_V$. Admissible set of the given task we shall designate through $\tilde{M}_V$ and to name as a relaxational polyhedron of Veber problem. It is obvious, that $M_V \subset \tilde{M}_V$.

Also it is obvious, that $\tilde{M}_V$ completely is contained in $(0,1)$ - cube, therefore each point from M_V is a top of polyhedron $\tilde{M}_V$. Further let us name

(1) the face with integer tops only as integer face;
(2) the polyhedron for which there is a path between any pair of integer tops taking place only through integer tops as quasi-integer polyhedron.

Theorem 3. Any two integer tops (y', v') and (y'', v'') of polyhedron $\tilde{M}_V$ belong to its some integer face ∎

PROOF. Let's consider a set of hyperplanes

$$y^i_m = y'^i_m, \quad (i,m) \in J \times V: \quad y'^i_m = y''^i_m; \qquad (11)$$

$$v^{i\,j}_{mn} = v'^{i\,j}_{mn}, \quad (\{i,j\}, m, n) \in E \times V^2: \quad v'^{i\,j}_{mn} = v''^{i\,j}_{mn}. \qquad (12)$$

which are supporting to a polyhedron $\tilde{M}_V$. The point set of a polyhedron $\tilde{M}_V$ satisfying to conditions 11 and 12 is a face. Let's designate this face through Φ. For a proof integerity of a polyhedron Φ it is enough to be convinced in absolute unimodularity of a matrix of a set of equations

$$A \begin{pmatrix} \hat{y} \\ \hat{v} \end{pmatrix} = b, \qquad (13)$$

obtained by an elemination from (1) - (3) all fixed variable and restrictions addressed an identities on an edge Φ. The system of restrictions (13) together with a condition of a nonnegativity sets a range not fixed variable $\hat{y}$ and $\hat{v}$ to face Φ. Point

$$\frac{1}{2}\left[\begin{pmatrix} \hat{y}' \\ \hat{v}' \end{pmatrix} + \begin{pmatrix} \hat{y}'' \\ \hat{v}'' \end{pmatrix}\right],$$

supplemented by fixed components belongs to face Φ. Besides for row of matrix A, appropriate to restrictions (2), we have

$$1 = \sum_{m:\ y'^i_m + y''^i_m \leq 1} a^i_m \frac{y'^i_m + y''^i_m}{2} = \frac{1}{2} \sum_{m:\ y'^i_m \neq y''^i_m} a^i_m. \qquad (14)$$

The remaining restrictions of an aspect (2), in which $y'^i_m = y''^i_m = 1$, are fulfilled on face Φ as identities.

Similarly, for row of a matrix A, appropriate to restrictions (3), we have

$$\sum_{n:\ v'^{i\,j}_{mn} \neq v''^{i\,j}_{mn}} a^{i\,j}_{mn} - 1 = 0, \quad (\{i,j\}, m) \in E \times V: \ y'^i_m \neq y''^i_m; \qquad (15)$$

$$\sum_{n:\ v'^{i\,j}_{mn} \neq v''^{i\,j}_{mn}} a^{i\,j}_{mn} - 2 = 0, \quad (\{i,j\}, m) \in E \times V: \ y'^i_m = y''^i_m = 1. \qquad (16)$$

The remaining restrictions of aspect (3) for which $y'^i_m = y''^i_m = 0$ or $v'^{i\,j}_{mn} = v''^{i\,j}_{mn} = 1$, are fulfilled on face Φ as identities.

From (14) - (16) follows, that the row of matrix A contain equally till two nonzero elements each of which is equal or 1 or -1.

Let's construct now two sets of columns of matrix A as follows

$$I_1 = \left\{\binom{i}{m} : y'^i_m = 1\right\} \bigcup \left\{\binom{i\,j}{mn} : v'^{i\,j}_{mn} = 1\right\},$$

$$I_2 = \left\{\binom{i}{m} : y''^i_m = 1\right\} \bigcup \left\{\binom{i\,j}{mn} : v''^{i\,j}_{mn} = 1\right\},$$

where $\binom{i}{m}$ is column appropriate y^i_m and $\binom{i\,j}{mn}$ is column appropriate $v^{i\,j}_{mn}$. It is clear that $I_1 \bigcap I_2 = \emptyset$ and the nonzero elements of each row of a matrix A belong to columns of one set if they have different signs, and columns of different subsets otherwise.

Thus, matrix A satisfies to a Heller's criterion (Heller, 1957) and, therefore, is absolutely unimodular.

The theorem 3 is proved ∎

It follows from the proved theorem that on the graph of a polyhedron $\tilde{M}_V$ there is a path between any pair of integer tops taking place only through integer tops. Thus it is proved

Theorem 4. The relaxational polyhedron of Veber problem is quasi-integer ∎

Problems with quasi-integer relaxational polyhedron is possible to solve by implementation of integer simplex-method. Let us construct the algorithm for solving Veber problem.

5. ALGORITHM FOR VEBER PROBLEM

Let φ be an univalent map $J \to V$. Then the point (y, v), defined in the correspondence with (7) - (8) is top of polyhedron $\tilde{M}_V$. It is easy to notice that nonzero coordinates of such top are only $x^i_{\varphi(i)} = 1$, $\forall i \in J$, and $v^{i\,j}_{\varphi(i)\varphi(j)} = 1$, $\forall i \in J$, i.e. is present only $|J| + |E|$ of nonzero coordinates. At the same time matrix of restrictions (2) - (3) has full rank equal $|J| + 2 \cdot |E| \cdot |V|$. Therefore, all integer tops of polyhedron $\tilde{M}_V$ are highly degenerate.

Dual solution (z, w) appropriate to primal solution (y, v) we shall search as a solution of linear programming problem $H(\varphi)$

$$\sum_{(i,m)\in J\times V} \left(\delta^i_m - \gamma^i_m\right) \to \max_{(w,\delta,\gamma)}, \qquad (17)$$

$$c^i_{\varphi(i)} - c^i_m = \sum_{j:\ \{i,j\}\in E} \left(w^{ij}_m - w^{ij}_{\varphi(i)}\right) - \delta^i_m,$$
$$(i, m) \in J \times V :\ m \neq \varphi(i); \qquad (18)$$

$$w^{ij}_{\varphi(i)} + w^{ji}_{\varphi(j)} = b\left(\{i, j\}, \varphi(i), \varphi(j)\right), \quad \{i, j\} \in E; \qquad (19)$$

$$w^{ij}_m + w^{ji}_n \leq b\left(\{i, j\}, m, n\right), \quad (\{i, j\}, m, n) \in E \times V^2 : \quad m \neq \varphi(i), n \neq \varphi(j); \qquad (20)$$

$$-\gamma^i_m \leq \delta^i_m \leq \gamma^i_m, \quad \gamma^i_m \geq 0, \quad i \in J,\ m \in V; \qquad (21)$$

The described above problem is constructed under the task

$$D(\varphi) = \sum_{i\in J} z^i \to \max_{(z,w)}, \qquad (22)$$

$$z^i \leq c^i_m + \sum_{j:\ \{i,j\}\in E} w^{ij}_m, \ (i, m) \in J \times V; \ (23)$$

$$w^{ij}_m + w^{ji}_n \leq b^{i\ j}_{mn}, \ \{i, j\} \in E, \ m, n \in V. \ (24)$$

which is dual task for $\tilde{\Theta}_V$ as follows. The group of restrictions ((18) is obtained from group of restrictions (23) by elemination of variable

$$z^i = c^i_{\varphi(i)} + \sum_{j:\ \{i,j\}\in E} w^{ij}_{\varphi(i)}, \qquad (25)$$

and inlet of variables δ^i_m supposing discrepancy. The equality (25) follows from $x^i_{\varphi(i)} = 1$ and conditions of a complementary slackness between direct and dual variables. Similarly, group of restrictions (24), appropriate to a variable $v^{i\ \ j}_{\varphi(i)\varphi(j)} = 1, \forall\{i, j\} \in E$, are replaced by equalities (19). The remaining restrictions of group (24) are the same as (24). A used target functional (17), and also the additional restrictions (21) guarantee, that in an optimum solution

$$\gamma^i_m = \left|\delta^i_m\right|, \quad (i, m) \in J \times V,$$

and the best value of a target functional is equal

$$H^*(\varphi) = \sum_{(i,m)\in J\times V} \min\left\{0,\ \delta^i_m\right\} \leq 0.$$

Thus, if $H^*(\varphi) = 0$ the current direct solution (y, v) has an appropriate dual admissible solution, and therefore is optimum. If $H^*(\varphi) < 0$ the set of defect restrictions of the task $H(\varphi)$

$$\tilde{B}(\varphi) = \left\{(i, m) \in J \times V :\ \delta^i_m < 0\right\} \neq \emptyset,$$

and there are tops with a smaller value of the functional.

It is obvious that not less than one of a variable $x^i_m :\ (i, m) \in \tilde{B}(\varphi)$ are nonzero for all adjacent tops with the best value of target function. If $x^i_m > 0,\ (i, m) \in \tilde{B}(\varphi)$ then it is follow from the integer condition that $x^i_m = 1$ and $x^i_n = y^{i\ j}_{n\ k} = 0$ if $n \neq m$. Thus task of searching of adjacent integer top is exchanged to installation of existence of $(i, m) \in \tilde{B}(\varphi)$ and univalent map $\tilde{\varphi}$:

$$(J \times \tilde{\varphi}(J)) \subset \left(\tilde{B}(\varphi) \bigcup B_0(\varphi)\right) \setminus B^i_m(\varphi),$$

where

$$B_0(\varphi) = \left\{(j, l) :\ \delta^j_l = 0\right\} \bigcup (J \times \varphi(J)),$$

$$B^i_m(\varphi) = \{(i, n) \in J \times V :\ n \neq m\}.$$

REFERENCES

[Balas E. and Padberg M., 1972.] On the set-covering problem. *Oper. Res.* **20. No 6.** P. 1153-161.

[Balas E. and Padberg M., 1975.] On the set-covering problem. An Algorithm for set partitioning. *Oper. Res.* **23. No 1.** P. 74-90.

[Francis R. L., White J. A., 1974.] *Facilities Layout and Location: an Analitical Approach.* Prentice-Hall, Englewood Cliffs. NJ.

[Garey M. R. and Jonson D. S., 1979.] *Computers and Intractability. A Guide to the Theory of $\mathcal{N}P$-Completeness.* Bell Laboratories, Murray Hill, New Jersey. W. H. Freeman and Company. San Fracisko.

[Grotschel M., Lovasz L., Schrijver A., 1981.] The Ellisoid Method and Its Conseqeuces in Combinatorial Optimization. *Combinatorica.* **1. No 2.** 169-197.

[Heller J., 1957.] On linear systems with integer valued solutions. *Pacif. J. Math.* **7. No 3.**

[Horst R. and Tuy H., 1993.] *Global Optimization: Deterministic Approaches.* – Heidelberg: Springer-Verlag.

[Khachiyan L. G., 1979.] Polynomial Algorithms to Linear Programming. *Soviet Math. Dokl.* **244. No 5.** P. 1093-1096.

[Panyukov A. V., Pelzwerger B. V., 1991.] Polynomial Algorithms to finite Veber problem for a tree network. *Journal of computational and Applied Mathematics.* **35.** P. 291-296.

[Trubin V. A., 1969.] On a method of solution of integer programing problems of a special kind. *Soviet Math. Dokl.* **10.** P. 1544-1546.

CONDITIONAL APPROXIMATION MINIMUM AND APPROXIMATION SADDLE POINTS OF CONVEX FUNCTIONS

V. E. Rolshchikov

Chelyabinsk State University
129, Br. Kashirinyikh str, Chelyabinsk, 454021, Russia

Abstract: Question on existence of conditional approximation minimum points are considered. Is shown, that the set of points conditional approximation minimum even in case of convex functions can not cut with set approximation saddle points. However in a limit on parameter r ($r \downarrow 0$) these sets converge to set of points of a conditional minimum. *Copyright © 1998 IFAC*

Keywords: Convex programming, Minimization, Multipliers.

In given work a problem of conditional minimization of function $f_0 : \mathbf{R}^n \to \mathbf{R}$ on set $Q \subset \mathbf{R}^n$, that is

$$f_0(x) \to \min , \ x \in Q;$$

$$Q = \bigcap_{i=1}^{m} Q_i, \ Q_i = \{x \in \mathbf{R}^n \mid f_i(x) \leq 0\} \ , (1)$$

where $f_i : \mathbf{R}^n \to \mathbf{R}$, $i = \overline{1,m}$ is considered. In general case of functions f_i, $\overline{0,m}$ are not smooth. We shall search the decision on the basis of entered and investigated in in works (Batukhtin and Maiboroda, 1984,1995; Batukhtin 1993) concept of approximate gradient. In general case on the basis of approximate gradient. it is possible to consider questions of optimization of discontinuous functions (Batukhtin 1993; Batukhtin and Maiboroda, 1984,1995; Rolshchikov 1988). Here we consider convex functions. Questions of conditional minimization of convex function at continuously differentiable restrictions (Batukhtin and Maiboroda, 1995; Rolshchikov 1991). In given work of a condition on function f_i, $\overline{1,m}$, are loosed up to convex it is possible of discontinuous functions. The necessity of such researches is justified by construction of numerical methods of optimization on the basis of approximate gradient (Batukhtin 1995 Batukhtin, et al., 1997; Bigil'deeva and Rolshchikov, 1994).

Let $f_i \ : \ R^n \to R$, $i = \overline{0,m}$, is convex eigenfuctions and $int(domf_i) \neq \emptyset$, $i = \overline{0,m}$.

Let $a(x;r,p;f)$ is approximation gradient ((Batukhtin and Maiboroda, 1984,1995; Batukhtin 1993)), equality determined from

$$a(x;r,p;f) = \frac{1}{d_r} \int\limits_{B_r} sf(x+s)p_r(s)ds \ ,$$

$$d_r = \int\limits_{B_r} s_i^2 p_r(s)ds = \frac{1}{n} \int\limits_{B_r} ||s||^2 p_r(s)ds \ ,$$

here

$$B_r = B_r(0), \ B_r(y) = \overline{V}(r,y),$$

$$V(r,y) = \{x \in R^n \mid ||x-y|| < r\};$$

$$p_r(s) = p^*(r,||s||), \ \forall s \notin B_r \ p_r(s) = 0,$$

$$\forall s \in B_r p_r(s) \geq 0 \ \int\limits_{B_r} p_r(s)ds = 1$$

Definition 1. The point $x_c \in Q$ is called the conditional approximation minimum point function $f_0 : R_n \to R$ at the condition $Q \subset R_n$ and fixed $r > 0$, p, if exists $\varepsilon > 0$ such, that for all point $x \in Q \bigcap B_\varepsilon(x_c)$ the inequality

$$\langle a(x;r,p;f_0), \ x - x_c \rangle \geq 0$$

will be executed.

We shall designate $CD(r, p, Q, f_0)$ set of all such points.

Theorem 1. Let $int(Q) \neq \emptyset$, function f_0 is bounded from below on Q (i.e. $\inf f_0(x) = c_* > -\infty$), and the set

$$M(f_0, c_*) = \{x \in R^n \mid f_0(x) \leq c_*\}$$

is bounded, then $\exists\ r^* > 0$ such, that for all $r \in (0, r^*]$, p the set $CD(r, p, Q, f_0)$ is not empty and is bounded.

beginthm Let $p_r(s) = \frac{1}{r^n} p^*(\frac{s}{r})$. Then the set $CD(r, p, Q, f_0)$ converges at $r \downarrow 0$ to set of conditional minimum's points of function f_0. endthm

We shall designate $\Lambda^{m+1} = \{\lambda \in R^{m+1} \mid \lambda_i \geq 0,\ i = \overline{0, m}\}$. As is known [2] the point $(x^*, \lambda^*) \in Q \times \Lambda^{m+1}$ is a saddle point of function

$$L(x,\ \lambda) = \sum_{i=o}^{m} \lambda_i f_i(x)$$

if and only if x^* is a minimum point of function $\psi_{\lambda^*}(x) = L(x,\ \lambda^*)$ and conditions complementary slackness $\lambda_i^* f_i(x^*) = 0, \quad i = \overline{1, m}$ are executed.

Definition 2. A point $(x_r,\ \lambda_r) \in Q \times \Lambda^{m+1}$ is called approximation saddle point of function L at fixed $r > 0$ and p, if $\exists \varepsilon > 0\ :\ \forall x \in B_\varepsilon(x_r)$

$$\langle a(x; r, p; \psi_{\lambda_r}),\ x - x_r \rangle \geq 0\ ,$$

$$\lambda_{r\ i} f_i(x_r) = 0, \quad i = \overline{1, m}\ .$$

We shall designate $SP(r, p, L) \subset Q \times \Lambda^{m+1}$ set of all such points.

Definition 3. A point $(x_*, \lambda^*) \in Q \times \Lambda^{m+1}$ is called generalized saddle point of function L,if $\exists r^* > 0\ :\ \forall r \in (0, r^*]\ \forall p\ \exists (x_r, \lambda_r) \in SP(r, p, L)$ and convergence

$$x_r \rightarrow x_*,\ \lambda_r \rightarrow \lambda^*$$

takes place at $r \rightarrow 0$.

We shall note, that sets $SP(r, p, L), CD(r, p, Q, f_0)$ can not coicide and even not to be cut.

Theorem 2. Let (x_*, λ^*) is generalized saddle point and $\lambda_0^* > 0$. Then x_* is a conditional minimum's points of function f_0.

REFERENCES

Batukhtin, V.D. and L.A. Maiboroda (1984). *Optimization of Discontinuous Functions.* Nauka, Moskow.

Batukhtin, V.D. and L.A. Maiboroda (1995). *Discontinuous Extremal Problems.* Gippocrat, St.-Peterburg.

Batukhtin, V.D. (1989). Problms of the Analysis of Discontinuous Functions and Nonsmoth Optimization. *Preprint Ural Branch of Sovet Union Academy of Science*

Batukhtin, V.D. (1993). On Solving Discontinuous Extremal Problems. *Journal of Optimization Theory and Applications,* **77**, 575-589.

Batukhtin, V.D., S.I. Bigil'deev and T.B. Bigil'deeva (1997). Numerical Methods for Solution of Discontinuous Extremal Problems. *Journal of Computer and System Sciences,* **3**, 113-120.

Vasilyev, F.P. (1980). *Numerical Methods of Solving Extremal Problems.* Nauka, Moskow.

Bigil'deeva, T.B. and V.E. Rolshchikov (1994). Numerical Methods of Optimization Discontinuous Functions. *News Russian Academy of Science, Technical Cybernetics,* **3**, 47-54.

Rolshchikov, V.E. (1988) Approximate Minimum of One Class of Discontinuous Function. *In proceedings: Nonsmooth Problems of Optimization and Control, Ural Branch Sovet Union Academy of Science* 41-56.

Rolshchikov, V.E. (1991) On Convergence of Sequence Points of Conditional Approximate Minimun *Vestn. Chelyabinsk State University,* **1**, 112-116.

A PIECEWISE LINEAR MINIMAX SOLUTION OF THE HAMILTON-JACOBI EQUATION

L.G. Shagalova

Institute of Mathematics and Mechanics,
S.Kovalevskaya str., 16, Ekaterinburg, 620219, Russia
e-mail: shag@imm.uran.ru

Abstract: The Cauchy problem for the Hamilton - Jacobi equation with Hamiltonian independent of time and the phase variable is considered under the assumption that the Hamiltonian and the boundary function are piecewise linear. A finite algorithm for the construction of the exact piecewise linear minimax (and/or viscosity) solution is developed in the case when the phase space is two-dimensional. The fact that a minimax solution is a piecewise linear function is established also for one special case when the phase space is three-dimensional. This results can be used in the researches of bifurcations of piecewise smooth solutions of PDEs of first order, and in the development of numerical methods.

Keywords: algorithms; differential equations; differential games; stability properties; numerical methods; piecewise linear analysis.

1. INTRODUCTION

It is known that Hamilton-Jacobi equations usually do not have classical solutions. By this reason many concepts of a generalized solution have been considered in the theory of these equations.

The investigations of many mathematicians are connected with the concept of a viscosity solution, introduced by Crandall and Lions (1983). Another approach to the definition of a generalized solution originates from the theory of positional differential games (Krasovskiĭ and Subbotin, 1974). In (Subbotin, 1980) a generalized solution of first - order PDE was defined by substituting the equation by a pair of differential inequalities. The term "minimax solution" was suggested in (Subbotin, 1991), where a study of such solutions was given. In particular, the equivalence of minimax and viscosity solutions was proved in this book. In monograph (Subbotin, 1995) the fact, that the theory of minimax solutions can be considered as a nonclassical method of characteristics, is substantiated.

In this paper the Cauchy problem for the Hamilton - Jacobi equation with Hamiltonian independent of time and the phase variable is considered. It is also assumed that the Hamiltonian and the boundary function are piecewise linear. For the case when the phase space is two-dimensional, and for one special case when the phase space is three-dimensional it is established that under these assumptions minimax solution also turns out to be a piecewise linear function.

If the phase space is two-dimensional, and the Hamiltonian is positively homogeneous, a finite algorithm for constructing of the exact minimax solution is elaborated and justified. The solution is constructed in the class of piecewise linear functions, that may be defined by the use of structural matrices. On the base of this algorithm the computing program and the program, drawing the level lines for corresponding minimax solutions were worked out.

The results of this paper can be used in researches

of bifurcations of piecewise smooth generalized solutions to the first-order PDEs in general form and in the development of corresponding numerical methods.

2. THE PROBLEM STATEMENT

The following Cauchy problem is considered

$$\frac{\partial u(t,x)}{\partial t} + H(D_x u(t,x)) = 0, t \in (0,1), x \in R^n \quad (1)$$

$$u(1,x) = \sigma(x), x \in R^n \quad (2)$$

where $D_x u$ is the gradient of the function u with respect to the variable x, and the Hamiltonian $H : R^n \to R$ and the terminal function $\sigma : R^n \to R$ are continuous and satisfy the conditions

$$|H(s^{(1)}) - H(s^{(2)})| \leq \lambda ||s^{(1)} - s^{(2)}||, \quad (3)$$

$$\sigma(\alpha x) = \alpha\sigma(x) \quad (4)$$

for any $s^{(i)} \in R^n, x \in R^n, \alpha \geq 0$.

It is required to construct a minimax (and/or viscosity) solution of this problem, which exists and is unique (Subbotin, 1991, 1995).

Use the relation

$$u(t,x) = (1-t)u(0, \frac{x}{1-t}), t \in [0,1), x \in R^n \quad (5)$$

following from the positive homogeneity condition (4), then the problem (1), (2) reduces to the problem of finding the function

$$\varphi(x) = u(0,x), x \in R^n. \quad (6)$$

The function φ is a minimax solution of the first order PDE

$$H(D\varphi(y)) + < D\varphi(y), y > - \varphi(y) = 0, y \in R^n, \quad (7)$$

which is considered along with the limit relation

$$\lim_{\alpha \downarrow 0} \alpha\varphi(y/\alpha) = \sigma(y), y \in R^n \quad (8)$$

The symbol $< l, s >$ denotes the scalar product of the vectors l and s. The minimax solution of (7) is the continuous function satisfying the pair of differential inequalities

$$\min_{f \in R^n} [d^-\varphi(y;f) - < s, f - y > + H(s) - \varphi(y)] \leq 0 \quad (9)$$

$$\max_{f \in R^n} [d^+\varphi(y;f) - < s, f - y > + H(s) - \varphi(y)] \geq 0 \quad (10)$$

for any $y \in R^n, s \in R^n$. The symbols $d^-\varphi(y;f)$ and $d^+\varphi(y;f)$ denote respectively the lower and upper derivatives of the function φ in the direction f evaluated at the point y:

$$d^-\varphi(y;f) = \lim_{\varepsilon \downarrow 0} \inf_{0<\delta<\varepsilon} \{[\varphi(y+\delta f) - \varphi(y)]\delta^{-1},$$

$$d^+\varphi(y;f) = \lim_{\varepsilon \downarrow 0} \sup_{0<\delta<\varepsilon} \{[\varphi(y+\delta f) - \varphi(y)]\delta^{-1}.$$

It should be noted, that the inequalities (9)-(10) can be written in other equivalent forms (Subbotin, 1991, 1995). Exact formulas are known for solution of problem (7), (8) ((1), (2)) in certain cases only. For example, if the Hamiltonian H or the boundary function σ are convex or concave, then the Hopf formulas (Hopf, 1965) can be applied directly. It follows from the results of (Pshenichnyi and Sagaidak, 1970; Bardi and Evans, 1984) that the generalized solutions defined by the Hopf formulas are viscosity (and/or minimax). However, there is no success in obtaining explicit formulas in the general case.

Below the problem (7), (8) will be considered under additional assumption that the functions H and σ are piecewise linear.

3. THE CONSTRUCTION OF AN EXACT SOLUTION ON THE PLANE

In this section the problem (7),(8) is considered under the following assumptions:

$$H(\cdot) \in PL, \sigma(\cdot) \in PL_+, n = 2 \quad (11)$$

Here the symbol PL denotes the set of positively homogeneous piecewise linear functions, and PL_+ stands for the set of all nonnegative functions in PL. It should be noted, that the assumption of the nonnegativity on the function σ does not lose the generality of consideration, and the positive homogeneity condition

$$H(\alpha s) = \alpha H(s) \quad (12)$$

for any $s \in R^2, \alpha \geq 0$, is essential.

A finite algorithm for the construction of the exact minimax solution of the problem (7), (8) ((1), (2)) was suggested (Subbotin and Shagalova, 1992). Now this algorithm describing below is justified in full.

3.1. The elementary problems.

In essence, the algorithm consists of successive solving elementary problems arising in a definite order. These problems can be formulated as follows. Let

$$\sigma^+(y) = \max\{< a, y >, < b, y >\},$$

$$\sigma^-(y) = \min\{< a, y >, < b, y >\},$$

where a, b and y are vectors in R^2.

Problems 1 and 2. Let some linearly independent vectors a and b be given. In problem 1 [in problem 2] it is required to construct the minimax solution of problem (7), (8), (11) with $\sigma = \sigma^+$ [with $\sigma = \sigma^-$].

The solutions of these problems are the functions

$$\varphi^+(y) = \max_{l \in [a,b]} \varphi_l(y),$$

$$\varphi^-(y) = \min_{l \in [a,b]} \varphi_l(y),$$

where $[a,b] = \{\lambda a + (1-\lambda)b | \lambda \in [0,1]\}$, $\varphi_l(y) =< l, y > +H(l)$.

Problems 3 and 4. Let some linearly independent vectors a, b and a number $r > 0$ be given. Let

$$\varphi^*(y) = \max\{\varphi_a(y), \varphi_b(y)\},$$

$$\varphi_*(y) = \min\{\varphi_a(y), \varphi_b(y)\},$$

It is required to construct a continuous functions φ^0 and φ_0 satisfying the relations

$$\{y \in R^2 | \varphi^0(y) = r\} = \{y \in R^2 | \varphi^*(y) = r\},$$

$$\{y \in R^2 | \varphi^0(y) < r\} = \{y \in R^2 | \varphi^*(y) < r\} = G^*,$$

$$\{y \in R^2 | \varphi_0(y) = r\} = \{y \in R^2 | \varphi_*(y) = r\},$$

$$\{y \in R^2 | \varphi_0(y) < r\} = \{y \in R^2 | \varphi_*(y) < r\} = G_*,$$

In the region G^* [respectively, G_*] the function φ^0 [φ_0] is to satisfy (9) and (10).

The solution of Problem 3 is the function $\varphi^0(y) = \max_s \varphi_s(y)$ for $s \in S_r(a,b)$, where $S_r(a,b) = \{s \in con(a,b) | \varphi_s(w_0) = r\}$,

$$con(a,b) = \{\lambda a + \mu b | \lambda \geq 0, \mu \geq 0\}$$

and the point w_0 is the solution of the system of two linear equations

$$\varphi_a(w_0) = r,$$

$$\varphi_b(w_0) = r.$$

In the general case (i.e., for arbitrary a, b, and r) the solution of Problem 4 may fail to exist. However, in the cases which arise in the construction of the solution of (7), (8), (11) the function φ_0 exists and has the form

$$\varphi_0(y) = \min\{\varphi_a(y), \varphi_b(y)\}.$$

3.2. The structure of the solution.

The function σ, according to (11), is formed by "sewing together" a finite collection of linear functions $\sigma_i(y) =< s_i, y >, i = 1, ..., n_\sigma$. Let $Z = \{s_i | i = 1, ..., n_\sigma\}$, $Z^* = co(Z \bigcup \{0\})$, where 0 and coN denote the zero vector and the convex hull of the set N. Let also the symbol Ω denote the set of points where the piecewise linear function H is not differentiable. Thus the set Ω consists of the point 0 and the points at which H is the sewing together of two linear functions. The following assertion is valid.

The solution φ of (6), (7), (11) is formed by sewing together the linear functions

$$\varphi_l(y) =< l, y > +H(l), l \in L,$$

where the set L consists of a finite number of elements, and $Z \subset L, (L \backslash Z) \subset (Z^* \bigcap \Omega)$.

The solution φ is constructed in the class of functions, that are formed by "sewing together" a finite collection of simple piecewise linear functions (SPLF). The main property of an SPLF is the following. If ψ is an SPLF, then for an arbitrary point y_* in its domain, there is a neighbourhood $O_\varepsilon(y_*)$ where ψ has one of three possible representations:

$$\psi(y) =< s_i, y > +h_i,$$

$$\psi(y) = \max\{< s_i, y > +h_i, < s_j, y > +h_j >,$$

$$\psi(y) = \min\{< s_i, y > +h_i, < s_j, y > +h_j > .$$

Here s_i and s_j are vectors in R^2, and h_i and h_j are numbers. Thus the domain of definitions of an SPLF contains no points in small neighbourhoods of which three or more linear functions are sewn together.

Structural matrices may be used for formal definition of SPLFs. The structural matrix (SM) contains an information about all linear functions, that are forming the corresponding SPLF. Given the SM, one can easy calculate the value of the corresponding SPLF in every point in its domain.

In the region $\{y \in R^2 | \varphi(y) > 0\}$ the solution φ of (6), (7), (11) can be written with the help of the sequence of structural matrices

$$M_1(c_1, \infty), M_2(c_2, c_1), ..., M_k(c_k, c_{k-1}), \quad (13)$$
$$0 \leq c_k < c_{k-1} < ... < c_1 < c_0 = \infty$$

The matrix $M_i(c_i, c_{i-1})$ defines the function φ in the region, where $c_i < \varphi < c_{i-1}$.

In the domain of φ there are finite number of singular points at which at least three linear functions are sewing together. All of such points are situated on the level lines $\{y \in R^2 | \varphi(y) = c_i\}, i = 1, ..., k$, and for every $i \in \{1, ..., k\}$ there is at least one singular point on the line $\{y \in R^2 | \varphi(y) = c_i\}$. To calculate any number $c_i, i = 1, ..., k$, called critical level, it is necessary to solve a finite number of Cramer systems of three linear equations.

Thus, formally the algorithm consists of a finite number of steps. Constructing of some SPLF is a matter of every step. The number of steps (the number of matrices in the sequence (13) is not known in advance. It will be defined in the process of the construction.

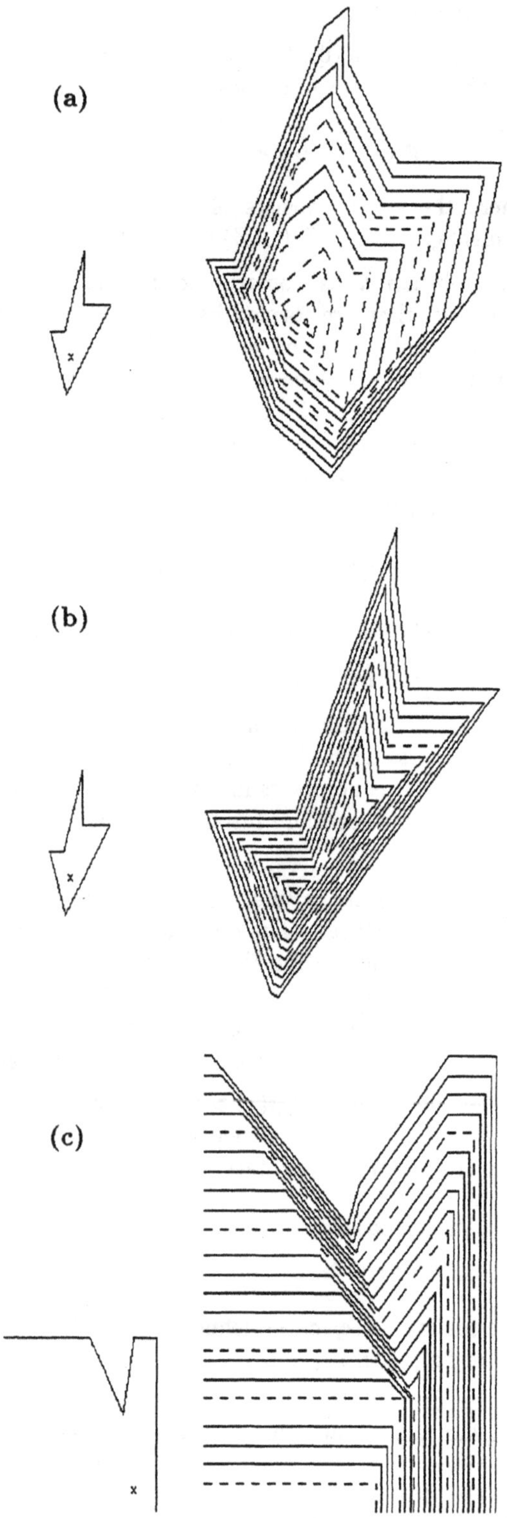

Fig. 1. Level lines of terminal functions and corresponding minimax solutions.

3.3. The computer realization.

On the base of the above algorithm the computing program and the program, drawing the level lines for corresponding minimax solutions were worked out for the case, when the Hamiltonian is the function of the form

$$H(s) = \min_{p \in P} < s, p > + \max_{q \in Q} < s, q >, s \in R^2,$$

where P and Q are convex polygons. Some illustrations, obtained with the help of these programs, are represented on Fig. 1.

On the left side of every picture on Fig. 1 there is the line $\{x \in R^2 | \sigma(x) = 1\}$. The origin is marked by cross. On the right side of the same picture there are level lines of corresponding minimax solution. The dotted lines denote the critical levels. Also it should be noted that the coordinate systems (the origins, and the scales on the axes) for the left and right parts of the picture are different, because the drawing program chooses them automatically.

3.4. The solution of non-reduced problem.

Constructing the solution φ of reduced problem (7), (8), (11), one can restore the solution u of original problem (1), (2), (11). This solution is a piecewise linear function defined in the space of variables t and x.

The following fact is valid. Let the sequence of structural matrices (13) define the solution problem (7), (8), (11). Then for any $t^* \in (0,1)$ the function $u(t^*, x)$ can be written in the form

$$M_1^*(c_1^*, \infty), M_2^*(c_2^*, c_1^*), ..., M_k^*(c_k^*, c_{k-1}^*),$$
$$0 \leq c_k^* < c_{k-1}^* < ... < c_1^* < c_0^* = \infty,$$

where $c_i^* = c_i/(1 - t_*)$, and the matrix M_i^* can be written formally with the help of simple transformation of the matrix M_i according with a definite rules.

4. THE STRUCTURE OF THE SOLUTION IN SOME SPECIAL CASES

Let the Hamiltonian H in problem (1), (2) be not positively homogeneous, and let $u(t, x)$ be the solution of this problem.

Define the positively homogeneous function

$$H^{\#}(s, r) = \begin{cases} |r| H\left(\frac{s}{|r|}\right), & if \quad r \neq 0 \\ \lim\limits_{r \downarrow 0} r H\left(\frac{s}{r}\right), & if \quad r = 0, \end{cases} \tag{14}$$

where $s \in R^n, r \in R$.

Consider the following Caushy problem

$$\frac{\partial u^{\#}}{\partial t} + H^{\#}(D_x u^{\#}, \frac{\partial u^{\#}}{\partial y}) = 0, \tag{15}$$

$$u(1, x, y) = \sigma(x) + y. \tag{16}$$

Here $t \in [0,1], x \in R^n, y \in R$.

The following assertion is valid (Subbotin, 1991).

The function $u(t, x)$ is the solution of (1), (2) iff the function $u^{\#}(t, x, y) = u(t, x) + y$ is the solution of (15), (16).

Let the functions H and σ be piecewise linear, and, besides, σ satisfy the condition (4). Using the above assertion, one can prove that minimax solution of (1), (2) is a piecewise linear function also in the cases
(a) $n = 2$, and H is not positively homogeneous;
(b) $n = 3$, H is positively homogeneous, and σ is a function of the form $\sigma(x) = \sigma(x_1, x_2) + x_3$, where $x = (x_1, x_2, x_3)^T \in R^3$, and the symbol T denotes transposition.

The proof is based on solving of elementary problems analogous those considered in section 3.1.

5. CONCLUSION

The Caushy problem for the Hamilton - Jacobi equation with Hamiltonian independent of time and the phase variable was considered in this article under the assumption that the data are piecewise linear. A finite algorithm for the construction of the exact minimax solution on the plane was described. Special case when minimax solution is piecewise linear was picked out when the phase space is three-dimensional. This results can be used in the researches of bifurcations of piecewise smooth solutions of first order PDEs in general form, and for approximations of such solutions.

ACKNOWLEDGEMENTS

This research was supported by the Russian Foundation of Basic Researches under Grant No. 96-01-00219 and Grant No. 96-15-96245.

REFERENCES

Bardi, M. and L.C. Evans (1984). On Hopf's formulas for solutions of Hamilton-Jacobi equations. *Nonlinear Analysis, Theory, Methods, Appl.*,**8 (11)**, pp. 1373-1381.

Crandall, M.G. and P.L. Lions (1983). Viscosity solutions of Hamilton-Jacobi equations. *Trans. Amer. Math. Soc.*, **277 (1)**, pp. 1-42.

Hopf, E. (1965). Generalized solutions of nonlinear equations of first order. *J. Math. Mech.* **14**, pp. 951-973.

Krasovskiĭ, N.N. and A.I. Subbotin (1974). *Positional differential games.* Nauka, Moscow. (in Russian; rev. English transl., (1988). *Game-theoretic control problems.* Springer-Verlag, Berlin.

Pshenichnyi, B.N. and M.I. Sagaidak (1970). Differential games of prescribed duration. *Kibernetika*, **2**, pp. 54-63. (in Russian; English transl., (1970). *Cybernetics*, **6**, pp. 72-83.)

Subbotin, A.I. (1980). A generalization of the basic equation of the theory of differential games. *Dokl. Akad. Nauk SSSR.* **254**, pp. 293-297. (in Russian; English transl., (1980). *Soviet Math. Dokl.*, **22**, pp. 358-362.)

Subbotin, A.I. (1991). *Minimax inequalities and Hamilton-Jacobi equations.* Nauka, Moscow.

Subbotin, A.I. (1995). *Generalized solutions of first order PDEs: the dynamical optimization perspective.* Birkhäuser, Boston.

Subbotin, A.I. and L.G. Shagalova (1992). A piecewise linear solution of the Cauchy problem for the Hamilton-jacobi equation. *Ross. Akad. Nauk Doklady,* **325 (5)**, pp. 932-936. (in Russian; English transl., (1993). *Russian Acad. Sci. Dokl. Math.*, **46 (1)**, pp. 144-148.)

ALGORITM OF OPTIMAL TIME MOVING OF A LINEAR SYSTEM TO THE CONVEX COMPACT

G.V.Shevchenko (Novosibirsk, Russia)

S.L.Sobolev Institute of Mathematics, Siberian Branch of Russian Academy of Sciences, 630090 Novosibirsk, av. of acad. Koptyug, 4,
e-mail: alexegor@math.nsc.ru

Abstract. In this paper an iterative algorithm for solving the problem of optimal time moving of a linear system from the initial state to the given convex compact has been proposed. Global convergence of the algorithm is proved.

Keywords. Optimal control, multivarible systems, computational method, differential equations, convex programming, convergence.

1. INTRODUCTION.

The problem of optimal time moving of a linear system from the initial state to the given convex compact is considered. A final state in the considered problem is unknown a priori in contrast to be same moving to the given final point (the classic problem of a time optimal control). It can be only maintained to satisfy the conditions of transversality. This fact hampers creating algorithms for solving the considered problem and explains, in our opinion, why for the classic problem of time optimal control it has been proposed an enouph great number of its, see, for example, (Gabasov, R. and Kirillova, F.M., 1966; Fedorenko, R.P., 1978; Akulenko, L.D., 1979; Kiselev, Ju.M., 1989; Kiselev, Ju.M. and Orlov, N.V., 1991; Shevchenko, G.V., 1990; Boldyrev, V.I., 1997).

For solving of this problem the iterative algorithm of the solution being a method of a sequential approximation of a boundary value of conjugate vector ψ^0 has been proposed. The algorithm synthesizes an idea of a multidimentional method of secants (Ortega G. and Reinboldt V., 1975) with the special way of rejecting the points.

2. THE STATEMENT AND GEOMETRICAL INTERPRETATION OF THE PROBLEM.

Let a controlled object be described by the system of linear ordinary differential equations

$$\dot{x}(t) = A(t)x + B(t)u(t), x(0) = x^0, \quad (1)$$

where $x \in R^n$ is a phase vector of a state of the object, $A(t)$ and $B(t)$ are continuous matrix-functions of dimensions $n \times n$ and $n \times s$, respectively, $u \in R^s$ is a piecewise continuous control hindered by restriction

$$u(t) \in U. \quad (2)$$

Here U is a convex corporal compact from R^s, the origin of coordinates belongs to U.

It is suggested that the system (1) is completely controllable and movable by bounded control (2) to the origin.

Problem. *To find an admissible control $u^0(t)$ ($t \in [0,T]$) (2) moving a system (1) from the initial state $x(0) = x^0$ to the given convex compact $\mathbf{G} \subset R^n$ in minimal time $T = T_{\text{opt}}$.*♣

Without loss of generality we may assumed that the origin belongs to the convex compact $0 \in \mathbf{G}$.

This problem is equivalent to the next problem: find a minimal time $T = T_{\text{opt}}$ under which region of attainability $\mathbf{R}(T)$ of the system (1) from the initial state $x(0) = x^0$ will have a point to that the boundary of the convex compact $\mathbf{G}$. A set $\mathbf{R}(T)$ is rigorously convex and bounded for every finite $T > 0$ if a set $B \cdot U$ is convex and bounded.

If $\mathbf{G} \bigcap \mathbf{R}(T) = \emptyset$ then $T < T_{\text{opt}}$. And then by the theorem of separability for convex sets there exists a hyperplane which rigorously separates the region of attainability $\mathbf{R}(T)$ from the set $\mathbf{G}$. Any hyperplane of support to $\mathbf{R}(T)$ (including a separable one of the set $\mathbf{G}$) can be determined by setting certain extreme points $z^i \in \mathbf{R}(T)$ ($i = \overline{1,n}$). An arbitrary choice of different extreme points z^i does not in general garantee that a hyperplane determed by these points will be rigorously separate $\mathbf{R}(T)$ from the set $\mathbf{G}$.

Let $\tilde{c}$ be a solution of the system of linear algebraic equations $\langle c, z^i \rangle = -1$ ($i = \overline{1,n}$) where $\langle .,. \rangle$ is a scalar product of vectors and the point z^{n+1} is determined in this way:

$$z^{n+1} = \arg \max_{x \in \mathbf{R}(T)} \langle \tilde{c}, x \rangle.$$

If $\langle \tilde{c}, z^{n+1} \rangle < \langle \tilde{c}, \tilde{y} \rangle$ where the vector $\tilde{y}$ is a solution of the convex programming problem

$$\min_{y \in \mathbf{G}} \langle \tilde{c}, y \rangle, \quad (3)$$

then the vector $\tilde{c}$ determines a hyperplane which rigorously separates $\mathbf{R}(T)$ from the set $\mathbf{G}$. Otherwise case in a collection of points $z^1, z^2, \ldots, z^{n+1}$ we retain such n points $z^1, z^2, \ldots, z^{i_0-1}, z^{i_0+1}, \ldots, z^{n+1}$ that $\rho(z^1, z^2, \ldots, z^{n+1}, \tilde{y}) = \rho(z^1, z^2, \ldots, z^{i_0-1}, z^{i_0+1}, \ldots z^{n+1}, \tilde{y})$ where $\rho(z^1, z^2, \ldots, z^{n+1}, \tilde{y})$ is the distance n-dimensional simplex $L(z^1, z^2, \ldots, z^{n+1})$ with vertices $z^1, z^2, \ldots, z^{n+1}$ from the point $\tilde{y}$. The described operations are made once again. As it will be shown later in setting an index i_0 in a definite way this process approaches to some rigorously separable hyperplane for sets $\mathbf{R}(T)$ and $\mathbf{G}$.

3. ALGORITHM OF SOLVING THE PROBLEM.

We denote: T_k is k-th approximation for an optimal time T_{opt}; $c^{(k)}$ is k-th approximation of an optimal value of the initial condition of the conjugate system

$$\dot{\psi} = -A^*(t)\psi; \quad (4)$$

$\Phi^{-1}(T_k)$ is an inverse matrix to the fundamental matrix of solutions of system (1), that is $\Phi^{-1}(T_k) = \Phi^{-1}(T_k, 0)$; $x(T_k, u^{\text{п}}(t))$ is the solution of the system (1) at the moment $t = T_k$ under the control $u^{\text{п}}(t)$ ($t \in [0, T_k]$) satisfying the L.S. Pontryagin maximum principle:

$$u^{\text{п}}(t) = \arg \max_{u \in U} \langle \psi^{\text{п}}(t), B(t)u \rangle,$$

where $\psi^{\text{п}}$ is an solution of system(4) with initial condition $\psi(0) = c^{(k)}$; k is an iteration number.

Algorithm

1. $k := 0$; $T_k := 0$; $\Phi^{-1}(T_k) := I_n$ (I_n is an identity matrix of dimension $n \times n$); $\tilde{c} = -x^0/\|x^0\|$; $p := 1$; $z^1 := x^0$; $m := 0$.

2. We find solution $\tilde{y}$ of the convex programming problem (3) and give $k := k+1$, $m := m+1$, $y^m := \tilde{y}$.

3. If $\max\limits_{i=\overline{1,p}} \langle \tilde{c}, z^i \rangle < \langle \tilde{c}, \tilde{y} \rangle$ we go to step 7. Otherwise to the next step. (Validity of this inequality means that $\tilde{c}$ determines a rigorously separable hyperplane for sets $L(z^1, \ldots, z^p)$ and G).

4. If $m \leq n$ then we find $\tilde{c}$ being a solution of the system of linear algebraic equations

$$\begin{array}{l} \langle c, y^i \rangle = -0.5 \, (i = \overline{1,m}), \\ \langle c, z^i \rangle = -1, (i = \overline{1, \min(p, n-m)}), \end{array} \quad (5)$$

and in case $\max\limits_{i=\overline{1,p}} \langle \tilde{c}, z^i \rangle \leq -0.5$ we norm of it and go to step 2. In case $\max\limits_{i=\overline{1,p}} \langle \tilde{c}, z^i \rangle > -0.5$ and $\min\limits_{i=\overline{1,p}} \langle \tilde{c}, z^i \rangle < -0.5$ we find solution c^* of system of linear algebraic equations

$$\langle c, z^i \rangle = -1, \, (i = \overline{1,p}),$$

norm it, give $\tilde{c} := c^*$ and go to step 2. In case $\min\limits_{i=\overline{1,p}} \langle \tilde{c}, z^i \rangle \geq -0.5$ we change the sign of vector $\tilde{c}$ and go to step 2.

5. We find $(\lambda_1^0, \ldots, \lambda_m^0)$ being solution of the system

$$\sum_{i=1}^{m} \lambda_i y^i = z^{j_0}, \; \sum_{i=1}^{m} \lambda_i = 1, \qquad (6)$$

where $j_0 = \arg \max\limits_{i=\overline{1,p}} \langle \tilde{c}, z^i \rangle$.

6. We give

$$i_0 := \begin{cases} \text{any } i \in \Lambda \bigcap \Omega, \text{if } \Omega \neq \emptyset \\ \arg \max\limits_{i \in \Lambda} (-\lambda_i^0/\omega_i), \text{in contrary case,} \end{cases}$$

where $\Lambda = \{i : \lambda_i^0 < 0\}$, $\Omega = \{i : \omega_i \leq 0\}$, $\omega_i = \langle Y_i^+, (z^{j_0}, 1) \rangle$, Y_i^+ is i-th row of the pseudoinverse matrix to the matrix of system (6) (Hantmakher, F.R., 1966). After that, giving $y^{i_0} := y^m$, $m := m-1$, we go to step 4.

7. If $T_k = 0$ then we give $p := 0$.

8. $c^{(k)} := \Phi^{-1}(T_k)\tilde{c}$;

$$T_k := \begin{cases} T_{k-1}, \text{ if } \langle \psi^{\Pi}(T_{k-1}), x(T_{k-1}, \\ \qquad u^{\Pi}(t)) \rangle \geq \langle \psi^{\Pi}(T_{k-1}), \tilde{y} \rangle \\ T^* \;\;, \text{ in contrary case,} \end{cases}$$

where T^* is such a minimal number $\tau > T_{k-1}$, that $\langle \psi^{\Pi}(\tau), x(\tau, u^{\Pi}(t)) \rangle = \langle \psi^{\Pi}(\tau), \tilde{y} \rangle$.

9. $p := p+1$; $u^p(t) = u^{\Pi}(t) (t \in [0, T_k])$, $c^p := \psi^{\Pi}(T_k)$, $z^p := x(T_k, u^{\Pi}(t))$.

10. If $z^p \in$ G then the problem has been solved and T_k is the optimal time, $u^p(t) \, (t \in [0, T_k])$ is the optimal control.

11. If $T_{k-1} < T_k$ then we determine controls $u^i(t)$ $(i = \overline{1, p-1})$ on the interval $[T_{k-1}, T_k]$ as:

$$u^i(t) = \arg \max_{u \in U} \langle \psi^i(t), B(t)u \rangle \; (t \in [T_{k-1}, T_k]),$$

where $\psi^i(t)$ is solution of system (4) with the initial condition $\psi(T_{k-1}) = c^i$. After that we give $z^i := x(T_k, u^i(t))$; $c^i = = \psi^i(T_k)$ $(i = \overline{1, p-1})$.

12. If $p \leq n$ then go to step 4.

13. We find $\lambda_1^0, \ldots, \lambda_p^0$ being solution of system

$$\sum_{i=1}^{p} \lambda_i z^i = \tilde{y}, \; \sum_{i=1}^{p} \lambda_i = 1. \qquad (7)$$

14. We give

$$i_0 := \begin{cases} \text{any } i \in \Lambda \bigcap \Omega, \text{ if } \Lambda \bigcap \Omega \neq \emptyset \\ \arg \max\limits_{i \in \Lambda} (-\lambda_i^0/\omega_i), \text{ in contrary case,} \end{cases}$$

where $\Lambda = \{i : \lambda_i^0 < 0\}$, $\Omega = \{i : \omega_i \leq 0\}$, $\omega_i = \langle Z_i^+, (\Phi(T, 0) \cdot x^0, 0) \rangle$, Z_i^+ is i-th row of the pseudoinverse matrix to the matrix of system (7). After that $z^{i_0} := z^p$; $c^{i_0} := c^p$; $u^{i_0}(t) := u^p(t) \, (t \in [0, T_k])$, $p := p-1$ and go to step 4.

4. PROOF OF A CONVERGENCE.

The whys and wherefores of the algorithm and the proof of its convegence rely on the following statements.

Lemma 1. *Let $\Lambda \neq \emptyset$. Then*
a) if $\Lambda \bigcap \Omega \neq \emptyset$ then

$$\rho(z^1, \ldots, z^p, v) = \rho(z^1, \ldots, z^{i_0-1}, z^{i_0+1}, \ldots, z^p, v) \quad (8)$$

for any $i_0 \in \Lambda \bigcap \Omega$;
b) if $\Lambda \bigcap \Omega = \emptyset$ then the equality (8) occurs for $i_0 = \arg\max_{i \in \Lambda}(-\lambda_i^0/w_i)$.

Lemma 2. *Let $\lambda_1^0, \ldots, \lambda_m^0$ be solution of system (6) and $\Lambda \neq \emptyset$. Then*
a) if $\Lambda \bigcap \Omega \neq \emptyset$ then

$$\rho(y^1, \ldots, y^m, z^{j_0}) = \rho(y^1, \ldots, y^{i_0-1}, y^{i_0+1}, \ldots, y^m, z^{j_0}) \quad (9)$$

for any $i_0 \in \Lambda \bigcap \Omega$;
б) if $\Lambda \bigcap \Omega = \emptyset$ then the equality (9) occurs for $i_0 = \arg\max_{i \in \Lambda}(-\lambda_i^0/w_i)$.

Now we proceed directly to proof of convergence of the algorithm. We introduce the next designations:

1. $z_k^1, \ldots, z_k^p$ is a collection of points $z^1, \ldots, z^p$ in k-th iteration.
2. $y_k^1, \ldots, y_k^m$ is a collection of points $y^1, \ldots, y^m$ in k-th iteration.
3. ζ_k is a point of intersection of a hyperplane passing through points $z_k^1, \ldots, z_k^{i_0-1}, z_k^{i_0+1}, \ldots, z_k^p$ (see step 14 of the algorithm) with a normal drawn to this hyperplane from the point y_k^m.
4. χ_k is a point of intersection of the hyperplane passing through points $y_k^1, \ldots, y_k^{i_0-1}, y_k^{i_0+1}, \ldots, y_k^m$ (see step 6 of the algorithm) with a normal drawn to this hyperplane from the point $z_k^{j_0}$.
5. $\rho_k = \rho(z_k^1, \ldots, z_k^p, y_k^1, \ldots, y_k^m)$ is the Howsdorf distance between simplicies $L(z_k^1, \ldots, z_k^p)$ and $L(y_k^1, \ldots, y_k^m)$ with vertices $z_k^1, \ldots, z_k^p$ and $y_k^1, \ldots, y_k^m$ respectively.

By virtue of lemmas 1, 2 the relationship takes place

$$\rho_{k+1} \leq \rho_k. \quad (10)$$

We suggest that the relationship (10) begining with some $k \geq k_0$, is fullfilled as an equality with $\rho_k \neq 0$, $k \geq k_0$. Otherwise, that is if $\rho_{k_0} = 0$, $\mathbf{R}(T_{k_0}) \bigcap \mathbf{G} \neq \emptyset$ is valid and the problem has been solved and T_{k_0} is an optimal time.

So, let $\rho_k = \text{const} \neq 0, \forall k \geq k_0$. We denote through $\widetilde{c}_k$ a solution of the system $\langle c, z^i \rangle = -1 \, (i = \overline{1,p})$ in k-th iteration. Two cases should be considered: a). $\max_{i=\overline{1,p}} \langle \widetilde{c}_k, z_k^i \rangle \geq \langle \widetilde{c}_k, y_k^m \rangle \, \forall \, k \geq k_0$; b). $\exists \, k_1 > k_0$ such that $\max_{i=\overline{1,p}} \langle \widetilde{c}_{k_1}, z_{k_1}^i \rangle < \langle \widetilde{c}_{k_1}, y_{k_1}^m \rangle$.

Consider the former. It is obvious that $z_k^i = z_{k_0}^i \, \forall \, k > k_0$. We suggest that $j_0 = \arg\max_{i=\overline{1,p}} \langle \widetilde{c}_k, z_{k_0}^i \rangle \equiv \text{const}$, that is, — independent of k. Let $y_k^* \in L(y_k^1, \ldots, y_k^m)$ be such that $\rho(y_k^1, \ldots, y_k^m, z_{k_0}^{j_0}) = \|y_k^* - z_{k_0}^{j_0}\|$. Since $\max_{i=\overline{1,p}} \langle \widetilde{c}_k, z_k^i \rangle \geq \langle \widetilde{c}_k, y_k^m \rangle \, \forall \, k \geq k_0$ then obviously $y_k^* = y_{k_0}^*, \forall \, k \geq k_0$. The point y_{k+1}^m, by the construction, lies on the same side from a hyperplane passing through the points $y_k^1, \ldots, y_k^m$ that the point $z_{k_0}^{j_0}$ does. Therefore, the inequality takes place

$$\|y_{k_0}^* - \chi_k\| > \|y_{k_0}^* - \chi_{k+1}\|.$$

Hence, $\chi_k \to y_{k_0}^*$ and there is founded such a $k_2 > k_0$ that $\langle \widetilde{c}_{k_2}, z_{k_0}^{j_0} \rangle < \langle \widetilde{c}_{k_2}, y_{k_2+1}^m \rangle$, which contradicts to the suggestion.

Let $j_0 = \arg\max_{i=\overline{1,p}} \langle \widetilde{c}_k, z_{k_0}^i \rangle \equiv j_0(k)$, that, is — dependent of k. Out of the sequence we single out $\{j_0(k)\}_{k=k_0}^{\infty}$ the subsequence $\{j_0(k_j)\}_{j=1}^{\infty}$, sush that $j_0(k_j) = j_0(k_{j+1}) \, \forall j \geq 1$. And again a contradictory takes place.

Now we consider the case b). Here we have (see step 8 of the algorithm) either $T_{k_1} = T_{k_1-1}$ and execution of step 9 of the algorithm for $k = k_1$

$$\max_{i=\overline{1,p-1}} \langle \widetilde{c}_k, z_k^i \rangle < \langle \widetilde{c}_k, y_k^m \rangle \text{ и } \langle \widetilde{c}_k, z_k^p \rangle \geq \langle \widetilde{c}_k, y_k^m \rangle, \quad (11)$$

or $T_{k_1} > T_{k_1-1}$.

We suggest that $T_k = T_{k-1} \, \forall k \geq k_1$. Then by virtue of the relationship (11) for $k = k_1$ it takes place the inequality $\|\zeta_{k-1}\| < \|\zeta_k\|$, since the case $\rho(z_k^1, \ldots, z_k^p, y_k^m) <$

$\rho(z_k^1, \ldots, z_k^{p-1}, y_k^m)$ contradicts to the suggestion: $\rho_k = \text{const} \neq 0$, $\forall k \geq k_0$.

Let relationship (11) be satisfyed for a sequence $\{k_j\}$. Then $\{||\zeta_{k_j}||\}$ is a strictlly increasing sequence. It is obviously bounded above by the value ρ_{k_0}. The sequence $\{||\zeta_{k_j}||\}$ has as its limit ρ_{k_0} by virtue of its strict increasity and the definitions of variables ζ_{k_j} and ρ_{k_0}. In a current iteration limit points $z_*^i = \lim_{j\to\infty} z_{k_j}^i$ $(i = \overline{1,p})$, $y_*^i = \lim_{j\to\infty} y_{k_j}^i$ $(i = \overline{1,m})$ are take as s z^i $(i = \overline{1,p})$, y^i $(i = \overline{1,m})$. Then for a some k^*, since the sequence $\{||\zeta_{k_j}||\}$ is strictly increasing we willl have relation $||\zeta_{k^*}|| > \rho_{k_0}$, which is inpossible. Hence, the suggestion what $T_k = T_{k-1}$ $\forall k \geq k_1$ is invalid.

Thus, there is a strictly increasing $\{T_k\}$ subsequence $\{T_{k_l}\}$, which can be singled out of $\{T_k\}$. It is obviously bounded above by the value T_{opt}. Hence, it is convergent. Let its limit be: $\lim_{l\to\infty} T_{k_l} = = T^*$. Since by the suggestion $\rho_k = \text{const} \neq 0$, $\forall k \geq k_0$ then $T^* < T_{\text{opt}}$. In a current iteration we take limit points $z_*^i = \lim_{l\to\infty} z_{k_l}^i$ $(i = \overline{1,p})$, $y_*^i = \lim_{l\to\infty} y_{k_l}^i$ $(i = \overline{1,m})$ as points z^i $(i = \overline{1,p})$, y^i $(i = \overline{1,m})$. Then such a iteration k' must occur when by virtue of strict increasing of the sequence $\{T_{k_l}\}$ the inequality $T_{k'} > T^*$ takes place. Contradiction. Hence, the suggestion that $\rho_k = \text{const} \neq 0$, $\forall k \geq k_0$ is invalid and $T^* = T_{\text{opt}}$.

Convergence has been proved.

REFERENCES

Gabasov, R. and Kirillova, F.M. (1966) *Building a sequential approximations for a some optimal control problems.* Automatics and telemechanics, No. 2. (in Russian)

Fedorenko, R.P. (1978) *Approximated solutions of an optimal control problems.* Moscow, Nauka. (in Russian)

Akulenko, L.D. (1979) *An Assimptotical solution of a some problems of the time optimal control type.* Applied mathematics and mechanics, v. 39, No. 4. (in Russian)

Kiselev, Ju.M. (1989) *Searching for an optimal solution of a time optimal control.* Systems: Mathematical methods of a description, SAPR and control, Kalinin, No. 29. (in Russian)

Kiselev, Ju.M. and Orlov, N.V., (1991) *A numerical algorithms of a linear time optimal control.* Journal of Computational Mathematics and Mathematical Physics, v. 31, No. 12. (in Russian)

Shevchenko, G.V. (1990) *Algorithms of Solving a some optimal control problem for a linear systems.* Novosibirsk. (Preprint/ AS USSR. Siberian Branch. Insnitute of mathematics, No. 11). 29 p. (in Russian)

Boldyrev, V.I. (1997) *Algorithm of solving a linear time optimal control problem.* Optimization, Control, Intellect. (Procedings of Russian Accociation of a mathematical programming, International Academy of Unlinear Sciences, Russian Accociation of Artifitial Intellect). Irkutsk, Ir. CC SB RAS, No. 2. pp. 119–125. (in Russian)

Ortega G. and Reinboldt V. (1975) *Iterative methods of solution for unlinear system of equations with most unknowns.* Moscow, Mir. (in Russian)

Hantmakher, F.R. (1966) *Theory of matrices.* Moscow, Nauka. (in Russian)

THE SYNTHESIS OF MINIMAX CONTROL OF LINEAR DYNAMICAL SYSTEMS WITH INCOMPLETE INFORMATION ABOUT PARAMETERS, DISTURBANCES AND INEXACT MEASUREMENTS *

Smolyansky N.Yu, Shiryaev V.I

Applied Mathematics Department of Southern Ural State University,
Lenin av. 76, Chelyabinsk, 454080, Russia

Abstract: Decision of identification problem for matrices' elements of system in the case of absent stochastic disturbances is proposed. The algorithm uses interval analysis procedures. *Copyright © 1998 IFAC*

Keywords: control theory, estimation, identification, uncertain dynamic linear system, intervals.

We offer the decision of identification problem for matrices elements of object as a result of partial and inexact measurements in the case of absent stochastic disturbances. The algorithm uses interval analysis procedures and geometrical operations on polytops for identification aims.

The linear dynamical system is described by equations of movement and measurement (Krasovsky, 1993)

$$\begin{cases} \mathbf{x_k} = \mathbf{A}\mathbf{x_{k-1}} + \Gamma \mathbf{w_k}\,; \\ \mathbf{y_k} = \mathbf{G}\mathbf{x_k} + \mathbf{H}\mathbf{v_k},\ k = 1, \dots\,, \end{cases} \tag{1}$$

where the vectors $\mathbf{x_k}$ and $\mathbf{y_k}$ are the state and the measurement of the system; the vectors $\mathbf{w_k}$ and $\mathbf{v_k}$ are the non-stochastic disturbances from the given convex sets W and V, respectively. $\mathbf{A}$ is the interval matrix of parameters with the elements in the form of the intervals $[\underline{a}(i,j), \overline{a}(i,j)]$. Another matrices of the system are known and defined. The physical sence of the interval matrix is contained in the following: the parameters of the dynamic system belong some intervals where they can vary in uncertain way for example in chaotic way (Mun, 1990).

In minimax case we have well-known equations for state vector $\mathbf{x_k}$ estimating with use of information sets X_k that are convex polytopes

$$\begin{cases} X_k = X_k(\mathbf{y_k}) \bigcap X_{k/k-1}\,, \\ X_k(\mathbf{y_k}) = \{\mathbf{x} :\ \mathbf{G}\mathbf{x} + \mathbf{H}\mathbf{v_k} = \mathbf{y_k}, \\ \qquad \mathbf{v_k} \in V\}\,, \\ X_{k/k-1} = \mathbf{A}X_{k-1} + \Gamma W\,, \\ k = 1, \dots\,. \end{cases} \tag{2}$$

*This work is supported by Russian Foundation of Basic Researches, grant 96-01-00460

Consider the identification algorithm for the matrix $\mathbf{A}$ that defines more exactly the interval elements of the matrix with use of current measurements and minimax filter's estimates. As a result we have better approximations of the point values of the parameters that are uncertain and are subject of the identification. For this problem the algorithm uses procedures of interval minimax estimation of the state vector. It transforms a priori information set on the basis of current measurement. Properties of the constructing interval transformation algorithm are important for identification. The construction of the interval transformation is based on the following statements.

Let $\mathbf{A}$ is an interval matrix, $\mathbf{x}$ is a vector in appropriate space. Then the exact transformation $\mathbf{Ax} = \bigcup_i \{\mathbf{A}_i\mathbf{x};\ \forall i:\ \mathbf{A}_i \in \mathbf{A}\}$ coincides with transformation of the vector in interval sence, image of that is the interval vector in same space.

Let $\mathbf{A}$ is an interval matrix, X is a convex polytope in appropriate space. Then the exact convex hull of the exact interval transformation is

$$conv(AX) = conv(\bigcup_i \{\mathbf{A}_i\mathbf{x};\ \forall i:\ \mathbf{x}_i \in V_X\}), \tag{3}$$

where V_X is the vertices set X.

So we need to know the vertices of the polytope X for constucting its interval transformation.

The identification algorithm (step K).

1. We obtain $X_{k/k-1}$ - a priori estimation of phase vector with help of filter (2) and improve it with help of measurements for finding X_k - a posteriori estimation. We use current value of identifiable intreval matrix $\mathbf{A}_{k-1}$.

2. Let's consider ratio from (2)

$$X_{k/k-1} = \mathbf{A}_{k-1}X_{k-1} + \Gamma W\ . \tag{4}$$

   Let $X_{k/k-1} = \mathbf{A}_{id}X_{k-1} + \mathbf{D}$, where $\mathbf{A}_{id} = \mathbf{A}_{k-1}$ our matrix, $\mathbf{D}$ - working set. $X_{k/k-1}$ is made with help of interval transformation and Minkovsky sum of polytops. We will memorize N vertices of X_{k-1} and corresponding hyperbars on each cycle when finding convex hull of interval transformation $\mathbf{A}_{k-1}X_{k-1}$ (where N is dimension of space).

3. $X_{k/k-1}$ is transformed with help of parallel transposition and similarity transformation for better approximation of X_k (a posteriori estimation). After transformations, we have set $X^{an}_{k/k-1}$. We will find interval matrix $\mathbf{A}_{id}$, so when we will calculate a priori estimation again, $X^{an}_{k/k-1} = X_{k/k-1}$ should be the truth.

   I.e. we have

$$\begin{aligned} X_{k/k-1} &= \mathbf{A}_{id}X_{k-1} + \mathbf{D}\ , \\ X^{an}_{k/k-1} &= \mathbf{A}^{H}_{id}X_{k-1} + \mathbf{D}\ , \\ X^{an}_{k/k-1} \overset{*}{-} \mathbf{D} &= \mathbf{A}^{i}_{id}X_{k-1}\ , \\ \mathbb{D} &= \mathbf{A}^{H}_{id}X_{k-1}\ , \end{aligned} \tag{5}$$

   geometrical subtraction is used in these relations.

4. When we obtained $\mathbf{A}_{k-1}X_{k-1}$, we used the algorithm for an interval transformation constructing, that uses vertices of polytop to hyperbars transformation and constructs convex hull of those hyperbars unite.

   We will use N hyperbars of $\mathbf{A}_{id}X_{k-1}$ for identification, that were memorized through step 2, but were transformed in according to transformation which was made on $X_{k/k-1}$ for obtaining $X^{an}_{k/k-1}$. These hyperbars will reside to polytop $\mathbb{D}$ from (4), i.e. we need to know N vectors of X_{k-1} and N corresponding hyperbars from $\mathbb{D}$ (that are oriented by axis). N hyperbars are N interval vectors points are transformed to which with help of multiplication on interval matrix.

   So, we have

$$C(i) = \mathbf{A}^{H}_{id}X_{k-1}(i), i = 1, \ldots, N,$$

   where $C(i)$ - hyperbar (interval vector), $X_{k-1}(i)$ corresponding to hyperbar vertice. We have $C = \mathbf{A}^{H}_{id}X_b$ where C is interval matrix with N interval vectors as columns, X_b is matrix columns of which are corresponding vertices of polytop X_{k-1}.

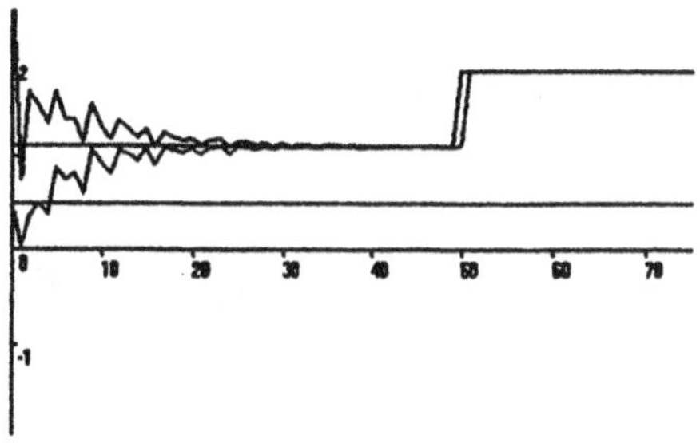

Fig. 1. Example of scalar identification

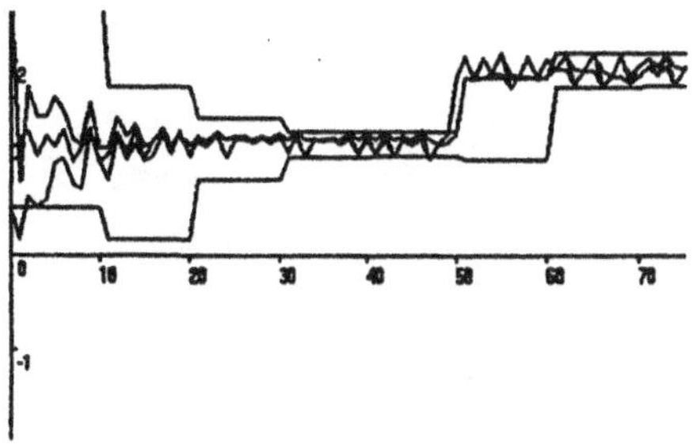

Fig. 2. Example of scalar identification

$$X_b{}^T(\mathbf{A}_{id}^H)^T = C^T \, , \qquad (6)$$

$$X_b{}^T(\mathbf{A}_{id}^H)^T(i) = C^T(i) \, .$$

I.e. we have obtained N interval linear systems with point left side and interval right side. We will solve them with help of S. Shary procedures(Shary, 1996) for solving such systems (interval linear system of equations).

So, we will find interval matrix $\mathbf{A}_{id}^H$ with help of solving of N interval linear systems. Each system (5) will give row of interval matrix.

5. Improves a posteriori estimation with help of $\mathbf{A}_{id}^H$ for better results in minimax filter.

6. Next step $(K+1)$ in filter (2) with using new matrix $\mathbf{A}_k = \mathbf{A}_{id}^H$.

If we apply this algorithm for dynamic system which have constant unknown parameters then it converges and gives us good identification results, see Fig. 1.

But in the case when parameters are chaotic and belong to intervals this algorithm shows bad results. For this case we use modification of algorithm with using of sliding window (see Fig. 2).

REFERENCES

Krasovsky, N.N. (1993). Control and stabilization under incomplete information. *Izvestiya RAN. Technical cybernetics*, **1**, 148-161.

Mun, F. (1990). *Chaotic Oscillations.* Mir, Moscow.

Shary, S.P. (1996). Algebraic approach to the interval linear static identification, tolerance and control problems or one more application of Kaucher arithmetic. *Reliable Computing*, **2**, 3-33.

DIFFERENTIAL GAME SOLUBILITY CONDITION IN H^∞-OPTIMIZATION

Vladimir Ya.Turetsky

The Ural State University, Ekaterinburg, Russia

Abstract: A solubility condition for the linear-quadratic differential game is applied to the H^∞ optimization problem. The game in rather general form is considered. Condition is obtained in a form of restriction on the maximal eigenvalue of some self-adjoint compact operator. *Copyright © 1998 IFAC*

Keywords: H^∞ optimization, differential game, operator eigenvalue.

1. INTRODUCTION

A well-known H^∞ optimization problem (Basar and Bernhard, 1991; Chu et al., 1986; Limebeer et al., 1991) is considered. In case of full state vector measurement it may be formulated as follows. Consider a controlled system

$$\dot{x} = A(t)x + B(t)u + C(t)v, t_0 \leq t \leq \vartheta, x(t_0) = x_0, \quad (1)$$

with $x \in R^n$, $u \in R^m, v \in R^k$ as state vector, control and disturbance correspondingly and the output $z^T = (x, u)$ (T denotes transposed vector or matrix); A, B, C are piecewise continuous matrices of corresponding dimensions. The problem is to decrease the influence of the disturbance, i.e to minimize the norm of operator $T_{vz} : v(\cdot) \rightarrow z(\cdot)$ via feedback inputs $u = Fx$. The well-known approach (Basar and Bernhard, 1991) gives a solution of the H^∞-problem in connection with associated linear-quadratic differential game for the (1) with the index

$$J_\gamma = \int_{t_0}^{\vartheta} |x(t)|^2 dt + \int_{t_0}^{\vartheta} |u(t)|^2 dt - \gamma^2 \int_{t_0}^{\vartheta} |v(t)|^2 dt, \quad (2)$$

where $|x|$ denotes an Euclidian vector norm. The solution

$$\gamma_0 = \inf ||T_{vz}|| \quad (3)$$

of the H^∞-problem is produced by the minimal $\gamma > 0$ with which the LQDG (1)-(2) is soluble, i.e the optimal value of the index is less than infinity. From this point of view a condition for LQDG (1)-(2) solubility leads directly to the solution of H^∞ problem. The aim of this paper is to apply a condition (Turetskij, 1989) obtained by the author in case of H^∞ problem.

2. LINEAR-QUADRATIC DIFFERENTIAL GAME

2.1. Problem statement

Consider a game in more general form than (1)-(2). Points $c^{(i)} \in [t_0, \vartheta], i = 1, \ldots, l$, and intervals $(a^{(j)}, b^{(j)}) \subset [t_0, \vartheta], j = 1, \ldots, s$, are prescribed and together with a given vector function $y(\cdot) \in L_2[t_0, \vartheta]$ generate a cost functional

$$J_{\alpha,\beta} =$$

$$= \sum_{i=1}^{l} |x(c^{(i)}) - y(c^{(i)})|^2 + \sum_{j=1}^{s} \int_{a^{(j)}}^{b^{(j)}} |x(t) - y(t)|^2 dt +$$

$$+ \alpha \int_{t_0}^{\vartheta} |u(t)|^2 dt + \beta \int_{t_0}^{\vartheta} |v(t)|^2 dt, \quad (4)$$

with $\alpha > 0, \beta > 0$. Index (4) may be rewritten in a form

$$J_{\alpha,\beta} = \int_{[t_0,\vartheta]} |x(t) - y(t)|^2 \mu(dt) +$$

$$+ \alpha \int_{t_0}^{\vartheta} |u(t)|^2 dt + \beta \int_{t_0}^{\vartheta} |v(t)|^2 dt, \quad (5)$$

where the first term is a Lebesgue-Stilties integral via corresponding Lebesgue measure $\mu(dt)$. Measure $\mu(\cdot)$ is constructed as follows. Let $T =$

$\cup_{j=1}^{s}[a^{(j)}, b^{(j)}]$, $\zeta(t)$ denotes indicator function of T set ($\zeta(t) = 1$ if $t \in T$ and $\zeta(t) = 0$ if $t \notin T$); $g([t_1, t_2])$ denotes number of points $c^{(i)}$ belonging $[t_1, t_2]$. Then

$$\mu([t_1, t_2]) = \int_{t_1}^{t_2} \zeta(t)dt + g([t_1, t_2]).$$

Note that functional (5) coinsides with the index (2) when $y(t) = 0, \mu(dt) = dt$ (i.e. $l = 1, a^{(1)} = t_0, b^{(1)} = \vartheta, s = 0$), $\alpha = 1, \beta = \gamma^2$.

The space $H = L_2([t_0, \vartheta], R^n, \mu)$ is a Hilbert space with the inner product $[h_1(\cdot), h_2(\cdot)] = \int_{[t_0,\vartheta]} h_1^T(t)h_2(t)\mu(dt)$ and norm $\|h(\cdot)\| = [h(\cdot), h(\cdot)]^{1/2}$. So the index (5) clearly may be represented in the following form:

$$J_{\alpha,\beta} = \|x(\cdot) - y(\cdot)\|^2 + \alpha\|u(\cdot)\|^2 - \beta\|v(\cdot)\|^2, \quad (6)$$

where the first norm is in the space H and the last two in $L_2([[t_0, \vartheta], R^m)$ and in $L_2([[t_0, \vartheta], R^k)$ correspondingly.

The players purposes are to minimize and maximize a functional (6). Optimal feedback strategies $u^0_{\alpha,\beta}(t, x), v^0_{\alpha,\beta}(t, x), t_0 \le t \le \vartheta, x \in R^n$, and game value $\rho^0_{\alpha,\beta}(t_0, x_0)$ are defined as usual. Below it is shown that the game value may be calculated as a program maximin which leads to the game solubility condition. To calculate a program maximin a dual representation of index is applied. To simplify a calculation a transformation $z = X[\vartheta, t]x$ is applied, where $X[t, \tau]$ is a Caushy matrix for system (1). The game (1)-(6) is rewritten in equivalent form:

$$\dot{z} = X[\vartheta, t](B(t)u + C(t)v), z(t_0) = X[\vartheta, t_0]x_0, \quad (7)$$

$$J_{\alpha,\beta} = \|w(\cdot) - y(\cdot)\|^2 + \alpha\|u(\cdot)\|^2 - \beta\|v(\cdot)\|^2, \quad (8)$$

where $w(t) = X[t, \vartheta]z(t)$.

The main result is as follows. Let introduce quasinilpotent Volterra operators

$$L_u u(\cdot) = \int_{t_0}^{t} X[\vartheta, \tau]B(\tau)u(\tau)d\tau,$$

$$L_v u(\cdot) = \int_{t_0}^{t} X[\vartheta, \tau]C(\tau)v(\tau)d\tau.$$

and compact self-adjoint non-negative operators $F_u = L_u L_u^*, F_v = L_v L_v^*$. Let $\lambda_{\alpha,\beta}$ denotes maximal eigenvalue of operator

$$F_{\alpha,\beta} = \frac{1}{\beta}F_v - \frac{1}{\alpha}F_u. \quad (9)$$

Theorem 2.1. *Suppose*

$$\lambda_{\alpha,\beta} < 1. \quad (10)$$

Then differential game (7)-(8) has the value

$$\rho^0_{\alpha,\beta}(t_0, z_0) = [w_0(\cdot), (I - F_{\alpha,\beta})^{-1}w_0(\cdot)], \quad (11)$$

where $w_0(t) = X[t, \vartheta]z_0 - y(t), \ t_0 \le t \le \vartheta$.

The proof of the theorem consists of several steps the main of which is a program maximin calculation. Explicit formulae for the game value and optimal strategies are obtained.

2.2. Program construction

A *program maximin* is defined as follows:

$$\rho_{\alpha,\beta}(t_0, z_0) = \max_{v(\cdot)} \min_{u(\cdot)} J, \quad (12)$$

where maximum and minimum are calculated via arbitrary functions $v(\cdot), u(\cdot) \in L_2[t_0, \vartheta]$. To calculate a program maximin (12) a dual representation of the first term in the index (8) is applied.

Lemma 2.1. *Let H is Hilbert space with inner product $[h_1, h_2]$ and norm $\|h\| = [h, h]^{1/2}$. Let R is linear self-adjoint positively defined operator mapping H to itself, $h_* \in H$ is a fixed element. Then*

$$[h_*, R^{-1}h_*] = \max_{h\in H}\left\{[h, h_*] - \frac{1}{4}[h, Rh]\right\}$$

and the unique maximizing vector is $h^* = 2R^{-1}h_*$.

Proof of the lemma is based on the obvious equality $[h, h_*] - \frac{1}{4}[h, Rh] = [h_*, R^{-1}h_*] - \frac{1}{4}[(Rh - 2h_*), R^{-1}(Rh - 2h_*)]$, where the last term is nonnegative and maximum is clearly achieved when $Rh - 2h_* = 0$ i.e. $h = h^* = 2R^{-1}h_*$. Particularly in case of $R = I$ (identity operator): $\|h_*\|^2 = \max_{h\in H}\left\{[h, h_*] - \frac{1}{4}\|h\|^2\right\}$ and with $h_* = w(\cdot) - y(\cdot)$ a program maximin (12) may be written as

$$\rho_{\alpha,\beta}(t_0, z_0) = \max_{v(\cdot)} \min_{u(\cdot)} \max_{h(\cdot)} \varphi(h(\cdot), u(\cdot), v(\cdot)), \quad (13)$$

where $\varphi(\cdot)$ denotes functional

$$\varphi(h(\cdot), u(\cdot), v(\cdot)) = [h(\cdot), w(\cdot) - y(\cdot)] - \frac{1}{4}\|h(\cdot)\|^2 + \alpha\|u(\cdot)\|^2 - \beta\|v(\cdot)\|^2.$$

Lemma 2.2. *For arbitrary fixed $v(\cdot) = v_*(\cdot) \in L_2([t_0, \vartheta], R^k)$*

$$\min_{u(\cdot)} \max_{h(\cdot)} \varphi(h(\cdot), u(\cdot), v_*(\cdot)) = \max_{h(\cdot)} \min_{u(\cdot)} \varphi(h(\cdot), u(\cdot), v_*(\cdot)). \quad (14)$$

Proof. The proof is based on the straightforward calculation of maximizing and minimizing elements. In all cases quadratic functionals in Hilbert spaces are maximized or minimized. So some details are omitted.

1. **Minimax**. It is clear from lemma 2.1 that with fixed $u(\cdot)$ maximum in the left side of (14) is achieved at $h^*(\cdot) = 2(w(\cdot) - y(\cdot))$. If $w_*(t) = X[t,\vartheta]z_0 + L_v v_*(\cdot) - y(t)$ then $w(\cdot) - y(\cdot) = w_*(\cdot) + L_u u(\cdot)$ and substitution $h(\cdot)$ with $2(w_*(\cdot) + L_u u(\cdot))$ gives

$$\varphi(h^*(\cdot), u(\cdot), v_*(\cdot)) = \|w_*(\cdot) + L_u u(\cdot)\|^2 +$$

$$+\alpha\|u(\cdot)\|^2 - \beta|v_*(\cdot)\|^2 =$$

$$= \|w_*(\cdot)\|^2 + 2[L_u^* w_*(\cdot), u(\cdot)] +$$

$$+[u(\cdot), (\alpha I + L_u^* L_u)u(\cdot)] - \beta|v_*(\cdot)\|^2,$$

fromwhere minimizing function is written by

$$u_*(\cdot) = -(\alpha I + L_u^* L_u)^{-1} L_u^* w_*(\cdot).$$

Then minimax is immedeately calculated by

$$\min_{u(\cdot)} \max_{h(\cdot)} \varphi(h(\cdot), u(\cdot), v_*(\cdot)) =$$

$$= [w_*(\cdot), (I - L_u(\alpha I + L_u^* L_u)^{-1} L_u^*)\, w_*(\cdot)] -$$

$$- \beta|v_*(\cdot)\|^2. \qquad (15)$$

2. **Maximin**. With arbitrary fixed $h(\cdot)$ functional

$$\varphi(h(\cdot), u(\cdot), v_*(\cdot)) = [h(\cdot), w_*(\cdot)] + [L_u^* h(\cdot), u(\cdot)] -$$

$$-\frac{1}{4}\|h(\cdot)\|^2 + \alpha\|u(\cdot)\|^2 - \beta|v_*(\cdot)\|^2$$

is minimized by $u^*(\cdot) = -\dfrac{1}{2\alpha} L_u^* h(\cdot)$ and

$$\varphi(h(\cdot), u^*(\cdot), v_*(\cdot)) = [h(\cdot), w_*(\cdot)] -$$

$$-\frac{1}{4}[h(\cdot), (I + \frac{1}{\alpha} L_u L_u^*) h(\cdot)] - \beta|v_*(\cdot)\|^2.$$

Maximizing function is given by $h_*(\cdot) = 2((I + \dfrac{1}{\alpha} L_u L_u^*)^{-1} w_*(\cdot)$ and

$$\max_{h(\cdot)} \min_{u(\cdot)} \varphi(h(\cdot), u(\cdot), v_*(\cdot)) =$$

$$= [w_*(\cdot), (I + \frac{1}{\alpha} L_u L_u^*)^{-1} w_*(\cdot)] - \beta|v_*(\cdot)\|^2. \quad (16)$$

To complete proof it is sufficient to show that

$$I - L_u(\alpha I + L_u^* L_u)^{-1} L_u^* = (I + \frac{1}{\alpha} L_u L_u^*)^{-1}$$

In fact
$(I - L_u(\alpha I + L_u^* L_u)^{-1} L_u^*)\ (I + \frac{1}{\alpha} L_u L_u^*) = I +$
$+\frac{1}{\alpha} L_u \left(I - (I + \frac{1}{\alpha} L_u L_u^*)^{-1} (I + \frac{1}{\alpha} L_u L_u^*)\right) L_u^* =$
$= I$. Left multiplication leads to the same result.

Calculating minimax instead of maximin helps to obtain expressions for optimal strategies and game value in more practical manner. This leads directly to the game solubility condition. From lemma 2.1 the following result is derived.

Lemma 2.3. *The program maximin is given by the following formula:*

$$\rho_{\alpha,\beta}(t_0, z_0) =$$

$$= [w_0(\cdot), (I + \frac{1}{\alpha} L_u L_u^* - \frac{1}{\beta} L_v L_v^*)^{-1} w_0(\cdot)], \quad (17)$$

where $w_0(t) = X[t,\vartheta]z_0 - y(t)$, $t_0 \le t \le \vartheta$, *if the inverse operator exists.*

Proof. Firstly it is clear that $w(\cdot) - y(\cdot) = w_0(\cdot) + L_u u(\cdot) + L_v v(\cdot)$. The proof is immedeately obtained by swapping min via $u(\cdot)$ and max via $h(\cdot)$ according to lemma 2.2 and then max via $v(\cdot)$ and max via $h(\cdot)$:

$$\rho_{\alpha,\beta}(t_0, z_0) =$$

$$= \max_{h(\cdot)} \max_{v(\cdot)} \min_{u(\cdot)} \{[w_0(\cdot) + L_u u(\cdot) + L_v v(\cdot), h(\cdot)] -$$

$$-\frac{1}{4}\|h(\cdot)\|^2 + \alpha\|u(\cdot)\|^2 - \beta\|v(\cdot)\|^2\}.$$

With fixed $h(\cdot)$ minimizing and maximizing elements are given by $u^0(\cdot) = -\dfrac{1}{2\alpha} L_u^* h(\cdot)$, $v^0(\cdot) = \dfrac{1}{2\beta} L_v^* h(\cdot)$. Substitution $u(\cdot), v(\cdot)$ with $u^0(\cdot), v^0(\cdot)$ leads to expression

$$\rho_{\alpha,\beta}(t_0, z_0) = \max_{h(\cdot)} \{[w_0(\cdot), h(\cdot)] -$$

$$-\frac{1}{4}[h(\cdot), (I + \frac{1}{\alpha} L_u L_u^* - \frac{1}{\beta} L_v L_v^*) h(\cdot)]\},$$

maximizing element is

$$h^0_{\alpha,\beta}(\cdot) = 2(I + \frac{1}{\alpha} L_u L_u^* - \frac{1}{\beta} L_v L_v^*)^{-1} w_0(\cdot)$$

and equality (17) is immediately derived. Optimal $u(\cdot), v(\cdot)$ are written as follows.

$$u^0_{\alpha,\beta}(\cdot) = -\frac{1}{2\alpha} L_u^* h^0_{\alpha,\beta}(\cdot),\ v^0_{\alpha,\beta}(\cdot) = -\frac{1}{2\beta} L_v^* h^0_{\alpha,\beta}(\cdot),$$

It may be easily shown that operators $F_u = L_u L_u^*, F_v = L_v L_v^*$ mapping H to H are self-adjoint, non-negative and compact. Operators are obviously self-adjoint and non-negative. Compactness is checked from explicit expressions for the operators:

$$F_u h(\cdot) = \int_{[t_0,\vartheta]} F_u(\eta, \nu) h(\nu) \mu(d\nu),$$

$$F_v h(\cdot) = \int_{[t_0,\vartheta]} F_v(\eta, \nu) h(\nu) \mu(d\nu),$$

with matrices

$$F_u(\eta,\nu) = = \frac{1}{\alpha} \int_{t_0}^{\min(\eta,\nu)} X[\eta,\tau]B(\tau)B^T(\tau)X^T[\nu,\tau]d\tau,$$

$$F_v(\eta,\nu) = = \frac{1}{\beta} \int_{t_0}^{\min(\eta,\nu)} X[\eta,\tau]C(\tau)C^T(\tau)X^T[\nu,\tau]d\tau.$$

Hence operator $F_{\alpha,\beta}$ (9) is compact too.

Lemma 2.3 and compactness of $F_{\alpha,\beta}$ operator together with the Fredholm alternative (Balakrishnan, 1976) lead to a following result.

Lemma 2.4. *If the condition (10) holds then program maximin exists and*

$$\rho_{\alpha,\beta}(t_0, z_0) = [w_0(\cdot), (I - F_{\alpha,\beta})^{-1} w_0(\cdot)],$$

The purpose of this paper is mainly to present a condition for game solubility. Nevertheless an explicit formulae for game value and optimal strategies are of interest itself. Let introduce matrices

$$Q(t) = = X[\vartheta,t]\left(\frac{1}{\beta}C(t)C^T(t) - \frac{1}{\alpha}B(t)B^T(t)\right)X^T[\vartheta,t],$$

$$G(t) = X^T[t,\vartheta]X[t,\vartheta].$$

Lemma 2.5. *If $\lambda_{\alpha,\beta} < 1$, then a program maximin is given by the following equality:*

$$\rho_{\alpha,\beta}(t_0, z_0) = z_0^T R_{\alpha,\beta}(t_0) z_0 + z_0^T r_{\alpha,\beta}(t_0) + g_{\alpha,\beta}(t_0), \tag{18}$$

where piecewise continuous functions $R_{\alpha,\beta}(t), r_{\alpha,\beta}(t), g_{\alpha,\beta}(t)$ (matrix, vector and scalar correspondingly) satisfy the following step-by-step differential equations

$$\frac{dR}{dt} = -RQ(t)R - \zeta(t)G(t),$$

$$R(c^{(i)}) = R(c^{(i)} + 0) + G(c^{(i)}), \tag{19}$$

$$\frac{dr}{dt} = -R(t)Q(t)r + 2\zeta(t)X^T[t,\vartheta]y(t),$$

$$r(c^{(i)}) = r(c^{(i)} + 0) - 2X^T[c^{(i)},\vartheta]y(c^{(i)}), \tag{20}$$

$$\frac{dg}{dt} = -\frac{1}{4}r^T(t)Q(t)r(t) - \zeta(t)|y(t)|^2,$$

$$g(c^{(i)}) = g(c^{(i)} + 0) + |y(c^{(i)})|^2, \tag{21}$$

with zero boundary conditions at $t = \vartheta + 0$.

The proof is based on the fact that maximizing element $h^0_{\alpha,\beta}(\cdot)$ (see lemma 2.3) satisfies a following integral equation in $L_2([t_0,\vartheta], R^n, \mu)$:

$$h(\cdot) - Fh(\cdot) = 2w_0(\cdot)$$

or

$$h(t) - \int_{[t_0,\vartheta]} (F_v(t,\tau) - F_u(t,\tau))h(\tau)\mu(d\tau) = 2w_0(t).$$

As usual in linear-quadratic control it may be shown that the integral equation is equivalent to two-point boundary problem

$$\frac{dp}{dt} = -\zeta(t)G(t)q + \zeta(t)2X^T[t,\vartheta]y(t),$$

$$p(c^{(i)}) = p(c^{(i)} + 0) + G(c^{(i)})q(c^{(i)}) - -2X^T[c^{(i)},\vartheta]y(c^{(i)}), \quad p(\vartheta + 0) = 0,$$

$$\frac{dq}{dt} = Q(t)p, \quad q(t_0) = 2z_0,$$

where $p(t) = \int_{[t,\vartheta]} X^T[\tau,\vartheta]h(\tau)\mu(d\tau)$,

$q(t) = \int_{t_0}^{t} Q(\tau)p(\tau)d\tau + 2z_0$. The solution of boundary problem may be searched in a form $p(t) = R(t)q(t) + r(t)$, where matrix and vector functions $R(t), r(t)$ satisfy equations (19), (20) which is demonstrated by straightforward calculation. Then $h(\cdot)$ is expressed via $p(\cdot)$ and $q(\cdot)$ and program maximin is calculated directly.

2.3. Game solution

The proof of theorem 2.1 is finished by the following lemma.

Lemma 2.6. *Suppose $\lambda_{\alpha,\beta} < 1$. Then differential game (7)-(8) has the value equal to the program maximin:*

$$\rho^0_{\alpha,\beta}(t_0, z_0) = \rho_{\alpha,\beta}(t_0, z_0).$$

The lemma is proven by means of standard technique of dynamic programming approach. From the equations (19) - (21) it is derived that program maximin satisfies the following step-by-step differential equation in partial derivatives (piecewise continuous version of Isaacs - Bellman equation):

$$\frac{\partial \rho}{\partial t} + \min_u \max_v [\operatorname{grad} \rho(t,z)^T X[\vartheta,t](B(t)u + C(t)v) + + \alpha|u|^2 - \beta|v|^2] + \zeta(t)|X[t,\vartheta]z - y(t)|^2 = 0 \tag{22}$$

with conditions

$$\rho(c^{(i)}, z) = \rho(c^{(i)} + 0, z) + |X[c^{(i)},\vartheta]z - y(c^{(i)})|^2,$$

$$i = 1,\ldots,l; \rho(\vartheta + 0, z) = 0.$$

Minimizing and maximizing functions are

$$u^*(t,z) = -\frac{1}{2\alpha}B^T(t)X^T[\vartheta,t]\operatorname{grad}\rho(t,z),$$

$$v^*(t,z) = -\frac{1}{2\beta}C^T(t)X^T[\vartheta,t]\,\mathrm{grad}\,\rho(t,z).$$

Suppose $z(t)$ is a solution generated by the strategy $u^*(\cdot)$ and arbitrary admissable function $v(t)$. The function $J^*(t)$ is defined as follows.

$$J^*(t) = \rho_{\alpha,\beta}(t, z(t))+$$

$$+\int_{t_0}^{\vartheta}[\alpha|u^*(t,z(t))|^2 - \beta|v(t)|^2+$$

$$+\varsigma(t)|X[t,\vartheta]z(t) - y(t)|^2]dt$$

Taking into account equation (22) it may be shown that piecewise continuous derivative of function $J^*(t)$ satisfies conditions

$$\frac{dJ^*(t)}{dt} \le 0, \tag{23}$$

$$J^*(c^{(i)}) = J^*(c^{(i)}+0)+|X[c^{(i)},\vartheta]z(c^{(i)})-y(c^{(i)})|^2. \tag{24}$$

Integrating inequality (23) from t_0 to ϑ and using conditions (24) the inequality $J \le \rho_{\alpha,\beta}(t_0,z_0)$ is obtained. Similarly it is shown that $J \ge \rho_{\alpha,\beta}(t_0,z_0)$ with $v^*(t,z)$ and arbitrary $u(t)$. These two conditions prove that program maximin $\rho_{\alpha,\beta}(t_0,z_0)$ is in fact a game value.

Combination of lemmas 2.5 and 2.6 gives the following formulae for the game value and optimal strategies.

$$\rho^0_{\alpha,\beta}(t_0,z_0) =$$

$$= z_0^T R_{\alpha,\beta}(t_0)z_0 + z_0^T r_{\alpha,\beta}(t_0) + g_{\alpha,\beta}(t_0), \tag{25}$$

$$u^0_{\alpha,\beta}(t,z) = -\frac{1}{2\alpha}B^T(t)d^0_{\alpha,\beta}(t,z), \tag{26}$$

$$v^0_{\alpha,\beta}(t,z) = \frac{1}{2\beta}C^T(t)d^0_{\alpha,\beta}(t,z), \tag{27}$$

where $d^0_{\alpha,\beta}(t,z) = X^T[\vartheta,t](R_{\alpha,\beta}(t)z+r_{\alpha,\beta}(t))$ and $R_{\alpha,\beta}(t), r_{\alpha,\beta}(t)$ functions satisfy (19) and (20) correspondingly.

Returning to the initial game (1)-(4) in condition $\lambda_{\alpha,\beta} < 1$ the game has a value

$$\rho^0_{\alpha,\beta}(t_0,x_0) = x_0^T X^T[\vartheta,t]R_{\alpha,\beta}(t_0)X[\vartheta,t]x_0+$$

$$+x_0^T X^T[\vartheta,t]r_{\alpha,\beta}(t_0) + g_{\alpha,\beta}(t_0)$$

and optimal strategies $u^0_{\alpha,\beta}(t,x)=u^0_{\alpha,\beta}(t,X[\vartheta,t]x)$, $v^0_{\alpha,\beta}(t,x)=v^0_{\alpha,\beta}(t,X[\vartheta,t]x)$,

The main result of this section is a LQDG solubility condition $\lambda_{\alpha,\beta} < 1$ which below is applied for the H^∞ problem.

3. H^∞ OPTIMIZATION

Let return to the initial H^∞ problem and associated linear-quadratic differential game (1)-(2). Let λ_γ denotes the maximal eigenvalue of compact operator $F_\gamma = \frac{1}{\gamma^2}L_v L_v^* - L_u L_u^*$. The results of section 2 evidently lead to the following statement.

Theorem 3.1. The solution of H^∞ problem is given by the value

$$\gamma_0 = \inf\{\gamma : \lambda_\gamma < 1\}. \tag{28}$$

Example. Consider a simple first-order system

$$\dot{x} = u + v.$$

In this case $X[t,\tau] = B = C = 1$ and operator F_γ is defined as

$$F_\gamma h(\cdot) = (\frac{1}{\gamma^2} - 1)Fh(\cdot) =$$

$$= (\frac{1}{\gamma^2} - 1)\int_{t_0}^{\vartheta}(\min(t,\tau) - t_0)h(\tau)d\tau.$$

Let find now eigenvalues of F operator: $Fh(\cdot) = \lambda h(\cdot)$. First of all 0 is evidently not an eigenvalue of F.

$$Fh(\cdot) = \int_{t_0}^{\vartheta}(\min(t,\tau) - t_0)h(\tau)d\tau =$$

$$= \int_{t_0}^{t}(\tau - t_0)h(\tau)d\tau + (t - t_0)\int_{t}^{\vartheta}h(\tau)d\tau$$

which leads to condition $h(t_0) = 0$. Then integrating by parts:

$$Fh(\cdot) = \int_{t_0}^{t}\left(\int_{\tau}^{\vartheta}h(\nu)d\nu\right)d\tau = \lambda h(t).$$

If $z(t)$ denotes $\int_{t}^{\vartheta}h(\tau)d\tau$ then $\lambda\frac{d^2z}{dt^2} + z = 0$ with boundary conditions $z(\vartheta) = 0, \dot{z}(t_0) = -h(t_0) = 0$ and $z(t) = c\sin\sqrt{\frac{1}{\lambda}}(\vartheta - t)$, $\dot{z}(t_0) = c\sqrt{\frac{1}{\lambda}}\cos\sqrt{\frac{1}{\lambda}}(\vartheta - t_0) = 0$. Hence $\sqrt{\frac{1}{\lambda}}(\vartheta - t_0) = \frac{\pi}{2}(2k+1), k = 0,1,2,\ldots$ and

$$\lambda_k = \frac{4(\vartheta - t_0)^2}{\pi^2(2k+1)^2}.$$

It is clear now that

$$\lambda_\gamma = \frac{4(\vartheta - t_0)^2}{\pi^2(1/\gamma^2 - 1)}$$

from which a solution of H^∞ problem is immedeately derived:

$$\gamma_0 = \frac{\pi}{\sqrt{4(\vartheta - t_0)^2 + \pi^2}} \tag{29}$$

This may be confirmed by explicit solution of equation (19) which in this case looks like as follows:

$$\dot{R} = -(\frac{1}{\gamma^2} - 1)R^2 - 1, \quad r(\vartheta) = 0$$

and

$$R(t) = R_\gamma(t) = \begin{cases} \frac{1}{\sqrt{b}} \operatorname{th} \sqrt{b}(\vartheta - t), & \gamma > 1; \\ \vartheta - t, & \gamma = 1; \\ \frac{1}{\sqrt{b}} \operatorname{tg} \sqrt{b}(\vartheta - t), & \gamma < 1; \end{cases}$$

where $b = |\frac{1}{\gamma^2} - 1|$. It is well seen that the game solution exists if

$$\sqrt{b}(\vartheta - t_0) < \pi/2$$

i.e.

$$\gamma > \frac{\pi}{\sqrt{4(\vartheta - t_0)^2 + \pi^2}}$$

and this condition gives the same critical γ value as (29).

In conclusion it may be noted that approach described may be applied in case of more general H^∞ problems in Hilbert spaces.

ACKNOWLEDGEMENTS

The work is financially supported by the Russian Foundation for Basic Researches (grant # 98-01-01160).

REFERENCES

Balakrishnan A.V. (1976) *Applied Functional Analysis* Springer-Verlag, New-York, Heidelberg, Berlin.

Basar T., P.Bernhard (1991). *H^∞ Optimal Control and Related Minimax Design Problems. A dynamic game approach.* Birkhauser.

Chu Ch., J.C.Doyle, E.B.Lee (1986). The general distance problem in H_∞ control theory. *Int. J. Control.* Vol. 44, No. 2, pp. 565 – 596.

Limebeer D.J.N., M.Green (1991). A game theoretic approach to H^∞ control. IMC Conf. "Robust Control System Design Using H^∞ and Related Methods". Cambridge.

Turetskij V.Ya. (1989). Solubility condition for a linear-quadratic game (English, Russian original). *Autom. Remote Control.* Vol. 50, No. 8, pp. 1045-1052; translation from *Avtomatika i Telemekhanika.* No. 8, pp. 55-64.

PROJECTION METHOD OF APPROXIMATION GRADIENT FOR CONVEX PROGRAMMING PROBLEMS

Vladimir A. Tyulyukin *

* *Ural State University of Economics, Ekaterinburg, Russia*

Abstract: In this report the problem of minimization of a convex function $f(x)$ on a convex closed and bounded set $Q \subset R^n$ is considered. The method described below concerns gradient methods of the search of extremum of convex functions. The discrete analogue of the approximation gradient plays here the role of the gradient (Batukhtin and Maiboroda, 1984; Batukhtin and Maiboroda, 1995). *Copyright ©1998 IFAC*

Keywords: Approximate analysis, Gradient methods, Convex optimization, Convex projections, Convex programming

1. DEFINITION OF THE APPROXIMATION GRADIENT

The discrete set is defined in the space R^n

$$\Omega(\nu) = \{s = (s_1, s_2, \ldots, s_n) \in R^n : s_i = \overline{-\nu, \nu}, i = \overline{1, n}\},$$

where ν is a fixed number ($\nu \in \mathbf{N}$, $\mathbf{N} = \{1, 2, \ldots\}$). It is designated for any number $\Delta > 0$:

$$\Omega_\Delta(\nu) = \{x \in R^n : x = \Delta s, s \in \Omega(\nu)\}.$$

Let $F(s, \Delta)$ is the function of the joint distribution of independent discrete random variables $s_i \in R^1, i \in \overline{1, n}$, their joint probabilities differ from zero on the set $\Omega_\Delta(\nu)$.

Let $G(\Omega_\Delta(\nu))$ is the class of symmetric functions of the distribution $F(s, \Delta)$ on the set $\Omega_\Delta(\nu)$.

The functions $a_0(x, \Omega_\Delta(\nu), F), a(x, \Omega_\Delta(\nu), F) = (a_1(x, \Omega_\Delta(\nu), F), \ldots, a_n(x, \Omega_\Delta(\nu), F))$ are defined by the condition of minimum of the functional

$$\gamma(x, a_0, a_1, \ldots, a_n, \Omega_\Delta(\nu), F) = \sum_{s_1=-\nu}^{\nu} \cdots \sum_{s_n=-\nu}^{\nu} [f(x + \Delta s) - a_0 - \langle a, \Delta s \rangle]^2 h_1(\Delta s_1) \ldots h_n(\Delta s_n).$$

Here $\langle x, y \rangle$ is the scalar product of vectors $x, y \in R^n$.

As $F(s, \Delta) \in G(\Omega_\Delta(\nu))$ it is received

$$a_0(x, \Omega_\Delta(\nu), F) = \sum_{s_1=-\nu}^{\nu} \cdots \sum_{s_n=-\nu}^{\nu} f(x + \Delta s) h_1(\Delta s_1) \ldots h_n(\Delta s_n),$$

$$a_i(x, \Omega_\Delta(\nu), F) = \frac{1}{\Delta \overline{s}_i^2} \sum_{s_1=-\nu}^{\nu} \cdots \sum_{s_n=-\nu}^{\nu} s_i \times f(x + \Delta s) h_1(\Delta s_1) \ldots h_n(\Delta s_n),$$

where $\overline{s}_i^2 = \sum_{s_i=-\nu}^{\nu} s_i^2 h_i(\Delta s_i), i = \overline{1, n}$.

The vector $a(x, \Omega_\Delta(\nu), F)$ is called the approximation gradient. Simplification of calculation of this gradient is passed from the approximation of the function $f(x + \Delta s)$ by hyperplane $y = a_0 + \langle a, \Delta s \rangle$ to the approximation for each coordinate separately. The modified approximation gradient is defined by the ratio:

$$a_i(x, \Omega_\Delta(\nu), F) = \frac{1}{\Delta \overline{s}_i^2} \sum_{s_i=-\nu}^{\nu} s_i f(x + \Delta s(i)) h_i(\Delta s_i),$$

where $s(i) = (0, 0, \ldots, 0, s_i, 0, \ldots, 0), s_i \in \overline{-\nu, \nu}$, $i = \overline{1, n}$. Hereinafter be based on a modified design. It will be kept to the initial designation of the approximation gradient .

2. THE DESCRIPTION OF THE PROJECTION METHOD OF THE APPROXIMATION GRADIENT

The map $\pi : R^n \mapsto Q$ is defined by the following relation

$$\pi(x) = \begin{cases} x, & x \in Q \\ P_Q(x), & x \notin Q \end{cases},$$

where $P_Q(x)$ is the projection of any point $x \in R^n$ on the convex set $Q \subset R^n$. The projection method of the approximation gradient means the constructing of a sequence $\{x^{(k)}\}_{k\in\overline{0,\infty}}$ according to the rule

$$x^{(k+1)} = \pi(x^{(k)} - \alpha_k l(x^{(k)}, \Delta_k)), \qquad (1)$$

where $l(x^{(k)}, \Delta_k)$ is the normalized approximation gradient of a convex function f(x) in the point $x^{(k)} \in Q$ for the chosen set of $\Omega_{\Delta_k}(\nu)$ and the chosen function $F(s, \Delta_k)$; numbers $\Delta_k \to 0$ for $k \to \infty$; α_k is the method step.

The point $x^{(0)} \in Q$, a sequences $\{\Delta_k\}_{k\in\overline{0,\infty}}$ and $\{\alpha_k\}_{k\in\overline{0,\infty}}$, the set $\Omega_{\Delta_k}(\nu)$, the function $F \in G(\Omega_{\Delta_k}(\nu))$ are given before the beginning of a process. The step α_k is chosen due to conditions

$$\lim_{k\to\infty} \alpha_k = 0, \sum_{k=0}^{\infty} \alpha_k = +\infty,$$

$$\sum_{k=0}^{\infty} \alpha_k^2 < +\infty. \qquad (2)$$

3. CONVERGENCE OF THE PROJECTION METHOD OF THE APPROXIMATION GRADIENT FOR PROBLEMS OF CONVEX PROGRAMMING

The following statements are need for the proof of convergence of an offered method.

Lemma 1. Let $f : R^n \mapsto R^1$ is the own convex function and $int(domf) \neq \emptyset$. Then for any point $x \in int(domf)$, any number $\varepsilon > 0$ and the finite number $\nu \in \mathbf{N}$ there is such number $\delta > 0$ that for any set $\Omega_\Delta(\nu)$ as $\Delta < \delta$ and any function $F \in G(\Omega_\Delta(\nu))$ the inclusion

$$a(x, \Omega_\Delta(\nu), F) \in V(\partial f(x), \varepsilon)$$

is carried out. Here $V(M, \varepsilon) = \{x \in R^n : \rho(x, M) \leq \varepsilon\}$; $\rho(x, M)$ is the distance from a point $x \in R^n$ up to the set $M \subset R^n$, $\partial f(x)$ is the subdifferential of a convex function $f(x)$ for a point x.

Lemma 2. Let $f : R^n \mapsto R^1$ is the own convex function,$int(domf) \neq \emptyset$ and the set M^* is not empty. Then for any number $\varepsilon > 0$, $\nu \in \mathbf{N}$ and all points $\overline{x} \notin V(M^*, \varepsilon)$ there will be numbers $\delta > 0$ and $\delta_1 > 0$ such that for any set $\Omega_\Delta(\nu)$ as $\Delta < \delta$ and any function $F \in G(\Omega_\Delta(\nu))$ the inequality

$$\langle l(\overline{x}, \Delta), \overline{x} - \overline{x}^* \rangle \geq \delta_1,$$

is true, where $M^* = \{x^* \in R^n : f(x^*) = \min_{x\in R^n} f(x) = f^*\}$, $\overline{x}^* \in P_{M^*}(\overline{x})$.

Proofs of these statements are like similar statements from (Batukhtin and Maiboroda, 1995).

Theorem. Let $f : R^n \mapsto R^1$ is the own convex function, $int(domf) \neq \emptyset$ and the sequence $\{\alpha_k\}_{k\in\overline{0,\infty}}$ of positive numbers satisfies to conditions (2), and the sequence $\{\Delta_k\}_{k\in\overline{0,\infty}}$ has the property: $\Delta_k \to 0$ for $k \to \infty$. Then for the sequence $\{x^{(k)}\}$, formed on the formula (1), for any point $x^{(0)} \in Q$ the condition

$$\lim_{k\to\infty} \rho(x^{(k)}, Q_*) = 0$$

is carried out, where $Q_* = \{x_* \in Q : f(x_*) = \min_{x\in Q} f(x)\}$.

Proof. At first validity of the theorem will be shown for the case, when there is the sequence of numbers $\{k_m\}_{m\in\overline{1,\infty}}$ such that $l(x^{(k_m)}, \Delta_{k_m}) = 0$. If in addition $M^* \bigcap Q \neq \emptyset$ then $Q_* = M^* \bigcap Q$, and if $M^* \bigcap Q = \emptyset$ thus there is such number $\varepsilon_* > 0$ that $V(M^*, \varepsilon_*) \bigcap Q = \emptyset$. Hence, taking into account the inclusion $x^{(k_m)} \in Q$, have that the sequence $\{k_m\}$ contains the finite number of elements.

For any number $\delta > 0$ the following set will be considered:

$$B(\delta) = \{x \in R^n : \exists c(x) \in \partial f(x), \\ c(x) \in V(0, \delta)\}.$$

It appears that $B(0) = Q_*$ and the property $B(\delta_1) \subset B(\delta_2)$ is fair for $\delta_2 > \delta_1$. Let Z_δ for any number $\delta > 0$ designates the intersection of the complement of the set $S(Q_*, \frac{\varepsilon}{2}) = \{x \in R^n : \rho(x, Q_*) < \frac{\varepsilon}{2}\}$ and the set $B(\delta)$. Sets Z_δ are closed and decrease as $\delta \to 0$. If in addition all sets Z_δ are not empty then there is the point y, belonging to all sets Z_δ. However such point satisfies to two relations contradicting each other: on the one hand $y \neq S(Q_*, \frac{\varepsilon}{2})$ and on the other hand - $y \in B(\delta)$ for all $\delta > 0$. But then there is the subgradient $c(y) \in \partial f(y)$ such that $c(y) \in V(0, \delta)$ for any number $\delta > 0$, i.e. $c(y) = 0$ for $y \in Q_*$. Therefore, sets Z_δ must be empty for some number $\delta = \delta_* > 0$ and the set $B(\delta_*)$ corresponding to this number δ_* is contained wholly in $V(Q_*, \frac{\varepsilon}{2})$.

Further, taking into account lemma 1, have that for any $\overline{x} \in Q$, $\delta_* > 0$ and $\nu \in \mathbf{N}$ there are the numbers $r(\overline{x}, \delta_*) > 0$ and $\zeta(\overline{x}, \delta_*)$ for which the relation

$$\forall x \in V(\overline{x}, r(\overline{x}, \delta_*)), \forall \Omega_\Delta(\nu) : \nu < \zeta(\overline{x}, \delta_*),$$

$$\forall F \in G(\Omega_\Delta(\nu)) \Rightarrow$$

$$\Rightarrow a(x, \Omega_\Delta(\nu), F) \in V(\partial f(\overline{x}), \delta_*) \qquad (3)$$

is carried out. Thus,the compact Q is covered by the infinite system of open solid spheres for each of them the ratio (3) is carried out. The finite subcover is selected from this infinite cover and let $x^{(i)} \in Q, i \in \overline{1,p}$ are centers of solid spheres of this finite subcover. Then the number $\zeta(\delta_*)$ is defined as follows: $\zeta(\delta_*) = \min\{\zeta(x^{(i)}, \delta_*), i \in \overline{1,p}\}$. As $\Delta_k \to 0$ for $k \to \infty$ then there is the number k_1 such that $\Delta_k < \zeta(\delta_*)$ for all $k \geq k_1$, and in addition for the number k_m, satisfying to the condition $k_m \geq k_1$, the inclusion

$$x^{(k_m)} \in V(Q_*, \varepsilon) \qquad (4)$$

is carried out for the infinite sequence $\{k_m\}_{m \in \overline{1,\infty}}$. Really, as $x^{(k_m)} \in Q$ there is such point $x^{(i_*)}$ for fixed $i_* \in \overline{1,p}$ that the inclusion

$$x^{(k_m)} \in S(x^{(i_*)}, r_1) \qquad (5)$$

is carried out. Here $r_1 \leq \frac{\varepsilon}{2}$. As $a(x^{(k_m)}, \Omega_{\Delta_{k_m}}(\nu), F) = 0$, then in accordance with (3) the inclusion $0 \in V(\partial f(x^{(i_*)}), \delta_*)$ is carried out, hence $x^{(i_*)} \in B(\delta_*) \subset V(Q_*, \frac{\varepsilon}{2})$. The inclusion (4) follows from last inclusions and from (5). The validity of the following ratio: $\forall \varepsilon > 0, \exists k^* : \forall k_m \geq k^* \Rightarrow x^{(k_m)} \in V(Q_*, \varepsilon)$ has been shown. The proof of the theorem is completed in the considered case.

Without bounding of a generality it will be supposed further that $l(x^{(k)}, \Delta_k) \neq \neq 0$ for any $k \in \mathbf{N}$. In this case it will be shown that for any number $\varepsilon > 0$ there is such number $k_\varepsilon \in \mathbf{N}$ that the inclusion $x^{(k_\varepsilon)} \in V(Q_*, \frac{\varepsilon}{2})$ is true. It will be proven ad absurdum, i.e. it will be assumed that for any number $k \in \mathbf{N}$ the condition $x^{(k)} \notin V(Q_*, \frac{\varepsilon}{2})$ is carried out.

The following function $\psi : R^n \mapsto R^1$ is determined by the ratio:

$$\psi(x) = \begin{cases} f(x), & x \notin H(f^*) \\ f^*, & x \in H(f^*) \end{cases},$$

which is the own convex function, $H(f^*)$ is the set of its minimums. Here $H(c) = \{x \in R^n : f(x) \leq c\}$. Taking into account property $\Delta_k \to 0$ for $k \to \infty$ and lemma 2, for function $\psi(x)$ the condition

$$\forall \varepsilon > 0, \exists \delta_1 > 0, k_1 \in \mathbf{N} : \forall k \geq k_1 \Rightarrow$$

$$\Rightarrow \langle l_\psi(x^{(k)}, \Delta_k), x^{(k)} - \overline{x}^{(k)} \rangle \geq \delta_1 \qquad (6)$$

is carried out, where $l_\psi(x^{(k)}, \Delta_k)$ is the normalized approximation gradient of the function $\psi(x)$ for the point $x^{(k)}$, $\overline{x}^{(k)} \in P_{H(f^*)}(x^{(k)})$, $x^{(k)} \in Q \setminus S(Q_*, \frac{\varepsilon}{2})$. As $x^{(k)} \notin \notin V(Q_*, \frac{\varepsilon}{2}), \varepsilon > 0$ and $\Delta_k \to 0$ for $k \to \infty$ then there is such number $k_2 \geq k_1$ that for all numbers $k \geq k_2$ the condition

$$\left(x^{(k)} + \Omega_{\Delta_k}(\nu)\right) \bigcap H(f^*) = \emptyset$$

is carried out. Thus $l(x^{(k)}, \Delta_k) \equiv l_\psi(x^{(k)}, \Delta_k)$ for any number $k \geq k_2$ and, taking into account (1), (6), for all numbers $k \geq k_2$ it gets that

$$\begin{aligned} \rho^2(x^{(k+1)}, Q_*) &= \| x^{(k+1)} - \overline{x}^{(k+1)} \|^2 \leq \\ &\leq \| x^{(k+1)} - \overline{x}^{(k)} \|^2 \leq \\ &\leq \| \pi(x^{(k)} - \alpha_k l(x^{(k)}, \Delta_k)) - \overline{x}^{(k)} \|^2 \leq \\ &\leq \| x^{(k)} - \alpha_k l(x^{(k)}, \Delta_k) - \overline{x}^{(k)} \|^2 \leq \\ &\leq \rho^2(x^{(k)}, Q_*) + \alpha_k^2 - 2\delta_1 \alpha_k. \end{aligned} \qquad (7)$$

As $\alpha_k \to 0$ for $k \to \infty$, there is the number $k_3 \geq k_2$ such that for all $k \geq k_3$

$$\alpha_k - 2\delta_1 < 0, \alpha_k - \frac{\varepsilon}{2} < 0. \qquad (8)$$

Then $\rho(x^{(k+1)}, Q_*) < \rho(x^{(k)}, Q_*)$ follows from (7), i.e. $x^{(k+1)} \neq x^{(k)}$. Summing (7) over k changing from k_3 up to $k_3 + m (m \geq 1)$ it is received that

$$\rho^2(x^{(k_3+m+1)}, Q_*) \leq \rho^2(x^{(k_3)}, Q_*) + \\ + \sum_{k=k_3}^{k_3+m} \alpha_k^2 - 2\delta_1 \sum_{k=k_3}^{k_3+m} \alpha_k,$$

but it is impossible for large $m \in \mathbf{N}$, because the series $\sum \alpha_k^2$ converges and the series $\sum \alpha_k$ diverges, i.e. there is the contradiction. Hence, there is such number $k_\varepsilon \geq k_3$ that $x^{(k_\varepsilon)} \in V(Q_*, \frac{\varepsilon}{2})$.

Then according to (8) it is received that $x^{(k)} \in V(Q_*, \varepsilon)$ for all $k \geq k_\varepsilon$.

Thus, for any number $\varepsilon > 0$ it is proved the existence of the number $k_\varepsilon \in \mathbf{N}$ such that for all $k \geq k_\varepsilon$ the inclusion $x^{(k)} \in V(Q_*, \varepsilon)$ is carried out. It proves the ratio

$$\lim_{k \to \infty} \rho(x^{(k)}, Q_*) = 0.$$

The theorem is proved.

The method, determined by relations (1)-(2), does not guarantee the monotonicity $f(x^{(k)}) > f(x^{(k+1)})$ at each iteration of process and converges rather slowly, but if the projection of the point to the set Q is calculated effectively, the method is very convenient and simple for the computer realization.

4. REFERENCES

Batukhtin, V.D. and L.A. Maiboroda (1984). *Optimization of discontinuous functions*. Nauka. Moscow.

Batukhtin, V.D. and L.A. Maiboroda (1995). *Discontinuous extremal problems*. Gippokrat. S-Peterburg.

CLASSIFICATION OF INDUSTRIAL ENTERPRISES ACCORDING TO LEVELS OF THEIR DANGER FOR ENVIRONMENT

D.G. Lokhnev, S.T. Zavalishchin [1]

Ural State Technical University - UPI

Abstract: A new approach based on Markov chains and multicriteria optimization to classify industrial enterprises according to levels of their danger for environment is suggested. *Copyright © 1998 IFAC*

Keywords: Enterprise, Events, Damage, Classification

1. INTRODUCTION

The report deals with the following problem. Given some totality of industrial enterprises, for example, nuclear power plants. The problem is to put in order these using two factors: (i) the industrial enterprise probability to meet with an extraordinary accident; (ii) the damage to be caused by such an accident. The main difficulties are the following ones. First, the damage has a multicomponent composition. This is caused by the necessity to take into account the damage for population, flora and fauna, pollution, economics and so on. Second, every damage of the above mentioned ones has an own scale of measurement. The considered problem solving the corresponding automation classification system of industrial enterprises according to inclination to meet with exstraordinary accidents and according to levels of their danger for environment may be developed. The system would enable decision makers to predict extraordinary accidents and to design look–ahead actions.

[1] The research reported here was carried out with the financial support of Ministry of Extreme Situations of Russian Federation (064/3.7.13)

2. ESTIMATING PROBABILITY TO MEET WITH AN ACCIDENT

In principle, the enterprise probability to meet with an accident can be calculated with the help of theory of reliability. However, practically it would be tedious to use such a way. So in what follows theory of random processes is applied. Namely, to solve the estimating problem it is supposed to characterize the enterprise by only two states: State 2 is an accident and State 1 is not so. Transitions from State 1 to State 2 and reverse these are suggested to be forced by Poisson flows of events with densities $\lambda(t)$ and $\mu(t)$ respectively. Then the current state of the enterprise forms a Markov chain.

Let $I_p = [t_0, t_p]$ be a time interval to observe the enterprise state. Let $N(t_p)$ denote the number of the accidents recorded for I_p, $T(t_p)$ denote the response time of the enterprise in State 2. With this, we have in mind both the accidents taken place and the accidents that could take place but have been prevented. The densities $\lambda(t_p)$ and $\mu(t_p)$ are abstract counterparts of the empirical quantities

$$\lambda(t_p) = \frac{N(t_p)}{t_p - t_0 - T(t_p)}, \; \mu(t_p) = \frac{N(t_p)}{T(t_p)}$$

respetively, with $t_p - t_0$ large enough. Let $P_i(t)$, $i = 1, 2$, be the probability of occurrence of the i–th state at the present moment t. The formulas

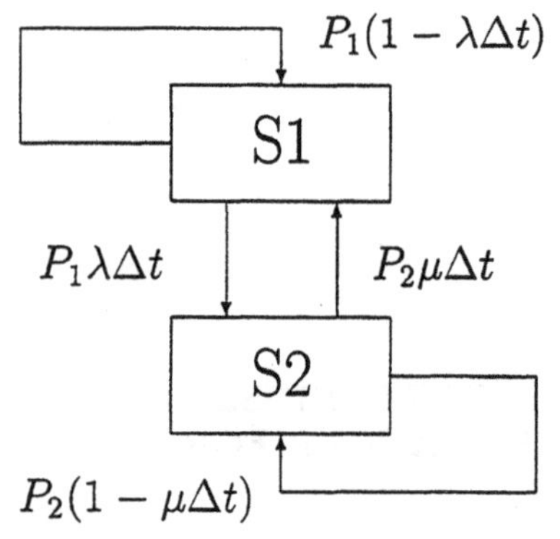

Fig. 1. The probability–transition diagram

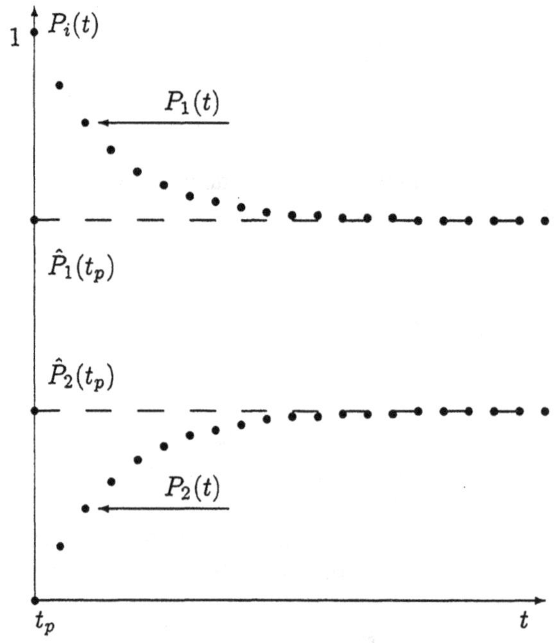

Fig. 2. Evolution of the probabilities in case $\lambda(t_p) = 1/3$, $\mu(t_p) = 2/3$

for densities $\lambda(t_p)$ $\mu(t_p)$ enable one to design the following prediction procedure for $P_i(t)$, $i = 1, 2$, $t \geq t_p$. The probability–transition diagram with Δt small enough is shown in Figure 1.

Hence, the Kolmogorov equations governing the evolution of probabilities $P_i(t)$, $i = 1, 2$ are

$$\frac{dP_1(t)}{dt} = -\lambda(t_p)P_1 + \mu(t_p)P_2\,,$$

$$\frac{dP_2(t)}{dt} = \lambda(t_p)P_1 - \mu(t_p)P_2\,.$$

It is well–known that

$$P_1(t) \to \hat{P}_1(t_p) = \frac{\mu(t_p)}{\lambda(t_p) + \mu(t_p)}\,,$$

$$P_2(t) \to \hat{P}_2(t_p) = \frac{\lambda(t_p)}{\lambda(t_p) + \mu(t_p)}$$

as $t \to \infty$. This is shown in Figure 2. Then the value $\hat{P}_i(t_p)$ is taken as asymptotically efficient estimate of $P_i(t_p)$, $i = 1, 2$.

3. CLASSIFICATION OF INDUSTRIAL ENTERPRISES

Let O_i, $i = 1, \ldots, n$, denote some enterprises, P_i be the probability to meet with an accident for the i–th enterprise. It is assumed that the damage to be caused by the accident for the i–th enterprise can be characterized by some vector of quantities: $D_{i1}, \ldots, D_{im}$, for example, relating to population, flora and fauna, pollution, economics and so on. Then the mean value of the above damage is the vector of quantities: $H_{ij} = P_i D_{ij}$, $j = 1, \ldots, m$.

Definition 1. The enterprise O_i is said to be more dangerous than an enterprise O_k if

$$H_{ij} \geq H_{kj}\,, \quad j = 1, \ldots, m$$

and such an inequality is strict for at least one j.

Definition 2. The enterprise O_{i_0} is said to be the most dangerous one if there is not another enterprise O_k that is more dangerous than the enterprise O_{i_0}.

The above enterprise O_{i_0} may be called Pareto optimal. The Pareto optimization theory makes possible to design procedures to solve the problem considered. For example, the following criterion, see (Podinovskii, Nogin, 1982), may be applied to find the most dangerous enterprises:

The enterprise O_{i_0} is the most dangerous one iff the equalities hold

$$H_{i_0 j} = \max_{O_i \in X_j(i_0)} H_{ij}\,, \quad j = 1, \ldots, m\,,$$

where, by definition,

$$X_j(i_0) = \{O_k \mid H_{kl} \geq H_{i_0 l}\,, \quad l = 1, \ldots, m, \quad l \neq j\}\,.$$

It is obvious that the enterprise O_{i_0} is in the set $X_j(i_0)$.

Sequential application of the above criterion enables one to divide the considered totality of enterprises into subsets $M_1, \ldots, M_q$ such that the following property holds. Let s be an integer, $1 < s \leq q$, then for every enterprise $O_i \in M_s$ there is a more danger enterprise $O_k \in M_{s-1}$.

It remains to put in order the enterprises which are in M_s, $s = 1, \ldots, q$. Such a problem may be solved as follows. Let

$$H^0_{ij} = (H_{ij} - m_j)/(M_j - m_j)\,,$$

where

$$m_j = \min_{1 \leq i \leq n} H_{ij}\,, \quad M_j = \max_{1 \leq i \leq n} H_{ij}\,.$$

Then the enterprises belonging to M_s may be ranked with the help of the following estimate

$$H_i = \left(\sum_{j=1}^{m}(H_{ij}^0)^a\right)^{1/a}, \quad i = 1,\ldots,n,$$

the parameter a ($1 \le a \le \infty$) being chosen by a decision maker. It is clear that the parametr a specifies the type of distance function in the space of relative estimates of damage, $H_{i1}^0, \ldots, H_{im}^0$. It should be noted that the case $a = 1$ is famous, see (Volkovich, 1969).

Example. The problem is to put into order eight enterprises. The corresponding indexes of damage are given in the following table

	H_{i1}	H_{i2}	H_{i3}
O_1	0,3	0,2	0,1
O_2	0,35	0,15	0,15
O_3	0,25	0,25	0,1
O_4	0,2	0,25	0,15
O_5	0,15	0,1	0,05
O_6	0,175	0,075	0,075
O_7	0,125	0,125	0,05
O_8	0,1	0,125	0,075

Step (a). Application of the above criterion makes possible to divide the considered set of enterprises into the two subsets $M_1 = \{O_1, O_2, O_3, O_4\}$ and $M_2 = \{O_5, O_6, O_7, O_8\}$. With this, for example, $X_1(1) = \{O_1, O_3, O_4\}$, $X_2(1) = \{O_1, O_2\}$, $X_3(1) = \{O_1\}$.

Step (b). If the parameter $a = 1$, then ranking the enterprises belonging to M_1, M_2 yields data combined into the following tables

	H_{i1}^0	H_{i2}^0	H_{i3}^0	H_i
O_1	0,666	0,5	0	1,166
O_2	1	0	1	2
O_3	0,333	1	0	1,333
O_4	0	1	1	2
M_j	0,35	0,15	0,1	
m_j	0,2	0,25	0,15	

	H_{i1}^0	H_{i2}^0	H_{i3}^0	H_i
O_5	0,666	0,5	0	1,166
O_6	1	0	1	2
O_7	0,333	1	0	1,333
O_8	0	1	1	2
M_j	0,175	0,125	0,075	
m_j	0,1	0,75	0,05	

As a result the two possible orders are obtained

$$O_2 \quad O_4 \quad O_3 \quad O_1 \quad O_6 \quad O_8 \quad O_7 \quad O_5$$
$$O_4 \quad O_2 \quad O_3 \quad O_1 \quad O_8 \quad O_6 \quad O_7 \quad O_5$$

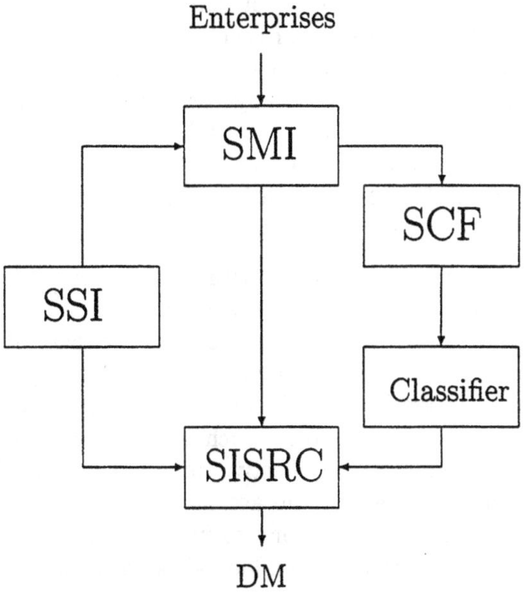

Fig. 3. The structure of the system

4. AN AUTOMATION SYSTEM OF CLASSIFICATION

On the basis of personal computer facilities, the automation classification system of industrial enterprises according to inclination to meet with exstraordinary accidents and according to levels of their danger for environment is developed. This system permits to create and manage information connected with enterprises, their characteristics and record–keeping accidents. By this information and the mathematical model developed above, the system creates and supports the classifier of enterprises. The structure of the system and data flows is shown in Figure 3.

The subsystem of standardization information creation and management (SSI) involves: (i) the list of categories of accidents; (ii) lists of regions; (iii) lists of branches.

The subsystem of management enterprises and accidents information (SMI) is intended to create and manage data concerning enterprises, their characteristics, past accidents. This subsystem makes use of standartization information and input information about enterprises and accidents.

Using present information base and the mathematical model developed the subsystem of the classifier formation (SCF) executes procedures of creating and recreating the classifier. This procedure must be started after each updating or correcting information connected with enterprises and accidents.

The subsystem of information search and reports creation (SISRC). This subsystem permits to execute the information search in the information base following various criteria and formats given. Information can be obtained in various aspects, for example: the level of danger of a given enterprise and the list of accidents in it; the list of the most dangerous enterprises in given regions; the list of the most dangerous enterprises from branches; lists of accidents taken place for given periods (in regions and branches as a whole). The regime of reports formation of arbitrary form at the request of user exists.

The system can operate with use one workstation or local networks and in a remote access mode. This enables decision makers to get necessary information concerning the industrial enterprises which have meet with an accident and the accidents by themselves. Such a convinience makes possible to do effective actions to relax the damage caused by the accident to environment. The system enables one to predict extaordinary accidents and to design look-ahead actions.

5. CONCLUSION

A new approach to classify industrial enterprises according to levels of their danger for environment is suggested. It makes use of theory of Markov chains and theory of multicriteria optimization.

On the basis of personal computer facilities, the automation classification system of industrial enterprises according to inclination to meet with extraordinary accidents and according to levels of their danger for environment is developed.

The automation classification system developed can operate in real–time applications provided that accidents may be registered in time.

Authors ought to note participating Dr. Sci. (Physics) Prof. B. Shul'gin, Dr. Sci. (Chem.) Prof. V. Berezyuk, Cand. Sci. (Physics) D. Shul'gin and D. Zavalishchin in the research presented above.

REFERENCES

Podinovskii V.V. and Nogin V.D. (1982). *Pareto optimal solutions of multicriteria problems.* Nauka, Moskow.

Volkovich V.L. (1969) *Multicriteria problems and methods of their solving.* In: *Proceedings of the seminar on cybernetics and computer techniques. Complicated control systems. Issue 1.* Naukova Dumka, Kiev.

ON CONTROLLABILITY OF ERGODIC SYSTEM LYAPUNOV EXPONENTS

Zaitsev V.A.

Udmurt State University, Izhevsk, Russia

Abstract: In this paper the linear differential time-dependent equation, that is defined by the ergodic dynamical system, is considered. This reseach focuses on investigation of conditions, under which Lyapunov exponents of this equation are locally controllable. *Copyright © 1998 IFAC*

Keywords: controllability, dynamical systems, stability, stochastic parameters.

1. INTRODUCTION

Investigations of Lyapunov exponents controllability problems are connected with motion separation and stabilization problems. A big number of research has been directed toward studying this property. The problem of Lyapunov exponents controllability was solved for autonomous systems (Popov, 1966) and for periodic systems (Brunovsky, 1969). It was proved, that the complete controllability of the system

$$\dot{x} = A(t)x + B(t)u, \quad (t, x, u) \in \mathbb{R}^{1+n+m}$$

is equivalent to Lyapunov exponents global controllability of the closed-loop control system

$$\dot{x} = (A(t) + B(t)U)x.$$

Time-dependent nonperiodical systems was studied by Popova and Tonkov (1994a, b, 1995a, b, 1997), and by Tonkov (1995). The recent advances one can find in (Makarov and Popova; Tonkov and Zaitsev).

In this work, it investigates the conditions, which permit to construct the control, ensuring "good environment" of the equation

$$\dot{x} = v_0(t, x) + u_1 v_1(t, x) + \dots + u_r v_r(t, x), \qquad (t, x) \in \mathbb{R}^{1+n} \quad (1)$$

trivial solution, where $u \doteq (u_1, \dots, u_r) \in U$, $0 \in U$. For example, let one need of finding the control $u(t) \in U$ such that the equation (1) is structurally stable (in the neighbourhood of zero). It turns out, that this problem is solvable if the linear equation

$$\dot{x} = A_0(t)x + u_1 A_1(t)x + \dots + u_r A_r(t)x, \qquad (t, x) \in \mathbb{R}^{1+n}, \quad (2)$$

where $A_i(t) = \partial v_i(t, 0)/\partial x$, is uniformly locally attainable and the matrix $A_0(t)$ is diagonalizable.

The case, which is considered there is the case of the equation with random parameters, i.e. the matrices $A_i(\cdot)$ are stationary processes. The main results are contained in the theorems 5, 6, 7.

2. NOTATIONS AND DEFINITIONS

$\mathbb{R}^n$ denotes the n-dimensional real Euclidean space equipped with the norm $|x| = \sqrt{x^* x}$ ($*$ is the transposition); $\mathcal{O}^n_\varepsilon(x) = \{y \in \mathbb{R}^n : |y - x| \leqslant \varepsilon\}$. Let $\operatorname{int} D$ and ∂D denote, respectively, the interior and the boundary of a set $D \subset \mathbb{R}^n$. Let the space $M_{n,m}$ of the linear operators $A : \mathbb{R}^m \to \mathbb{R}^n$ be identified with the space of the $n \times m$ matrices ($M_n \doteq M_{n,n}$); $|A| = \max\{|Ax| : |x| = 1\}$ is the norm, $\rho(A, B) = |A - B|$ is the metric in $M_{n,m}$. Set $\mathcal{B}_\varepsilon(I) = \{H \in M_n : |H - I| \leqslant \varepsilon\}$, I is the identity matrix, $\mathcal{B}_\varepsilon = \{H \in M_n : |H| \leqslant \varepsilon\}$. For any nonempty compact sets $D_1, D_2 \subset M_n$ let $\rho(A, D_1) = \min_{B \in D_1} \rho(A, B)$ denote the distance between the point A and the set D_1; $\operatorname{dist}(D_1, D_2) = \max\{\max_{A \in D_1} \rho(A, D_2), \max_{B \in D_2} \rho(B, D_1)\}$ is the hausdorffian distance between the sets D_1, D_2.

Let a topological dynamical system (Ω, f^t) be given. Ω is a metric complete separable space, f^t is a flow on Ω, that is an one-parameter group of transformations of Ω, which is continuous with respect to $(t,\omega) \in \mathbb{R} \times \Omega$. Suppose that Ω is compact. Let $\gamma(\omega)$ denote the trajectory of the motion $t \to f^t\omega$, $\overline{\gamma}(\omega)$ denote the completion of $\gamma(\omega)$ by metric of the space Ω. The set $E \subset \Omega$ is said to be invariant if $f^tE \subset E$ for all $t \in \mathbb{R}$, and E is said to be minimal if it is invariant, compact and $\overline{\gamma}(\omega) = E$ for every $\omega \in E$.

To each pair $(\mathbb{A}, \omega)$, where $\mathbb{A} \doteq (A_0, A_1, \ldots, A_r) : \Omega \to M_{n,m}$, $m = n(r+1)$, and vector $u = (u_1, \ldots, u_r)$ one can associate the equation

$$\dot{x} = A_0(f^t\omega)x + u_1A_1(f^t\omega)x + \cdots \cdots + u_rA_r(f^t\omega)x, \quad x \in \mathbb{R}^n. \quad (3)$$

$\mathbb{A}$ assumed to be satisfying the following condition: for each fixed $\omega_0 \in \Omega$, the function $t \to |\mathbb{A}(f^t\omega_0)|$ is Lebesgue measurable, bounded on $\mathbb{R}$, and for any $\varepsilon > 0$ and $N > 0$ there exists $\delta > 0$ such that $\rho_\Omega(\omega, \omega_0) < \delta$ implies $\max_{|t|\leqslant N} \int_t^{t+1} |\mathbb{A}(f^s\omega) - \mathbb{A}(f^s\omega_0)|\, ds < \varepsilon$, where ρ_Ω is the metric in Ω.

The equation (3) and $(\mathbb{A}, \omega)$ are identified.

By $(\mathbb{A}, \omega)$ one can construct the matrix differential equation

$$\dot{Z} = A_0(f^t\omega)Z + \Big(u_1A_1(f^t\omega) + \ldots \cdots + u_rA_r(f^t\omega)\Big)X_0(t,0,\omega) \quad (4)$$

with respect to $Z \in M_n$ ($X_0(t,s,\omega)$ is the Cauchy function of the equation $\dot{x} = A_0(f^t\omega)x$). Let $\mathfrak{L}_\vartheta(\omega)$ be the attainable set of the equation (4) from zero at time ϑ under measurable functions $u : [0,\vartheta] \to \mathbb{R}^r$. Thus, $G \in \mathfrak{L}_\vartheta(\omega)$ if and only if there exists $u_G : [0,\vartheta] \to \mathbb{R}^r$ such that for $u = u_G(t)$ the equation (4) has the solution, which satisfies the boundary conditions $Z(0) = 0$, $Z(\vartheta) = G$. The equation (4) is linear, that is why $\mathfrak{L}_\vartheta(\omega)$ is the linear subspace in M_n.

Definition 1. The equation $(\mathbb{A}, \omega)$ is called *consistent*, if there exists $\vartheta > 0$ such that $\mathfrak{L}_\vartheta(\omega) = M_n$ and *uniformly consistent*, if there exist $\vartheta > 0$ and $\beta > 0$ such that $\mathfrak{L}_\vartheta(\omega_0) = M_n$ for all $\omega_0 \in \overline{\gamma}(\omega)$ and inequality $|u_G(t,\omega_0)| \leqslant \beta|G|$ holds for all $t \in [0,\vartheta]$ (where $u_G(t,\omega_0)$ is the control, which translates the solution of the equation (4) from zero to G at time ϑ for $\omega = \omega_0$).

The definitions of the consistency and the uniform consistency were introduced by Popova and Tonkov (1994a) for the equation $\dot{x} = (A(f^t\omega) + B(f^t\omega)UC^*(f^t\omega))x$, where U is the matrix of control parameters, for problems of Lyapunov exponents controllability, and were investigated by Popova and Tonkov (1995b, 1997), by Tonkov (1995), and by Tonkov and Zaitsev.

3. CONSISTENCY

Let

$$\gamma_{ijps}(\vartheta,\omega) = \int_0^\vartheta \sum_{l=1}^r e_i^*\widehat{A}_l(t,\omega)e_pe_s^*\widehat{A}_l^*(t,\omega)e_j\, dt,$$

$i,j,p,s = 1,\ldots,n$, $e_k \in \mathbb{R}^n$ be the kth column of the identity matrix, $\widehat{A}_l(t,\omega) = X_0(0,t,\omega)A_l(f^t\omega)X_0(t,0,\omega)$, $l = 1,\ldots,r$. One can consider the $(n \times n)$-matrices $\Gamma_{ij}(\vartheta,\omega) = \{\gamma_{ijps}(\vartheta,\omega)\}_{p,s=1}^n$, $i,j = 1,\ldots,n$, and the $(n^2 \times n^2)$-matrix $\Gamma(\vartheta,\omega) = \{\Gamma_{ij}(\vartheta,\omega)\}_{i,j=1}^n$. The matrix $\Gamma(\vartheta,\omega)$ is said to be *the consistency matrix* of the equation (3) on $[0,\vartheta]$.

Lemma 1. *The consistency matrix satisfies the following conditions:*

a) $\Gamma(\vartheta,\omega) = \Gamma^*(\vartheta,\omega)$;

b) $\nu(\vartheta,\omega) \geqslant 0$, *($\nu$ is the least eigenvalue of the matrix Γ);*

c) $\nu(\vartheta_1,\omega) \geqslant \nu(\vartheta,\omega)$, *if* $\vartheta_1 \geqslant \vartheta$.

Let the functions $u_{ip}^l : [0,\vartheta] \times \Omega \to \mathbb{R}$ define by the equations

$$u_{ip}^l(t,\omega) = e_i^*\widehat{A}_l(t,\omega)e_p, \quad i,p = 1,\ldots,n, \quad l = 1,\ldots,r,$$

and

$$u_{ip}(t,\omega) = \mathrm{col}(u_{ip}^1(t,\omega),\ldots,u_{ip}^r(t,\omega)), \quad i,p = 1,\ldots,n. \quad (5)$$

Lemma 2. *The vector functions* (5) *are linearly independent on* $[0,\vartheta]$ *if and only if* $\nu(\vartheta,\omega) > 0$.

Theorem 1. *The following statements are equivalent:*

a) *the equation* (3) *is consistent;*

b) *there exists* $\vartheta > 0$ *such that the matrix* $\Gamma(\vartheta,\omega)$ *is positive definite* $(\nu(\vartheta,\omega) > 0)$;

c) *the vector functions* (5) *are linearly independent on* $[0,\vartheta]$ *for some* $\vartheta > 0$.

Theorem 2. *The equation* $(\mathbb{A},\omega)$ *is uniformly consistent if and only if there exist* $\vartheta > 0$ *and* $\varepsilon > 0$ *such that inequality* $\nu(\vartheta,\omega_0) \geqslant \varepsilon$ *holds for any* $\omega_0 \in \overline{\gamma}(\omega)$.

4. ATTAINABILITY

Let $U \subset \mathbb{R}^r$ be given convex compact set and $0 \in U$. The set $\mathcal{U} \doteq \{u : \mathbb{R} \to U\}$, where u is Lebesgue measurable, is said to be *the set of permissible controls*; $\mathcal{U}_\vartheta \doteq \mathcal{U}|_{[0,\vartheta]}$. Let the equation $(\mathbb{A},\omega)$ be fixed. To each permissible control $u = u(t)$ one can associate the Cauchy function $X_u(t,s,\omega)$ of the equation (3). It is easy to prove that the function $\omega \to X_u(t,s,\omega)$ is continuous at each point $\omega_0 \in \Omega$ uniformly with respect to (t,s) on a compact in $\mathbb{R}^2$. Note that if the control is stationary over flow, i.e. $u(t,\omega) = u(f^t\omega)$, then $X_u(t+\tau, s+\tau, \omega) = X_u(t,s,f^\tau\omega)$.

Let us consider the matrix equation

$$\dot{X} = A_0(f^t\omega)X + u_1A_1(f^t\omega)X + \cdots \\ \cdots + u_rA_r(f^t\omega)X, \quad X \in M_n, \quad (6)$$

which corresponds to the equation (3). For any $\vartheta > 0$ one can construct the attainable set $\mathfrak{D}_\vartheta(\omega) = \{X \in M_n : X = X_u(\vartheta, 0, \omega),\ u(\cdot) \in \mathcal{U}\}$ of the equation (6).

Definition 2. The equation $(\mathbb{A}, \omega)$ is called

a) *locally attainable* if there exist $\vartheta > 0$ and $\varepsilon > 0$ such that

$$\mathcal{B}_\varepsilon(I)X_0(\vartheta, 0, \omega) \subset \mathfrak{D}_\vartheta(\omega);$$

b) *uniformly locally attainable* if there exist $\vartheta > 0$ and $\varepsilon > 0$ such that the inclusion

$$\mathcal{B}_\varepsilon(I)X_0(\vartheta, 0, \omega_0) \subset \mathfrak{D}_\vartheta(\omega_0)$$

holds for all $\omega_0 \in \overline{\gamma}(\omega)$.

In other words, the local attainability of the equation (3) means that there are $\vartheta > 0$ and $\varepsilon > 0$ such that for any matrix $H \in \mathcal{B}_\varepsilon(I)$ there exists the control $u(\cdot, H) \in \mathcal{U}_\vartheta$ with

$$X_u(\vartheta, 0, \omega) = HX_0(\vartheta, 0, \omega), \quad (7)$$

and the uniform local attainability means that there are $\vartheta > 0$ and $\varepsilon > 0$ such that for any continuous function $H : \overline{\gamma}(\omega) \to \mathcal{B}_\varepsilon(I)$ there exists the control $u(\cdot, \omega_0) \in \mathcal{U}_\vartheta$ with

$$X_u(\vartheta, 0, \omega_0) = H(\omega_0)X_0(\vartheta, 0, \omega_0).$$

for all $\omega_0 \in \overline{\gamma}(\omega)$.

Lemma 3. *Let E be the invariant compact set in Ω and the equation* (3) *be uniformly locally attainable for all $\omega \in E$. Then it is uniformly locally attainable* on E, *i.e. there exist the common constants $\vartheta > 0$ and $\varepsilon > 0$ for all $\omega \in E$, that are in the definition* 2b).

The consistency and the local attainability are closely related. It is true, that the uniform local attainability in the first approximation is the uniform consistency in the noncritical case ($0 \in \operatorname{int} U$).

Let the substitution $Y(t, \omega) = X_u(t, 0, \omega) - X_0(t, 0, \omega)$ be made in the equation (6). Then the equation will be

$$\dot{Y} = A_0(f^t\omega)Y + \Big(u_1A_1(f^t\omega) + \cdots \\ \cdots + u_rA_r(f^t\omega)\Big)Y + \Big(u_1A_1(f^t\omega) + \cdots \\ \cdots + u_rA_r(f^t\omega)\Big)X_0(t, 0, \omega), \quad Y \in M_n \quad (8)$$

with respect to Y. The equality (7) implies the following conditions: $Y(0, \omega) = 0$, $Y(\vartheta, \omega) = GX_0(\vartheta, 0, \omega)$, where $G = H - I$. Let $\mathbb{D}_\vartheta(\omega)$ be the attainable set of the equation (8) from zero at time $\vartheta > 0$ under controls $u(\cdot) \in \mathcal{U}$. Note that $0 \in \mathbb{D}_\vartheta(\omega) \subset M_n$. The inclusion $0 \in \operatorname{int} \mathbb{D}_\vartheta(\omega)$ holding for some $\vartheta > 0$ is equivalent to the local attainability of the equation (3), the inclusion $\mathcal{B}_\varepsilon \subset \mathbb{D}_\vartheta(\omega_0)$ holding for some $\vartheta > 0$ and $\varepsilon > 0$ and for all $\omega_0 \in \overline{\gamma}(\omega)$ is equivalent to the uniform local attainability.

Theorem 3. *If $0 \in \operatorname{int} U$ and the equation* (3) *is consistent (uniformly consistent) then it is locally attainable (uniformly locally attainable).*

The attainable set of the equation (8) is

$$\mathbb{D}_\vartheta(\omega) = \{\int_0^\vartheta X_u(\vartheta, t, \omega)(u_1(t)A_1(f^t\omega) + \cdots \\ \cdots + u_r(t)A_r(f^t\omega))X_0(t, 0, \omega)\,dt,\ u(\cdot) \in \mathcal{U}_\vartheta\}.$$

Lemma 4. *For every $\vartheta \geqslant 0$ and $\omega \in \Omega$ it holds:*

a) *the set $\mathbb{D}_\vartheta(\omega)$ is compact in M_n;*

b) *the function $(\vartheta, \omega) \to \mathbb{D}_\vartheta(\omega)$ is continuous in hausdorffian metric* dist;

c) *the inclusion $\mathbb{D}_\vartheta(\omega) \subset \mathbb{D}_{\vartheta+\tau}(\omega)$ is valid for every $\tau \geqslant 0$;*

d) *the inclusion $\mathbb{D}_\vartheta(f^\tau\omega)X_0(\tau, 0, \omega) \subset \mathbb{D}_{\vartheta+\tau}(\omega)$ is valid for every $\tau \geqslant 0$.*

Let $\vartheta > 0$ and $\omega \in \Omega$ be fixed. Lemma 4 implies that $\partial\mathbb{D}_\vartheta(\omega)$ is compact in M_n. The inclusion $0 \in \mathbb{D}_\vartheta(\omega)$ holds for any ϑ и ω, that is why the inclusion $0 \in \operatorname{int} \mathbb{D}_\vartheta(\omega)$ is equivalent to the inequality $\rho(0, \partial\mathbb{D}_\vartheta(\omega)) > 0$; $\mathcal{B}_\varepsilon \subset \mathbb{D}_\vartheta(\omega)$ is equivalent to $\rho(0, \partial\mathbb{D}_\vartheta(\omega)) \geqslant \varepsilon$.

Thus, the equation $(\mathbb{A}, \omega)$ is locally attainable if and only if there exists $\vartheta > 0$ such that $\rho(0, \partial\mathbb{D}_\vartheta(\omega)) > 0$, and the equation $(\mathbb{A}, \omega)$ is uniformly locally attainable if and only if there exist $\vartheta > 0$ and $\varepsilon > 0$ with the inequality $\rho(0, \partial\mathbb{D}_\vartheta(\omega_0)) \geqslant \varepsilon$ holding for all $\omega_0 \in \overline{\gamma}(\omega)$. Note that lemma 4 implies the continuity of the function $(\vartheta, \omega) \to \rho(0, \partial\mathbb{D}_\vartheta(\omega))$.

Theorem 4. *Let $\overline{\gamma}(\omega)$ be minimal. Then the equation $(\mathbb{A}, \omega)$ is locally attainable if and only if it is uniformly locally attainable.*

Let a probabilistic Borel measure μ be given, $\mathfrak{B}$ be the completion of the Borel σ-algebra by this measure. Let μ be invariant over the flow f^t: $\mu(\Omega_0) = \mu(f^{-t}\Omega_0)$, $(t, \Omega_0) \in \mathbb{R} \times \mathfrak{B}$. Thus, the function $\mathbb{A}$ generates the stationary (in the restricted sense) process $(t, \omega) \to \mathbb{A}(f^t\omega)$. It is assumed that the measure of any nonempty open set is positive. Moreover, let the dynamical system be ergodic with respect to measure μ, i.e., the measure of any invariant set is either 0 ore 1.

Let Σ_0 denote the set of the locally attainable equations. Thus, $\Sigma_0 = \bigcup_{\vartheta > 0} S_\vartheta$, where $S_\vartheta = \{\omega \in \Omega : \rho(0, \partial\mathbb{D}_\vartheta(\omega)) > 0\}$. By lemma 4 $\Sigma_0 = \bigcup_{\vartheta \in \mathbb{N}} S_\vartheta$, hence, Σ_0 is open.

Lemma 5. *For any $\tau \geqslant 0$ the inclusion $f^{-\tau}\Sigma_0 \subset \Sigma_0$ holds.*

It implies that Σ_0 is invariant(mod0), that is $\Sigma_0 = f^t\Sigma_0$ (μ-almost everywhere) for all t. Consequently, if $\Sigma_0 \neq \varnothing$, then $\mu(\Sigma_0) = 1$. Let $\widehat{\Sigma}$ be the set of the uniformly locally attainable equations, that is

$$\widehat{\Sigma} = \bigcup_{\vartheta>0}\bigcup_{\varepsilon>0}\bigcap_{\omega_0\in\overline{\gamma}(\omega)} \{\omega\in\Omega : \rho(0,\partial\mathbb{D}_\vartheta(\omega_0)) \geqslant \varepsilon\}.$$

This equality one can rewrite as $\widehat{\Sigma} = \bigcup_{\vartheta\in\mathbb{N}}\bigcup_{n\in\mathbb{N}}\bigcap_{t\in\mathbb{R}} f^{-t}S_\vartheta^n$, where $S_\vartheta^n = \{\omega \in \Omega : \rho(0,\partial\mathbb{D}_\vartheta(\omega)) \geqslant \frac{1}{n}\}$. Let $\widehat{\Sigma}$ be compact. Hence, by lemma 3 there exist the common constants ϑ and n for all $\omega\in\widehat{\Sigma}$, that are in the definition of the uniform local attainability. In fact, it means that the union on ϑ and n is finite, that is there are $\vartheta_0\in\mathbb{N}$ and $n_0\in\mathbb{N}$ such that the equality $S_\vartheta^n = S_{\vartheta_0}^{n_0}$ holds for all $\vartheta\geqslant\vartheta_0$ and $n\geqslant n_0$.

Lemma 6. *One can rewrite* $\bigcap_{t\in\mathbb{Z}}$ *instead of* $\bigcap_{t\in\mathbb{R}}$ *in the set* $\widehat{\Sigma}$.

Thus,

$$\widehat{\Sigma} = \bigcup_{\vartheta\in\mathbb{N}}\bigcup_{n\in\mathbb{N}}\bigcap_{t\in\mathbb{Z}} f^{-t}S_\vartheta^n = \bigcup_{\vartheta=1}^{\vartheta_0}\bigcup_{n=1}^{n_0}\bigcap_{t\in\mathbb{Z}} f^{-t}S_\vartheta^n =$$

$$= \bigcap_{t\in\mathbb{Z}} f^{-t}S_{\vartheta_0}^{n_0} = \bigcap_{t\in\mathbb{Z}} f^{-t}(\bigcup_{\vartheta=1}^{\vartheta_0}\bigcup_{n=1}^{n_0} S_\vartheta^n) =$$

$$= \bigcap_{t\in\mathbb{Z}} f^{-t}(\bigcup_{\vartheta\in\mathbb{N}}\bigcup_{n\in\mathbb{N}} S_\vartheta^n) =$$

$$= \bigcap_{t\in\mathbb{Z}} f^{-t}(\bigcup_{\vartheta\in\mathbb{N}} S_\vartheta) = \bigcap_{t\in\mathbb{Z}} f^{-t}\Sigma_0.$$

Consequently, by lemma 5 $\mu(\widehat{\Sigma}) = \mu(\Sigma_0) = 1$, therefore $\widehat{\Sigma} = \Sigma_0 = \Omega$.

The converse is true.

Theorem 5. *Let the dynamical system be ergodic and* $\Sigma_0 \neq \varnothing$. *Then* $\widehat{\Sigma}$ *is compact if and only if* $\mu(\widehat{\Sigma}) = \mu(\Sigma_0)$ *(and hence* $\widehat{\Sigma} = \Sigma_0 = \Omega$*).*

5. CONTROLLABILITY OF LYAPUNOV EXPONENTS

Let the whole spectrum $\lambda_1(\omega,u), \ldots, \lambda_n(\omega,u)$ of the equation (3) Lyapunov exponents (Demidovitch, 1967, p.145) be given.

Definition 3. The equation $(\mathbb{A},\omega)$ possesses *the uniform local controllability of Lyapunov exponents* if there is $\delta > 0$ such that for any $\beta = (\beta_1,\ldots,\beta_n) \in O_\delta^n(0)$ and $\omega_0\in\overline{\gamma}(\omega)$ there exists the permissible control $u : \mathbb{R}\times\overline{\gamma}(\omega) \to U$, which implies $\lambda_j(\omega_0,u) = \lambda_j(\omega_0,0) + \beta_j$, $j = 1,\ldots,n$ for all $\omega_0\in\overline{\gamma}(\omega)$.

Definition 4. The equation

$$\dot{x} = A_0(f^t\omega)x \tag{9}$$

is said to be *diagonalizable*, if we can reduce the equation (9) by Lyapunov transformation (Demidovitch, 1967, p.154) to an equation with the diagonal matrix.

Theorem 6. *Let the equation (9) be diagonalizable and* $(\mathbb{A},\omega)$ *be uniformly locally attainable. Then* $(\mathbb{A},\omega)$ *possesses the uniform local controllability of Lyapunov exponents.*

Theorem 7. *Let the dynamical system be ergodic,* Σ_0 *be nonempty,* $\widehat{\Sigma}$ *be compact and the equation (9) be diagonalizable for all* $\omega\in\Omega$. *Then* $(\mathbb{A},\omega)$ *possesses the uniform local controllability of Lyapunov exponents for all* $\omega\in\Omega$.

6. ACKNOWLEDGEMENT

The work is supported by Russian Foundation for Basic Reseach (grant 97–01–00413) and by Competition Center of Fundamental Natural Science (grant 97–0–1.9). The author thanks Professor E.L. Tonkov for help in proving some results.

REFERENCES

Brunovsky, P. (1969). Controllability and linear closed-loop controls in linear periodic systems. *Journal of Differential Equation*, **6**, 296—313.

Demidovitch, B.P. (1967). *Lecture on mathematical stability theory.* Nauka, Moscow (in Russian).

Makarov, E.K. and S.N. Popova. On global controllability of two-dimensional linear system Lyapunov invariants totality. *Differentsial'nye Uravneniya*, (to appear, in Russian).

Popov, V.M. (1966). *Hiperstabilitatea sistemelor automate.* Editura Academiei Republicii Socialiste Romania, Bucuresti (in Romanian).

Popova, S.N. and E.L. Tonkov (1994a). Control over the Lyapunov exponents of consistent systems. I. *Differential Equations*, **30**, 1556—1564.

Popova, S.N. and E.L. Tonkov (1994b). Control of the Lyapunov exponents of consistent systems. II. *Differential Equations*, **30**, 1800—1807.

Popova, S.N. and E.L. Tonkov (1995a). Control over Lyapunov exponents of consistent systems. III. *Differential Equations*, **31**, 209—218.

Popova, S.N. and E.L. Tonkov (1995b). Uniform consistency of linear systems. *Differential Equations*, **31**, 672—674.

Popova, S.N. and E.L. Tonkov (1997). Consistent systems and control of the Lyapunov exponents. *Differentsial'nye Uravneniya*, **33**, 226—235 (in Russian).

Tonkov, E.L. (1995). Control problems for Lyapunov exponents. *Differential Equations*, **31**, 1646—1651.

Tonkov, E.L. and V.A. Zaitsev. Attainability, consistency and Millionshchikov rotation method. *Izvestiya VUZov. Matematika*, (to appear, in Russian).

EVOLUTIONARY ALGORITHMS FOR MULTIPLE CRITERIA DECISION MAKING IN CONTROL

Peter J. Fleming and Andrew J. Chipperfield
Department of Automatic Control and Systems Engineering
The University of Sheffield
Mappin Street, Sheffield S1 3JD, UK

ABSTRACT

Many problems arising in control and systems engineering require the simultaneous optimisation of multiple, often conflicting, design criteria, such as performance, reliability, environmental impact, and cost. Unlike in single-objective optimisation, the global solution to such problems is seldom a single point, but a family of compromise solutions known as the Pareto-optimal set. These solutions are optimal in the sense that improvement in any objective can only be achieved at the expense of degradation in at least one of the remaining objectives. Here, it is shown that multiobjective genetic algorithms (MOGAs) can be a powerful decision-making aid. It is possible to search for many Pareto-optimal solutions concurrently, while concentrating on relevant regions of the Pareto set. Also, a human decision maker may interactively supply preference information to the algorithm as it runs. Applications of MOGAs being pursued include controller design, model selection in system identification, on-line scheduling, and multidisciplinary optimisation problems.

1 INTRODUCTION

Real-world control problems usually involve the satisfaction of multiple performance measures, or objectives, which should be solved simultaneously. In some situations, the objective functions may be solved in isolation from one another and an insight concerning the best that may be obtained in each performance domain obtained. However, when considering the overall problem, suitable solutions will seldom be found in this way. An optimal performance in one objective domain will often imply an unacceptably low performance in one or more of the remaining objectives, necessitating the need for some sort of compromise solution to be reached. Suitable solutions to problems posing such conflicts should offer an ``acceptable'' (though possibly sub-optimal solution in a single objective sense) performance across all the objectives. In this case ``acceptable'' is then a problem-dependent and subjective notion.

The simultaneous solution of multiple, possibly competing, objective functions is unlikely to yield a single Utopian solution. Instead, the solution of a multiobjective optimisation (MO) problem is a set of Pareto-optimal solutions which in most practical situations is likely to be very large. Subsequently, there is a difficulty in representing the set of Pareto-optimal solutions and in choosing a suitable solution from this set when there is no information regarding the relative performance of each objective. The size of the solution set can, however, be reduced by including a set of objective function goals which must also be satisfied.

Various non-linear programming methods have been developed to solve the MO problem (see, for example, Becker et al, 1979). However, this is not a trivial task as practical problems are generally non-convex, multimodal and frequently non-smooth or exhibit discontinuities. These traditional approaches use deterministic transition rules, generally to implement a form of hill climbing, and as such can only be expected to work well if the problem is small and has few local minima, i.e. distinct regions in decision variable space that yield Pareto-optimal solutions. Additionally, they will require a good estimate of the solution if they are not to converge to some local, sub-optimal solution. For larger, more realistic problems or ones that may have many local minima, algorithms with probabilistic transition rules offer greater potential for success.

Here, after an introduction to the basic components of an evolutionary algorithm, the notion of a decision maker is presented and used as the basis for constructing a multiobjective genetic algorithm (MOGA). Application examples are cited which demonstrate the versatility and utility of the approach.

2 EVOLUTIONARY ALGORITHMS

Evolutionary algorithms are based on computational models of fundamental evolutionary processes such as selection, reproduction and mutation, as shown in Fig. 1. Individuals, or current approximations, are encoded as strings composed over some alphabet(s), e.g. binary, integer, real-valued etc., and an initial population of chromosomes, Chrom, in Fig. 1, is produced by randomly sampling these strings. Once a population has been produced it may be evaluated using an objective function or functions that characterise an individual's performance in the problem domain. Where the encoding of chromosomes uses a mapping from the decision variables to some other alphabet, e.g. real-values encoded as binary strings, it will be necessary to decode the chromosomes before the objective function may be evaluated and a cost vector, Cost, assigned to the population. The objective function(s) is also used as the basis for selection and determines how well an individual performs in its environment. A fitness value is then derived from the raw performance measure given by the objective function(s) and is used to bias the selection process towards promising areas of the search space. Highly fit individuals will be assigned a higher probability of being selected for reproduction than individuals with a lower fitness value. Therefore, the average performance of individuals can be expected to increase as the fitter individuals are more likely to be selected for reproduction and the lower fitness individuals get discarded. Note that individuals may be selected more than once at any generation (iteration) of the EA and that the temporary vector of selected individuals, Sel, may therefore contain more than one copy of any individual in the original population.

Selected individuals are then reproduced, usually in pairs, through the application of genetic operators and these new individuals may then overwrite their parents in the vector, Sel. These operators are applied to pairs of individuals with a given probability and result in new offspring that contain material exchanged from their parents. The offspring from reproduction are then further perturbed by mutation. These new individuals then make up the next generation, Chrom. These processes of selection, reproduction and evaluation are then repeated until some termination criteria are satisfied.

```
procedure EA {
   initialise(Chrom);
   while not finished do {
      Cost = objv_fun(decode(Chrom));
      Sel = select(Chrom, Cost);
      Sel = reproduce(Sel);
      Chrom = mutate(Sel);
      Gen = Gen + 1;
   }
}
```

Fig. 1 An Evolutionary Algorithm

Although similar in concept, many variations exist in EAs. A comprehensive discussion of the differences between the various EAs can be found in Spears et al (1993).

3 MULTIOBJECTIVE OPTIMISATION AND DECISION MAKING

The use of multiobjective optimisation (MO) recognises that most practical problems require a number of design criteria to be satisfied simultaneously, viz.:

$$\min_{\mathbf{p} \in \Omega} \mathbf{F}(\mathbf{p}) \quad \ldots(1)$$

where $\mathbf{p}=[p_1, p_2, \ldots, p_q]$ and Ω defines the set of free variables, $\mathbf{p}$, subject to any constraints and $\mathbf{F}(\mathbf{p}) = [f_1(\mathbf{p}), f_2(\mathbf{p}), \ldots, f_n(\mathbf{p})]$ are the design objectives to be minimised.

Clearly, for this set of functions, $\mathbf{F}(\mathbf{p})$, it can be seen that there is no one ideal 'optimal' solution, rather a set of Pareto-optimal solutions for which an improvement in one of the design objectives will lead to a degradation in one or more of the remaining objectives. Such solutions are also known as *non-inferior* or *non-dominated* solutions to the multiobjective optimisation problem.

Generally, members of the Pareto-optimal solution set are sought through solution of an appropriately formulated non-linear programming problem. A number of approaches are currently employed including the ε-constraint, weighted-sum and goal attainment methods (Hwang and Masud, 1979). However, such approaches require precise expression of a, usually not well understood, set of weights and goals.

If the trade-off surface between the design objectives is to be better understood, repeated application of such methods will be necessary. In addition, non-linear programming methods cannot handle multimodality and discontinuities in function space well and can thus only be expected to produce local solutions.

Evolutionary algorithms, on the other hand, do not require derivative information or a formal initial estimate of the solution region. Because of the stochastic nature of the search mechanism, genetic algorithms (GAs) are capable of searching the entire solution space with more likelihood of finding the global optimum than conventional optimisation methods. Indeed, conventional methods usually require the objective function to be well behaved, whereas the generational nature of GAs can tolerate noisy, discontinuous and time-varying function evaluations, see Goldberg (1989). Moreover, EAs allow the use of mixed decision variables (binary, *n*-ary and real-values) permitting a parameterisation that matches the nature of the design problem more closely. Single objective GAs, however, do still require some combination of the design objectives although the relative importance of individual objectives may be changed during the course of the search process.

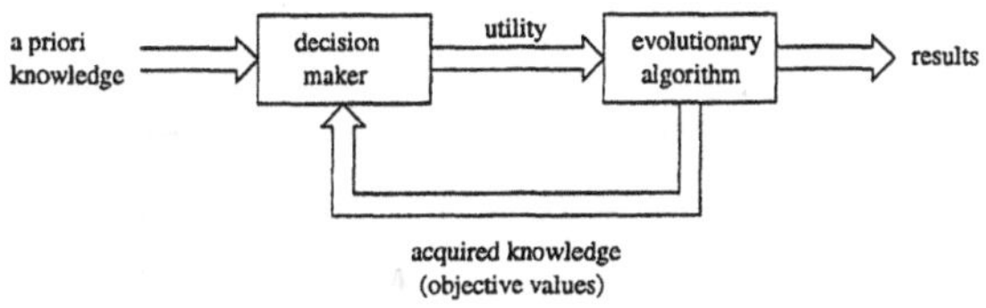

Fig. 2 A General Multiobjective Evolutionary Optimiser

A general view of multiobjective evolutionary optimisation has been proposed by Fonseca and Fleming (1995) and is illustrated in Fig. 2. The decision-maker block represents a utility assignment strategy, which may be anything from a straightforward weighted-sum approach to an intelligent decision maker or human operator. The EA is employed to generate a set of candidate solutions according to the utility level assigned by the decision maker to the current set of solution estimates. The decision-maker thus influences the production of new solution estimates and as these are evaluated they provide new trade-off information which can be used by the decision maker to refine its current goals and preferences. The effect of any changes in the decision process, perhaps arising from taking recently acquired information into account, is seen by the EA as a change in environment. In the next section, a multiobjective decision making process, based on a Pareto-ranking approach, is described and a multiobjective evolutionary algorithm developed.

4 MULTIOBJECTIVE GENETIC ALGORITHMS

The notion of fitness of an individual solution estimate and the associated objective function value are closely related in the single objective GA described earlier. Indeed, the objective value is often referred to as fitness although they are not, in fact, the same. The objective function characterises the problem domain and cannot therefore be changed at will. Fitness, however, is an assigned measure of an individual's ability to reproduce and, as such, may be treated as an integral part of the GA search strategy.

As Fonseca and Fleming (1995) describe, this distinction becomes important when performance is measured as a vector of objective function values as the fitness must necessarily remain scalar. In such cases, the scalarisation of the objective vector may be treated as a multicriterion decision making process over a finite number of candidates - the individuals in a population at a given generation. Individuals are therefore assigned a measure of utility depending on whether they perform better, worse or similar to others in the population and, possibly, by how much. The remainder of this section describes the main differences between the simple EA outlined earlier and MOGAs.

4.1 Decision Strategies

In the absence of any information regarding the relative importance of design objectives, Pareto-dominance is the only method of determining the relative performance of solution estimates. Non-dominated individuals are all therefore considered to be `best' performers and are thus assigned the same fitness (Goldberg, 1989), e.g. zero. However, determining a fitness value for dominated individuals is a more subjective matter. The approach adopted here is to assign a cost proportional to how many individuals in a population dominate a given individual, Fig. 3. In this case, non-dominated individuals are all treated as desirable.

If goal and/or priority information is available for the design objectives then it may be possible to differentiate between some non-dominated solutions. For example, if degradations in individual objectives still allow those goals to be satisfied whilst also allowing improvements in other objectives that do not already satisfy their design goals, then these degradations should be accepted. In cases where different levels of priority may be assigned to the objectives then, in general, it is only important to improve the high priority objectives, such as hard constraints, until the corresponding design goals are met, after which improvements may be sought in the lower priority objectives.

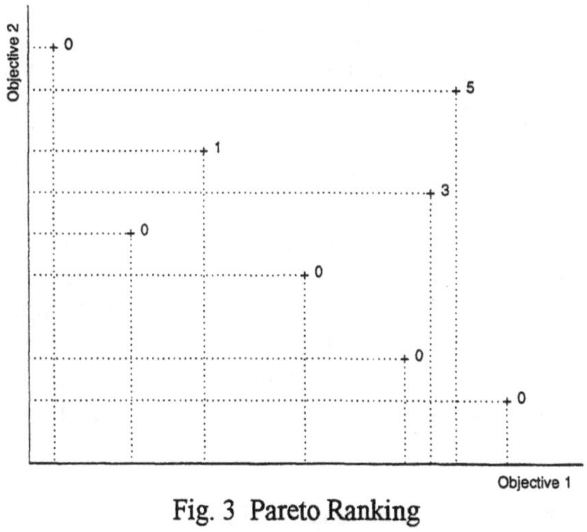

Fig. 3 Pareto Ranking

These considerations have been formalised in terms of a transitive relational operator, *preferability*, based on Pareto-dominance, which selectively excludes objectives according to their priority and whether or not the corresponding goals are met (Fonseca, 1995). For simplicity only one level of priority is considered here. Consider two objective vectors **u** and **v** and the corresponding set of design goals, **g**. Let the smile, $\smile$ denote the components of **u** that meet their goals and the frown $\frown$ those that do not. Assuming minimisation, one may then write

$$\mathbf{u}^{\smile} \le \mathbf{g}^{\smile} \wedge \quad \mathbf{u}^{\frown} > \mathbf{g}^{\frown} \qquad \ldots(2)$$

where the inequalities apply component-wise. This is equivalent to

$$\forall i \in \smile, u_i \le g_i \quad \wedge \quad \forall i \in \frown, u_i > g_i, \quad \ldots(3)$$

where u_i and g_i represent the components of **u** and **g**, respectively. Then, **u** is said to be preferable to **v** given **g** if and only if

$$(\mathbf{u}^{\frown} \mathrel{p\!<} \mathbf{v}^{\frown}) \vee \left\{ \begin{array}{l} (\mathbf{u}^{\frown} = \mathbf{v}^{\frown}) \wedge \\ \left[(\mathbf{v}^{\smile} \not\le \mathbf{g}^{\smile}) \vee (\mathbf{u}^{\smile} \mathrel{p\!<} \mathbf{v}^{\smile}) \right] \end{array} \right\} \qquad \ldots (4)$$

where $p\!<$ is a dominance operator such that $\mathbf{u} \mathrel{p\!<} \mathbf{v}$ denotes that **u** dominates **v**, i.e.

$$\forall i \in \{1,\ldots,n\}, u_i \le v_i \wedge \exists i \in \{1,\ldots,n\}: u_i < v_i \qquad \ldots (5)$$

Hence **u** will be preferable to **v** if and only if one of the following is true:

1. The violating components of **u** dominate the corresponding components of **v.**
2. The violating components of **u** are the same as the corresponding components in v, but v violates at least one other goal.
3. The violating components of **u** are equal to the corresponding components of v, but **u** dominates v as a whole.

4.2 Fitness Mapping and Selection

After a cost has been assigned to each individual, selection can take place in the usual way. Suitable schemes include rank-based cost to fitness mapping (Baker, 1985), followed by stochastic universal sampling (Baker, 1987) or tournament selection, also based on cost, as described by Ritzel et al (1994).

Exponential rank-based fitness assignment is illustrated in Fig. 4. Here, individuals are sorted by their cost - in this case the values from Fig. 3 - and assigned fitness values according to an exponential rule (determined by the particular ranking method employed) in the first instance, shown by the narrow bars in Fig. 4. A single fitness value is then derived for each group of individuals sharing the same cost, through averaging, and is shown in the figure by the wider bars.

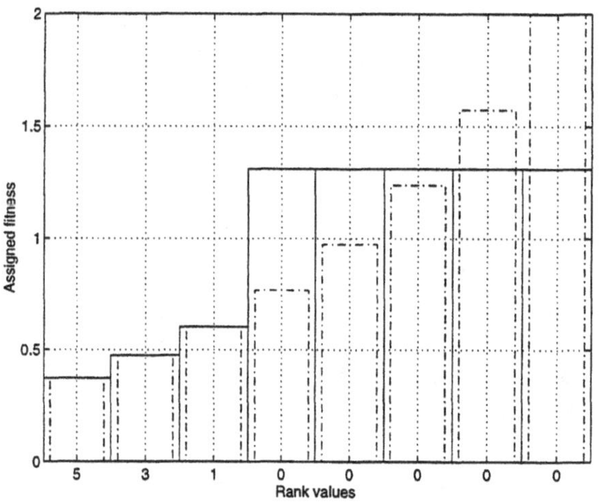

Fig. 4 Rank-Based Fitness Assignment

4.3 Fitness Sharing

Even though all preferred individuals in the population are assigned the same level of fitness, the number of offspring that they will produce, which must obviously be integer, may differ due to stochastic nature of EAs. Over generations, these imbalances may accumulate resulting in the population focusing on an arbitrary area of the trade-off surface, known as *genetic drift* (Goldberg and Segrest, 1987). Additionally, recombination and mutation may be less likely to produce individuals at certain areas of the trade-off surface, e.g. the extremes, giving only a partial coverage of the trade-off surface.

Originally introduced as an approach to sampling multiple fitness peaks, fitness sharing (Goldberg and Richardson, 1987 helps counteract the effects of genetic drift by penalising individuals according to the number of other individuals in their neighbourhood. Each individual is assigned a *niche count*, initially set to zero, which is incremented by a certain amount for every individual in the population, including itself. A *sharing function* determines the contribution of other individuals to the niche count as a function of their mutual distance in genotypic, phenotypic or objective space. Raw fitness values are then weighted by the inverse of the niche count and normalised by the sum of the weights prior to selection. The total fitness in the population is re-distributed, and thus shared, by the population. However, a problem with the use of fitness sharing is the difficulty in determining the niche size, σ_{share} , i.e. how close together individuals may be before degradation occurs.

An alternative, but analogous, approach to niche count computations are kernel density estimation methods (Silverman, 1986), as used by statisticians. Instead of a niche size, a smoothing parameter, h, whose value is also ultimately subjective, is used. However, guidelines for the selection of suitable values for h have been developed for certain kernels, such as the standard normal probability density function and Epanechnikov kernels. The Epanechnikov kernel may be written as

$$K_e(d/h)=\begin{cases}\frac{1}{2}c_n^{-1}(n+2)[1-(\frac{d}{h})^2] & \text{if } \frac{d}{h}<1 \\ 0 & \text{otherwise}\end{cases} \quad \ldots(6)$$

where n is the number of decision variables, c_n is the volume of the unit n-dimensional sphere and d/h is the normalised Euclidean distance between individuals. Apart from the constant factor, $\frac{1}{2}c_n^{-1}(n+2)$, this kernel is a particular case of the family of power law sharing functions proposed by Goldberg and Richardson (1987).

Silverman (1986) gives a smoothing factor that is approximately optimal in the least mean integrated squared error sense when the population follows a multivariate normal distribution for the Epanechnikov kernel $K_e(\mathrm{d})$ as

$$h=[8c_n^{-1}(n+4)(2\sqrt{\pi})^n / N]^{1/(n+4)} \quad \ldots(7)$$

for a population with N individuals and identity covariance matrix. Where populations have an arbitrary sample covariance matrix, S, this may simply be `sphered', or normalised, by multiplying each individual by a matrix **R** such that $\mathbf{RR}^T = \mathbf{S}^{-1}$. This means that the niche size, which depends on S and h, may be automatically and constantly updated, regardless of the cost function, to suit the population at each generation and used directly to perform sharing in Euclidean decision variable spaces.

4.4 Mating Restriction

Mating restrictions are employed to bias the way in which individuals are paired for reproduction (Deb and Goldberg, 1989). Recombining arbitrary individuals from along the trade-off surface may lead to the production of a large number of unfit offspring, called lethals, that could adversely affect the performance of the search. To alleviate this potential problem, mating can be restricted, where feasible, to individuals form within a given distance of each other, σ_{mate}. A common practice is to set $\sigma_{mate} = \sigma_{share}$ so that individuals are allowed to mate with one another only if they lie within a distance h from each other in the `sphered' space used for sharing (Fonseca and Fleming, 1995).

4.5 Progressive Preference Articulation

As the population of the MOGA evolves, trade-off information will be acquired. In response to the optimisation so far, the operator may wish to investigate a smaller region of the search space or even move on to a totally new region. This can be achieved by resetting the goals supplied to the MOGA which, in turn, affects the ranking of the population and modifies the fitness landscape concentrating the population on a different area of the search space. The priority of design objectives may also be changed interactively using this scheme.

The introduction of a small number of random individuals at each generation, say 10-20\%, has been shown to make the EA more responsive to sudden changes in the fitness landscape as occurs when the optimisation is changed interactively (Grefenstette, 1992). This technique may also be employed by a MOGA and is used in the example presented in the next section.

4.6 Algorithm Description

A pseudo-code outline of the multiobjective genetic algorithm is shown in Fig. 5. The population is initialised and the chromosomes are decoded, if necessary, and then evaluated according to the multiple objective functions. Preference-based ranking, pref_rank in Fig. 5, assigns a non-unique cost to each individual dependent on its dominance in the population such that all non-dominated individuals are ranked zero, as described in Section 4.1. As well as the vector of performance goals, GoalV, an additional vector of objective priority levels, PriV may also be specified, although this is not used in the example here.

The niche counts, Share, are calculated using a kernel estimator based on the Epanechnikov kernel. The decoded decision variables, DVar are passed to the function twice as they are both the sample data and the points where the population density should be estimated. The default smoothing parameter Sigma (h) and a matrix **R**, such that DVar * R has an identity covariance matrix, are also returned by the estimation function for use later during mating restriction.

The function ranking uses Share to perform fitness sharing between individuals of equal cost as part of the fitness assignment procedure. Individuals can now be selected for reproduction, in this case by stochastic universal sampling, and allowance should be made at this point if random chromosomes are to be inserted into the population after mutation so that only the required number of individuals are selected. Mating restriction is implemented by reordering the selected individuals in Sel so that consecutive pairs correspond to individuals within the required distance Sigma of one another within normalised decision variable space wherever possible (restrict in Fig. 5).

Recombination of individuals may now proceed as normal and the resulting population mutated. If random chromosomes are to be appended to the population then this should occur after mutation so that they will have to survive selection before they can reproduce with the main population. This is most likely to occur when the fitness landscape changes, as a result of changes in GoalV or PriV, and the population is no longer well adapted to it.

```
procedure MOGA {
   initialise(Chrom);
   while not finished do {
      DVar = decode(Chrom);
      ObjV = multi_obj_fun(DVar);
      Cost = pref_rank(ObjV, GoalV);
      [Share, Sigma, R] =
         epanechnikov(DVar, DVar);
      Fitn = ranking(Cost, Share);
      Sel = select(Chrom, Fitn);
      Sel = reproduce    (restrict(decode(Sel)*R,
Sigma));
      Chrom = mutate(Sel);
   }
}
```

Fig. 5 A Multiobjective Genetic Algorithm

5 CONTROL AND SYSTEMS ENGINEERING APPLICATIONS

To conclude, the following range of applications provide an indication of the versatility of the use of MOGAs in multiple criteria decision making in control.

5.1 Gas Turbine Engine Control System Design

In an early study based on a non-linear model of a Rolls-Royce gas turbine engine, Fonseca and Fleming (1998) demonstrated how parameters for an existing fixed-structure controller could be found to satisfy a set of control system design requirements, extracted from the original stability and response requirements provided by the manufacturers. Frequency-domain measures were computed from a model linearised around an operating point. Time-domain characteristics were derived from the response of a full non-linear model to a (small) step input, obtained through simulation. The objective functions were not required to be conveniently mathematically tractable; rather they were common criteria used by practising control engineers, such as gain and phase margin, rise and settling time.

The several objective functions, the multiobjective ranking algorithm and all GA routines were implemented as MATLAB M- and MEX-files. A graphical interface was also written as an M-file, making use of the graphical facilities of MATLAB. The model was simulated with SIMULINK.

A typical set of design trade-offs resulting from a MOGA design exercise is shown in Fig. 6. In this "parallel co-ordinates representation", each line in the graph represents a potential solution to the design problem, indicating the achieved objective values for that solution. All solutions are both non-dominant **and** satisfy the prescribed *goals* as represented by the "×" marks. The decision-maker (DM) must select a suitable compromise from this set of solutions. DM may interact with the MOGA as it runs to "tighten" or "slacken" the *goals*, in order to target a specific compromise solution.

Through such a representation (Fig. 6), the DM is informed of conflicts, or otherwise, between objectives. For example, in Fig. 6, solution lines between Objectives 2 and 3 clearly cross one another, indicating that improvement in one objective can only be achieved at the expense of the other objective. Other refinements at the disposal of DM include the ability to specify "hard" constraints for objectives.

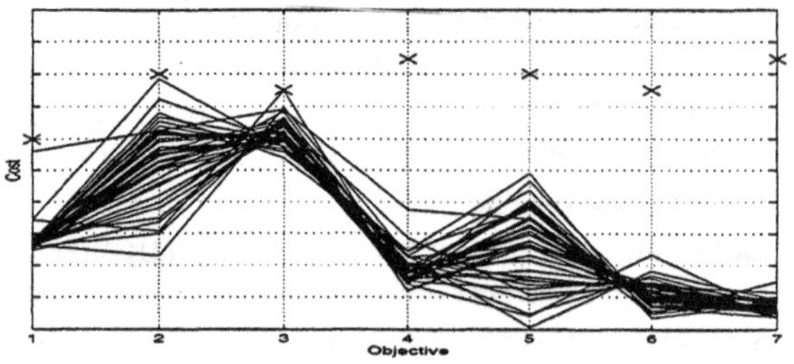

Fig. 6 Design Objective Trade-Offs

5.2 Control mode analysis for advanced concept aero-engines

Chipperfield et al (1997) extended this approach to evaluate a *set* of candidate control *modes* in a study to consider some problems likely to be associated with new variable cycle engine concepts. The nine objectives considered included typical control system design criteria along with physical constraints, such as maximum turbine blade temperature. Simple controller implementations were evaluated for a variety of loop configurations and a selection of alternative actuators and sensors. This provided valuable insights into the characteristics of the alternative control mode solutions which satisfied the design criteria. For example, it was possible to identify control modes which offered solutions using reliable sensors or, for example, those which placed less stress on physical components at the expense of a less exacting performance.

5.3 System Architectures for Distributed Aero-Engine Control Systems

Recently, Thompson et al (1998) have explored the potential of the increasing use of embedded intelligence through deployment of smart sensors and actuators in future distributed control architectures for aero-engines. Use of MOGA in assessing the potential of alternative architectures is dramatically reducing design times, while providing an opportunity to contemplate a rich set of potential solutions. Metrics such as technological risk, fault diagnosis capability and cost are introduced. Variables include bus interconnection topology, number of smart interface units and "mix" of dumb/smart sensors and actuators. In 100 generations, the MOGA was able to consider 4000 architectures, assisting DM (expected, in this case, to be an expert systems architecture designer) to identify characteristics of classes of solution such as simplex, duplex and triplex architectures and to identify key conflicts between objectives.

5.4 Multidisciplinary Optimisation

Multidisciplinary optimisation (MDO) is needed for increasingly complex design problems where system performance characteristics are influenced by more than one discipline, for example, the aerodynamic-structural optimisation of an aircraft wing. The MOGA approach is proving effective in addressing the non-commensurate objectives which necessarily arise in such problems. Moreover, Khatib and Fleming (1997) have successfully applied the methodology to examples derived from the NASA MDO Test Suite (Padula et al 1996). Current work is successfully matching the use of MOGA with the industrial design environment's organisation and practice.

5.5 Related Work

Other interesting applications and developments grow apace. Dakev et al (1997) have harnessed MOGA to produce H_∞ designs for an electromagnetic suspension control system for a Maglev vehicle, simultaneously satisfying specific performance criteria within the framework of an H_∞ loop-shaping design procedure.

Schroder et al (1998) have successfully applied MOGA *on-line* to tune PID controllers for an active magnetic bearing application with an immediate 50% improvement in performance. Recent work has similarly placed this work within the framework of an H_∞ loop-shaping design procedure.

Rodriguez-Vazquez et al (1998) have advanced the contemplation of multiple design criteria into the domain of *multiobjective genetic programming* and demonstrated how non-linear system identification may be achieved through attention to competing criteria such as minimum residual error, long-term prediction error and model parsimony, while allowing the genetic program to evolve an appropriate model structure.

Finally, Shaw and Fleming (1997) have applied MOGA to a scheduling problem in a chilled ready meal industry in order to improve optimisation of production schedules, where objectives include minimisation of rejected orders, staffing balance and lateness of batches within orders. This particular technique is being currently extended to batch scheduling problems arising in the chemical process industry.

ACKNOWLEDGEMENTS

The authors wish to acknowledge the support of the UK Engineering and Physical Sciences Research Council (Grant GR/J70857) and companies such as Rolls-Royce plc, Rolls-Royce & Associates, British Aerospace and Pennine Foods. They also wish to acknowledge the excellent related work of members of the Evolutionary Computing Group and the Rolls-Royce UTC, both at AC&SE, University of Sheffield, especially Carlos Fonseca, Nick Dakev, Haydn Thompson, Wael Khatib, Jane Shaw and Peter Schroder, whose work is cited here.

REFERENCES

Baker JE(1985): Adaptive selection methods for genetic algorithms, In: John J. Grefenstette (ed): *Genetic Algorithms and Their Applications: Proc. of the First International Conference on Genetic Algorithms*, Lawrence Erlbaum, 1985, pp. 101-111.

Baker JE (1987): Reducing bias and inefficiency in the selection algorithm". In: John J. Grefenstette (ed): *Genetic Algorithms and Their Applications: Proc. of the Second International Conference on Genetic Algorithms*, Lawrence Erlbaum, 1987, pp 14-21.

Becker RG, Heunis AJ and Mayne DQ (1979): Computer-aided design of control systems via optimization., *IEE Proc. D*, vol 126, pp 573-578.

Chipperfield AJ, Fleming PJ and Betteridge HC (1997): Control mode analysis for advanced concept aero-engines, *Proc 33rd AIAA/ASME/SAE/ASEE Joint Propulsion Conference*, Seattle USA.

Dakev NV, Whidborne JF, Chipperfield AJ and Fleming PJ (1997): Evolutionary H_∞ design of an EMS control system for a Maglev vehicle, *Proc Inst Mech Engrs, Part I: J of Sys and Contr. Eng*, vol 211, pp 345-355.

Deb K and Goldberg DE (1989): An investigation of niche and species formation in genetic function optimization". In: J. David Schaffer (ed): *Proc. of the Third International Conference on Genetic Algorithms, Morgan Kaufmann*, San Mateo, CA, pp 42-50

Fonseca C.M (1995): Multiobjective Genetic Algorithms with Application to Control Engineering Problems, *PhD Thesis*, University of Sheffield.

Fonseca CM and Fleming PJ (1995): An overview of evolutionary algorithms in multiobjective optimization, *Evolutionary Computation*, vol 3, no 1, pp 1-16.

Fonseca CM and Fleming PJ: Multiobjective optimization and multiple constraint handling with evolutionary algorithms - Part II: Application example, *IEEE Trans Systems, Man and Cybernetics*, vol 28, no 1, 1998, pp 38-47.

Goldberg DE and Richardson J (1987): Genetic algorithms with sharing for multimodal function optimization, In: In: John J. Grefenstette (ed): *Genetic Algorithms and Their Applications: Proc. of the Second International Conference on Genetic Algorithms*, Lawrence Erlbaum, 1987, pp. 41-49.

Goldberg DE and Segrest P (1987): Finite Markov chain analysis of genetic algorithms, In: In: John J. Grefenstette (ed): *Genetic Algorithms and Their Applications: Proc. of the Second International Conference on Genetic Algorithms*, Lawrence Erlbaum, 1987, pp. 1-8.

Goldberg DE (1989): *Genetic Algorithms in Search, Optimization and Machine Learning*, Addison-Wesley, Reading, Massachusetts.

Grefenstette JJ (1992): Genetic algorithms for changing environments, In: R. Manner and B. Manderick (eds): *Parallel Problem Solving From Nature 2*, North-Holland, pp. 137-144

Hwang C-L and. Masud ASM (1979): Multiple Objective Decision Making - Methods and Applications, Vol. 164 of *Lecture Notes in Economics and Mathematical Systems*, Springer-Verlag, Berlin.

Khatib W and Fleming PJ (1997): Evolutionary computing for multidisciplinary optimisation, *Proc 2nd IEE/IEEE International Conference on Genetic Algorithms in Engineering Systems: Innovations and Applications GALESIA 97*, Glasgow, pp 7-12.

Padula S, Alexandrov N and Green L (1996): NASA MDO Test Suite at NASA Langley Research Center, *6th AIAA/USAF/NASA/ISSMO Symposium on Multidisciplinary Analysis and Optimisation*, AIAA Paper 96-4028, Bellevue, WA, USA.

Rodríguez-Vázquez K and Fleming PJ (1998): Multi-objective genetic programming for nonlinear system identification, *Electronics Letters*, vol 34 no 9, pp 930-931.

Ritzel BJ, Eheart JW and Ranjithan S (1994): Using genetic algorithms to solve a multiple objective groundwater pollution containment problem, *Water Resources Research*, vol 30, pp 1589-1603.

Schroder P, Green B, Grum N and Fleming PJ (1998): On-line genetic auto-tuning of PID controllers for an active magnetic bearing application, *Preprints of the 5th IFAC Workshop on Algorithms and Architectures for Real-Time Control AARTC 98*, Mexico.

Shaw KJ and Fleming PJ (1997): Including real-life problem preferences in genetic algorithms to improve optimisation of production schedules, *Proc 2nd IEE/IEEE International Conference on Genetic Algorithms in Engineering Systems: Innovations and Applications GALESIA 97*, Glasgow, pp 239-244.

Silverman BW (1986): Density Estimation for Statistics and Data Analysis, *Vol. 26 of Monographs on Statistics and Applied Probability*, Chapman and Hall, London.

Spears WM, De Jong KA, Back T, Fogel DB and De Garis H (1993): An overview of evolutionary computation". In: Machine Learning: ECML-93 European Conference on Machine Learning, Lecture notes in Artificial Intelligence, Vol. 667 of Lecture notes in Artificial Intelligence, Springer-Verlag, pp. 442-459

Thompson HA, Fleming PJ and Chipperfield AJ (1998): Multiobjective optimisation of systems architectures for distributed aero-engine control systems, to appear in *Joint Propulsion Conference*, Paper AIAA-98-3727.

AUTHOR INDEX

www.ingramcontent.com/pod-product-compliance
Lightning Source LLC
Chambersburg PA
CBHW081149250326
R18032300001B/R180323PG41598CBX00003B/5
* 9 7 8 0 0 8 0 4 3 2 3 7 3 *